Social Inequalities and Cancer

Edited by M. Kogevinas, N. Pearce, M. Susser and P. Boffetta

IARC Scientific Publications No. 138

International Agency for Research on Cancer, Lyon, 1997

Published by the International Agency for Research on Cancer,
150 cours Albert Thomas, F-69372 Lyon cedex 08, France

© International Agency for Research on Cancer, 1997

Distributed by Oxford University Press, Walton Street, Oxford, UK OX2 6DP (Fax: +44 1865 267782) and in
the USA by Oxford University Press, 2001 Evans Road, Carey, NC 27513, USA (Fax: +1 919 677 1303).
All IARC publications can also be ordered directly from IARC*Press*
(Fax: +33 4 72 73 83 02; E-mail: press@iarc.fr).

Publications of the World Health Organization enjoy copyright protection in
accordance with the provisions of Protocol 2 of the Universal Copyright Convention.
All rights reserved.

The designations used and the presentation of the material in this publication do not imply the
expression of any opinion whatsoever on the part of the Secretariat of the World Health Organization
concerning the legal status of any country, territory, city, or area or of its authorities,
or concerning the delimitation of its frontiers or boundaries.

The mention of specific companies or of certain manufacturers' products does not imply
that they are endorsed or recommended by the World Health Organization in preference to others
of a similar nature that are not mentioned. Errors and omissions excepted,
the names of proprietary products are distinguished by initial capital letters.

The authors alone are responsible for the views expressed in this publication.

The International Agency for Research on Cancer welcomes requests for permission to
reproduce or translate its publications, in part or in full. Applications and enquiries should be addressed
to the Editorial & Publications Service, International Agency for Research on Cancer,
which will be glad to provide the latest information on any changes made to the text, plans for new editions,
and reprints and translations already available.

IARC Library Cataloguing in Publication Data

Social Inequalities and Cancer/
    editors, Manolis Kogevinas ... [et al.]
    (IARC Scientific Publication; 138)

    1. Neoplasms – epidemiology
    2. Socioecomonic factors
    I. Kogevinas, Manolis II. Title III. Series

ISBN 92 832 2138 9   (NLM Classification: W1)
ISSN 0300–5085

Foreword

Differences in morbidity and mortality across socioeconomic groups have been documented since the beginning of this century. IARC has previously published an overview of the data from England and Wales up to the early 1980s (Cancer Mortality by Occupation and Social Class 1851–1971, IARC Scientific Publications No. 36). This new publication brings together data on socioeconomic differences in cancer incidence and survival from many countries. In addition, detailed information on the social distribution of the most important causes of cancer is presented. This book shows that in both industrialized and less-developed countries, cancer incidence and cancer survival are related to socioeconomic status. Lower classes tend to have higher cancer incidence and poorer cancer survival than higher social classes, although this pattern differs for specific cancers. It is shown that social class differences in cancer incidence can, in part, be explained by known risk factors, particularly tobacco smoking, occupational exposures, reproductive behaviour, diet and chronic infections.

This book is intended to stimulate research in understanding the causes of socioeconomic differences in cancer and will help in developing appropriate effective measures for the prevention of social inequalities in cancer control.

P. Kleihues
Director, IARC

Acknowledgements

We are grateful to all our colleagues who contributed to this book by providing published or unpublished data, or who helped in other ways: B.C. Balram (Health and Community Services, New Brunswick, Canada); L. Barlow (National Board of Health and Welfare, Stockholm, Sweden); R. Bourbonnais (Laval University, Quebec, Canada); M. Dosemeci (National Cancer Institute, Bethesda, Washington DC, USA); S. Fincham (Alberta Cancer Board, Edmonton, Alberta, Canada); M.A. Fingerhut (National Institute of Occupational Health and Safety, Cincinnati, USA); J. Fox (Office of Population Censuses & Surveys, London, UK); M. Giraldez (National School of Public Health, Lisbon, Portugal); P. Jozan (Hungarian Central Statistical Office, Budapest, Hungary); H. Kromhout (National Institute of Public Health and Environmental Protection, Bilthoven, The Netherlands); A. Leclerc (INSERM, Paris, France); E. Lynge (Danish Cancer Society, Copenhagen, Denmark); H. Malker (National Institute of Occupational Health and Safety, Stockholm, Sweden); A. Mielck (Institute for Medical Informatics and Systems Research, Neuherberg, Germany); C.E. Minder (Institute of Social and Preventive Medicine, Bern, Switzerland); I. Plesko (Slovak Academy of Sciences, Bratislava, Slovak Republic); E. Pukkala (Finnish Cancer Registry, Helsinki, Finland); M. Rahu (Estonian Institute of Experimental and Clinical Medicine, Tallinn, Estonia); E. Regidor (Ministry of Health and Consumption, Madrid, Spain); J. Siemiatycki (University of Quebec, Quebec, Canada); M. Susser (Columbia University, New York, USA); M. Thun (American Cancer Society, Atlanta, USA), R. Wilkins (Statistics Canada, Ottawa, Canada); and W. Zatonski (The Maria Sklodowska-Curie Memorial Cancer Center and Institute of Oncology, Warsaw, Poland). We also thank Montse Ginesta, Barcelona, for the painting used on the cover of the book.

The Editors

Contributors

O. Andersen
Center for Research in Health and Social Statistics
The Danish National Research Foundation
Sejerøgade 11
2100 Copenhagen Ø, Denmark

A. Auvinen
Finnish Cancer Registry
Lisankatu 21B
00170 Helsinki, Finland

V. Beral
Cancer Epidemiology Unit
Imperial Cancer Research Fund
Gibson Building
Radcliffe Infirmary
Oxford OX2 6HE, UK

L.F. Berkman
Department of Health and Social Behaviour and Epidemiology
Harvard School of Public Health
677 Huntington Avenue
Boston, MA 02115, USA

P. Boffetta
Unit of Environmental Cancer Epidemiology
IARC
150 cours Albert Thomas
69372 Lyon cedex 08, France

F.X. Bosch
Servei d'Epidemiologia
Hospital Duran i Reynals
Autovia de Castelldefels, Km 2.7
08907- Hospitalet de Llobregat
Barcelona, Spain

C. Boschi-Pinto
Department of Epidemiology
Harvard School of Public Health
677 Huntington Avenue
Boston, MA 02115, USA

F. Faggiano
Department of Hygiene and Community Medicine
University of Torino
Via Santena 5bis
10126 Torino, Italy

A. Feeney
University College London
Medical School

Department of Epidemiology and Public Health
1–19 Torrington Place
London WC1E 6BT, UK

S. Karjalainen
Medical Research Council
Academy of Finland
P.O. Box 57
00551 Helsinki, Finland

M. Kogevinas
Department of Epidemiology and Public Health
Institut Municipal d'Investigació Mèdica (IMIM)
80 Doctor Aiguader Rd
Barcelona 08003, Spain

E. Lynge
Danish Cancer Society
Strandboulevarden 49, Box 839
2100 Copenhagen Ø, Denmark

S. Macintyre
MRC Medical Sociology Unit
6 Lilybank Gardens
Glasgow G12 8R2, UK

M. Marmot
University College London
Medical School
Department of Epidemiology and Public Health
1–19 Torrington Place
London WC1E 6BT, UK

H. Møller
Center for Research in Health and Social Statistics
The Danish National Research Foundation
Sejerøgade 11
2100 Copenhagen Ø, Denmark

N. Muñoz
Unit of Field and Intervention Studies, IARC
150 cours Albert Thomas
69372 Lyon cedex 08, France

T. Partanen
Institute of Occupational Health
Topeliuksenkatu 41aA
00250 Helsinki, Finland

N. Pearce
Wellington Asthma Research Group

Department of Medicine
Wellington School of Medicine
P.O. Box 7343
Wellington South, New Zealand

M. Porta
Institut Municipal d'Investigació Mèdica (IMIM)
80 Doctor Aiguader Rd
Barcelona 08003, Spain

J.D. Potter
Cancer Prevention Research Program
Fred Hutchinson Cancer Research Center
1124 Columbia Street MP 702
Seattle, WA 98104, USA

K. Resnicow
Emory University
Rollins School of Public Health
1518 Clifton Road
Atlanta, GA 30322, USA

K.V. Shah
The Johns Hopkins University
School of Hygiene and Public Health
Baltimore, MD 21205, USA

S. de Sanjosé
Servei d'Epidemiologia
Hospital Duran i Reynals
Autovia de Castelldefels, Km 2.7
08907- Hospitalet de Llobregat
Barcelona, Spain

I. dos Santos Silva
Epidemiology Monitoring Unit
Department of Epidemiology and Population Sciences
London School of Hygiene and Tropical Medicine
Keppel Street
London WC1E 7HT, UK

R. Saracci
Istituto Fisiologia Clinica CNR
Epidemiology Section
Via Trieste
56100 Pisa, Italy

N. Segnan
Unit of Cancer Epidemiology
Department of Oncology
S. Giovanni Hospital
Via S. Francesco da Paola 31
10123 Torino, Italy

S.D. Stellman
American Health Foundation,
320 East 43rd Street
New York, NY 10017, USA

S.O. Stuver
Center for Cancer Prevention and Department of Epidemiology
Harvard School of Public Health
677 Huntington Avenue
Boston, MA 02115, USA

I. Susser
Department of Anthropology
Hunter College
City University of New York
695 Park Avenue
New York, NY, USA

M. Susser
Gertrude H. Sergievsky Center
Columbia University
P&S Box 16
630 West 168th Street
New York, NY 10032, USA

L. Tomatis
Scientific Director
Istituto dell' infanzia
63 via dell'Istria
34137 Trieste, Italy

H. Tønnesen
Department of Surgery
Copenhagen County Hospital
Herlev Ringvej 75
2730 Herlev, Denmark

D. Trichopoulos
Center for Cancer Prevention and
Department of Epidemiology
Harvard School of Public Health
677 Huntington Avenue
Boston, MA 02115, USA

P. Westerholm
National Institute of Occupational Health
Solna, Sweden

A. Woodward
Department of Public Health
Wellington School of Medicine
University of Otago
PO Box 7343
Wellington South, New Zealand

Contents

Social inequalities and cancer ... 1
The Editors: M. Kogevinas, N. Pearce, M. Susser and P. Boffetta

General Considerations

Chapter 1: Why study socioeconomic factors and cancer? ... 17
N. Pearce

Chapter 2: Poverty and cancer ... 25
L. Tomatis

Chapter 3: Social theory and social class ... 41
I. Susser

Chapter 4: The measurement of social class in health studies: old measures and new formulations ... 51
L.F. Berkman and S. Macintyre

Evidence of Social Inequalities in Cancer

Chapter 5: Socioeconomic differences in cancer incidence and mortality ... 65
F. Faggiano, T. Partanen, M. Kogevinas and P. Boffetta

Chapter 6: Socioeconomic differences in cancer survival: a review of the evidence ... 177
M. Kogevinas and M. Porta

Explanations for Social Inequalities in Cancer

Chapter 7: General explanations for social inequalities in health ... 207
M. Marmot and A. Feeney

Chapter 8: Tobacco smoking, cancer and social class ... 229
S.D. Stellman and K. Resnicow

Chapter 9: Alcohol drinking, social class and cancer ... 251
H. Møller and H. Tønnesen

Chapter 10: Diet and cancer: possible explanations for the higher risk of cancer in the poor ... 265
J.D. Potter

Chapter 11: Socioeconomic differences in reproductive behaviour ... 285
I. Dos Santos Silva and V. Beral

Chapter 12: Social differences in sexual behaviour and cervical cancer ... 309
S. de Sanjosé, F.X. Bosch, N. Muñoz and K. Shah

Chapter 13:	**Infection with hepatitis B and C viruses, social class and cancer** S.O. Stuver, C. Boschi-Pinto and D. Trichopoulos	319
Chapter 14:	**Infection with *Helicobacter pylori* and parasites, social class and cancer** P. Boffetta	325
Chapter 15:	**Exposure to occupational carcinogens and social class differences in cancer occurrence** P. Boffetta, P. Westerholm, M. Kogevinas and R. Saracci	331
Chapter 16:	**Unemployment and cancer: a literature review** E. Lynge	343
Chapter 17:	**Unemployment and cancer in Denmark 1970–1975 and 1986–1990** E. Lynge and O. Andersen	353
Chapter 18:	**Environmental exposure, social class and cancer risk** A. Woodward and P. Boffetta	361

Socioeconomic Differences in Health Care

Chapter 19:	**Socioeconomic status and cancer screening** N. Segnan	369
Chapter 20:	**Possible explanations for social class differences in cancer patient survival** A. Auvinen and S. Karjalainen	377

Social inequalities and cancer

A summary by the Editors

Why study socioeconomic factors and cancers?
Inequalities in health reflect social inequalities in society; they provide perhaps the most convincing index of inequality (Chapter 2). Despite attempts to change the social structure and to arrive at a more egalitarian society, social inequalities have not disappeared and seem even to be increasing worldwide. At the global level, socioeconomic differences in health are stark. They are apparent in the worse sanitary conditions, higher mortality, lower life expectancy and lower cancer survival rates of the populations of developing countries compared with those of industrialized countries. Differences in cancer risk are also seen within industrialized countries between the socioeconomically less and more favoured population groups. In certain areas of industrialized countries, social and environmental conditions comparable with those existing in the poorest countries of the world have been recreated. However, social inequalities in health are not limited to those of lowest socioeconomic status but operate across the whole of society.

The occurrence of cancer within a population can be studied at many different levels, including forms of social entities, 'the individual', a particular organ system, or a particular molecule (Chapter 1). The causes of cancer can also be studied at these different levels, including socioeconomic factors, lifestyle, and genetic alterations in a clone of cells. Clearly, there are advantages in understanding disease causation at all of the different levels at which it can be analysed. Although cancer risk factors such as tobacco smoke may appear to operate mainly at the individual level, exposure may occur due to a wide range of political, economic and social factors; conversely, tobacco smoke ultimately has effects at the cellular and molecular levels, including the production of mutations in crucial genes. Of course, it is important to gain information, and take action, at all possible levels, but the history of public health shows that changes at the population level are usually more fundamental and effective than changes at the individual level, even when a single risk factor accounts for most cases of disease. In this sense, a risk factor such as smoking can be regarded as a secondary symptom of deeper underlying features of the social and economic structure of society. Thus, just as a variety of health effects in various organ systems (for example, various types of cancer) may have a common contributing cause (for example, tobacco smoking) at the level of the individual, a variety of individual exposures (for example, smoking and diet) may have common socioeconomic causes at the population level.

This volume
This volume is organized in four parts and 20 chapters. The first part, 'General considerations', contains four chapters presenting an overview of issues of poverty and health, and also discussing theoretical and methodological issues on the definition and measurement of social class in epidemiological studies. (Regional, gender or ethnic differences in health, important in their own right, are beyond the scope of this book.) The second part, 'Evidence of social inequalities in cancer', includes two chapters summarizing international data on social class differences in cancer incidence and mortality, and in cancer survival. The third part, 'Explanations for social inequalities in cancer', contains 12 chapters. It starts with a discussion of general explanations for social inequalities and cancer, and then international data on the prevalence of major cancer risk factors in different social strata are presented, particularly for tobacco, alcohol, diet, reproductive patterns, sexual behaviour, infectious agents, environmental and occupational exposures, and the effects of unemployment. The extent to which these risk factors explain socioeconomic differences in cancer incidence is discussed. The fourth part comprises two chapters on socioeconomic differences in health care, which present and discuss differences in access to and use of health services, particularly in relation to the early diagnosis of cancer.

Theories of social class and measurement of social inequality (Chapters 3 and 4)

Concepts of class theory developed with the emergence of industrial society in the nineteenth century (Chapter 3). For an understanding of current divisions, however, theories must reflect the advances of capitalism and the global economy that characterize the late-twentieth century. In industrialized societies, reductions in the industrial workforce and the growth of finance, investment and real-estate industries worldwide have added a new service workforce that is largely female. Large sectors of industry have departed in search of cheaper labour in poorer countries. As a result, in those areas too, a new industrial workforce has emerged. Concomitantly, accumulation of land used for cultivation for the world market in less developed agricultural regions has led to an increase in mobile agricultural labour and a shift of landless labourers to the cities of less developed countries. In addition, both upward and downward mobility have occurred for individuals and groups in specific populations as well as for particular diseases in developed and less developed countries. All these changes have precipitated fundamental changes in class, gender and family relationships and transformed the living conditions of populations in both developed and less developed societies. These changes have major implications for the patterns of health and disease in the world today.

The measurement of socioeconomic status requires that we think more precisely about both conceptual issues and issues more traditionally thought of as measurement issues (Chapter 4). Progress in this area rests on our ability to identify those aspects of socioeconomic status that are most closely related to health, human development, and life expectancy. Measures of socioeconomic status have been based on characteristics of the individual as well as on characteristics of the environment or more ecologically based measures. Each of these types of socioeconomic status measures has strengths and weaknesses and in all likelihood taps somewhat different aspects of class. In measuring socioeconomic status across diverse populations, it is also crucial to be sensitive to the ways in which measurement varies across different cultures, ethnic and demographic groups. It is likely that more refined research in this area will clarify more fully why socioeconomic status is so profoundly related to health status. In order to understand this relationship, efforts will have to be focused on identifying not only those psychosocial or biological processes that occur 'downstream' as a result of socioeconomic status (for example, occupational exposures as a mediator of the higher cancer risk of manual social classes) but also the nature of the social experience itself and those 'upstream' forces that place so many individuals at risk (for example, the reasons why at this time manual social classes take up smoking more frequently than non-manual classes, although the cigarette smoking habit originated in the high classes).

Socioeconomic differences in cancer incidence and mortality (Chapter 5)

Data on the presence, magnitude and consistency of socioeconomic differentials in mortality and incidence of all malignant neoplasms and 24 individual types of neoplasms in 35 populations from 20 countries are reviewed in Chapter 5. Reasonably consistent excess risks in men in lower social strata were observed for all respiratory cancers (nose, larynx and lung) and cancers of the oral cavity and pharynx, oesophagus, stomach, and, with a number of exceptions, liver, as well as for all malignancies taken together. For women, low-class excesses were consistently encountered for cancers of the oesophagus, stomach, cervix uteri and, less consistently, liver. Men in higher social strata displayed excesses of colon and brain cancers and skin melanoma. In the two Latin American populations for which data were available, lung cancer was more frequent in higher social strata. Excesses in high socioeconomic strata were seen in women in most populations for cancers of the colon, breast, ovary, and skin melanoma. Data for the United Kingdom, Denmark, Italy and New Zealand are shown in Figure 1 for men and Figure 2 for women. Longitudinal data from England and Wales suggest widening over time of social class differences in men for all cancers combined (Figure 3) and for cancers of the lung, larynx and stomach, and in women for all cancers combined and for cervical cancer.

Socioeconomic differences in cancer survival (Chapter 6)

In the discussion of social inequalities in health there has been much debate on the role of medical

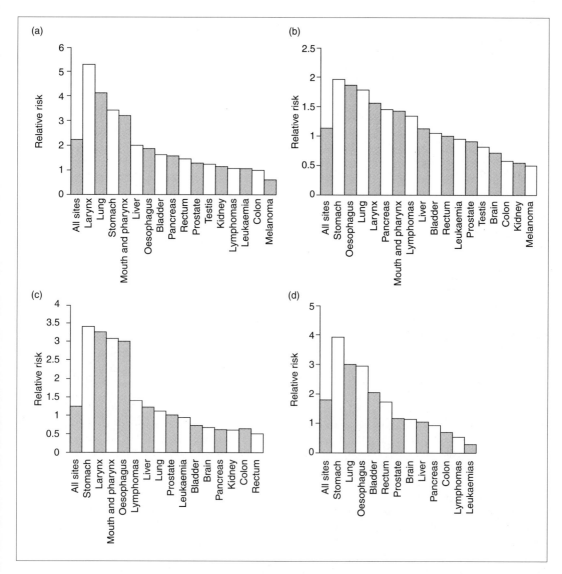

Figure 1. (a) Cancer mortality in men (aged 20–64) in social class V versus class I, in Great Britain during the years 1979–1980 and 1982–1983 (OPCS, 1986). (b) Cancer incidence in unskilled men versus employees group I (all ages), in Denmark during the years 1979–1980 (Lynge & Thygesen, 1990). (c) Cancer mortality in illiterate men versus men with university education (men aged 18–74), in Italy during the years 1981–1982 (Faggiano et al., 1994). (d) Cancer mortality of men (aged 15–64) in social class V versus class I, in New Zealand during the years 1984–1987 (Pearce & Bethwaite, in press).

care. To understand the potential importance of socioeconomic differences in prompt detection and treatment of cancer, data on cancer survival are essential. These have been examined less extensively than differences in cancer incidence. Forty-two studies on social class differences in cancer survival, covering 12 cancer sites in 14 different countries, are reviewed in Chapter 6. Social class differences in cancer survival appear remarkably general (Figure 4). Patients in low social classes had consistently poorer survival than those in high social classes. The magnitude of the differences for most cancer sites is fairly narrow, with most relative risks falling between a range of 1 and 1.5. The widest differences were observed for cancers of good prognosis and specifically cancers of the female breast, corpus uteri, bladder and colon. Social differences in cancer survival were present in both genders and in most countries and were found consistently whichever socioeconomic indicator was used.

General explanations for social inequalities in health (Chapter 7)

Life expectancy has always differed according to status in society, with a higher mortality among those of lower social status. Although cancer and cardiovascular diseases are proportionally more common as causes of death in rich societies, in industrialized countries the major causes of death

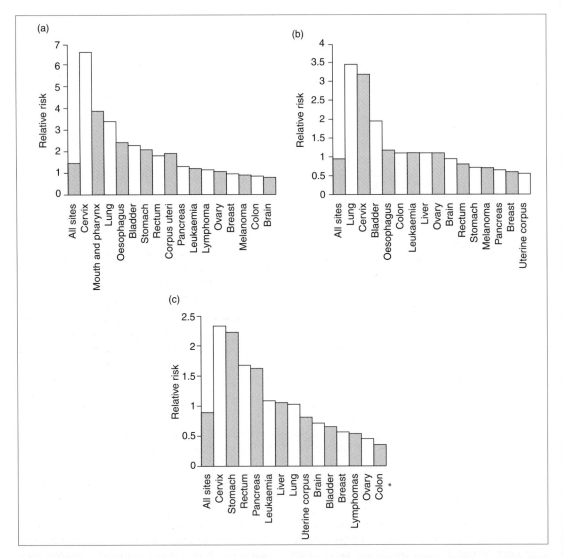

Figure 2. (a) Cancer mortality in women (aged 18–74) of social class V versus class I, in Great Britain during the years 1979–1980 and 1982–1983 (OPCS, 1986). (b) Cancer incidence in unskilled women versus employees group I (all ages), in Denmark during the years 1970–1980 (Lynge & Thygesen, 1990). (c) Cancer mortality in illiterate women versus women with university education (women aged 18–74), in Italy during the years 1981–1982 (Faggiano et al., 1994).

are more common in those of lower social status. Much of the discussion about social inequalities in health has been focused on the health disadvantage of those of lowest socioeconomic status. Data from the Whitehall studies show that the social gradient in morbidity and mortality exists across employment grades in British civil servants, none of whom is poor by comparison with people in developing countries, suggesting that there are factors that operate across the whole of society. The magnitude of socioeconomic differences in health varies between societies, and over time within societies. This suggests that identification of factors that influence socioeconomic status and health, and the pathways by which they operate, is an important public health task that could lay the basis for a reduction in inequalities in health.

Tobacco smoking (Chapter 8)

Consumption of tobacco products is causally connected with many types of cancer – mainly lung, larynx, mouth and pharynx, oesophagus and bladder cancers. Tobacco is the main specific contributor to total mortality in many developed countries and has become a major contributor in developing countries as well. In most industrialized countries, prevalence of cigarette smoking is currently higher in low than in high social classes, the differences being more pronounced in men than in women. The pattern shown in Table 1 for Spain is characteristic of the pattern observed in industrialized countries in the last decades. This pattern of tobacco consumption may not be typical for developing countries. In some industrialized countries, smoking was more frequent in high social classes during the first half of this century. Trends in prevalence of smoking in the United States of America (Figure 5) and many other countries indicate that the proportion of current smokers has fallen more rapidly in high than in low social classes. To formulate and carry out effective tobacco control activities it is important to assess the relative incidence of tobacco-related cancers in different social strata and the prevalence of tobacco use across strata. Despite many years of data gathering, the information base is far from complete, especially in developing countries where tobacco use is increasing rapidly and where aggressive marketing by the transnational tobacco industry is occurring. A key question is the extent to which tobacco usage can 'explain' the ob-

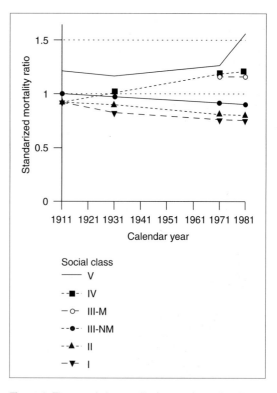

Figure 3. Time trends in mortality for men from all malignant neoplasms, by social class in England and Wales during the period 1911–1981. SMR, standardized mortality ratio.

served social class differences in cancer risk. Class differences in lung cancer are likely to be mostly related to the unequal distribution of tobacco smoking between social classes, and in some fairly simple situations this has been satisfactorily demonstrated. Nevertheless, there are many unresolved issues, especially with regard to the role of collateral exposures, such as hazardous occupations, poor diet, and limited access to health care.

Alcoholic-beverage drinking (Chapter 9)

Alcohol drinking causes cancers of the upper gastrointestinal and respiratory tracts and liver cancer. Patterns of alcohol drinking by socioeconomic status are not consistent between countries and between genders. A role of alcohol drinking in the observed negative social class gradients for alcohol-related cancers is very likely in men in France, Italy and New Zealand. Evidence that is less strong but suggestive of a role of alcohol drinking is seen for men in Brazil, Switzerland, the United Kingdom and Denmark. Although a role of alcohol drinking in cancer causation is likely or possible in certain populations, other factors may contribute as well, most notably tobacco smoking and dietary habits.

Diet (Chapter 10)
There are a variety of ways in which diet may influence the development of human cancers. In Chapter 10, a theoretical framework is proposed in which a main feature is a dietary pattern to which humans are well adapted – an 'original diet'. This original dietary pattern had specific features, which included regular exposure to a variety of substances on which human metabolism is dependent but that are not usually explicitly labelled as 'essential nutrients'. The theory suggests that the higher risk of cancer in the low social classes at this time, in both the developed and developing world, is related, to an as yet unknown degree, to the fact that the amount of variation from the diet to which we are well adapted is greater in that portion of the population who have less access to the world's goods and services. This is particularly true regarding the intake of fresh vegetables and fruit, which are almost universally consumed in smaller quantities among the poor in most parts of the world. Some diet-related cancers, particularly breast cancer, run counter to the general trend towards higher risks in poorer people; it is probable that social class differences in other risk factors, particularly reproductive history, explain this discrepancy at least in part.

Reproductive factors (Chapter 11)
Socioeconomic variations in the risk of female reproductive cancers are marked. Data from the World Fertility Surveys, the Demographic and Health Surveys, and other national surveys are examined in Chapter 11 to assess whether these variations in cancer risk might be explained, at least in part, by socioeconomic variations in reproductive behaviour. Marked socioeconomic differentials in reproductive pattern were present in almost all settings: countries with low and high levels of modernization,

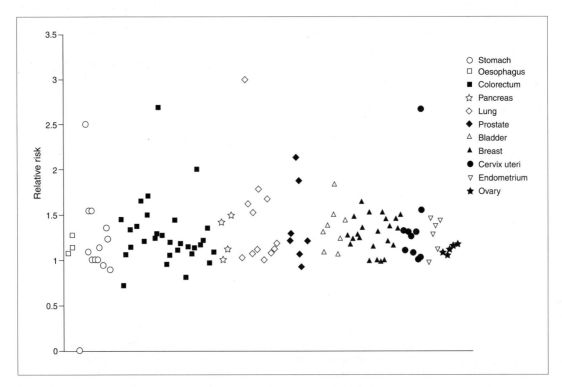

Figure 4. Socioeconomic differences in cancer survival: relative risks for patients in low versus high socioeconomic status in 33 studies.

Table 1. Prevalence of smoking (percentage) according to age and education in Spain, 1987

Education	Males		Females	
	16–24 years	65+ years	16–24 years	65+ years
Less than primary school	67%	34%	43%	1%
Primary school	62%	32%	46%	4%
High school	51%	39%	48%	6%
University	53%	37%	62%	5%

and countries with low and high levels of fertility (Table 2). In general, women of higher socioeconomic status and with more education had lower fertility and later age at first birth, a greater prevalence of childlessness, shorter duration of breast feeding and later age at menopause. The direction and size of these differences varied markedly from country to country according to level of economic development and, within each country, from generation to generation of women. In Western countries, some of these socioeconomic differences may possibly be narrowing in recent generations. There was little evidence of socioeconomic variations in age at menarche. The observed socioeconomic differentials in most aspects of reproductive behaviour account for some of the socioeconomic variation in risk of female reproductive cancers. However, this relationship could not be assessed directly because such analysis would require unavailable birth-cohort-specific data on socioeconomic variations in reproductive behaviour and in cancer risks.

Sexual behaviour and infection with human papillomavirus (Chapter 12)

Information on social class differences in sexual behaviour is available only for a limited number of, mostly industrialized, countries. According to population-based surveys in industrialized countries, men of low socioeconomic status report fewer sexual partners than men of high status. There is no clear indication that the same is true of women (Table 3). Cervical cancer is the most important cancer linked with sexual behaviour. It is the most common cancer in women in developing countries and the sixth most common in developed countries. In all areas, it is more frequent among women of low socioeconomic status, and is associated with multiple sexual partners and early age at first sexual intercourse. Both incidence and mortality are reduced by screening.

The human papillomavirus (HPV) has been shown to be the main biological agent causing cervical cancer. The extent to which infection with HPV and other sexually transmitted diseases relates to the occurrence of socioeconomic differences in cervical cancer incidence was examined in two parallel case–control studies in Spain and Colombia. The results, presented in Chapter 12, indicate that socioeconomic differences in the incidence of cervical cancer can, in part, be explained by differences in the prevalence of HPV DNA. Male sexual behaviour, and particularly contacts with prostitutes, may be a major contributor to the higher prevalence of HPV DNA among the poor.

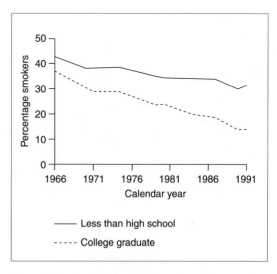

Figure 5. Time trends in the prevalence of smoking: percentage of current smokers in the USA by education.

Chronic infections (Chapters 13 and 14)

Various infectious agents, in addition to HPV, have been associated with the occurrence of cancer. Relations of such organisms to social class are available only for *Helicobacter pylori* and the hepatitis viruses (Table 4). The hepatitis B and C viruses (HBV and HCV) are major etiological factors in the occurrence of hepatocellular carcinoma worldwide, but most especially in developing countries where the majority of liver cancer cases can be found. *H. pylori* has been associated with stomach cancer.

In parallel with the geographic distribution of hepatocellular carcinoma, high levels of HBV endemicity also are concentrated in the developing world (Chapter 13). Low educational attainment, lower social stratum, and crowded urban residence have been reported to predict higher HBV chronic carrier prevalence in both developed and developing countries. More importantly, the effect of poverty on HBV endemicity is clearly evident among younger age groups, and earlier chronic HBV infection seems to increase the risk of development of hepatocellular carcinoma. The limited number of studies of the seroepidemiology of HCV also report an association between higher prevalence of antibodies to HCV and indicators of low social class. It would appear that the striking correlation between hepatocellular carcinoma and low socioeconomic status, both within industrialized societies and when comparing industrialized with less developed countries, is largely related to the impact of poverty on the spread of HBV and probably HCV.

Studies in the United Kingdom and USA strongly suggest that social class factors, in particular those acting during childhood, are determinants of infection with the bacterium *H. pylori* (Chapter 14). The odds ratio of seroprevalence are of the order of 1.5–5 for lower social classes compared with higher social classes. A conservative estimate of the role of social class, acting through an increased prevalence of *H. pylori* infection, on the burden of stomach cancer resulted in an estimated number of over 50 000 stomach cancer cases per year worldwide – or approximately 8% of all stomach cancers.

Table 2. Reproductive behaviour by socioeconomic status in developed and developing countries

	Developed countries		Developing countries	
	General tendency	Comments	General tendency	Comments
Parity	Higher in manual social classes	U-shaped relation in some countries	Higher in manual social classes	Differences more pronounced in Central and South America
Age at birth of first child	Earlier in manual social classes		Earlier in manual social classes	Less consistent data than in developed countries
Childlessness	No consistent pattern	Little data	No data	International data indicate a reduction of childlessness at the first phases of modernization, but this trend may reverse subsequently
Age at menarche	No data		Little variation	
Age at menopause	Earlier in manual social classes	Little data	No data	
Breastfeeding	Shorter duration in manual social classes		Longer duration in manual social classes	

Table 3. Sexual partnership as reported in population-based studies

Author (country)	Sex, age	Sexual partnership	Socioeconomic status		
			Low (%)	High (%)	P value[a]
Leigh et al., 1993 (USA)	Men & women: 42.8	≥5 partners in last 5 years	5.5	12.2	0.016
Seidman et al., 1992 (USA)	Women: 15–44	≥ 2 partners 3 months before interview	6.5	2.6	<0.01
Laumann et al., 1994 (USA)	Men & women: 18–59	>10 partners in adult lifetime	14.6	22.2	<0.01
Johnson et al., 1994 (UK)	Men: 35–44	≥2 partners 1 year before interview	6.8	10.1	<0.01
	45–59		2.8	7.3	<0.01
	Women: 35–44	>1 partner 1 year before interview	4.4	4.1	0.89
	45–59		2.4	2.2	0.91
Spira et al., 1994 (France)	Men: 18–69	>1 partner 1 year before interview	7.4	15.9	<0.01
	Women: 18–69	>1 partner 1 year before interview	3.0	7.6	<0.01
Melbye & Biggar, 1992 (Denmark)	Men: 18–59	>1 partner 1 year before interview	5.4	25.7	<0.01

[a] P values refer to the comparison between high and low socioeconomic status.

Occupational factors (Chapter 15)

Occupational exposures are responsible for about an estimated 4% of the total of human cancers in industrialized countries. These cancers, however, are concentrated among manual workers and in the lower social classes, thus contributing to the social class gradient in cancer incidence and mortality. An estimate from 1971 cancer mortality data for England and Wales assigns to occupational cancer about a third of the total difference between high (I, II and III-NM) and low (III-M, IV and V) social classes, and about a half of the differences for lung and for bladder cancer (Table 5). Direct evidence on the extent of the contribution of occupational exposure to carcinogens to social class differences is lacking. Several problems, such as possible interaction between carcinogens, and the effect of extraoccupational confounding factors, add further elements of uncertainty.

Unemployment (Chapters 16 and 17)

With a tenth of the labour force involuntarily out of work, unemployment has become an important element among the socioeconomic determinants of health in the rich countries (Chapter 16). Unemployed men have an excess cancer mortality of close to 25% compared with that of all men in the labour force (Table 6). The available data from England and Wales, Finland and Denmark (including recent data presented in Chapter 17) indicate that this excess risk is found both in periods when the unemployment rate is about 1% and in periods when it is around 10%. Furthermore, it persists long after the start of unemployment and the risk does not disappear when social class, smoking, alcohol intake, and previous sick days are controlled for. The excess risk comes mainly from lung cancer, and the excess risk of lung cancer does not disappear when social class and number of previous

Table 4. Prevalence of infection with *H. pylori* and prevalence of HBsAg by socioeconomic status – results of selected studies

Study	Population	Social class indicator	Category	Prevalence (%)
Sitas et al., 1991 Prevalence of *H. pylori*	749 adults, UK	Registrar General's social class	I, II III IV, V	49 57 62
Fiedorek et al., 1991 Prevalence of *H. pylori*	245 children, USA	Income of family	<US$ 5000/year >US$ 25 000/year US$ 5000–25 000/year	39 27 16
Patel et al., 1994 Prevalence of *H. pylori*	554 children, UK	Persons per room	<0.5 0.5–1.0 >1.0	10 9 23
Awidi et al., 1984 Prevalence of HBsAg	Volunteer blood donors, Jordan	Residence	Non-crowded urban areas Intermediate level urban areas Poor, crowded urban areas, refugee camps, rural areas	0.7 1.7 6.9

HBsAg, hepatitis B surface antigen.

Table 5. Ratios of cancer mortality between manual and non-manual social classes with and without excluding cancers attributable to occupational exposures (England and Wales, 1971)[a]

Cancer site	Crude rate ratio (R_c)[b]	Rate ratio for the proportion of cancers not attributable to occupation (R_a)[c]	Excess risk (%) attributable to occupation[d]
Liver	1.16	1.09	42
Larynx	1.76	1.71	5
Lung	1.71	1.37	48
Nose	1.38	0.90	100
Skin (non-melanoma)	1.77	1.55	29
Prostate	1.19	1.17	9
Bladder	1.36	1.17	52
All cancers	**1.40**	**1.27**	**32**

[a]Based on 25–64 years cumulative rates reported by Logan (1982). Only cancer sites that have been strongly related with occupational exposures are reported. Proportions of cancers attributable to occupation were derived from Doll & Peto (1981); all cancers related to occupation were assumed to occur among manual workers.
[b]Ratio of the rate among manual workers to the rate among non-manual workers.
[c]As crude rate ratio, after excluding cancers attributable to occupation (see Chapter 15 for details).
[d]Percentage of the crude rate ratio accounted for by cancers attributable to occupation, or $[(R_c - R_a) / (R_c - 1)] \times 100$.

Table 6. Total cancer mortality in unemployed men

Country, population	Study period	Age	SMR/RR	Reference[a]
Italy, 1981 census, mortality	1981–1985	15–59	1.75	Costa & Segnan, 1987
UK, 1971 census, mortality	1971–1981	15–64	1.44	Moser et al., 1990
UK, 1981 census, mortality	1983	16–64	1.38	Moser et al., 1987
UK, 1971 census, incidence	1971–1981	15–64	1.29	Kogevinas, 1990
UK, Regional Heart Study, mortality	1978/80–89	40–59	1.74	Morris et al., 1994
Finland, 1980 census, mortality	1981–1985	30–54	1.39	Martikainen, 1990
Denmark, 1970 census, mortality	1970–1980	20–64	1.33	Iversen et al., 1987
Denmark, 1970 census, incidence	1970–1975	20–64	1.25	Lynge & Andersen, 1996
Denmark, 1986 census, mortality	1986–1990	20–64	1.23	Lynge & Andersen, 1996
USA, 1979–1983, Current Population Survey, mortality	1979–1983	25–64	0.86	Sorlie & Rogot, 1990

SMR, standardized mortality ratio; RR, relative risk/rate ratio.
[a]See complete list of references in chapters by Lynge and by Lynge & Andersen in this volume.

sick days are controlled for. Unemployment does not increase smoking, but unemployed men have a slightly higher smoking prevalence before unemployment. However, as the excess lung cancer risk among unemployed men remains after controlling for social class it seems unlikely that it can be explained only by differences in smoking prior to unemployment.

Environmental factors (Chapter 18)
Exposure to a variety of environmental factors associated with cancer occurrence varies by social class. These factors include air pollutants (SO_2, NO_2, total suspended particles and so on), toxic waste hazards, and ionizing and other radiation. Heavy environmental pollution has been associated with an increased risk of some cancers and in particular lung cancer, and limited evidence suggests that individuals from low social classes are exposed to higher levels of environmental pollutants than individuals from high social classes. This may be due to the placement of new sources of pollution or of toxic processes in disadvantaged areas, or to the s elective migration of the poorer sectors of society to these areas. The available data do not allow any conclusion on the contribution to social class differences in cancer occurrence of exposure to environmental pollution. Exposure to ultraviolet (UV) radiation is due principally to sunlight, and is modified strongly by personal behaviours such as choice of recreation and use of protective clothing. Those in outdoor occupations are likely to receive the highest cumulative exposure to UV radiation. There is no clear evidence from recent surveys in Australia and North America that socioeconomic factors are strongly related to nonoccupational exposure to UV radiation. Information is lacking on the influence of socioeconomic status on sun exposure in other parts of the world. There is little information on the social distribution of exposure to ionizing radiation.

Socioeconomic status and cancer screening (Chapter 19)
The only widely applied cancer screening programmes are those for cervical and female breast cancer. Participation in breast cancer screening has been shown to depend on income and education, health insurance and type of health service. Women of low social classes tend to have lower screening participation rates than those in higher classes (Table 7). Socioeconomic differences in screening practices tend to decrease when participation is promoted, cultural and economic barriers are removed and social support is offered. In both developed and developing countries, women of low socioeconomic status have a higher than average risk of cervical cancer, and a lower than average participation in Papanicolaou smear screening.

Table 7. Sociodemographic differences in breast cancer screening – results from selected studies in the United States

Reference	Setting	Socioeconomic classification	Screening participation rate	Comments
Lane et al., 1992	Telephone survey, Suffolk County, NY, USA; random sample of women aged 50–75 living in the community (n= 404) and of women using health centres (n = 795)	*Education* <12 years High school Post high school ≥College graduate <12 years High school Post high school ≥College graduate	17% 25% 24% 49% 27% 31% 27% 41%	Percentage use of mammography in community sample Percentage use of mammography in the health care centres sample
Fletcher et al., 1993	Controlled intervention study, 1987–1990, in two communities of N. Carolina, USA; 1000 women aged 50–74	*Education* ≤High school >High school ≤High school >High school	26% 39% 33% 53%	Mammography utilization in the control community 1987 Mammography utilization in the control community 1989 Narrower differences in the intervention community
Kang et al., 1994	Household Survey, 1986; 670 African-American women	*Education* <High school High-school grad. 1–3 years college 4+ years college	0.84 (0.46–1.54) 1.0 0.79 (0.39–1.59) 1.31 (0.49–3.47)	Logistic regression; odds ratios and 95% CIs for the use of routine mammography
Urban et al., 1994	Washington state, 1984; telephone survey of 1538 women aged 50–75	*Income* <US$ 15 000 >US$ 15 000 Unknown <US$ 15 000 >US$ 15 000 Unknown	27% 48% 37% 41% 18% 29%	Percentage of women aged 50–64 having mammography within one year of interview Percentage of women aged 50–64 never having a mammography
Reeves et al., 1995	Wisconsin Tumour Registry, USA, 1988–1990; 3197 women with breast cancer	*Education* <High school High-school grad. Some college College graduate	1.0 1.2 (0.9–1.6) 1.3 (0.9–1.7) 1.5 (1.0–2.1)	Logistic regression; odds ratios and 95% CIs. Mammographic detection compared with self-detected cancer. No differences observed for CBE versus self-detected cancer.
Anderson et al., 1995	National Health Interview Survey, Cancer Control Supplement, USA; 5052 women older than 40 years in 1987 and 2709 women in 1992	*Education* <12 years 12 years >12 years <12 years 12 years >12 years	15% 25% 31% 36% 52% 61%	Reported use of mammography, 1987 Reported use of mammography, 1992

CBE, clinical breast examination; grad., graduate.

Explanations for social class differences in cancer patient survival (Chapter 20)

Social class differences in cancer patient survival have been reported for most cancer types and in a number of countries. The source of these differences has been studied less thoroughly and less systematically than social class differences in cancer occurrence. Stage of disease at diagnosis appears to be the most important factor contributing to the social class differences in cancer patient survival, although the evidence is not always consistent (Table 8). This has been observed most clearly for gastrointestinal and gynaecological cancers. Differences in survival are generally wider with localized than with advanced stages of disease. The reasons why cancers are more frequently diagnosed at a local stage in high than low social classes is not fully understood at the moment. Of other potential contributing factors, the role of treatment and psychosocial factors has scarcely been studied.

Table 8. Proportion of cases of various cancers diagnosed at a local stage, by socioeconomic status

Reference; country	Socioeconomic classification	Colon	Rectum	Lung	Female breast	Uterine cervix	Uterine corpus	Bladder	Prostate
Linden, 1969; USA	County				82%				
	Private				83%				
Lipworth et al., 1970; USA	<US$ 5000	9%	9%	2%		35%	35%	39%	27%
	>US$ 5000	10%	16%	8%		45%	46%	36%	16%
Lipworth et al., 1972; USA	Non-private	34%	53%	13%	31%	38%	48%	66%	57%
	Private	30%	36%	24%	40%	35%	60%	63%	52%
Berg et al., 1977; USA	Indigent	35%		15%	35%	69%		70%	49%
	Non-indigent	37%		19%	38%	75%		75%	44%
Keirn & Metter, 1985; USA	Indigent	34%		34%	45%				
	Non-indigent	29%		29%	35%				
Dayal et al., 1987; USA	Low	34%	34%						
	Medium	42%	42%						
	High	44%	44%						
Karjalainen & Pukkala, 1991; Finland	I (high)				51%				
	II				51%				
	III				48%				
	IV (low)				47%				
Brenner et al., 1991; Germany (colorectal)	Low	54%							
	Medium	58%							
	High	51%							
Kato et al., 1992; Japan	Non-employed	33%							
	Service	36%							
	Production	42%							
	Clerical	46%							
	Professional	38%							
Auvinen, 1992; Finland	I (high)	44%	44%						
	II	41%	41%						
	III	42%	42%						
	IV (low)	37%	37%						

Biological indicators of tumour aggressiveness have failed to explain the social class differences.

Conclusions

Clear evidence from industrialized and less developed societies shows that both cancer incidence and cancer survival are related to socioeconomic factors. Lower social classes tend to have higher cancer incidence and poorer cancer survival overall rates than higher social classes, although this pattern differs for specific cancers. Social class differences in cancer incidence can, in part, be explained by known risk factors. Tobacco smoking appears as the single most important mediating factor for the occurrence of socioeconomic differences in cancer. The extent to which tobacco smoking 'explains' socioeconomic differences in one or more of the cancers that it causes has rarely been directly addressed in epidemiological studies. Occupation, reproductive behaviour and biological agents (HPV, *H. pylori*, HBV and HCV) have also been shown to be important mediating factors for the occurrence of socioeconomic differences in cancer incidence. The main factor associated with the poorer survival of cancer patients of low socioeconomic status is stage of the cancer at diagnosis, although the evidence showing that cancer patients of low social classes present at a later stage has not always been consistent. For breast and cervical cancer, the differences in stage at diagnosis may, in part, be attributed to the differential use of cancer screening programmes. Studying the magnitude of socioeconomic differences and the mediating factors for the occurrence of these differences provides valuable information for the prevention of health inequalities. At this time, however, there is insufficient evidence to discriminate between socioeconomic factors, the social distribution of specific cancer risk factors and the overall 'package' of social inequality.

References

Anderson, L.M. & May, D.S. (1995) Has the use of cervical, breast, and colorectal cancer screening increased in the United States? *Am. J. Public Health*, 85, 840–842

Auvinen, A. (1992) Social class and colon cancer survival in Finland. *Cancer*, 70, 402-409

Awidi, A.S., Tarawneh, M.S., El-Khateeb, M., Hijazi, S. & Shahrouri, M. (1984) Incidence of hepatitis B antigen among Jordanian volunteer blood donors. *Publ. Hlth. Lond.*, 98, 92–96

Berg, J.W., Ross, R. & Latourette, H.B. (1977) Economic status and survival of cancer patients. *Cancer*, 39, 467–477

Brenner, H., Mielck, A., Klein, R. & Ziegler, H. (1991) The role of socioeconomic factors in the survival of patients with colorectal cancer in Saarland, Germany. *J. Clin. Epidemiol.*, 44, 807–815

Costa, G. & Segnan, N. (1987) Unemployment and mortality. *Br. Med. J.*, 294, 1550–1551

Dayal, H., Polissar, L., Yang, C.Y. & Dahlberg, S. (1987) Race, socioeconomic status and other prognostic factors for survival from colo-rectal cancer. *J. Chron. Dis.*, 40, 857–864

Doll, R. & Peto, R. (1981) *The causes of cancer: quantitative estimates of avoidable risk of cancer in the United States today*. New York, Oxford University Press.

Faggiano, F., Zanetti, R. & Costa, G. (1994) Cancer risk and social inequalities in Italy. *J. Epidemiol. Community Health*, 48, 447–452

Fiedorek, S.C., Malaty, H.M., Evans, D.L., Pumphrey, C.L., Casteel, H.B., Evans, D.J. Jr. & Graham, D.Y. (1991). Factors influencing the epidemiology of *Helicobacter pylori* infection in children. *Pediatrics*, 88, 578–582

Fletcher, S.W., Harris, R.P., González, J.J., Degnan, D., Lannin, D.R., Strecher, V.J., Pilgrim, C., Quade, D., Earp, J.A. & Clark, R.L. (1993) Increasing mammography utilization: a controlled study. *J. Natl Cancer Inst.*, 85, 112–120

Giovino, G.A., Schooley, M.W., Zhu, B.P., Chrismon, J.H., Tomar, S.I., Peddicord, J.P., Merrit, R.K., Husten, C.G. & Eriksen, M.P. (1994) Surveillance for selected tobacco-use behaviors – United States, 1900–1994. *MMWR*, 43, (Suppl. 3), 1-43

Iversen, L., Andersen, O., Andersen, P.K., Christoffersen, K., & Keiding, N. (1987) Unemployment and mortality in Denmark 1970-80. *Br. Med. J.*, 295, 879–884

Johnson, A. & Wadworth, J. (1994) Heterosexual partnership. In: Johnson, A.M., Wadsworth, J., Wellings, K., Field, J., ed., *Sexual attitudes and lifestyles*, Oxford, Blackwell Scientific Publications, pp. 110–182

Kang, H.S., Bloom, J.R. & Romano, P.S. (1994) Cancer screening among African–American women: their use of tests and social support. *Am. J. Public Health*, 84, 101–103

Karjalainen, S. & Pukkala, E. (1990) Social class as a prognostic factor in breast cancer survival. *Cancer*, 66, 819–826

Kato, I., Tominaga, S. & Ikari, A. (1992) The role of socioeconomic factors in the survival of patients with gastrointestinal cancers. *Jpn. J. Clin. Oncol.*, 22, 270–277

Keirn, W. & Metter, G. (1985) Survival of cancer patients by economic status in a free care setting. *Cancer*, 55, 1552–1555

Kogevinas, M. (1989) *Longitudinal study: Socio-demographic differences in cancer survival 1971–1983*. London, HMSO

Lane, D.S., Polednak, A.P. & Burg, M.A. (1992) Breast cancer screening practices among users of county-funded health centers vs women in the entire community. *Am. J. Public Health*, 82, 199–203

Laumann, E.O., Gagnon, J.H., Michael, R.T. & Michaels, S. (1994) *The social organization of sexuality*. Chicago, The University of Chicago Press, pp. 172–224

Leigh, B.C., Temple, M.T. & Trocki, K.F. (1993) The sexual behavior of US adults: Results from a national survey. *Am. J. Public Health*, 83, 400–408

Linden, G. (1969) The influence of social class in the survival of cancer patients. *Am. J. Public Health*, 59, 267–274

Lipworth, L., Abelin, T. & Connelly, R.R. (1970) Socioeconomic factors in the prognosis of cancer patients. *J. Chronic Dis.*, 23, 105–115

Lipworth, L., Bennett, B. & Parker, P. (1972) Prognosis of nonprivate cancer patients. *J. Natl Cancer Inst.*, 48, 11–16

Logan, W.P.D. (1982). *Cancer mortality by occupation and social class 1851–1971* (IARC Scientific Publications No. 36). Lyon, IARC

Lynge, E. & Thygesen, L. (1990) *Occupational cancer in Denmark*. Copenhagen

Martikainen, P. (1990) Unemployment and mortality among Finnish men. *Br. Med. J.*, 301, 407–411

Melbye, M., Biggar, R.J. (1992) Interactions between persons at risk for AIDS and the general population in Denmark. *Am. J. Epidemiol.*, 135, 593–602

Morris, J.K., Cook, D.G. & Shaper, A.G. (1994) Loss of employment and mortality. *Br. Med. J.*, 308, 1135–1139

Moser, K., Goldblatt, P., Fox, J. & Jones, D. (1990) Unemployment and mortality. In: Goldblatt P.O., ed., *Longitudinal study: mortality and social organisation*, London, HMSO, pp. 81–97

Moser, K.A., Goldblatt, P.O., Fox, A.J. & Jones, D.R. (1987) Unemployment and mortality: comparison of the 1971 and 1981 longitudinal study samples. *Br. Med. J.*, 294, 86–90

Office of Population Censuses and Surveys (1919) Supplement to 75th Annual Report. London, HMSO.

Office of Population Censuses and Surveys (1938) Decennial Supplement, Part IIa. London, HMSO.

Office of Population Censuses and Surveys (1977) Decennial Supplement, London, HMSO.

Office of Population Censuses and Surveys (1986) Occupational Mortality 1979–80, 1982–83, Decennial suppplement, London, HMSO.

Patel, P., Mendall, M.A., Khulusi, S., Northfield, T.C. & Strachan, D.P. (1994) *Helicobacter pylori* infection in childhood: risk factors and effect on growth. *Br. Med. J.*, 309, 1119–1123

Pearce, N., & Bethwaite, P. (1997) Social class and male cancer mortality in New Zealand, 1984–1987. *N.Z. Med. J.* in press

Reeves, M.J., Newcomb, P.A., Remington, P.L. & Marcus, P.M. (1995) Determinants of breast cancer detection among Wisconsin (United States) women, 1988-90. *Cancer Causes Control*, 6, 10–111

Regidor, E., Gutiérrez-Fisac, J.L. & Rodríguez, C. (1994) *Diferencias y desigualdades en salud en Espana*. Madrid, Ediciones Diaz de Santos

Seidman, S.N., Mosher, W.D. & Aral, S.O. (1992) Women with multiple sexual partners: United States, 1988. *Am. J. Public Health*, 82, 1388–1394

Sitas, F., Forman, D., Yarnell, J.W.G., Burr, M.L., Elwood, P.C., Pedley, S. & Marks, K.J. (1991) *Helicobacter pylori* infection rates in relation to age and social class in a population of Welsh men. *Gut.*, 32, 25–28

Sorlie, P.D. & Rogot, E. (1990) Mortality by employment status in the national longitudinal mortality study. *Am. J. Epidemiol.*, 132, 983–992

Spira, A., Bajos, N. & ACSF group (1993) *Les comportements sexuels en France*. Paris, La documentation Française.

Urban, N., Anderson, G.L. & Peacock, S. Mammography screening: how important is cost as a barrier to use?. *Am. J. Public Health*, 84, 50–55

Why study socioeconomic factors and cancer?

N. Pearce

The occurrence of cancer within a population can be studied at many different levels, including forms of social organization, the individual, a particular organ system, or a particular molecule. The causes of cancer can also be studied at these different levels, including socioeconomic factors, lifestyle, the organ burden of a carcinogen, or DNA adducts. Clearly, there are advantages in understanding disease causation at all of the different levels at which it operates. Although cancer risk factors such as tobacco smoke may appear to operate at the individual level, exposure may occur due to a wide range of political, economic and social factors; conversely, tobacco smoke ultimately also has effects at the cellular and molecular levels, including the production of mutations in DNA. Of course, it is important to gain information, and take action, at all possible levels, but the history of public health shows that changes at the population level are usually more fundamental and effective than changes at the individual level, even when a single risk factor accounts for most cases of disease. In this sense, a risk factor such as smoking can be regarded as a secondary symptom of deeper underlying features of the social and economic structure of society. Thus, just as a variety of health effects in various organ systems (for example, various types of cancer) may have a common contributing cause (for example, tobacco smoking) at the level of the individual, a variety of individual exposures (for example, smoking and diet) may have common socioeconomic causes at the population level. In many instances there is clear evidence that cancer is related to socioeconomic factors, but this does not appear to be fully explained by known risk factors. More importantly, there is little evidence as to which socioeconomic factors are of most importance, or whether it is the overall 'package' of social inequality that is responsible for the differences in cancer risk. The aim of this book is therefore to summarize what is already known, and to identify gaps in our knowledge.

Socioeconomic factors include education, income, assets, housing and occupation. 'Social class' can be used as a convenient summary term for various socioeconomic factors, but can also be used to denote more profound divisions within society.

The primary goal of public health is the prevention of disease in human populations, and socioeconomic factors are of major importance in this context. Epidemiology is the field of scientific investigation that attempts to discover the major causes of disease in the population so that public health action can be taken, although a great deal of epidemiological knowledge has been gained by researchers in related fields, particularly in the social sciences. For example, traditional definitions of epidemiology commonly refer to 'the study of the distribution and determinants of health-related states or events in specified populations, and the application of this study to control of health problems' (Last, 1988), although some recent definitions of epidemiology ignore the population perspective (for example, Miettinen, 1985; Rothman, 1986).

Of course, most cancer epidemiologists know about the importance of social class and socioeconomic factors. In the first week of their epidemiological training they learn about the work of Virchow, Chadwick, Engels and others who exposed the appalling social conditions during the industrial revolution, and the work of Farr (1860) and others who developed methods of social class classification and revealed major socioeconomic differences in death rates in the nineteenth century. They also learn about the subsequent dramatic decline in infectious diseases that occurred before the development of modern pharmaceuticals and has been attributed to improvements in nutrition, sanitation and general living conditions (McKeown, 1979), although specific public health interventions on factors such as urban congestion probably also played a major role (Szreter, 1988). They may also learn that there are now major social class differences in incidence and mortality from cancer and other chronic diseases, and that social class differences in mortality are increasing and are greater now (in relative terms) than they were in the nineteenth century (Marmot & McDowell, 1986; Pappas et al., 1993). After this, they usually forget about socioeconomic factors during the rest of their careers, except perhaps to occasionally adjust for social class in multivariate analyses of risk factors such as tobacco smoking, diet and other 'lifestyle' factors.

Nowadays, cancer epidemiologists typically study factors that have very low relative risks (for example, studies of new occupational carcinogens, or dietary studies of low levels of intake of micronutrients, usually reveal relative risks of less than 1.5) and that account for a small proportion of cancer cases. In contrast, in most industrialized countries, studies have repeatedly found strong associations between social class and cancer (Logan, 1982), with a nearly twofold relative risk for cancer when comparing the most disadvantaged group with the most advantaged group (although there are some specific cancer types for which the differential is in the opposite direction). However, few cancer epidemiologists study social class as an issue of major importance in itself, and it did not feature (except for a brief mention as a confounder) in the most comprehensive and authoritative review of the causes of cancer in the United States of America (Doll & Peto, 1981).

So why is social class so often just a footnote in cancer epidemiology studies?

Epidemiologists

Some of the reasons for this lack of interest may lie in the personal and professional situations of epidemiologists. Most epidemiological studies nowadays require substantial funding, and in most countries the main sources of funding are governmental or voluntary agencies that have little interest in, or sympathy for, studies of socioeconomic factors and health. Epidemiologists, either through choice or through necessity, tend to go 'where the money is'. Moreover, they tend to be most interested in risk factors that they can relate to, or may even be exposed to. Epidemiologists are frequently at risk from factors such as cigarette smoke, alcohol, diet, viruses and some occupational exposures, but they are rarely at risk of being poor. The poor may be occasionally encountered in random population surveys, or after taking the wrong exit from the autoroute, but in daily life they are invisible.

A further issue is that the study of socioeconomic factors is often seen as 'too political' and not a proper subject for scientific investigation. This argument reveals more about the proponent than about the issue. If the goal of epidemiology (and public health in general) is to discover (and ultimately take action on) the major causes of disease, then any factor that is a major cause of disease should warrant study; the decision not to study socioeconomic factors is itself a political decision to focus on what is 'politically acceptable' rather than what is most important in scientific and public health terms.

The art of the possible

A related argument is that socioeconomic factors are 'not easily modifiable'. Public health, like politics, can be viewed as 'the art of the possible' and socioeconomic factors are often placed into the 'too hard' basket. However, governments have repeatedly shown that social and economic differences are not 'God-given' but, for better or worse, are directly affected by government policies, often in unexpected ways (Black, 1993; Hewlett, 1993). Even when governments can have little effect on the overall gross national product (GNP), they can have major effects on how it is distributed by changes in the money supply, the level of inflation, the level of employment, the minimum wage and the average wage, taxation, and the level and availability of social services and social security benefits. There is some preliminary evidence that inequitable distribution of the GNP can have a more significant impact on overall national mortality rates than the actual level of GNP (Wilkinson, 1992, 1994). For example, in some countries, a large increase in GNP has been accompanied by little benefit in terms of health, whereas some relatively poor countries (for example, China, Jamaica and Costa Rica) have made major improvements in health care and life expectancy (Sen, 1980). Thus, the way in which the GNP is 'shared' is as important as its absolute level. Public health measures that aim to address the health problems of poverty may ultimately find themselves in conflict with government policies (or may even have the ultimate policy impact of changing the government), but this does not make the role of socioeconomic factors any less important, or less worthy of study.

Single risk factors

However, perhaps the main reason why socioeconomic factors have received little attention in cancer epidemiology is that they do not appear to be 'real causes', or at least are not as straightforward as factors such as tobacco smoke. Modern epidemiology became widely recognized with the discovery of tobacco smoking as a cause of lung cancer in the early 1950s (for example, Doll & Hill, 1952) and subsequent decades have seen major discoveries relating to other causes of cancer such as asbestos, ionizing radiation, hepatitis B, and dietary factors. These epidemiological successes have in some cases led to successful preventive interventions without the need for major social or political change. For example, occupational carcinogens such as asbestos can (with some difficulty) be controlled through regulatory measures, and exposures to known occupational carcinogens have been reduced in industrialized countries in recent decades. Another example is the successful World Health Organization campaign for the elimination of smallpox (Tesh, 1988). More recently, some countries have passed legislation to restrict advertising of tobacco and smoking in public places and have adopted health promotion programmes aimed at changes in 'lifestyle'.

These successes of 'risk factor' epidemiology have been striking and have undoubtedly prevented many cases of cancer. However, in recent years epidemiologists have struggled to find new major risk factors for cancer, and studies have increasingly focused on rare exposures or weak risk factors. As a result, epidemiologists have increasingly resorted to high-technology solutions (including new molecular markers of exposure) in an attempt to measure the risks associated with these weak risk factors (Pearce et al., 1995).

A more fundamental problem is that the success of 'risk factor' epidemiology has been more temporary and more limited than might have been expected (Loomis & Wing, 1991; Wing, 1994). It is one thing to discover that tobacco smoke is the major cause of lung cancer, but redressing this situation is a different problem entirely. For example, Graham (1989) suggests that smoking can be viewed as a strategy enabling women to cope with stress, while at the same time undermining their health and that of their children (Power et al., 1991). Why do manual workers smoke more than non-manual workers (and find it more difficult to give up)? Why have most physicians taken notice of the epidemiological evidence and given up smoking whereas nurses continue to smoke in great numbers?

Moreover, it can be argued that the fundamental problem of tobacco lies in its production rather than in its consumption (Tesh, 1988). As long as tobacco is produced (and governments provide subsidies and incentives to tobacco farmers) then someone somewhere is going to smoke it. In my own country (New Zealand/Aotearoa), tobacco was unknown before the arrival of European explorers in the eighteenth century. Tobacco was one of the 'gifts' given by the English representatives before the signing of the Treaty of Waitangi, which was followed by extensive settlement (and colonization) of New Zealand by Europeans. Nowadays, Maori women have some of the highest smoking rates and the highest lung cancer rates in the world. More generally, the (limited) success of legislative measures in industrialized countries has led the tobacco industry to shift its promotional activities to developing countries so that more people are exposed to tobacco smoke than ever before (Barry, 1991; Tominaga, 1986).

Levels of causality

Thus, the link between tobacco and lung cancer is as much a social, economic and political problem as it is a problem of individual 'lifestyle'. In the context of this book, smoking and socioeconomic factors are not alternative explanations for disease; rather smoking is one mechanism by which socioeconomic factors cause disease, and smoking is therefore an intermediate factor in the causal pathway leading from socioeconomic factors to disease.

In this sense, the apparently competing explanations for disease causation (for example, 'tobacco smoking' or 'socioeconomic factors') can be reconciled by recognizing that these explanations operate at different levels of analysis (Pearce, 1996). The occurrence of cancer (and other diseases) within a population can be studied at many different levels (Susser, 1973), including forms of social organization, 'the individual', a particular organ system, or a particular molecule. The causes of cancer can also be studied at these different levels, including socioeconomic factors, 'lifestyle', the organ burden of a carcinogen, or DNA adducts. For example, Potter (1992) argues that:

> 'A question relevant to the etiology of cancer that is seldom asked is: What gets cancer – the genes, the cell, the organism, or perhaps even the population? The potential answers are not necessarily exclusive, even given reductionist tendencies and the genuine and justified excitement over discoveries in the molecular biology of cancer. Rather these are levels of explanation that may be more or less coherent within themselves but provide even more information when they exist in a framework provided by all of the explanatory modes.'

Clearly, there are advantages in understanding disease causation at all of the different levels at which it operates. Although cancer risk factors such as tobacco smoke may appear to operate at the individual level, exposure may occur due to a wide range of political, economic and social factors; conversely, tobacco smoke ultimately also has effects at the cellular and molecular levels including the production of mutations in DNA.

So what is the most appropriate level at which to commence the study of the causes of cancer? Most researchers will immediately answer that their own discipline has it right, and all of the others have got it wrong. Usually this is presumed to be so obvious that no supporting arguments are necessary. Molecular biologists will focus on the carcinogenic process at the molecular level, with the belief that this will ultimately explain the major causes of cancer. In recent years, much of public health activity (including epidemiological research and some social science research) has focused on aspects of individual 'lifestyle' (perhaps mirroring economic and political trends, which have placed greater emphasis on individual responsibility in recent years) and the targeting of specific 'risk factors'. In contrast, some social scientists and epidemiologists emphasize that the major improvements in health status have come from social and economic changes and their influence on factors such as housing, income and nutrition (McKeown, 1979; Szreter, 1988).

'Top-down' and 'bottom-up'

These various pathways to understanding the disease process fall into two main approaches, which mirror wider scientific debates in recent centuries.

The 'bottom-up' approach [variants of which include reductionism, positivism, or the downstream approach (McKinlay, 1993)] focuses on understanding the individual components of a process at the lowest possible level and using this information as the 'building blocks' to gain knowledge about higher levels of organization. One current example is molecular epidemiology, which attempts to understand disease at the molecular level and then (ultimately) to use this knowledge in public health policy (for example, by screening populations for susceptibility to specific carcinogens). This approach stems from the clinical tradition and is typified by an emphasis on specific risk factors and the use of the randomized clinical trial as a paradigm. It certainly yields useful information about the level under study (for example, the molecular level), but it is debatable whether it is an effective and efficient long-term strategy for gaining knowledge or preventing disease at the population level. As Smith (1985) notes, this approach lacks distinctive theory regarding the occurrence of disease at the population level, and its products can be likened to 'a vast stockpile of almost surgically clean data untouched by human thought' (Anonymous, 1994). Although it has an air of scientific purity, this approach is in fact rarely used in other sciences or related disciplines; for example, nobody would attempt to predict the weather or the motion of the planets from measurements of individual molecules. Such an approach is not only impossible in practice (because of the infinitely large amount of information required), but recent work in chaos theory has shown that such an approach is also impossible in theory because small inaccuracies can produce huge effects in non-linear systems (Firth, 1991).

In contrast, the 'top-down' approach [variants of which include the structural approach (Tesh, 1988), the dialectical approach (Levins & Lewontin, 1985), and the upstream approach (McKinlay, 1993)] starts at the population level so as to ascertain the main factors that influence health status within the population. Studying disease at the population level usually requires a greater emphasis on observational (epidemiological) studies rather than experimental studies, and may also involve a greater use of 'ecological' studies of 'sick populations' rather than 'analytical' epidemiological studies of 'sick individuals' (Rose, 1992). Thus, the 'top-down' approach stems from the demographic/social science tradition (rather than the clinical trial paradigm). The study of socioeconomic differences in cancer primarily belongs to this tradition, which has been supported in a recent editorial in *The Lancet* (1994) that argued for the 'need to move away from the almost exclusive focus of research on individual risk, toward the social structures and processes within which ill-health originates, and which will be more amenable to modification' (McKinlay, 1993).

Links between levels

It should be emphasized that, even though it is important to start at (and return to) the population level, it is also important to conduct studies at other levels so as to explain the mechanisms by which these population factors operate. In particular, it is of interest to ascertain to what extent the observed effects at the population level are explained by known risk factors. For example, in the Whitehall study, Marmot *et al.* (1984) found extensive social class differences in coronary heart disease, which were only partially explained by known coronary risk. Syme and Berkman (1979) and Cassel (1976) proposed a more general explanation in which psychosocial factors (stress) influence susceptibility to various specific risk factors.

Even when social class differences for a particular cancer site are explained by the operation of known risk factors this does not mean that socioeconomic factors are not of importance. Of course, it is important to gain information, and take action, at all possible levels, but the history of public health shows that changes at the population level are usually more fundamental and effective than changes at the individual level, even when a single risk factor accounts for most cases of disease. In this sense, a risk factor such as smoking can be regarded as a secondary symptom of deeper underlying features of the social and economic structure of society (Townsend & Davidson, 1982). Thus, just as a variety of health effects in various organ systems (for example, various types of cancer) may have a common contributing cause (for example, tobacco smoking) at the level of the individual, a variety of individual exposures (for example, smoking and diet) may have common socioeconomic causes at the population level.

Furthermore, the 'populations' that epidemiologists study are not just collections of individuals conveniently grouped for the purposes of study, but are instead historical entities. Every population has its own history, culture, organization, and economic and social divisions, which influence how and why people are exposed to particular factors. For example, Terris (1979) argues that:

'The causes of cholera in India today go back hundreds of years in India's history, to the British invasion and destruction of once-flourishing textile industries; the maintenance of archaic systems of land ownership and tillage; the persistence of the caste system and the unbelievable poverty, hunger, and crowding; the consequent inability to afford the development of safe water supplies and sewage disposal systems; and, almost incidentally, the presence of cholera vibrios.'

As a result of such historical considerations, the strength, and even the direction, of socioeconomic disease gradients will vary between populations and over time; for example, coronary heart disease was at one time a disease of the affluent, but has become a disease of the poor as smoking and eating habits have changed over time (Wing, 1988). Furthermore, although specific cancer risk factors will play an important role in any population, their contribution to disease risk will be modified by the baseline disease risk and the presence of various co-carcinogens and cancer promoters, making it impossible to assume a universal dose–response relationship (Wing, 1994). Thus, generalization of study findings is much more difficult in the population sciences than in the physical and biological sciences, and appropriate interventions will differ widely between populations.

Socioeconomic factors and cancer

This book on socioeconomic factors and cancer has been prepared with these issues in mind. Just as mortality from most infectious diseases primarily declined due to general improvements in housing, income and nutrition (rather than treatment or prevention aimed at specific viruses or bacteria), it is likely that the greatest advances in cancer prevention will come from social and economic changes that in turn affect 'lifestyle' and exposure to specific risk factors.

In many instances there is clear evidence that cancer is related to socioeconomic factors, but this does not appear to be fully explained by known risk factors. More importantly, there is little evidence as to which socioeconomic factors are of most importance, or whether it is the overall 'package' of social inequality that is responsible for the differences in cancer risk. The aim of this book is therefore to summarize what is already known, and to identify gaps in our knowledge.

The book is intended to cover the major groups of risk factors that may contribute to socioeconomic differences in cancer, but it is not intended to be exhaustive. In particular, we do not address genetic factors; these undoubtedly play a role in most cancers, but are unlikely to play a major role in social class differences in cancer risk. It should be noted, however, that genetic factors may make a minor contribution by causing early disease (for example, Down's syndrome) that may itself affect both social class selection and subsequent cancer risk, or by acting as determinants (or susceptibility factors) for cancer in combination with various exposures arising in polluted environments.

In most countries there are major ethnic differences in socioeconomic status, and ethnic differences in cancer risk are therefore undoubtedly affected by the various socioeconomic factors considered in this book. However, a comprehensive review of ethnic differences in cancer (including cancer in indigenous peoples, migrants, and other ethnic groupings) is beyond the scope of this book. We therefore do not directly consider ethnic differences in cancer risk except for countries (for example, the United States) where data on ethnic differences may be the only information that is available as a surrogate for data on socioeconomic differences in cancer risk.

In the following three introductory chapters, we present an overview of issues of poverty and health and methods of measuring social inequality including income, education, housing, assets, occupation and employment status (we do not consider regional, gender or ethnic differences in health, which are important in their own right but are beyond the scope of this book).

In Part I of the main body of the book, we then summarize current knowledge regarding socioeconomic differences in cancer incidence, survival and mortality. In Part II we discuss general explanations for social inequalities in cancer, and then consider to what extent the socioeconomic differences in cancer risk may be explained by specific risk factors and aspects of health care. Finally, in the concluding chapter we attempt to draw conclusions regarding what is already known, and to make recommendations for further research.

Acknowledgements

This work was funded in part by a Visiting Scientist Award of the International Agency for Research on Cancer and in part by a Senior Research Fellowship of the Health Research Council of New Zealand. I thank Phillippa Howden-Chapman, Peter Davis, John McKinlay and Steve Wing for their comments on the draft manuscript.

References

Anonymous (1994) Population health looking upstream. *Lancet*, 343, 429–430

Barry, M. (1991) The influence of the U.S. tobacco industry on the health, economy, and environment of developing countries. *New Engl. J. Med.*, 324, 917–920

Black, D. (1993) Deprivation and health. *Br. Med. J.*, 307, 1630–1631

Cassel, J. (1976) The contribution of the social environment to host resistance. *Am. J. Epidemiol.*, 104, 107–123

Doll, R. & Hill, A.B. (1952) A study of the aetiology of carcinoma of the lung. *Br. Med. J.*, 2, 1271–1286

Farr, W. (1860) On the construction of life tables, illustrated by a new life table of the healthy districts of England. *J. Inst. Act.*, IX

Firth, W.J. (1991) Chaos – predicting the unpredictable. *Br. Med. J.*, 303, 1565–1568

Graham, H. (1989) Women and smoking in the United Kingdom: the implications for health promotion. *Health Promotion*, 3, 371–382

Hewlett, S.A. (1993) *Child neglect in rich nations*. New York, UNICEF

Last, J.M., ed. (1988) *A dictionary of epidemiology*. New York, Oxford University Press

Levins, R. & Lewontin, R. (1985) *The dialectical biologist.* Cambridge, MA, Harvard University Press

Logan, W.P.D., ed. (1982) *Cancer mortality by occupation and social class, 1951–1971* (IARC Scientific Publications No. 36). Lyon, International Agency for Research on Cancer

Loomis, D. & Wing, S. (1991) Is molecular epidemiology a germ theory for the end of the twentieth century? *Int. J. Epidemiol.*, 19, 1–3

McKeown, T. (1979) *The role of medicine.* Princeton, NJ, Princeton University Press

McKinlay, J.B. (1993) The promotion of health through planned sociopolitical change: challenges for research and policy. *Soc. Sci. Med.*, 36, 109–117

Marmot, M.G. & McDowell, M.E. (1986) Mortality decline and widening social inequalities. *Lancet*, ii, 274–276

Marmot, M.G., Shipley, M.J. & Tose, G. (1984) Inequalities in health – specific explanations of a general pattern? *Lancet*, i, 1003–1006

Miettinen, O.S. (1985) *Theoretical epidemiology: principles of occurrence research.* New York, Wiley

Pappas, G., Queen, S., Hadden, W. & Fisher, G. (1993) The increasing disparity in mortality between socioeconomic groups in the United States, 1960 and 1986. *New Engl. J. Med.*, 329, 103–109

Pearce, N. (1996) Traditional epidemiology, modern epidemiology, and public health. *Am. J. Public Health* 86, 678–683

Pearce, N., Davis, P.B., Smith, A.H. & Foster, F.H. (1985) Social class, ethnic group, and male mortality in New Zealand, 1974–8. *J. Epidemiol. Community Health*, 39, 9–14

Pearce, N., san José, S., Boffetta, P. et al. (1995) Limitations of biomarkers of exposure in cancer epidemiology. *Epidemiology*, 6, 190–194

Potter, J.D. (1992) Reconciling the epidemiology, physiology, and molecular biology of colon cancer. *J. Am. Med. Assoc.*, 268, 1573–1577

Power, C., Manor, O. & Fox, J. (1991) *Health and class: the early years.* London, Chapman & Hall

Rose, G. (1992) *The strategy of preventive medicine.* Oxford, Oxford University Press

Rothman, K.J. (1986) *Modern epidemiology.* Boston, Little Brown

Sen, A. (1980) *Levels of poverty: policy and change* (World Bank Staff Working Paper No. 401). Washington, DC, World Bank

Smith, A. (1985) The epidemiological basis of community medicine. In: Smith, A., ed., *Recent advances in community medicine 3*. Edinburgh, Churchill Livingstone. pp. 1–10

Susser, M. (1973) *Causal thinking in the health sciences.* New York, Oxford University Press

Syme, S.L. & Berkman, L.F. (1976) Social class, susceptibility and sickness. *Am. J. Epidemiol.*, 104, 1–8

Szreter, S. (1988) The importance of social intervention in Britain's mortality decline c. 1850–1914: a re-interpretation of the role of public health. *Soc. Hist. Med.*, 1, 1–37

Terris, M. (1979) The epidemiologic tradition. *Public Health Rep.*, 94, 204

Tesh, S.N. (1988) *Hidden arguments: political ideology and disease prevention policy.* London, Rutgers

Tominaga, S. (1986) Spread of smoking to the developing countries. In: Zaridze, D. & Peto, R., eds, *Tobacco: a major international health hazard* (IARC Scientific Publications No. 74). Lyon, International Agency for Research on Cancer. pp. 125–133

Townsend, P. & Davidson, N., eds (1982) *Inequalities in health: the Black report.* London, Penguin

Wilkinson, R. (1992) National mortality rates: the impact of inequality. *Am. J. Public Health*, 82, 1082–1084

Wilkinson, R. (1994) Divided we fall: the poor pay the price of increased social inequality with their health. *Br. Med. J.*, 308, 1113–1114

Wing, S. (1988) Social inequalities in the decline of coronary mortality. *Am. J. Public Health*, 78, 1415–1416

Wing, S. (1994) Limits of epidemiology. *Med. Global Survival*, 1, 74–86

N. Pearce
Wellington Asthma Research Group, Department of Medicine, Wellington School of Medicine, PO Box 7343, Wellington, New Zealand

Poverty and cancer

L. Tomatis

Despite the attraction of certain utopias and the convincing strength of some of the social and philosophical theories underlying attempts to change the social structure and to achieve a more egalitarian society, social inequalities have not disappeared and seem even to be increasing worldwide. Inequalities in health are part of the social inequalities present in our society and one of their most convincing indices. Sanitary conditions are worse, mortality higher, survival rates of cancer patients lower, and life expectancy shorter in developing countries than in industrialized countries. Similar if not identical differences can be seen within industrialized countries between socioeconomically less and more favoured population groups. In many areas of the industrialized countries social and environmental conditions comparable with those existing in the poorest countries last century have been recreated. Occupational risks are becoming a serious problem in developing countries, largely as a consequence of the transfer of hazardous industries from industrialized countries where certain industries are judged to be unacceptable. A similar double standard is applied to tobacco advertising and sales in the industrialized and developing countries. The projections of the total number of cancer cases in the next decades indicate a generalized increase, proportionally greater in developing than in industrialized countries.

Despite many attempts in various periods of human history to achieve an egalitarian society by distributing wealth and reducing poverty, and despite the attraction of certain utopias and the strength of several social and philosophical theories, social inequalities have not disappeared and seem even to be increasing worldwide. Inequalities in health are just part of the social inequalities present in our society and are one of their most convincing indices (Susser *et al.*, 1985).

The identification of causes and mechanisms behind the origin and persistence of poverty in Western society has perhaps been an achievement for sociologists and philosophers, but it has not led to a comparable achievement in the prevention of social disparities around the world (Kosa & Zola, 1976). The incapacity of our society to eliminate poverty is indeed one of the most blatant examples of failure in prevention. Similarly, the identification of the causes of a considerable proportion of human cancers, including natural substances like tobacco and asbestos, as well as industrial chemicals, medical drugs, radiation and viruses (Tomatis, 1990; Tomatis *et al.*, 1989), has been a great achievement by laboratory scientists and epidemiologists, but up to now has been followed by only limited victories for cancer prevention. The accumulated knowledge of the etiology of cancer has in fact not been applied with the efficiency necessary to achieve a substantial reduction in the incidence of at least certain human cancers. By some perverse mechanism this failure has been attributed more to inadequacy of etiological knowledge than to inefficiency in its practical use in prevention. In this way, the role of the environmental carcinogenic agents so far identified, with the single notable exception of tobacco, has been unjustifiably downgraded.

Poverty can be defined as a form of inequality below certain limits, thus implying that not all conditions that are just below the average within a given social environment can be considered as poverty. It may also be interpreted as a relative concept. If poverty is defined in relation to the average conditions in a given society, and what is measured is the extent of the inequality, then the proportion of people experiencing such poverty in a poor country is lower than in a rich country. In a poor country, most people generally share conditions close to the average for the country, but in a rich country a greater proportion of people are likely to share living conditions distant from the average and below the threshold of poverty, or at least of what can be called relative poverty.

In today's society the phenomenon of relative deprivation appears to be shifting the border of poverty upwards by including people who must renounce non-essential goods, like for instance vacations or a car. At the same time, the gap between the very rich and the very poor is widening and the proportion of the very poor – those who lack essentials – is increasing.

Poverty measured purely by its economic dimension (which is likely to underestimate its true dimensions) is still well rooted even in rich parts of the world. In 1985, 15.4% of Europeans (14.4% of all families) were considered poor, with Portugal and Ireland at one extreme with 32.7% and 19.5%, respectively, considered poor, and Belgium and Denmark at the other extreme with 5.9% and 8.0%, respectively, considered poor. In Italy in 1985, the proportion was 14.2%, but it had risen by 1988 to 15.2% (Commissione d'Indagine sulla Povertà e l'Emarginazione, 1992). In the United States of America in 1993, 15.1% (39.3 million) of the population (an increase of 0.3% from 1992) were living below the poverty line, which was set at US$ 14 763 a year for a family of four (Anonymous, 1994).

In discussing poverty and health, one enters a territory in which the borders between public health, the social sciences and politics are indistinct. Sigerist (1956) stated, on different occasions, that 'in any given society the incidence of illness is largely determined by economic factors' and that 'the problem of public health is ultimately political'. The great scientist of the last century Rudolf Virchow (Ackerknecht, 1981) must have shared this view when he was fighting for his idea of public health and social justice, at the risk of appearing an outmoded anticontagionist. He and another eminent scientist, Max von Pettenkofer (Hume, 1927), achieved an improvement in living and sanitary conditions in Berlin and Munich that resulted in a decrease of infant and infectious disease mortality. Such a decrease would not have been obtained if the purely scientific view of the contagionists, who were nevertheless scientifically correct, had totally prevailed at that time. McKeown (1988), in a more guarded statement that would have been shared by Winslow (1980), proposed that 'poverty is not a direct cause of disease, but it is the main determinant of influences that lead to disease'.

In this context, the German physician Johann Peter Frank from Göttingen deserves a special mention. Frank won the Chair of Clinical Medicine in Pavia in 1784, and in 1786 became Director General of Public Health of Lombardy, which was then under Austrian domination. In 1790 at the University of Pavia, he delivered the famous speech entitled 'De Populorum miseria: morborum genitrice' (Frank, transl. 1941). Under any form of government, he said, we may expect that the rich and the poor will have diseases peculiar to them, following the inevitable law of social disparity. He was deeply concerned, however, by the 'tremendous consequences for the public health of the extreme poverty from which the greater and most useful section of the population is being crushed'. Frank was not a revolutionary, but rather a loyal subject of King Joseph II, and tried, passionately but without success, to persuade authorities of the need for social reform to improve the people's health – however, all that he obtained was a call to return to Göttingen.

Economics and health

There has been a tendency recently to consider health as being purely, or mainly, an economic problem. There are various and severe limitations in the contributions that economics can make to public health (Fuchs, 1993). Economics by itself will not explain why different people give different emphasis to health care or to public health, nor clarify the criteria by which public health may become a priority, nor identify the sociocultural characteristics of a society or of an era within which the preferred choices are not necessarily in harmony with economics *sensu stricto*. While economics will be able to explain the costs that an equitable health system may imply, it will never explain why some people may prefer such a system, even if it would mean a slowing down of technological advances.

Civil rights originated and developed from deep moral and religious roots. These original foundations still exist and must continue to exist to support the moral commitment, which has been modified, but certainly not eliminated, by the secularization of society. While the extension of rights to all citizens should be connected with a due emphasis on reciprocal obligations, the definition of how much equality current society would like to achieve and the degree of social differences it would be ready to accept remains very difficult. The difficulties actually begin when we try to define what it is that separates a difference from an inequality and, furthermore, where and when a need becomes a right and a legitimate right (Pennacchi, 1994).

Health expenditures account for a considerable and increasing proportion of the gross national product (GNP) in industrialized countries (La Santé en France, 1989; Saracci, 1990; Geddes, 1991; Nau, 1991; Anonymous, 1991). This has not resulted in proportionate gains in public health. One of the major reasons for the lack of correspondence between increase in expenditure and improvement in well-being is probably that no more than 2–3% of the budget for health is usually earmarked for prevention; another is that it is generally the segment of the population that most needs health care that has the least access to it. A third, and perhaps the main, reason is that investment in health has been seen too often in terms similar to investment in other sectors of the economy and has thus been conditioned by the same imperatives of profit that drive industries or private enterprises. As the real 'profit' produced by health services – that is, the improvement of health status – is not easily quantifiable, all expenditures in public health are seen as low priorities.

The USA spends about 13% of its GNP on health (Fuchs, 1993) despite the fact that about 15% of its population is not covered by health insurance and therefore does not have access to a health protection system (Gibbons, 1991). As a comparison, the United Kingdom spends about 7% of its GNP on the National Health Service, which covers the entire population.

General and infant mortality

Several attempts to improve public health were made in some countries at the time of the Renaissance (Cipolla, 1976; Cipolla et al., 1992), but the first official recognition of the necessity to intervene in favour of the poor in general, and of the working class in particular, was made in the last century (The Chadwick report on the sanitary condition of the labouring population, 1971; Rosen, 1993).

In 1828, the French physician Villermé recorded the sharp contrasts between death rates in the rich and the poor and noted that infant and childhood mortality was almost twice as high among the poor as among the wealthy (Villermé, 1928; Coleman, 1979). Benoiston de Châteauneuf, a friend and contemporary of Villermé, provided unassailable evidence that differences in age-specific mortality increase with age, with practically none of the poor reaching old age (Benoiston de Châteauneuf, 1830). About 10 years later similar differences in mortality among people living in rich and poor residential areas were observed in the United Kingdom (Shryok, 1979; Strong, 1990).

Disease and destitution may have been considered part of the inscrutable plan of the Almighty, but when the injury and death of the workers interfered with industrial production and put profit in jeopardy, and when infectious diseases spread from poor to rich districts, it was time to take action. It was indeed pointed out that some sanitary and hygienic measures had to be taken because they were 'necessary not less for the welfare of the poor than the safety of property and the security of the rich' (Rosen, 1993). The famous Chadwick report on the 'sanitary condition of the labouring population' of Great Britain (1971) was published in 1842, largely as a result of the initiative and perseverance of Edwin Chadwick. Chadwick was a lawyer by education and a disciple of Jeremy Bentham. He was appointed secretary to a newly created Poor Law Board in 1834 and, after it was dissolved, became a member of the new National Board of Health in 1839. His report marked the beginning and formed the basis of the wave of sanitary reforms initiated towards the middle of the last century.

The first compulsory insurance covering the costs of care in cases of diseases and maternity of workers was instituted in Germany by Bismarck in 1883. Austria followed in 1888, Denmark in 1892, Luxembourg in 1902, Norway in 1909, the United Kingdom and Switzerland in 1911, and the Netherlands in 1913; other countries followed after World War I – for instance, France in 1928 and Sweden in 1948. In 1919, the International Labour Organization, of which the International Labour Office is a component, was created with the official and advertised goal of favouring the implementation of social justice worldwide. In 1948, the General Assembly of the United Nations approved the Universal Declaration of Human Rights in which article 25 proclaims that every individual has the right to a standard of living sufficiently high to guarantee his health, as well as full protection against unemployment, diseases, invalidity, widowhood and age, or in any other case of loss of the means of subsistence. This brief, necessarily incomplete, historical survey of how and when governments have tried to deal with certain social issues provides some elements on which to measure the distance that separates good intention and rhetorical statements from the reality. Such stated intentions seem more likely to have been realized when the egalitarian motivation is subsumed with a precise political goal that has little or only an indirect connection with altruism, as in the case of the initiative taken by Bismarck.

In the twentieth century, United Kingdom mortality rates for tuberculosis, heart diseases, bronchitis and stomach and other cancers were higher among the less favoured social classes IV and V than among the more favoured classes I and II in the 1930s and the same gap existed and had become even wider 50 years later (Greenwood, 1935; Logan, 1954; Wilkinson, 1986; Townsend & Davidson, 1982). A considerable overall improvement in infant survival rates occurred in the United Kingdom after 1921, but the social gradient indicating a higher mortality within classes IV and V did not change in the next 50 years (Rosen, 1993). In the United States, infant mortality has in recent years decreased among Black and White infants, but the disparity in mortality between Black and White infants has persisted or even increased, even when the comparison is made between infants with the same socioeconomic conditions. The chances of a Black infant dying in 1950 and 1988 were 1.6 and 2.1 times greater, respectively, than for a White infant (Schoendorf et al., 1992). This increasing disparity is observed while the absolute infant mortality rates are declining in both White and Black infants, suggesting that the causes of the decline do not substantially affect the disparity (Wise & Pursley, 1992).

The *ad hoc* Working Group on Inequalities in Health that the United Kingdom Labour Government set up in 1977 confirmed that while overall death rates had continued to fall, the difference between the classes remained proportionally the same or had even widened in all age groups. At the time the report of the group (known as the Black report, from the name of its chairman) was completed and ready for distribution, the Labour Government had been replaced by a Conservative Government, which tried to prevent its publication. A limited number of copies were nevertheless later distributed, and the Black report subsequently became available to the public because it was published independently as a paperback (Townsend & Davidson, 1982). The new government claimed that the pursuit of equality in health would hinder the much needed economic growth and that even greater inequality, by permitting faster growth, would give more real benefit to the less favoured than a policy of equal shares for all (Strong, 1990). Although no one has ever been able to provide solid evidence for this conjecture, similar policies still seem to be supported in certain circles in the United Kingdom and elsewhere. A report that was prepared as a follow-up of the Black report several years later also encountered serious difficulties in its publication, as the government tried again, but again unsuccessfully, to prevent it (Whitehead, 1987). A further updating of the study confirmed the persistence and even widening of the differences in health between the more and less favoured socioeconomic groups and re-emphasized the link between health differentials and material conditions, rather than individual behaviours (Phillimore et al., 1994; Wilkinson, 1994).

In most industrialized countries, the distribution of economically active individuals by occupational class has changed with time. The percentage of economically active men in the United Kingdom assigned to classes I and II increased considerably from a combined total of 13.8 in 1931 to 23.2 in 1971, while the percentage assigned to classes IV and V decreased from 38.4 to 26.4 (Townsend, 1982). It might be inferred that inequalities had been reduced, since the segment of society with higher incomes and lower death rates had increased and the one with lower incomes and higher death rates had shrunk. What was also true, however, was that the death rate had fallen much faster in class I than in class V, so that the gap between the rich and the poor grew even greater despite of, and in parallel to, the changes in the sizes of the two classes. The persistence and even widening of such a gap have been further confirmed by a more in-depth and detailed analysis of the relationships between trends in mortality and relative poverty and class differences (Wilkinson, 1989). A relatively new phenomenon is the growing proportion of jobless individuals. Several studies in different countries point to this condition as being associated with particularly severe consequences for health status (Costa & Segnan, 1988; Morris et al., 1994).

Marxist theory, which was seen as forming the background to the Black report, has been heavily criticized for being reductionist, as it gives overriding importance to economic circumstances and, by doing so, underestimates cultural factors. Non-Marxist sociologists and economists have suggested that other modes of class stratification would be more meaningful than the manual/nonmanual distinction. Thus a division has been proposed between, on the one hand, a large middle class whose members are employed, most often own their residences and can afford highly privatized consumption and, on the other hand, an underclass (representing about 25% of the population) that is low-waged or unemployed, lives in rented accommodation, and is highly dependent on public services. This division between a more affluent employed population and the less advantaged underclass has seemed more significant than the conventional Marxist division between manual and non-manual occupation (Pahl, 1984). However, while it is understandable that Marxist theories have incited strong adverse reactions, alternative theories have not yet offered satisfactory criteria to explain and eliminate social and health inequalities. Nevertheless, it is clear that in industrialized countries the original sharp division between manual and non-manual employment has lost much of its original significance. It is thought a sort of modern fraud to use 'the term "work" to cover what for some is... dreary, painful or socially demeaning and what for others is enjoyable, socially reputable and economically rewarding' (Galbraith, 1992). The main reason why privileged members of society oppose all reforms, including a sanitary reform, which while ensuring a general improvement of social conditions would imply an unavoidable increase in taxes, is the obtuse and pertinacious preference given to immediate advantage as compared with a hypothetical better future in a more equitable society, a future that would be better also in the sense that it could prevent a catastrophe from which not even the rich would be protected (Galbraith, 1992).

Marxist-inspired theories have also been accused of ignoring biology, and in particular a possible genetic explanation for the persistence of differences in class-related mortality. While there has been no great support for an explanation of the gap between classes based only, or mainly, on genetic differences, more attention has been paid to the suggestion that there might be a heightened general susceptibility to disease in particular groups. For example, an increased predisposition to cancer and other diseases might go with a lower socioeconomic position, perhaps due to immune suppression related to stress from adverse socioeconomic conditions. However, against this hypothesis stands the heterogeneity in the associations of socioeconomic conditions and specific cancer sites, as well as the clear evidence of the role of environmental/hygienic and behavioural factors, which strongly suggest that no single factor could account for the association seen and for the alleged increased general susceptibility to disease (Susser et al., 1985; Smith et al, 1991).

Inequalities in health in industrialized countries

Most of the studies on the relationship between socioeconomic class and cancer have been carried out in industrialized countries and have consistently shown that the total incidence, as well as mortality, of cancer at all sites is higher in the lower socioeconomic groups and is due mainly to an increased incidence and mortality at certain sites. From the classical study of Clemmesen and Nielsen (1951) to the more recent of Kogevinas (1990), it appears that the sites where the differences are the highest are the stomach, lung and cervix uteri. Three large studies in the United Kingdom have consistently found mortality to be higher in low socioeconomic groups for cancers of the lung, stomach, liver and oesophagus, while in one of the studies higher rates were reported also for the bladder, rectum and pancreas (Smith et al., 1991). A study in Italy found the greatest differences for cancers of the lung, pharynx, larynx, stomach, bladder and cervix uteri (Faggiano et al., 1994), and a study in Argentina for cancers of the lung, bladder, larynx and pancreas in males, and for cervix uteri (Matos et al., 1994).

The inescapable conclusion from all these studies is that in developed countries such as the United Kingdom there has been little progress in the reduction of inequalities in health between the different socioeconomic groups in the last 50 years (Whitehead, 1987; Townsend et al., 1986; Marmot & McDowell, 1986; Smith & Jacobson, 1989; Acheson, 1990; Marmot et al., 1991). In this context it is perhaps useful to remember that the income of chief executive officers of the major American companies was 29 times greater than that of the average manufacturing worker in 1980, and 93 times greater 10 years later (Galbraith, 1992).

There is little wonder that inequalities in health have been found in every country in which they have been sought and in which adequate information is available. The relationships between socioeconomic conditions and health have in recent years been the object of more studies in the United Kingdom than anywhere else, but results reflecting similar contrasts have also been obtained in the United States, Denmark, France, Italy, Argentina and Australia (Susser et al., 1985; Clemmesen & Nielsen, 1951; Matos, 1994; Desplanques, 1984, 1991; McMichael, 1985; Wilkins et al., 1990; Williams et al., 1991; Terris, 1990; Costa et al., 1994; Iscovich, 1989). In several studies the importance of differences in educational level has been stressed. Lower educational-level groups have higher death rates, which can to a large extent be explained by a higher prevalence of risk factors related to working conditions, material living conditions, lifestyles and ways to cope with stress (a spectrum that should be extended to include cultural factors) (Kunst & Mackenbach, 1994). From a study in Italy, it appears that in the 1980s there would have been 40 000 deaths fewer per year if people without a university education had had the same mortality as people with a university degree (Costa et al., 1994). Inequalities in education seem therefore to be an indicator of social and health inequalities that is as or more important than differences in occupations (Bouchardy et al. 1993; Vineis & Capri, 1994).

In the last 50 years, lung cancer mortality has continued to increase in the lower socioeconomic groups but has started to decrease in the socio-economically more favoured groups. The usual explanation is that within the lower socioeconomic groups, smoking (as well as drinking and other unhealthy behaviours) is more and more frequent than among the favoured groups, where smoking was more frequent in earlier times. Differences in smoking, however, cannot entirely explain the difference in mortality among the socioeconomic groups, as lung cancer is more frequent in lower socioeconomic groups even among non-smokers (Smith et al., 1991). The mortality from gastric cancer is decreasing in all groups, but the gap between the more and the less favoured socioeconomic groups not only has not been reduced but has slightly widened (Desplanques, 1991) in a way similar to what is observed for mortality of Black and White infants (Schoendorf et al., 1992), stressing once more that the factors affecting overall mortality may be different from those at the origin of the class disparities. In order to decrease such disparities, prevention should be concerned not only with the universal availability and delivery of services, but also with sociocultural differences in which disparities may be deeply rooted.

An element that can certainly influence mortality, at least at some cancer sites, is access to early diagnosis and to adequate therapy. In the study of Kogevinas (1990) this was seen in the higher mortality for endometrial cancer in the less favoured socioeconomic groups. Cancer patients of more favoured socioeconomic groups treated in private clinics were shown to survive their disease better than patients of less favoured socioeconomic groups treated in public hospitals (Linden, 1969). This differential in survival between socioeconomic groups has been further confirmed for cancers of relatively good prognosis, for which the duration of survival could be expected to depend partially on the timing and quality of treatment (Vineis & Capri, 1994; Kogevinas et al., 1991).

Cancer in developing countries

There is a striking relationship between per capita GNP and life expectancy. The world maps of the distribution of GNP and of life expectancy provide impressive evidence of their direct relationship, to the point that the maps could be easily mistaken for each other (The World Bank Atlas, 1990; The Commission on Health Research for Development, 1990). The disparity in wealth between the nations is such that an average individual in a less developed country earns 50 times less (in certain countries hundreds of times less) than an average individual in an industrialized country (The World Bank Atlas, 1990).

It is useful to mention the origin of the term 'developing countries' and the particular meaning it conveys. The term, which apparently derives from the view expressed by the USA President Harry Truman in 1949 that most areas of the world were underdeveloped, reflects entirely a Western concept of development. This implies that 'developing' countries will necessarily have to go through the same phases of development that Western countries have experienced during the industrial revolution, only at a delayed pace (Pearce *et al.*, 1994). The incapacity, or unwillingness, of the Western world to conceive a development that may follow other paths and pursue goals other than those of the industrialized countries may be seen as almost equivalent to the perpetuation of neocolonialism, which grants an apparent political independence but also guarantees a very tenacious economic dependence.

Occupational risks in developing countries are becoming a very serious problem (Kogevinas *et al.*, 1994; Rantanen *et al.*, 1994) largely as the consequence of the transferring of hazardous industries from highly industrialized countries, where certain industries are now judged unacceptable because of the risks for health and the environment, to poor countries, where adequate legislation protecting the workers and the environment does not yet exist (Simonato, 1986; La Dou, 1992; Jeyaratnam, 1994). The problem is further exacerbated by the little time given for the adjustment and training of the workers for new technologies (Kogevinas *et al.*, 1994). A pertinent example is the so-called 'maquiladora factories' located near the border of Mexico and the USA. In 1970, there were 16 such assembly plants; in 1991 the number had climbed to probably over 1 900 (25% of them located in Tijuana and 18% in Juarez), employing about 500 000 people at a salary that is much lower than in the USA and also lower than Mexican industry pays in the interior of the country (Moure-Eraso *et al.*, 1994). Acute health effects are reported to be common, but given the high labour turnover rates and the lack of health care structures, it will be extremely difficult to survey the long-term health effects. For instance, gross exposure to asbestos of workers, as well as the neighbouring population, in an asbestos textile plant has been reported (Abrams, 1979).

The differences in ranking of cancer sites between the industrialized and developing countries are similar to those seen between different socioeconomic groups within industrialized countries. Cancers of the stomach, cervix uteri, liver and oesophagus are most common both in developing countries and in the less favoured socioeconomic groups in industrialized countries (Parkin *et al.*, 1993). However, there is a notable discrepancy in interclass differences between the industrialized and the developing countries. In the developing countries, lung cancer mortality is highest in the socioeconomically more privileged groups, even though the rates remain considerably lower than those seen in industrialized countries (Cuello *et al.*, 1982). This can be explained by the fact that as cigarettes are expensive, only the well-to-do people in developing countries have been able to afford them and they have smoked more than the poor. It is relevant to note the different rates of growth in tobacco consumption in industrialized and developing countries, with the former showing decreases and the latter high rates of growth, providing good evidence for the success of the tobacco multinationals' efforts to open new profitable markets. Immediate and effective measures to prevent the massive introduction of the habit of smoking tobacco in developing countries, where the habit does not exist or where it has only recently been introduced, could avoid an epidemic of major proportions of lung cancer and other tobacco-related cancers and diseases. However, like the double standard with regard to occupational risks, there is a quite different attitude toward tobacco publicity and sales in the industrialized and developing countries. The concern for the damage to health that tobacco causes appears to have some difficulty crossing the borders of rich countries, which do not seem to be concerned if their tobacco corporations promote tobacco sales in the poor countries. The attributable fraction of cancers at certain sites (namely lung, oral cavity, oesophagus, pancreas, larynx and bladder) is directly related to the frequency of the smoking habit and therefore higher in African regions where the habit was introduced earlier – that is South Africa and several northern African countries (Sasco *et al.*, 1994). Pertinent action now could still prevent future epidemics of tobacco-related diseases in a large part of Africa (Chapman *et al.*, 1994; Peto, 1994).

Another inverse trend between industrialized and developing countries is seen in the use of certain hazardous chemicals: over 50 million pounds of pesticides that are either banned from use or unregistered or restricted in the USA (such as chlordane, mirex, dicofol, ziram and dibromochloropropane) were still shipped in 1990 from the USA, mainly to developing countries and in particular those in Latin America and Africa (Smith & Beckmann, 1991; Boffetta *et al.*, 1994).

A further example of the ambiguity that governs the relationships between industrialized and developing countries concerns the attitude towards environmental protection. There is little doubt that industrialized countries have contributed enormously more than developing countries to the downgrading of the world environment. Still now a significant proportion of the environmental pollution that occurs in developing countries is due to the exploitation of resources by multinational corporations. Industrialized countries now intend to impose their new environmental standards to poor countries, which may see this imposition as a continuation of the politics of colonial domination. They could not afford such standards economically and technically without substantial help. The pretension of the rich countries that, when convenient, there is one and the same law for the rich and the poor, closely resembles a virtuous hypocrisy (Anonymous, 1993).

Countries in central and eastern Europe that were part of the Soviet empire, Russia included, are experiencing health problems to a large extent similar to those of the developing countries. The conditions are obviously not the same – the competence and technological know-how of central and eastern Europe is rather more similar to those of Western countries than developing countries. Nevertheless, hazardous industrial productions are in operation and it is not evident that they will rapidly be modified or shut down. Occupational risks are therefore high and extensive pollution creates problems not only for the occupationally exposed workers but also for the general population (Pereira et al., 1992). Life expectancy is not increasing as it is in other European countries and adult male mortality rates are much higher than in the rest of Europe. Age-standardized male mortality rates have continued to decline in western Europe, but they rose since 1960 between 2% and 13% in countries of central and eastern Europe. In the age group 45–49 years, mortality rates in males increased between 1965 and 1989 by 7% in East Germany and by 131% in Hungary (Feachem, 1994). A considerable part of the increase seems to be attributable to a steep increase of ischaemic heart disease. In addition to having hazardous industries, unbalanced diets and inadequate health care structures, central and eastern European countries, like developing countries, are faced with the attempt of tobacco multinationals to conquer their markets with all sorts of incentives and publicity.

The increasing burden of cancer

Projections of cancer incidence in future years indicate that the number of cancer cases is almost certain to increase everywhere in the world. Figures 1–3 show projections of the total number of cancer cases in the year 2010 in North and South America, Europe, Africa and Asia. The expected increases in the number of cancer cases are 27% in Europe, 116% in Africa, 44% in North America, 101% in South America, 92% in all of Asia and 68% in Japan. Part of this increase, proportionally much more conspicuous in developing countries, can be attributed to demographic changes (the aging of the population); the rest is a real increase (Parkin et al., 1993). In as much as available health structures are insufficient to cope with the present demand, it is difficult not to worry about the disastrous situation we may run into in the near future.

As the absolute number of cases will almost inevitably increase over the next decades, the cost of cancer therapy becomes an even more relevant concern. It has been estimated that the average cost per patient of cancer treatments within the European Union was, in 1991, 3 000 ECU (1 ECU = about 1.3 US$) for conventional radiotherapy, 7 000 ECU for surgery, 12 000 ECU for chemotherapy, and at least 40 000 ECU for bone marrow transplantation. It is highly unlikely that developing countries, but perhaps also countries of central and eastern Europe, will be in a position to provide adequate treatment for most of their cancer patients, as even the richest countries are already unable to guarantee it. This is perhaps the most powerful argument in support of primary prevention of cancer.

Primary prevention will not be a panacea for all evils, but it is an inescapable conclusion that only a decrease in the incidence of chronic degenerative diseases, and/or a considerable delay in their manifestation, would avoid a further increase in the costs of health.

In addition, unless any extraordinary breakthrough in therapy is made soon, the rates of success of therapy will remain low for cancers that are inoperable and/or which show diffuse metastases at the time of diagnosis. The situation in this respect is already critical in developing countries, where the average survival time for cancer patients from the time of diagnosis is on average less than a third of that in industrialized countries.

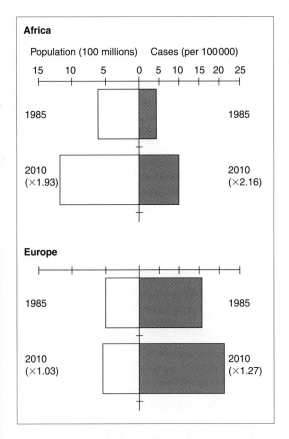

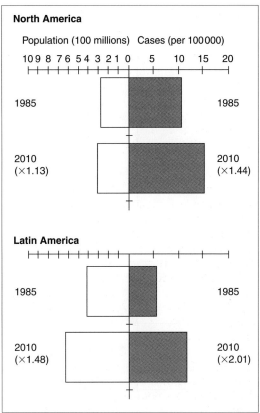

Figure 1. Cancer projections 1985–2010 for Africa and Europe based on synthetic age-specific incidence (assuming that there will be no change in age-specific incidence rates over time).

Figure 2. Cancer projections 1985–2010 for North America and Latin America based on synthetic age-specific incidence (assuming that there will be no change in age-specific incidence rates over time).

Projections of cancer costs, even with the caution with which projections of this type must be taken, show an upward trend. In 1988, the Netherlands spent 1 894 million Dutch Guilders (MDF) on the care of cancer patients, of which 61%, or 1 146 MDF, were spent on patient hospital care. The projected costs for the year 2005 and 2020 are 2 312 and 2 778 MDF, respectively – a 18% and 31% increase over those of 1988 (Koopmanshap et al., 1994).

Medical consumption and relative costs show two peaks during the course of cancer, namely during the first year following diagnosis and in the last year of life. The question is often raised whether health care costs can be substantially reduced by reducing the cost of care of terminal patients. Measures taken with the intent of reducing unnecessary medical services, and among these are rarely included expenditures for the care of patients known in advance to be dying, have resulted in a reduction of both efficient and wasteful services (Emanuel & Emanuel, 1994). It is not at all clear, therefore, that ethics and economics can be reasonably linked to reduce what can be only very imprecisely defined as 'futile care', and that such reduction would in any case be substantial.

Therefore, if governments have to find a way to reduce the cost of health, other measures must be sought. Within countries of the OECD (Organization for Economic Cooperation and Development), government spending to maintain the welfare system (pensions are first in the ranking of expenditures, and health comes second) has almost doubled between 1960 and 1990, passing from 28.1% to 43.8% of the GNP. Of the underlying pressures

Social Inequalities and Cancer

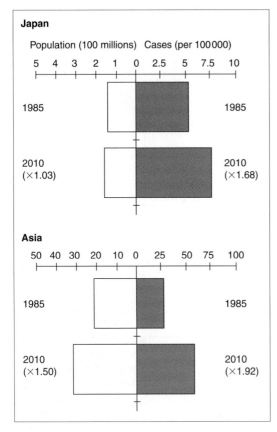

Figure 3. Cancer projections 1985–2010 for Japan and Asia based on synthetic age-specific incidence (assuming that there will be no change in age-specific incidence rates over time).

on welfare spending, the most important is the aging of the population. The number of people over 65 years of age in the OECD countries has grown from 61 million in 1960 to 100 million in 1990 and is expected to increase further in the next decades (Willman, 1994). Increased longevity may not necessarily mean an increase in medical care expenses if there is a decrease in the incidence of chronic diseases in middle- aged individuals. However, the development of new diagnostic and therapeutic devices could increase expenses, as the demand for them will grow even if they are shown not to be cost-effective.

It is rather encouraging, however, that several poor countries have been able to improve their educational and health systems more and faster than their economies. One reason for this is probably that health services and educational expansion are both labour intensive and therefore relatively inexpensive in poor economies (Sen, 1994), but the most important reason is the genuine will to improve health and educational services. Such will has been shown more clearly by several developing countries than by most industrialized countries.

The rich and the poor have different health histories

It is unrealistic at present to try to give a molecular explanation for the differences in incidence and mortality for most cancers between more and less favoured socioeconomic classes. Understanding of the multistage progression towards the clinical manifestation of malignancy has evolved a great deal since its earliest description (Berenblum & Shubik, 1949; Foulds, 1969), but it is likely that many more genetic and epigenetic alterations than have been identified so far are required to complete the process of carcinogenesis (Loeb, 1991, 1994). Thus we cannot yet have a complete and satisfactory translation into molecular terms of the effect(s) of environmental exposures than can be demonstrated epidemiologically, although recent results have already permitted a certain proportion of cancer cases to be attributed to specific environmental factors (Jones et al., 1991; Hollstein et al., 1993; Harris, 1994; Greenblatt et al., 1994). Every individual's 'health history' is characterized by life-long influences and superimposed short-term factors, but health biographies of the rich and the poor show divergences that are the result of the cumulation and interaction of a series of events, some of which can already be identified (others seem likely to become identifiable in the near future), and are qualitatively and quantitatively different. Schematically, this could, for instance, mean that certain individuals and certain segments of the population are exposed more frequently and to more hazardous agents than others and/or less frequently to protective agents.

How far one has to go back into a person's history to trace the beginning of the long development of most cancers we do not know precisely, but it is already clear that events that take place early in life, perinatally and even before conception, may contribute to an increased risk of cancer later in life (Tomatis et al., 1992). In this context, it is relevant to note that the strongest correlation between exposure of parents before conception and an increased cancer risk in the progeny has been reported in relation to exposure to occupational carcinogens and radiations (Tomatis, 1994).

Conclusions

Sanitary conditions are worse, mortality higher, survival rates of cancer patients lower, and life expectancy shorter in developing countries than in industrialized countries. Similar if not identical differences can be seen within industrialized countries between the socioeconomically less and more favoured population groups. Until some years ago, one could have perhaps corrected the above sentence by saying 'can *still* be seen', but in the present economic situation and with the present low priority given to social interventions, it would rather appear that such differences not only will not be reduced but also will affect an increasing proportion of the population within many industrialized countries. The growing phenomenon of lack or loss of employment is also directly related to a worsening of life expectancy. Losing a job may result in a doubling of the risk of death. In many areas of the industrialized world social and environmental conditions comparable with those existing in the poorest countries last century have been recreated (Drucker, 1993). This can be seen as the result of the progressive establishment and consolidation of a 'two-thirds society' as predicted in 1985 (Glotz, 1987). In a two-thirds society, the marginalization and social degradation of the weakest third of the society is accepted and programmed, while selfishness and individualism are encouraged and rewarded. The weakest third of the society, which has characteristics of the 'functional underclass' described by Galbraith (1992), is composed of the unemployed (that is, those who have lost and those who have not found employment), the unqualified or semiskilled labourers, the migrant workers, the handicapped, the less gifted and the young who cannot find their place in the professional system.

The projections of the total number of cancer cases in the next decades indicate a generalized increase, proportionally greater in developing than in industrialized countries. Given the inadequacy of the currently available sanitary structures and of the preventive measures that may reduce the anticipated increase, a potentially disastrous situation has to be feared.

Progress in science has had a negligible effect on improving the usual behaviour of people. Today's world provides a clear indication that the trend is towards increasing selfishness of the rich countries – that is, those in which science is better developed. It almost appears as if the progress of science has committed us to rigid obedience to an economic system that is prone to condemn the weakest portion of the world's population to starvation and poor health – a sort of aberrant neocolonialism. As Bertrand Russell said 'science enables the holders of power to realize their purposes more fully than they would otherwise do', but 'science is no substitute for virtue' (Russell, 1924). Even if the proportion of the GNP spent for health worldwide had been slightly increased in recent years, it would still have remained smaller than the proportion spent for defence. It cannot be ignored that almost one trillion dollars were spent in 1988 worldwide for military purposes (Sivard, 1991). Although there has been a tendency to reduce the production of arms for export and some reduction in military expenditures has been announced and has probably taken place, it is not clear that a real inverse, and hopefully irreversible, trend in military expenditures has been initiated (Barnet, 1991).

Most countries and governments justify the severe limitations of health resources by maintaining that there is an apparent infinity of demands that could in any case never be satisfied. Such a claimed infinity of demands is in fact surely in large part the result of an accumulation of unsatisfied demand over a long period. One can ask whether health resources are unavoidably and irremediably scarce, or whether such scarcity is a political choice. In our society, the imperative of profitability conditions even dominates research priorities and the resulting medical care system, of which prevention today represents a minimal part. Whether one day we shall be able to reverse the present trend so as to achieve a more equal distribution of resources and an orientation of research toward prevention as a main goal remains to be seen, but much will depend on our commitment as scientists and citizens of this world. In today's society, we could at least demand, as a first step, that health measures and intervention of undoubted efficacy not be rationed (Frankel, 1991). We should then also press for acceptance of the principles that a basic and irreplaceable element of public health policy is the improvement of living standards, that much improvement is within the reach of government policy (Wilkinson, 1986; Terris, 1990), and that a real and stable improvement in the health of populations depends on, and can only occur through the reduction of inequalities (Bartley, 1994) and the respect of human rights (Mann *et al.*, 1994; Susser, 1993; Leary, 1994).

References

Abrams, H.K. (1979) Occupational and environmental health problems along the U.S.–Mexico border. *Southwest Econ. Soc.*, 4, 3–20

Acheson, E.D. (1990) Edwin Chadwick and the world we live in. *Lancet*, 336, 1482–1485

Ackerknecht, E.H. (1981) *Rudolf Virchow*. New York, Arno Press

Anonymous (1991) How affordable is public health? *Nature*, 353, 587–588

Anonymous (1993) Environmental protection or imperialism? *Nature*, 363, 657–658

Anonymous (1994) The number of Americans living below poverty line. *International Herald Tribune*, Oct. 7, p. 3

Barnet, R.S. (1991) L'Amérique de Bush ne renonce pas à la puissance militaire. *Le Monde diplomatique*, Nov., p. 8

Bartley, M. (1994) Health costs of social injustice. *Br. Med. J.*, 309, 1177–1178

Benoiston de Châteauneuf, L.F. (1830) De la duré de vie chez le riche et chez le pauvre. *Ann. Hyg. Publique Méd. Légale*, 3, 5–15

Berenblum, I. & Shubik, P. (1949) An experimental study of the initiating stage of carcinogenesis and a re-examination of the somatic mutation theory of cancer. *Br. J. Cancer*, 3, 109–118

Boffetta, P., Kogevinas, M., Pearce, N. & Matos, E. (1994) Cancer. In: Pearce, N., Matos, E., Vainio, H., Boffetta, P. & Kogevinas, M., eds, *Occupational cancer in developing countries* (IARC Scientific Publication No. 129). Lyon, International Agency for Research on Cancer. pp. 11–16

Bouchardy, C., Parkin, D.M., Khlat, M., Mirra, A.P., Kogevinas, M., De Lima, F.D. & De Cravalho Ferreira, C.E. (1993) Education and mortality from cancer in Sao Paulo, Brazil. *Ann. Epidemiol.*, 3, 64–70

The Chadwick report on the sanitary condition of the labouring population with the local reports for England and Wales and other related papers (1971) Shannon, Ireland, Irish University Press

Chapman, S., Yach, D., Saloojee, Y. & Simpson, D. (1994) All Africa conference on tobacco control. *Br. Med. J.*, 308, 189–191

Cipolla, C.M. (1976) *Public health and the medical profession in the Renaissance*. Cambridge, MA, Cambridge University Press

Cipolla, C.M. (1992) *Miasmas and disease*. New Haven, CO, Yale University Press

Clemmesen, J. & Nielsen, A. (1951) The social distribution of cancer in Copenhagen, 1943 to 1947. *Br. J. Cancer*, 5, 159–171

Coleman, W. (1979) *Death is a social disease*. Madison, WI, University of Wisconsin Press

The Commission on Health Research for Development (1990) *Health research – essential link to equity in development*. Oxford, Oxford University Press

Commissione d'Indagine sulla Povertà e l'Emarginazione (1992) *Secondo rapporto sulla povertà in Italia*. Milano, Italy, Franco Angeli Editore

Costa, G. & Segnan, N. (1988) Mortalità e condizione professionale nello studio longitudinale torinese. *Epidemiol. Prev.*, 10, 48–57

Costa, G., Faggiano, F., Cadum, E., Lagorio, S., Arcà, M., Farchi, G., De Maria, M. & Pagnelli, F. (1994) Le differenze sociali nella mortalità in Italia. In: Costa, G. & Faggiano, F., eds, *L'equità nella salute in Italia*. Milano, Italy, Franco Angeli

Cuello, L., Correa, P. & Haenszel, W. (1982) Socioeconomic class, differences in cancer incidence in Cali, Columbia. *Int. J. Cancer*, 29, 637–643

Desplanques, G. (1984) L'inégalité sociale devant la mort. *Econ. Stat.*, 162, 29–51

Desplanques, G. (1991) *Les cadres vivent plus vieux*. Paris, Institut National de Statistiques et d'Etudes Economiques No. 158

Drucker, E. (1993) Molecular epidemiology meets the fourth world. *Lancet*, 342, 817–818

Emanuel, E.J. & Emanuel, L.L. (1994) The economics of dying. The illusion of cost savings at the end of life. *New Engl. J. Med.*, 330, 540–544

Faggiano, F., Zanetti, R. & Costa, G. (1994) Le differenze sociali nell'incidenza dei tumori a Torino negli anni '80. In: Costa, G. & Faggiano, F., eds, *L'equità nella salute in Italia*. Milano, Italy, Franco Angeli

Feachem, R. (1994) Health decline in eastern Europe. *Nature*, 367, 313–314

Foulds, L. (1969) *Neoplastic development*. London, Academic Press

Frank, J.P. (transl. Singer, H., 1941) The people's misery: mother of disease. *Bull. Hist. Med.*, 9, 81–100

Frankel, S. (1991) Health needs, health-care requirements, and the myth of infinite demand. *Lancet*, 337, 1588–1590

Fuchs, V.R. (1993) *The future of health policy.* Cambridge, MA, Harvard University Press

Galbraith, J.K. (1992) *The culture of contentment.* Boston, MA, Houghton Mifflin Co.

Geddes, M. (1991) *La salute degli Italiani.* Rome, La nuova Italia Scientifica

Gibbons, A. (1991) Does war on cancer equal war on poverty? *Science,* 253, 260

Glotz, P. (1987) *Manifeste pour une nouvelle gauche européenne.* Paris, Editions de l'Aube

Greenblatt, M.S., Bennett, W.P., Hollstein, M. & Harris, C.C. (1994) Mutations in the p53 tumor suppressor gene: clues to cancer etiology and molecular pathogenesis. *Cancer Res.,* 54, 4855–4878

Greenwood, M. (1935) *Epidemics and crowd diseases.* London, William and Norgate

Harris, C.C. (1994) p53: at the crossroads of molecular carcinogenesis and risk assessment. *Science,* 262, 1980–1981

Hollstein, M.C., Wild, C.P., Bleicher, F., Chutimataewin, S., Harris, C.C., Srivatanakul, P. & Montesano, R. (1993) p53 mutations and aflatoxin B1 exposure in hepatocellular carcinoma patients from Thailand. *Int. J. Cancer,* 53, 51–55

Hume, E.E. (1972) *Max von Pettenkofer.* New York, Paul B. Hoeber

Iscovich, J.M. (1989) Mortalidad por cancer y su relacion con las clases socio-ocupacionales en la provincia de Buenos Aires, Argentina, 1938–1943: un enfoque historico. *Med. Segur. del Trabajo,* 36, 74–82

Jeyaratnam, J. (1994) Transfer of hazardous industries. In: Pearce, N., Matos, E., Vainio, H., Boffetta, P. & Kogevinas, M., eds, *Occupational cancer in developing countries* (IARC Scientific Publication No. 129). Lyon, International Agency for Research on Cancer. pp. 23–29

Jones, P.A., Buckley, J.D., Henderson, B.E., Ross, R.K. & Pike, M.C. (1991) From gene to carcinogen: a rapidly evolving field in molecular epidemiology. *Cancer Res.,* 51, 3617–3620

Kogevinas, M. (1990) *Longitudinal study: socio-demographic differences in cancer survival, 1971–1983* (Office of Population Censuses and Surveys: Series LS No. 5). London, Her Majesty's Stationery Office

Kogevinas, M., Marmot, M.G., Fox, A.J. & Goldblatt, P.O. (1991) Socio-economic differences in cancer survival. *J. Epidemiol. Community Health,* 45, 216–219

Kogevinas, M., Boffetta, P. & Pearce, N. (1994) Occupational exposure to carcinogens in developing countries. In: Pearce, N., Matos, E., Vainio, H., Boffetta, P. & Kogevinas, M., eds, *Occupational cancer in developing countries* (IARC Scientific Publication No. 129). Lyon, International Agency for Research on Cancer. pp. 63–95

Koopmanshap, M.A., van Roijen, L., Bonneux, L. & Barendregt, J.J. (1994) Current and future costs of cancer. *Eur. J. Cancer,* 30, 60–65

Kosa, J. & Zola, I.K., eds (1976) *Poverty and health.* Cambridge, MA, Harvard University Press

Kunst, A.E. & Mackenbach, J.P. (1994) The size of mortality differences associated with educational level in nine industrialized countries. *Am. J. Public Health,* 84, 932–937

La Dou, J. (1992) The export of industrial hazards to developing countries. In: Jeyaratnam, J., ed., *Occupational health in developing countries.* Oxford, Oxford University Press. pp. 340–358

Leary, V. (1994) The right to health in international human rights law. *Health Hum. Rights,* 1, 24–56

Linden, G. (1969) The influence of social class in the survival of cancer patients. *Am. J. Public Health,* 59, 267–274

Loeb, L.A. (1991) Mutator phenotype may be required for multistage carcinogenesis. *Cancer Res.,* 51, 3075–3079

Loeb, L.A. (1994) Microsatellite instability: marker of a mutator phenotype in cancer. *Cancer Res.,* 54, 5059–5063

Logan, W.P.D. (1954) Social class variations in mortality. *Br. J. Prev. Soc. Med.,* 8, 128–137

McKeown, T. (1988)*The origins of human disease.* Oxford, Basil Blackwell

Mancina, C. (1994) Giustizia, eguaglianza et solidarietà. In: Pennacchi, L., ed., *Le Ragioni dell'Equità.* Bari, Italy, Edizioni Dedalo

Mann, J., Gostin, L., Gruskin, S., Brennan, T., Lazzarini, Z. & Fineber, G.H.V. (1994) Health and human rights. *Health Hum. Rights,* 1, 6–23

Marmot, M.G. & McDowall, M.E. (1986) Mortality decline and widening social inequalities. *Lancet,* 2, 274–276

Marmot, M.G., Smith, D.G., Stansfeld, S., Patel, C., North, F., Head, J., White, I., Brunner, E. & Ferney, A. (1991) Health inequalities among British civil servants: the Whitehall study II. *Lancet,* 337, 1387–1393

Matos, E.L., Loria, D.I. & Vilensky, M. (1994) Cancer mortality and poverty in Argentina: a geographical correlation study. *Cancer Epidemiol. Biomarkers Prev.,* 3, 213–218

Morris, J.K., Cook, D.G. & Shaper, A.G. (1994) Loss of employment and mortality. *Br. Med. J.,* 308, 1135–1139

Moure-Eraso, R., Wilcox, M., Punnett, L., Copeland, L. & Levenstein, C. (1994) Back to the future: sweatshop conditions on the Mexico–U.S. border. I. Community health impact of maquiladora industrial activity. *Am. J. Ind. Med.,* 25, 311–324

Nau, J.Y. (1991) Les dépenses de santé pourraient représenter près de 10% du PIB en l'an 2000. *Le Monde*, May 12–13

Pahl, R. (1984) *Divisions of labour*. Oxford, Basil Blackwell

Parkin, D.M., Pisani, P. & Ferlay, J. (1993) Estimates of the worldwide incidence of eighteen major cancers in 1985. *Int. J. Cancer*, 54, 594–606

Pearce, N., Matos, E., Koivusalo, M. & Wing, S. (1994) Industrialization and health. In: Pearce, N., Matos, E., Vainio, H., Boffetta, P. & Kogevinas, M., eds, *Occupational cancer in developing countries* (IARC Scientific Publication No. 129). Lyon, International Agency for Research on Cancer. pp. 7–22

Pennacchi, L. (1994) Processi, principi e politiche nella riprogettazione del welfare state. In: Pennacchi, L., ed., *Le Ragioni dell'Equità*. Bari, Italy, Edizioni Dedalo

Pereira, F.P., Hemminki, K., Gryzbowska, E., Motykiewicz, G., Michalk, A.J., Santella, R.M., Young, T., Dickey, C., Brandt-Rauf, P., De Vivo, I., Blaner, W., Tsai, W. & Chorazy, M. (1992) Molecular and genetic damage in humans from environmental pollution in Poland. *Nature*, 360, 256–258

Peto, R. (1994) Smoking and death: the past 40 years and the next 40. *Br. Med. J.*, 309, 937–939

Phillimore, P., Beattie, A. & Townsend, P. (1994) Widening inequality of health in northern England, 1981–91. *Br. Med. J.*, 308, 1125–1128

Rantanen, J., Lehtinen, S., Kalimo, R., Nordman, H., Vainio, H. & Viikari-Juntura, E. (1993) *New epidemics in occupational health*. Helsinki, Finland, Finnish Institute of Occupational Health

Rosen, G. (1993) *A history of public health*. Baltimore, MD, The Johns Hopkins University Press

Russell, B. (1924) *Icarus or the future of science*. New York, E.P. Dutton

La Santé en France (1989) Paris, La Documentation Française

Saracci, R. (1990) Pour en finir avec l'inégalité face à la santé. *Le Monde*, Oct. 10

Sasco, A., Grizeau, D., Pobel, D., Chatard, O. & Danzon, M. (1994) Tabagisme et classe sociale en France de 1974 à 1991. *Bull. Cancer*, 81, 355–359

Schoendorf, K.C., Hogue, C.J.R., Kleinman, J.C. & Rowley, D. (1992) Mortality among infants of black as compared with white college-educated parents. *New Engl. J. Med.*, 326, 1522–1526

Sen, A. (1994) Population: delusion and reality. *NY Rev. Books*, 41, 62–71

Shryok, R.H. (1979) *The development of modern medicine*. Madison, WI, University of Wisconsin Press

Sigerist, H.E. (1956) *Landmarks in the history of hygiene*. London, Oxford University Press

Simonato, L. (1986) Aspects of occupational cancer in developing countries. In: Khogali, M., Omar, Y.T., Gjorgor, A. & Ismail, A.S., eds, *Proceedings of the Second International Union Against Cancer (UICC): conference on cancer prevention*. London, Pergamon Press. pp. 101–106

Sivard, R.L. (1991) *World military and social expenditures*. Washington, DC, World Priorities

Smith, A. & Jacobson, B., eds (1989) *The nation's health*. London, King Edward's Hospital Fund for London

Smith, C. & Beckmann, S.L. (1991) *Export of pesticides from U.S. Ports in 1990*. Los Angeles, Foundation for Advancement in Science and Education

Smith, D.G., Leon, D., Shylley, M.J. & Rose, G. (1991) Socioeconomic differentials in cancer among men. *Int. J. Epidemiol.*, 20, 339–345

Strong, P.M. (1990) Black on class and mortality: theory, method and history. *J. Public Health Med.*, 12, 168–180

Susser, M.W. (1993) Health as a human right: an epidemiologist's perspective on the public health. *Am. J. Public Health*, 83, 418–426

Susser, M.W., Watson, W. & Hopper, K. (1985) *Sociology in medicine*. New York, Oxford University Press

Terris, M. (1990) Public health policy for the 1990s. *J. Public Health Policy*, 11, 281–295

Tomatis, L., ed. (1990) *Cancer: causes, occurrence and control* (IARC Scientific Publication No. 100). Lyon, International Agency for Research on Cancer

Tomatis, L. (1994) Transgeneration carcinogenesis: a review of the experimental and epidemiological evidence. *Jpn. J. Cancer Res.*, 85, 443–454

Tomatis, L., Aitio, A., Wilbourn, J. & Shuker, L. (1989) Human carcinogens so far identified. *Jpn. J. Cancer Res.*, 80, 795–807

Tomatis, L., Narod, S. & Yamasaki, H. (1992) Transgenerational transmission of carcinogenic risk. *Carcinogenesis*, 13, 145–151

Townsend, P. & Davidson, M., eds (1982) *Inequalities in health: the Black report*. London, Penguin

Townsend, P., Phillimore, P. & Beattie, A. (1986) *Inequalities in health in the Northern Region*. Bristol, England, University of Bristol and Northern Region Health Authority

Villermé, L.R. (1828) Mémoire sur la mortalité en France dans la classe aisée et dans la classe indigente. *Mém. Acad. R. Méd.*, 7, 51–98

Vineis, P. & Capri, S. (1994) *La salute non è una merce*. Torino, Bollati Boringhieri

Whitehead, M. (1987) *The health divide*. London, The Health Education Authority

Wilkinson, R.G. (1986) Socioeconomic differences in mortality: interpreting the data on their size and trends. In: Wilkinson, R.G., ed., *Class and health*. London, Tavistock Publications. pp. 1–20

Wilkinson, R.G. (1989) Class mortality differentials, income distribution and trends in poverty 1921–1981. *J. Soc. Policy*, 307, 307–335

Wilkinson, R.G. (1994) Divided we fall. *Br. Med. J.*, 308, 1113–1114

Willman, J. (1994) Welfare versus wealth of nations. *Financial Times*, Oct. 25, p. 13.

Winslow, C.E.A. (1980) *The conquest of epidemic diseases*. Madison, WI, University of Wisconsin Press

Wise, P.H. & Pursley, D.M. (1992) Infant mortality as a social mirror. *New Engl. J. Med.*, 326, 1558–1559

The World Bank Atlas (1990) Washington DC, The World Bank

Corresponding author:
L. Tomatis
Istituto dell'infanzia, 63 via dell'Istria,
34137 Trieste, Italy

Social theory and social class

I. Susser

Concepts of class developed with the emergence of industrial society in the nineteenth century. For an understanding of current divisions, theories must reflect the advances of capitalism and the global economy that characterize the late twentieth century. In industrialized societies, reductions in the industrial workforce and the growth of finance, investment and real-estate industries worldwide have produced a new, largely female, service workforce. Large sectors of industry have departed in search of cheaper labour in poorer countries, which also have a rising number of women workers. In those areas, as a result, a new industrial workforce has emerged. Concomitantly, accumulation of land in less developed agricultural regions for production for the world market has led to an increase in mobile agricultural labour and a shift of landless labourers to the cities of less developed countries. In addition, both upward and downward mobility have occurred for individuals and groups in specific populations, as well as for particular diseases in developed and less developed countries. All these processes have precipitated fundamental changes in class, gender and family relationships and transformed the living conditions of populations in both developed and less developed societies. These changes have major implications for the patterns of health and disease in the world today. Objective measures of social change may be difficult to construct and use in epidemiological cancer research. Since questions of class and shifting social relations are directly implicated in the patterns of disease, they must be assessed in future research as accurately as possible.

'To the social scientist, uneven distributions of disease, illness and sickness in society are manifestations of social structure and culture that reveal variations in culture, disparities in resources, or differences among subgroups in the condition of daily life. For epidemiologists, discovery of such variations is a starting point; they must elucidate their medical significance, and through preventive medicine attempt to eliminate them.'
Susser et al., 1985

Readers ensconced in the fortress of empiricism and accustomed to the use of a variety of indices may perhaps ask what need there is for a theory to guide the choice and the use of epidemiological indices for social class. To begin to answer this question, whether in the biomedical or the social sciences, we must recognize that concepts and theory underlie many and probably most indicators or measures of material states. Indicators misinterpreted because of mistaken theory can lead to false conclusions and misguided practice. In the biomedical sciences, theories of immunity underlie indicators of the immune state, and theories of haemodynamics and electrophysiology underlie indicators of blood flow and cardiac impulses. In social science, political, economic or ethnic concepts may underlie administrative or geographical boundaries. Social class and its indicators differs from these, however. It is a construct, something more than a notion, which describes an aspect of social reality that has no observable material presence. The existence of the reality must be inferred. It need not for that reason be a phantasmagoria. During the century (and more) since Mendel, before genes could be physically located, they were a construct that the notorious Lysenko could challenge and thereby gain Stalin's ear. Cells, molecules and atoms were also constructs before they had more than an assumed presence.

In these instances, and many others, appropriate theory at once led to rapid scientific advance. It is fair to assume that sound social class theory will lead not only to a better understanding of society but also to a better understanding and measurement of how health and disease are distributed within and across societies. To achieve this, we need to comprehend the underlying forces that create and separate social strata. We must discover what holds the strata in place and what promotes and prevents the flow of groups and individuals between them. In the following section, I review the historical foundations of theories of class. Next, I consider contemporary developments in class theory and issues of social mobility and social change. I then examine the economy and class structure of the United States of America as a historical example of changing class relations in the global economy of advanced capitalism. Finally, I discuss shifting class relations in less developed countries and the contrasting impact of the global economy on living conditions in less developed countries. A foundation for class theory is the work of Karl Marx in the mid-nineteenth century. His successor, elaborator and challenger, in the nineteenth century and beyond, is Max Weber. I shall deal briefly with the founding theories of both Marx and Weber.

Theories of social class

Hierarchy and stratification have been features of all known human societies, whether on the minor scale of kinship or the major scale of caste or class. Hunters and gatherers are differentiated biologically and socially by age and gender at the least. In more complex societies, strata are differentiated by privately owned land and property. The resulting constraints of ownership on common access to property also differentiate group interest. Ownership entails the potential for class conflict and the exercise of power to defend it. The consequences of stratification for health have long been noted. In *The Republic* (Book III), Plato reports a dialogue that describes how social position governs the assumption of the sick role:

> 'When a carpenter is ill, he expects to receive a draught from his doctor that will expel the disease by vomiting or purging, or else to get rid of it by cauterizing, or a surgical operation; but if any one were to prescribe to him a long course of diet, and to order bandages for his head, with other treatment to correspond, he would soon tell such a medical adviser that he has not time to be ill, and hint that it was not worth his while to live in this way, devoting his mind to his malady, and neglecting his proper occupation: and then, wishing his physician a good morning, he would enter upon his usual course of life, and either regain his health and live in the performance of his business; or, should his constitution prove unable to bear up, death puts an end to his troubles. Yes, and for a man in that station of life, this is thought the proper use to make of medical assistance...'

The state of health of a population is one facet of social structure, a facet that reflects the form of the structure and the elements contained within it. An understanding of the full extent of the relations of states of health to social structure, however, did not emerge for considerably more than two millennia after Plato. With Jean Jacques Rousseau, we see clearly expressed the idea that human capacities are moulded by society. After the French Revolution, and as the industrial revolution gathered momentum, writers and researchers began to adumbrate this idea and extend it to health. True insight into the pertinence of class for health awaited the means to measure the health of populations. These means, first devised by John Graunt and William Petty in the seventeenth century, like much else in the founding of modern science, were gained at the height of mercantile capitalism (Merton, 1973). Graunt documented the flight of the higher classes from London during the Great Plague and the mortality of those who remained. In Germany, a century later, Johann Peter Frank observed that 'every social group has its own type of health and disease determined by the mode of living' (Frank, transl. 1941), and poverty was 'the mother of diseases'. By the early nineteenth century, capital and industrialization in England, and the Revolution in France, had broken down the remnants of feudalism. Contractual relations and a class system finally superseded the primacy of kinship relations in economic production. The owners of capital, in contract with a newly emerged urban working class, were now the economic driving force. Thus in the industrial town of Mulhouse in 1826, the conservative Pierre-Louis Renee Villerme first set out the facts of a social class gradient in mortality (Villerme, 1826).

In England, in the next decade, William Farr began to follow Villerme's lead, and Edwin Chadwick and Friedrich Engels, each in his own distinctive way, drew on Farr. This work presaged and influenced the theory of the formation and the consequences of the class system of Karl Marx, developed while he was a political refugee in England. Marx saw that the manner in which people produce, exchange and consume goods defines the nature and scope of their social and political or power relations.

In summary, the central feature of Marx's theory is the definition of classes as discrete entities that are not independent but in dynamic relation with each other. Classes that are the result of processes of production that assemble occupations in groups unequal in power and status, and that therefore have inherently conflicting interests (Marx, transl. 1977).

Subsequent to Marx, the most enduring and influential theorist of class is Max Weber, another great nineteenth-century intellectual (Weber, 1947). Weber did not disagree with Marx on the economic stratification of society, but he took the view that three distinct systems of stratification exist in industrial society. To economic interest he added prestige (which he called honour) and political power. Certain groups could and did acquire prestige independently of wealth and economic interest. Some, like the clergy, the nobility and the learned professions, derive prestige from the preindustrial past. Others acquire prestige by virtue of especially valued occupational or artistic skills. Status groups share styles of life, education and culture, and are bound together by common interests. Such status groups have proliferated in industrial societies, and they are linked in a continuum of prestige that runs from the nobility to the illiterate poor. Weber also treats power as an issue separate from economic and occupational structure. The effect of his analysis was to temper Marx's view of classes locked in inevitable economic and political conflict over ownership of property and capital. He did not deny that the three systems of stratification might overlap to a degree that might produce the appearance of discrete class entities. Thus his concept readily allowed either for gradients in each dimension, or for a system that derived a single class gradient by combining selected indicators, whether of one, two or three dimensions.

Contemporary formulations of class

In contemporary sociology, in the United States of America (USA) particularly, the use of statistical methods designed to reflect a more or less Weberian view of class has produced a variety of complex measures of socioeconomic status. More often, however, those following a quasi-Weberian definition of class have been content to use the census categories. Thus to capture the stratification of industrialized societies, some analyse ethnic categories of Black, Hispanic-non-White, Hispanic-White and White. Others adopt such categories as 'underclass' or rely simply on income or poverty levels, occupational groupings or residential neighbourhoods.

In terms of Marxist definitions of class, two contrasting ways of conceptualizing class have emerged over the past decades. One derives from the structural Marxists (Althusser & Balibar, 1970; Poulantzas, 1975) and has been systematically formulated in the USA by Eric Olin Wright. Wright sees capitalism dividing the USA population into different classes – service workers, industrial workers, corporate executives – and he argues that these groups can be categorized objectively. He maintains that their political perspectives, cultures, behaviours and health can be expected to differ according to these divisions. However, he argues that many groups find themselves in contradictory locations, such as managers who are paid less than many workers, and that their perspectives reflect these contradictions. Wright's formulation has been useful for sociologists and others who have wished to implement this view of class status by statistical analysis of such objective groupings as type of occupation and income. (For a discussion and debate on Wright's theories, see Wright, 1978, 1989). The British social historian E.P. Thompson took a different view from that elaborated by Althusser, Wright and others. Thompson saw class emerging through social conflict. As groups come into conflict, and as their interests are found to be contradictory, politically conscious classes emerge in opposition to one another. Thus with the rise of industrial capitalism in England, manufacturers began to recognize a common interest in controlling both the government and the workforce. In analysing eighteenth-century England, Thompson documented the rise of this new capitalist upper class. He then reconstructed the parallel emergence of a self-conscious working class, in response to the attacks on workers' wages by this unified ruling class (Thompson, 1966, 1978).

The significant divergence of Thompson from Wright and the structural Marxists has to do with his emphasis on process. Thompson envisaged classes as emerging through changing historical circumstances. Groups gradually begin to make conscious political sense of the daily experiences of opposing interests in capitalist society. Such class formations are not predictable and can be explained only through the analysis of detailed historical processes. Wright depends not on an empirical analysis of historical change and the experiences of each group in the population, but rather on theoretical concepts of the structure of advanced capitalism and the divisions among workers and capitalists. Such a formulation facilitates objective measures and comparisons of health and class without reference to attitudes, political consciousness, the development of social movements or even daily experiences. However, whether such concrete and objective formulations accurately capture the complexities of historical change in advanced capitalist societies remains problematic. Interaction, exchange and the dynamic interpretation of class that flows from social mobility in modern society are unlikely to be captured.

Social mobility

The peoples of all the industrialized societies of the Western world share a high degree of social mobility, both upward and downward. In the application of social class theory, it is essential to recognize that social systems are complex and made more so because they are ever shifting. In the course of the history of the most developed economies – from feudalism through mercantile capitalism and industrial capitalism to the postindustrial present – societies have grown more dynamic. Social change has accelerated, and so too has the mobility of both individuals and groups.

We must rid ourselves of the image of class societies stratified like the layers of a cone-shaped Neapolitan ice cream or a wedding cake. Such metaphors may have been an adequate reflection of feudal society but are not useful for a socially mobile class society. Within class societies there is mobility both upward and downward. However, generally, individuals move singly in successive operations from one stratum to the next. For groups, too, there is mobility and change in social position. As we see in caste societies, even the ranking of castes changes over time (Srinivas, 1966). In class societies, the same has happened to various occupations: some have risen and others have declined in position and prestige. Furthermore, the shape and size of occupational strata have changed with time. In developed societies, the manual unskilled working class has declined and what has been viewed as the middle class has grown larger in the course of the twentieth century. In less developed societies, the elite and the poor are separated by a smaller middle class, and the gap between them is even greater than in the older industrial societies. Recent developments in the USA, however, suggest shrinkage of the middle class, with polarization once again increasing between the wealthy and the working class and poor (Mollenkopf & Castells, 1991).

Thus, as Thompson's model suggests, in the struggle for power, classes can win and lose position within different historical periods. This struggle for position, which includes the struggle for working conditions, working hours, days of work, health benefits and pensions, has far-reaching implications for the life opportunities and health of members of each class over time.

The Weberian model of stratification allows for a structure of vertical segments, each with its own horizontal strata and with more or less prestige, status and power. This model is the basis of the pluralist view of a society comprising interest groups that are constantly vying for power and shifting in relative status. Analyses of the mobility of ethnic groups have been built on this model, both in the USA and the Caribbean. From the standpoint of a Marxian concept of class, however, such shifts in the status of ethnic or interest groups may promote the mobility of a small group of leaders and ethnic entrepreneurs but do not necessarily facilitate class mobility for large numbers of people. In addition, the focus on ethnic identity and competition is seen as dividing the working class and diverting workers from recognizing their common interests. Social mobility can be lateral or horizontal as well as vertical, as in the case of migration either within countries or between countries. In all these cases mobility is a key element in health patterns. Sometimes individuals move upward or especially downward and carry their diseases with them, as they do their distinctive cultures and genes. That is reflected by the theory of social drift and has been one explanation for the accretion of schizophrenia in the lower social classes.

A pertinent factor in disease distribution is the fact that cultural forms and behaviour too have social mobility independently of persons; they are adopted and move upwards or downwards or indeed laterally from group to group.

The idea has long persisted that not only does poverty carry with it miseries and diseases, but civilization too engenders its own ills. Henry Sigerist gave one of his works the very title *Civilization and disease* (1943). Peptic ulcer and coronary heart disease, which among other diseases rose dramatically in the first third of this century, were characterized as 'diseases of civilization'. In the late 1950s, however, it became apparent from cohort analysis that although peptic ulcer and its subforms (gastric and duodenal ulcer) had first been recognized as serious afflictions among the upper classes, mortality from the disease had steadily diffused downwards through the social classes until it was predominantly a disease of the lower classes (Susser & Stein, 1960). Later, a similar picture could be discerned with regard to coronary heart disease (Susser et al., 1985).

Smoking is a factor in both diseases and it is possible that the diffusion of smoking related to patterns of both peptic ulcer and coronary heart disease. The smoking of cigarettes began in the late nineteenth century as an upper-class habit, first among men. During World War I it became a widespread habit – and was in fact encouraged by the distribution of cigarettes – among working-class men and, to some extent, among upper-class women. Gradually the habit of cigarette smoking became predominantly a habit of the lower classes. The diffusion downwards was made more apparent by the fact that in Great Britain and the USA and some other societies, the gradual decline of smoking followed the same track as the adoption of the smoking habit. The upper-class groups that had begun smoking first were also the first to desist.

HIV infection in the USA has followed a pattern similar to those of heart disease and peptic ulcer. Identified originally among upper- and middle-class gay men, members of this group were also the first to mobilize politically and to adopt preventive strategies effective in slowing the spread of infection. The rate of increase of HIV infection in the USA is now fastest among the poor, and specifically among poor and minority women. As the prevalence of infection is highest in poor inner city communities, they face high risks of infection transmitted through unprotected sex and intravenous drug use with shared needles. Thus, factors of class and life opportunities have once again steered the course of an epidemic to centre among the most disadvantaged groups in a society.

Thus individuals move up and down and across class divisions, groups shift in power and position and the relative life chances of entire classes change over time. Diseases follow these shifts and changes. To illuminate the distributions and the causes of the shifts in health and disease, the complex and changing patterns of stratification of each society must first be understood. These patterns in turn reflect different historical experiences and the differential access to strategic resources.

Changing class structure in the global economy

In the USA and other Western societies in recent decades, global economic changes have forced a further re-evaluation of concepts of class, both Weberian and Marxist. Since current social theory of whatever persuasion conceptualizes some form of a global economy, both developed and less developed countries must now be viewed in this context. It is in fact the dynamics of the interactions between developed and less developed societies that on both fronts have lent the impetus to many economic and social changes in the past 20 years.

The USA provides an example of the development of an advanced capitalist society that has had a dominant influence on the global economy. Although each industrialized or postindustrial state has been created within unique historical circumstances and in recent decades other countries have industrialized at a rapid rate and entered global capitalist competition at a different pace, the USA can reasonably be taken to provide an empirical test for theories of social class. In order to explore concepts of class further, therefore, we begin with an examination of categorical divisions and changes in USA society since the 1950s.

A number of characteristics observed since the 1950s complicate discussion of the 'working class' in the USA. One major area of change surrounds the dramatic increase in the service economy in the sectors of civil service, health services, retail and restaurant services, and also support services for the expanding industries of finance, insurance and real estate. Jobs in these sectors have been filled disproportionately by women. As a result, employment patterns, gender relations and families have been restructured (for a discussion of these changes, see

Sassen, 1991; Castells, 1989; Bluestone & Harrison, 1982; Gordon et al., 1982).

Social class, whether defined by Marx, by Weber, Talcott Parsons and their followers, or by the British Registrar General, referred to men and their families. Women were classified by the employment of their husbands or fathers. When women worked infrequently outside the home, this classificatory criterion appeared less problematic than it now does. Since the 1960s, however, theorists have been debating how to define class for women, whether working or not. Discussion has centred around the value of women's unpaid work in the home, the hierarchies of domestic relations, the segregation of women workers in the labour force, and the lower pay for women than for men in equivalent jobs. Each of these issues raises questions about how to define 'working-class women'. For example, are they working class regardless of their jobs because (as previously defined) their husbands have working-class jobs? When married and not earning income, are they 'unemployed' or homemakers? When working in the service economy are they working class even if their husbands are lawyers and executives? (For a review, see Susser, 1986, 1989.)

Questions about women's work and the growing service sector have been further complicated by the export of industrial jobs from the north-eastern to the southern and south-western USA, and the departure of industry to poorer countries in search of cheaper labour (Nash & Kelly, 1983). These changes have forced many working-class men out of the once typical well-paid non-unionized service sector. Such major shifts have resulted in the need for the two-worker family (Bluestone & Harrison, 1982; Gordon et al., 1982). Since the 1980s it has taken two pay cheques to support a working-class household in a manner for which the man's pay cheque alone was sufficient in the 1950s and 1960s. Thus, in the USA, a typical working-class household in the 1990s may look very different from a working-class household 20 years earlier.

The families that form households have changed in structure, with many fewer nuclear households and many more single women supporting children. Where the two-parent nuclear family does exist, we are likely to find two adults with employment in the service sector; neither job may be unionized and neither may carry health benefits for the family. In order to describe accurately the health chances of individuals and families in different sectors of the USA economy in terms of class, the concept must be flexible enough to take these major changes into account. (For discussions of the changing economy and workforce of the USA, see Bluestone & Harrison, 1982; Gordon et al., 1982; Wilson, 1987; Susser, 1996b; Nash, 1989; Sidel, 1986; Smith, 1984).

Thus, as noted, the shift to service work and the departure of industrial work has been accompanied by a general decline in the unionization of American workers and a concomitant reduction in the benefits of health and pensions gained by unions in a continuing struggle over the course of the past one hundred years. Since 1971, workers' wages have been declining. Men have been retiring earlier, in many cases forced out of the workforce by closing factories, and women have been working not only more often but also longer, certainly in part to replace the loss of men's wages. Neighbourhoods that previously housed industrial workers with stable jobs were built around heavy industry with extensive pollution problems. Now such factories have gone elsewhere. The surrounding residents are underemployed, but the health and social consequences of industrial waste in abandoned neighbourhoods remain. The economic shifts described above have affected the overall stratification patterns of the USA. Results of the 1990 census reveal an increasing gap between the earnings of American workers and of people with higher incomes. While the incomes of USA residents earning over US$ 200 000 per year increased in the past decade, those earning under US$ 25 000 did not improve their relative status. This increasing gap between high- and low-income Americans has been reflected in changing mortality rates (Pappas et al., 1993). Nationally, between 1960 and 1986 mortality rates for men and women with incomes above US$ 25 000 improved more than did those of people with lower incomes. Indeed, the mortality rates of men with incomes under US$ 25 000 declined hardly at all.

A new poverty has followed the departure of industry and dependence on lower-wage, less-unionized service work in the USA. Some conceive this change as the creation of an 'underclass'. In the early 1980s the concept of underclass was introduced to describe and explain the persistence of extreme poverty in the USA (for further discussion of this term see Wilson, 1987; Jones & Susser, 1993). This concept, like that of the culture of poverty, involved a pattern of behaviours attributed to the poor. Characteristics used frequently as indices of underclass status included teenage pregnancies, female-headed households, and substance abuse. These conditioned behaviours were seen as perpetuating the poverty of poor people and creating the very problems that prevented them from benefiting from the opportunities in USA society.

While some moderate exponents of underclass theories placed their explanations of the social organization of the poor within the context of a changing economy, the emphasis on measurement and the policy applications focused on a static view of culture and socialization. The model implies that people are trapped in patterns of behaviour that they teach to their children and are unable to change even when the situation changes. On the contrary, overwhelming evidence from anthropology and history documents the ability of human beings to adjust to new circumstances and to create new cultural strategies. The static view of culture reflected in underclass theory tends to reinforce negative stereotypes of the poor without illuminating the causes for the multiple problems documented.

Workers who belong to so-called 'minority groups' are most affected by shifts in the USA economy. The most recently hired workers in the civil service and health bureaucracy – African-American and Latino workers – entered the northeastern USA workforce just as the shift from industrial to service work was taking place. When the service bureaucracy is reduced, as occurs periodically in response to fiscal crises and political pressure, a disproportionate number of members of minority populations lose their jobs. Thus, minorities not only have maintained the merest fingerhold in the industrial workforce but also, lacking seniority, have often been ousted even from the service sector. Unemployment, homelessness and high mortality rates at younger ages have accompanied their exclusion from declining industries and their insecure hold on any kind of employment. This has contributed to the stereotype of an unchangeable underclass as well as to continuing racial discrimination and segregation in employment and housing conditions (Wilson, 1987, 1996).

Complementing the decline of unions and well-paying jobs has been a partial expansion of peripheral industries, sweat shops and what has come to be known as the 'informal economy' (for extensive discussions of this concept, see Portes et al., 1989). The informal economy operates outside the legal requirements of USA institutions. By its very definition it is difficult to trace and measure. Undocumented immigrants and others work for cash, with few records, and without reporting income either to the Internal Revenue Service or the Social Security Administration. Workers are not organized and have almost no recourse against management policy. Such conditions make health measures particularly difficult to implement. Workers receive no health benefits and have no basis on which to demand healthy working conditions.

Since the dramatic change in the immigration laws of 1965, a new immigrant population (equivalent to the great immigration waves of the turn of the century) has entered the USA. Immigrants provide a low-paid workforce for restaurants, the fashion industry and other small manufacturers, domestic services, car repairs and the informal sector in general. Immigrants who have entered the USA without adequate documentation are likely to avoid medical care for fear that the disclosures of information required in clinics and hospitals might lead to legal problems or deportation. (For discussions of the new immigration, see Portes & Rumbaut, 1990; Kasinitz, 1992; Hing, 1993.)

Not all immigrants work in the informal economy. Some have found manufacturing employment (as, for instance, Dominicans and Chinese in the garment industry of New York City – see Kwong, 1987; Waldinger, 1986). Others enter the USA with high levels of education or wealth and fall into an entirely different category. Indeed, immigrants earn more than many members of minority groups born in the USA, at least among those who do report income (Waldinger, 1986). Nevertheless, evaluations of socioeconomic status and class position by health researchers are complicated by the existence of a poorly documented informal economy that engages a high proportion of a poor immigrant population, some without legal status, as well as a number of poor Americans. Any analysis of class and health in the USA, it is plain, must take into account a shifting economy that affects all classes, the export of industry on a global scale, changing employment for men and women, changing families, declining unions and health benefits, declining wages of workers, an increasing gap between rich and poor with decreasing possibilities for the children of the middle class, the creation of a new poverty and new manifestations of racial discrimination, a dramatic influx of immigration, and the growth of an informal economic sector. Only in the context of these changes can we understand the emerging class divisions in terms of daily experience, life opportunities and health (Harvey, 1990; Castells, 1996; Susser, 1996a,b).

Perspectives on class in less developed countries
Since the colonial period, many cities in less developed countries have been stratified by class interlocking with race (Epstein, 1958; Srinivas, 1966). Discussions of class have centred around the stratification among colonized peoples as well as between the colonial and neocolonial elites and the general population. When in colonial times the colonized were counted at all, as in India, stratification was frequently neglected in analysis among both the colonized population and the colonialists. During the last three decades, however, differentiation within the general population has become both obvious and significant. Theories of the 1960s concerning the genesis of underdevelopment as well as ideas of a world system have been extensively criticized for not attending to the historical emergence of classes within previously colonized societies (Wolf, 1982). As within industrialized societies, issues of race and gender complicate in different ways analyses of class divisions (Stoler, 1989; Nash & Kelly, 1983).

In many less developed societies, a large peasantry still exists. Peasants were defined by Marx as producers of goods for the cities and not as separate from urban development (Wolf, 1966). In this way they are differentiated from what used to be known as tribal groups, subsistence farmers or pastoralists. Peasants have been viewed as tied to state societies and dependent upon selling their surplus goods to the cities. Extensive discussion has focused around differentiation of the peasantry. In Latin America and parts of Africa, the more prosperous peasants have acquired land and advanced technology, while others have lost their land and been forced to search for agricultural wage labour or to migrate to the cities (Roseberry, 1989; Lennihan, 1990). This process has been accelerated over the past two decades. Higher-income peasants were able to invest in the fast-growing seeds, fertilizers, insecticides and high technology – the so-called Green Revolution – often to the exclusion of the poorer peasantry who were then edged out of the markets. These economic changes have produced a landless and mobile working class in contrast to the landed peasantry. Plantation workers and agricultural wage labourers have come to be seen as a growing rural proletariat and migrant labour patterns have been seen as the movement of that rural proletariat (Mintz, 1985; Vincent, 1984).

Thus, land accumulation for cultivation for the world market in less developed countries has led to the eviction of peasants from their land and to displacements either into agricultural wage labour or to the cities where they form the massive informal settlements of urban squatters. Informal urban settlements often house millions of people. They lack sanitation, electricity, running water or paved roads. The sources of poor health are manifest. Such situations still breed plagues. The settlements are untouched by city regulations for fire protection or control of industrial pollution. In Mexico City, such settlements burned down when liquid natural gas in storage sites surrounded by millions of squatters caught fire. Similarly, with the Union Carbide dioxin disaster in Bhopal, India, thousands of people were living, contrary to city regulation, in informal settlements surrounding the factory.

The departure of industry from the USA, which has changed the structure of its own workforce, has also transformed the world of workers in the less developed countries to which these industries were transplanted (Rothstein & Blim, 1992; Susser 1985, 1992). Wages in the new factories of the less developed world generally much exceed those in other available jobs. In fact, in countries such as the Dominican Republic, few jobs exist apart from those in the transnationally owned industries. Other work is to be found in the low-paid informal sector, which serves as a peripheral economy. For instance, street vendors sell cigarettes and provide other supplies for those employed in the elite sector and able to afford them. The new industrial development is largely unregulated, frequently uses old and unsafe technology, and brings with it the environmental and occupational hazards familiar in early industrialization. Unions seldom exist and those that do are usually powerless to defend against hazards (Susser, 1985). Workers in these better-paid industries, however, unlike those in the informal sector, may be provided with health benefits. Thus, the divisions between workers in the formal and informal sectors have a variety of implications for differential health outcomes.

In the face of major recessions and world trade deficits in less developed countries, the International Monetary Fund (IMF) has provided economic assistance. Such assistance has been proffered in the context of required 'structural adjustment' policies. These policies have generally stipulated a reduction of services financed by national governments, including health and social services. Thus, international assistance in the 1980s and 1990s brought with it a restructuring that decreased the health and human services available to the population. Any analysis of health and stratification in poor nations needs to take account of such policies. Of particular interest is the fact that low national per capita income does not correlate with mortality. However, the greater the income disparities between rich and poor in a country, the higher the levels of mortality. Mortality is thus associated with inequality rather than poverty alone (Wilkinson, 1986). Just as in the USA migration is a major factor in stratification, routes of labour migration between less developed countries and from there to highly industrialized countries are equally significant. Remittances to home countries are an important consideration in evaluating internal differentiation among the local population. Residents may in fact raise both social position and health status with assistance from kin abroad.

Thus, in the formulation of concepts of class and health in less developed countries, account needs to be taken of large global economic trends with significant local consequences. These include ongoing transformations with accumulation of peasant holdings in agriculture, the increasing migration to the cities and growth of informal settlements, the movement of industry and the consequent differentiation among wage workers, and the global labour migration and patterns of remittances.

Conclusion

An examination of class division is essential for illuminating the distribution of health and disease in modern societies. Fundamental changes in class, gender and family relationships and transformed the living conditions of populations in both developed and less developed societies. These changes have major implications for the patterns of heath and disease in the world today. Research concerning the epidemiology of cancer will have to rely on sophisticated statistical measures of the distribution of disease. However, an analysis of emerging patterns of class and shifting economies and populations provides the context for effective sampling and interpretation of the data. Although objective measures of some aspects of social change may be difficult to construct, this should not discourage researchers from examining questions of class and shifting social relations as accurately as is currently possible. As these are the parameters that limit or open life possibilities, they are also directly implicated in the patterns of disease and must be assessed in future research.

References

Althusser, L. & Balibar, E. (1970) *Reading capital*. New York, Pantheon

Bluestone, B. & Harrison, B. (1982) *The deindustrialization of America*. New York, Basic Books

Castells, M. (1989) *The informational city*. London, Blackwell

Castells, M. (1996) The net and the self: working notes for a critical theory of the informational society. In: Susser, I., ed., *Anthropological perspectives on the informational society. Critique Anthropol.* 16, 9–38

Epstein, A.L. (1958) *Politics in an urban African community*. Manchester, Manchester University Press

Frank, J.P. (transl. Singer, H., 1941) People's misery – mother of disease. *Bull. Hist. Med.*, 9, 81–100

Gordon, D., Edwards, R. & Reich, M. (1982) *Segmented work, divided workers: the historical transformation of labor in the United States*. Cambridge, Cambridge University Press

Harvey, D. (1990) *The condition of postmodernity*. London, Blackwell

Hing, B.O. (1993) *Making and remaking Asian America through immigration policy: 1850–1990*. New York, Stanford University Press

Jones, D. & Susser, I., eds (1993) *The widening gap between rich and poor. Critique Anthropol.* 13, 11

Kasinitz, P. (1992) *Caribbean New York*. Ithaca, N.Y., Cornell University Press

Kwong, P. (1987) *The new Chinatown*. New York, Hill and Wang

Lennihan, L. (1990) Wage contracts in Northern Nigeria. In: Roseberry, W. & O'Brien, J., eds, *Golden ages, dark ages.* Berkeley, University of California Press. pp. 107–126

Marx, K. (transl. Fowkes, B., 1977) *Capital.* New York

Merton, R.K. (1973) *The sociology of science: theoretical and empirical investigations* (revised edn). Chicago, University of Chicago Press

Mintz, S. (1985) *Sweetness and power.* New York, Penguin Press

Mollenkopf, J. & Castells, M., eds (1991) *The dual city.* London, Sage Publications

Nash, J. (1989) *From tank town to high tech.* Albany, NY, State University of New York Press

Nash, J. & Kelly, P., eds (1983) *Women, men and the new inter-national division of labor.* Albany, NY, State University of New York Press

Pappas, G., Hadden, W. & Fisher, G. (1993) The increasing disparity of morality between socioeconomic groups in the United States: 1960–1986. *New Engl. J. Med.*, 329, 103–109

Portes, A. & Rumbaut, R., eds (1990) *Immigrant America.* Berkeley, University of California Press

Portes, A., Castells, M. & Benton, L., eds (1989) *The informal economy.* Baltimore, Johns Hopkins University Press

Poulantzas, N. (1975) *Classes in contemporary capitalism.* London, Verso

Roseberry, W. (1989) *Anthropologies and histories.* New Brunswick, NJ, Rutgers University Press

Rothstein, F. & Blim, M., eds (1992) *Anthropology and the global factory.* New York, Bergin and Garvey

Sassen, S. (1991) *The global city.* Princeton, NJ, Princeton University Press

Sidel, R. (1986) *Women and children last.* New York, Viking

Sigerist, H. (1943) *Civilization and disease.* Chicago

Smith, N. (1984) *Uneven development.* New York, Basil Blackwell

Srinivas, M. (1966) *Social change in modern India.* Berkeley, CA, University of California Press

Stoler, A. (1989) Making an empire respectable: the politics of race and sexual morality in the twentieth century colonial cultures. *Am. Ethnologist*, 16, 634–660

Susser, I. (1982) *Norman Street.* Oxford, Oxford University Press

Susser, I. (1985) Union Carbide and the community surrounding it: the case of a community in Puerto Rico. *Int. J. Health Services*, 15, 561–583

Susser, I. (1986) Work and reproduction in sociologic context. In: *Reproduction and the workplace* (*Occupational medicine: state of the art review*). pp. 345–359

Susser, I. (1989) Gender in the anthropology of the United States. In: *Gender and Anthropology.* American Anthropological Association

Susser, I. (1991) The separation of mothers and children. In: Mollenkopf, J. & Castells, M., eds, *The dual city.* Sage Publications. pp. 207–225

Susser, I. (1992) Women as political actors in rural Puerto Rico: continuity and change. In: Rothstein, F. & Blim, M., eds, *Anthropology and the global factory.* New York, Bergin and Garvey. pp. 206–220

Susser, I. (1996a) The shaping of conflict in the space of flows. In: Susser, I., ed., *Anthropological perspectives on the informational society. Critique Anthropol.* 16, 39–49

Susser, I. (1996b) The construction of poverty and homelessness in US cities. *Ann. Rev. Anthropol.* 25, 411–435

Susser, M. & Stein, Z. (1962) Civilization and peptic ulcer. *Lancet*, 1, 115–119

Susser, M., Watson, W. & Hopper, K. (1985) *Sociology in Medicine.* Oxford, Oxford University Press

Thompson, E.P. (1966) *The making of the English working class.* New York, Vintage Press

Thompson, E.P. (1978) *The poverty of theory.* London, Merlin Press

Villerme, I.R. (1826) Rapport. *Archives générales de médicine*, 10, 216–245

Vincent, J. (1984) *Teso in transformation.* Berkeley, University of California Press

Waldinger, R. (1986) Ethnicity and musical chairs: ethnicity and opportunity in post-industrial New York. *Politics and Society*, 15, 369–402

Weber, M. (transl. Henderson, A.R. & Parsons, T., 1947) *The theory of social and economic organization.* London

Wilkinson, R., ed. (1986) *Class and health.* London, Tavistock

Wilson, W. (1987) *The truly disadvantaged.* Chicago, IL, Chicago University Press

Wilson, W. (1996) *When work disappears: the world of the new urban poor.* New York, NY, Knopf

Wolf, E. (1966) *Peasants.* Englewood Cliffs, NJ, Prentice Hall

Wolf, E. (1982) *Europe and the people without history.* Berkeley, University of California Press

Wright, E. (1978) *Class, crisis and the state.* London, Verso

Wright, E., ed. (1989) *The debate on classes.* London, Verso

I. Susser
Department of Anthropology, Hunter College, City University of New York, 695 Park Ave, New York, NY 10021, USA

The measurement of social class in health studies: old measures and new formulations

L.F. Berkman and S. Macintyre

The measurement of socioeconomic status (SES) is a serious matter that requires us to think more precisely about both conceptual issues and issues more traditionally thought of as measurement issues. Progress in this area rests on our ability to identify those aspects of SES that are most closely related to health, human development, and life expectancy. In this chapter we review measures of SES based on characteristics of the individual as well as on characteristics of the environment or more ecologically based measures. Each of these types of SES measures has strengths and weaknesses and in all likelihood taps somewhat different aspects of class. In measuring SES across diverse populations, it is also crucial to be sensitive to the ways in which measurement varies across different cultural, ethnic and demographic groups.

It is likely that as we conduct more refined research in this area we will understand more fully why SES is so profoundly related to health status. However, so as to understand this relationship, we will need to expand efforts to identify not only those psychosocial or biological processes that occur 'downstream' as a result of SES but also the nature of the social experience itself and those 'upstream' forces that place so many individuals at risk.

Over the past decade there has been considerable interest in social class inequalities in health and length of life in industrialized societies. An extensive literature now covers empirical manifestations of such inequalities, trends in these over time, explanations for them, and methods of measuring socioeconomic status (SES), health, premature death, and the magnitude of inequalities (Macintyre, 1996). The aim of this chapter is to review some common and some uncommon ways of measuring SES or social class. By doing this, we hope to arrive at a deeper and more penetrating understanding of what it is about one's social class position that is so closely related to health, development and life expectancy.

Our objective in this review is to lead epidemiologists towards a richer understanding of the measurement of social class and the underlying reasons for it having been so consistently observed over the last century and a half to be related to health status. Our aim is not to document social class differentials in health status (for reviews on this topic, see Macintyre, in press; Marmot et al., 1987; Bunker et al., 1989; Davey Smith et al., 1994; Adler et al., 1994); nor is our aim to discuss the theoretical and conceptual underpinnings of the meaning of social class developed in the social sciences (for more information on this topic, see the chapter by Susser in this book). Rather, we hope that by taking a social epidemiological perspective that incorporates an understanding of the social dimensions and implications of social class (focusing 'upstream') as well as a biological understanding of ways in which social position influences the onset and progression of disease (focusing 'downstream'), we may help to clarify the processes that generate and maintain socioeconomic differentials in health.

Studies of social class and health show poorer health and shorter life expectancy the lower one's position in the social class scale, in all industrialized countries studied so far. However, the magnitude of the differences varies by the measures of SES used, the measures of the health outcome used, the social group being studied, and the particular setting. There is a tendency in the literature on inequalities to assume that the 'best' measure of SES is the measure that produces the steepest SES gradients for the particular group in question. This can involve a circularity in the argument about the relationship between SES and health – a circularity that can interfere with precise thinking about, and study of, the processes producing social gradients in health. The usefulness of a measure depends on the analytical purpose at hand. The question 'what measures of social differentiation produce the greatest observed differentiation in health?' is not the same as the question 'what is the relationship between a specific measure of social differentiation, developed to capture a particular dimension of social experience, and health?', but the two are often confused (for example in debates about whether it is 'better' to classify women by their own or their husbands' occupations). In this chapter we describe properties of various methods of measuring social class without assuming that the 'gold standard' against which they should be measured is the strength of their association with health.

Which measurements are used in particular countries is dependent on the type of socioeconomic information commonly available in those countries. This in turn relates to deep-rooted political and cultural understandings about the nature of social stratification, and the axes of differentiation that are assumed to be significant and that it is politically feasible to collect, in those countries. In the United States of America (USA), race/ethnicity is routinely recorded in vital statistics, cancer registries, and social surveys, whereas occupation tends not to be so collected (Krieger, 1992). In Great Britain, by contrast, race/ethnicity is rarely recorded in these sorts of datasets (the 1991 census was the first ever to try to elicit this information), whereas occupation is a key social signifier and is routinely collected in all official datasets and surveys; data on income are also less frequently collected. In Great Britain such a high proportion of the population has only the lowest educational qualifications [for example, in a six-country comparison of years of education in relation to mortality conducted by Valkonen, 81% of the men and 86% of the women from England and Wales had left school at the statutory minimum school leaving age (Valkonen, 1989)] that years of education is rarely used as a useful measure of SES because of its lack of variance. In some European countries, such as the Netherlands, health-related data at an individual level were not generally available until recently and ecological data had to be used as a proxy [a standardized procedure for measuring SES on the basis of education, occupational class or income has now been developed and will be incorporated into routine data including hospital admission data (Mackenbach, 1994)]. By contrast, the Nordic countries have for some time been able to obtain a number of SES measures (education, occupation and income) from official records and link these with mortality and other health-related measures. As much social epidemiological research on cancer uses large-scale datasets, often derived from record linkage of official social and medical statistics, investigators must often rely on the types of socioeconomic data routinely available in their countries rather than on those measures that they might regard as the most conceptually appropriate.

This chapter is divided into two major sections. In the first section, we review traditional measures of SES, which are based on characteristics of the individual. We then discuss three major issues related to ascertainment of SES based on these indicators: the precision with which they are measured; their appropriateness for women, minorities and older people; and the need to articulate a clearer set of hypotheses about the characteristics of SES that are related to health. In the second section, we discuss assessments of SES that are based not on individual characteristics but on characteristics at a household or community level. The advantages and disadvantages of such measures are identified.

Traditional measures of SES: assessment at the level of the individual

While social class and SES have somewhat different meanings in the sociological literature, based largely on theories developed by Marx and Weber, we have elected to use them interchangeably here (Lipset, 1968).

Social class as described by Weber (1946) had three domains: (1) class, by which he meant ownership and economic resources; (2) status, by which he meant prestige, community ranking or honour; and (3) political power. This tripartite definition has led many social scientists to identify multiple indicators of social class that may be used individually or cumulatively to represent these different dimensions. In an excellent and detailed review of this material and its relevance for those working in the field of public health, Liberatos et al. (1988) discuss the three common indicators of SES – wealth

Table 1. Summary of social class measures by three characteristics

Measures	Categories/scores	Advantages	Disadvantages
Occupation			
Edwards (Haug, 1977)	12 categories; data comparable for 1940–1980 census; 13 categories for 1980	Used by Census Bureau Provides comparability over time since 1940 Widely used since 1940	Each occupational category contains wide variations in income and education
Nam–Powers OSS (Nam & Terrie, 1986)	Scores range 0–100; available for 1960, 1970 and 1980 census	Each score interpretable as a cumulative percentile Data available for male, female, Black and total labour forces	Not sufficiently used to provide empirical evidence of its performance
Siegel (Siegel, 1971)	Scores range 0–100; available for 1960 census only	One of few scales utilizing prestige scores	Based on prestige data collected 20 years ago. Not updated to 1980. Available for male labour force only
Treiman (Treiman, 1977)	Scores range 0–100; can be grouped into eight occupational levels	Only occupational scale that applies internationally Applies to both industrialized and developing countries	Based on prestige data collected 20 years ago. Not updated to 1980. Available for male labour force only
Income	Categories vary depending on population	May measure unique aspects of social class	Varies within occupations and is inconsistent with education Sensitive to changes in life circumstances. Increases with age up to age 65. Not comparable across different years of family sizes unless adjusted. Sensitive topic in USA – 9% refusal rate
Education	Usual category range 2–5; sometimes used as a quantitative variable	Stable over life course Good predictor of mortality from all causes	Fixed early in adult life. Decreasing variability over time. Status does not rise monotonically with years
Composites			
Duncan SEI; (Duncan, 1961; Stevens & Featherman, 1981)	Scores range 0–99; available for 1950, 1970 and 1980 census	Most frequently used in social science research	Positively skewed distribution Original scale based on 1950 male labour force; updates use studies from 1960s to supplement
Hollingshead (Hollingshead, 1975)	Original scores range 11–77 subdivided into 5 classes; available for 1950 census Revision scores range 8–66 subdivided into 5 classes; available for 1970 census	Widely used during 1960s and early 1970s	**Original** Based on 1950 census. Validated in one small Connecticut city No update for 1980 **Revision** Scores for each working spouse are averaged. Census categories used in revision have been modified, requiring additional questions of respondents

Table 1. (Contd) Summary of social class measures by three characteristics			
Measures	**Categories/scores**	**Advantages**	**Disadvantages**
Nam–Powers SES (Nam & Terrie, 1986; Nam & Powers, 1983)	Scores range 0–100; available for 1960, 1970 and 1980 census	Each score interpretable as a cumulative percentile Data available for male, female, Black and total labour forces. Scores are normally distributed	Not sufficiently used to provide empirical evidence of its performance. Potentially redundant if used in combination with individual's education and income
Warner ISC (Miller, 1983)	Scores range 12–84		Difficult to rate dwelling area and house type. Limited applicability since validated on small communities in 1940s
Indices combining income and education	Ad hoc measures	Can be specifically tailored to study population	No systematic validation. Each scale specific to a given study making cross-study comparisons difficult

Modified from Liberatos et al., 1988.

(or income), occupation and education – and the ways in which they are measured or combined to form composite indices. Wealth is clearly most directly related to Weber's idea of class based on ownership and access to economic resources. Occupational rankings based on prestige tap Weber's domain of status (Nam & Terrie, 1982), whereas those occupational rankings based on income may also tap his domain of class. Education, perhaps the most commonly used measure of class in North America, is an indicator of both class and status. As on an individual level completed education generally precedes employment and the ability to earn income, it may influence social position in a powerful way. Table 1 outlines the scales of SES based on occupation, income and education most commonly used in North America, along with their major advantages and disadvantages.

Work from outside North America also uses occupation, income and education, although as noted earlier the availability of individual-level data on these three dimensions, and their perceived relevance, varies from country to country. Investigators in the USA tend to use education and income (Kitagawa & Hauser, 1973; Pappas et al., 1993); those in Great Britain, and in countries such as New Zealand with previous close links to Great Britain, tend to use occupational social class (Townsend et al., 1992; Pearce et al., 1993); and education has commonly been used in several European countries (Valkonen, 1989).

As noted in virtually all reviews on SES gradients and health, the consistency and strength of the associations between SES and morbidity and mortality, both within and across countries, attest to the validity of the indicators themselves and the degree to which these relatively crude indicators must be tapping some underlying construct of social stratification that powerfully influences health (Syme & Berkman, 1976; Williams, 1990). However, from a measurement perspective there are certain troublesome aspects of these brief scales, which pose problems worthy of consideration. The three issues outlined below are particularly relevant, and are discussed in more detail in the following sections.

• Differences in the slope of the SES gradient or in the magnitude of differences between different social categories are reported both between studies and between different groups within studies. A major question is whether these differences reflect imprecision in measurement of either SES or the health outcome of interest or reflect real variations in risk.
• Measures of SES were most often developed using middle-aged employed male populations.

The validity of such measures for women, older populations and ethnic minorities is unclear.
• Investigators rarely articulate precisely what it is about SES, or about the particular measure of it they are using, that they hypothesize to influence health status. For instance, most scientists currently argue that it is not poverty alone that conveys disease risk, as there is a steady gradient of risk all the way up the social scale. However, they often fail to articulate whether they are using education, income or occupation simply as an indicator of an underlying SES gradient (and if so, how that gradient influences health) or whether (and if so, how) they see education, income and occupation as directly influencing health.

Imprecision in measurement of SES

Social scientists whose major efforts are directed towards the measurement and study of social stratification often argue that crude indicators of SES such as income, occupation and educational level are inadequate measures of SES.

One concern relates to the use of income as a measure of wealth or economic status. Economists point out that income captures economic status only partially since income measures do not include assets such as inherited wealth, savings, benefits, or ownership particularly of homes or motor vehicles. More detailed information must be collected to identify these other sources of wealth. Excellent examples of how this information has been incorporated into studies with important health outcomes are the Health and Retirement Study funded by the National Institute of Aging to the University of Michigan and cross-cultural studies conducted by Rand.

Not only do traditional measures of income fail to capture wealth but they also often fail to measure income earned from the 'informal economy'. As Susser points out in the chapter in this book, many recent immigrants and minorities work in an informal economy for cash with no job security or benefits. Additionally, many people – rich and poor – exchange goods and services, and barter. Such informal transfers are poorly documented and rarely included in measures of income.

Finally, many investigators from both epidemiology and public health have remarked that measures of income must be adjusted to account for the number of people supported by the income. Clearly, an income of US$ 25 000 for a family of two or a single person is not equivalent to the same income supporting a family of six. When we add to these measurement issues the fact that of all the measures of SES, individuals are least likely to report their income, regarding it as a highly sensitive and private topic, we can see why income is the least used of SES indicators. This is unfortunate because no other measure of SES has the psychometric properties of being continuous and spread along such a broad range from low (the depths of poverty) to high (wealth). Furthermore, no other indicator so clearly taps the dimension of potential access to material goods and services as unambiguously as does economic resources assessed from income and wealth.

Measures of occupational status or prestige are commonly used in epidemiological studies. There are several occupationally based classifications of social class in Great Britain, mainly based on the Registrar General's classification of occupations that was first used to examine social gradients in births, infant mortality and adult mortality around the time of the 1921 British census (Stevenson, 1928). This grouped occupations into six social classes (three non-manual and three manual) according to a combination of skill levels and general standing in the community. It was explicitly not based simply on the average income of the occupations. In a lecture to the Royal Statistical Society in 1928, Stevenson, the Registrar General who had developed the classification, described a method of examining infant mortality by family income that had recently been used in the USA, and commented (Stevenson, 1928):

'So far as this method can be applied it is of course ideal for estimation of the effects of wealth as such... But its drawback is that it may fail altogether as in index of culture, probably the more important influence. The power of culture to exert a favourable influence on mortality, even in the complete absence of wealth, is well illustrated by the clergy. The income test, if it could be applied, would certainly place them well down the list yet their mortality is remarkably low... The method suggested, therefore, as on the whole best meeting the various conditions which have to be considered is that of inferring social position (largely but by no means exclusively a matter of wealth or poverty, culture also having to be taken into account) from occupation.'

When Stevenson first applied his classification to birth and death rates he was pleased to see that they produced steady gradients, and commented that this was 'at the same time an indication both of success in the social grading in the population and of the association of mortality with low status' (Stevenson, 1928). Thus from the start there has been some circularity about this classification of occupations when applied to birth or death rates – the validity of the classification being assessed by its correlation with these rates, and the strength of social influences on these outcomes being assessed by the linear gradients produced by this occupational social classification (for a critique, based on this issue, of the use of the classification, see Jones & Cameron, 1984).

Nevertheless, scales similar to the British one are widely used in other countries and have been used for a number of between-country comparisons of inequalities (Vagero & Lundberg, 1989; Leclerc et al., 1990; Kunst & Mackenbach, 1994)

In the USA, occupational scales have usually been based on prestige or income (Siegel, 1971; Treiman, 1977; Duncan, 1961; Featherman et al., 1975). Apart from the issue of the conceptual underpinning of these various scales, the greatest problem with them is that the job rankings have proved to be relatively unstable over time; that is, new requirements and economic needs have changed job standing in terms of both income and status over time. For instance, when scales were initially developed in the 1950s and 1960s, white-collar, office jobs were almost always ranked higher than blue-collar jobs. Over time, however, with the entry of women into the workforce and the growth of 'pink-collar' office jobs (secretarial, sales and so on), some blue-collar jobs have gained much higher earning power as well as more prestige and job characteristics associated with control, independence and skill than these white-collar jobs have. These recent status differentials are not always reflected in job rankings. Investigators must take care to use scales that reflect the characteristics of the population being studied. Some scales, including those developed by Edwards, Nam–Powers and Duncan, have been updated to the 1980 census and reflect some of these changes.

Prestige-ranked scales of occupation vary by country, making international comparisons difficult. For instance, in France intellectuals and artists rank high on prestige-based occupational scales. Thus, teachers at both the secondary and university level and visual artists are ranked high even though their income is not among the highest in the country. In the USA, such professionals are ranked lower. In some countries it is considered legitimate to have an unambiguous ranking from the top to the bottom of the occupational scale (as in the British system) but in others certain occupational groups (such as farmers, the self-employed, or the armed forces) stand outside the occupational ranking system and their placement in a scale is therefore problematic. Occupational scales that rank occupations uniformly across countries will obscure these differences yet scales based on national norms may make comparisons difficult to interpret (Fox, 1989; Kunst & Mackenbach, 1994).

Occupational classes are made up of heterogeneous occupations and there is considerable variation within each class in education, income, prestige and risks. Studies that examine more homogeneous occupational groups within specific industries or employment settings [for example, studies in Great Britain of the army, the National Health Service, and the civil service in London (Lynch & Oelman, 1981; Balarajan, 1989; Davey Smith et al., 1990)], find much bigger differences between these groups in mortality than are found for the occupational classes in which they are normally classified [for example, standard mortality ratios for coronary heart disease among men in the British army were 33 for direct-entry officers and 205 for private soldiers, a sixfold difference that is greater than the difference between all social class I men and all social class V men (Lynch & Oelman, 1981)]. It has been argued on the basis of findings such as these that conventional occupational class measurements tend to underestimate the impact on mortality of socioeconomic position because of imprecision of measurement (Davey Smith & Egger, 1992).

One of the reasons that education is used most frequently as an indicator of social class is the ease and consistency with which it is measured. While distributions in educational level may vary by region, country, age and gender, years of education remains one of the most reliable and valid indicators of SES. It may be used as a continuous measure or categorized at meaningful cut points, such as completion of high school or university. Furthermore, education is often used as an indicator because of all common measures it is least likely to be influenced by disease, as most people complete their education before they reach the age of 20 or 25 years. With the exception of a few psychiatric disorders, there are few diseases that threaten to disrupt the educational process in European and North American countries (Kitagawa & Hauser, 1973). [However, as pointed out by Goldberg and Morrison, achieved educational level is not a good measure of social drift following illness. They found that many schizophrenics had educational records commensurate with their social background but then held jobs that were of lower skill level than their background or education would predict. Many also remained living with their parents. For these reasons, using educational attainment and residence as measures of SES could underestimate the intergenerational downward social drift experienced by schizophrenics (Goldberg & Morrison, 1963)].

A technique used to circumvent absence of individual-level income, education or occupational data is to classify people according to household assets such as whether the home is owned or rented, and whether there is a car or garden. These have been shown both to be independently predictive of mortality and to add to the predictive power of other measures such as occupational social class and grade of employment in the civil service (Fox & Goldblatt, 1982; Davey Smith et al., 1990). It has been argued that such household measures of assets create a more finely grained hierarchy of socioeconomic position and thus demonstrate that conventional measures understate the power of SES to influence life chances (Davey Smith & Egger, 1992). However, they represent some of the least characterized measures of SES and little research has been conducted into their social meaning and implications or the processes by which they influence health. Some have taken them as simply more refined measures of material well-being (Davey Smith & Egger, 1992). But the finding that top-rank civil servants in London who do not own cars have higher mortality than car-owning colleagues in the same grades raises as many questions as it answers.

It is likely that top-rank civil servants can all afford cars, so why do some not have cars? Because they prefer to walk from their elegant town houses, because they have had their driver's licence revoked, or because they use taxis and a work-provided chauffeur? Car ownership may actually be directly health promoting (by enhancing social contacts, and providing convenient access to health services, recreation, and food shopping), but it is often regarded simply as a marker of wealth. Similar sorts of questions are raised about the social meaning of home ownership as opposed to home rental: is it directly health promoting in some way, a marker of wealth, or confounded with other variables (such as region of the country, or employment in the armed forces or other occupations that involves frequent moves)?

Generational and aging effects need to be taken into account in using individual- or household-level measures of SES. The social meaning of education, income and occupation will vary between different birth cohorts who grew up in, and now inhabit, very different social contexts. The significance of a college-level education will for example differ between someone now 75 years old and someone now 35 years old because a much higher proportion of the latter's age cohort will have attended college. What may be relevant for a 75 year old is not how their absolute income compares with that of her 35-year-old grandchild but how it compares with the income of their age peers. The occupational structure, and with it the occupational class structure, has changed greatly in all industrialized societies such that, in general, succeeding generations are likely to appear to have higher social class position. The social meaning and significance for health of other measures such as car or house ownership are also likely to differ both by generations and by age. It is thus important when examining SES relationships with health both to standardize for age and to think clearly about the applicability for different ages and cohorts of the measures and underlying construct of SES being used.

The use of indicators of SES among women, minorities and older populations

As we stated earlier, most measures of social class were developed and subsequently validated on men, primarily men in the labour force. Extrapolating to other populations has proved to be quite problematic. Even after 30–40 years of research in this area in the USA (and 70 years in Great Britain), there are no completely successful resolutions as to how to classify housewives, retirees, or minorities who may hold the same job as White males but do not gain the same benefits. The robust and consistent relationships between SES and health or mortality found for men of working age (whether measured by education, income or occupation) are often not found for women, older people and ethnic minorities. For example, in cross-national comparisons the relationship between education and mortality is less consistent (and less likely to be linear) among women than among men (Valkonen, 1989; Koskinen & Martelein, 1994), and occupational class is much less predictive of mortality among Maori compared with non-Maori men in New Zealand (Pearce et al., 1993) and among migrants from the Indian subcontinent in Great Britain compared with the general population (among Afro-Caribbeans in Great Britain, mortality is actually higher among higher occupational classes) (Marmot et al., 1984).

Only a small part of the problem lies in imprecision in measurement. For instance, while it is worthwhile to rescale occupation to incorporate occupations held predominantly by women (for example, nursing and clerical work) and carefully consider where 'pink-collar' occupations fit in occupational rankings, such readjustments to the scales do not inform us about how to deal with two-occupation families or how to classify women who do not work in the labour force. These problems pose larger challenges and force us to confront more directly the conceptual underpinnings of measures of SES. More serious consideration should be given to classifying couples who share households by the highest occupational ranking between them or by developing new indicators of SES that are not gender specific.

In Great Britain steeper gradients in mortality have been observed when women have been classified by their husbands' occupations than by their own, and within any own occupational social class defined by the women's own occupations there are gradients by husbands' occupations. For example, among women whose own occupations place them in social class III non-manual, SMRs range from 72 among those whose husbands are in class I to 117 among those whose husbands are in class V. These differences are even greater for economically inactive women (SMRs by husbands' classes ranging from 55 to 130) (Fox & Goldblatt, 1982). Various attempts have been made to improve social classifications among women by adding in other social characteristics such as marital status, economic activity and indicators of household wealth, and it has been argued that 'accurately to reflect the relation between a woman's life circumstances and mortality it is necessary to utilize other measures than those based solely on occupation' (Moser et al., 1988). As with evaluations of the arguments that specific occupations provide sharper differentiation in mortality than broad groupings (see above) and that household- or asset-based measures add to predictions of mortality, it is important to be clear about whether the aim is to produce the most accurate social predictors of mortality risk (which might contain a number of elements including marriage and specific occupational exposures) or whether the intention is to clarify the relationship between occupational class and risk. It is also important to note an asymmetry in discussions of women's and men's social classifications; researchers have rarely looked to see whether the educational levels, incomes or occupations of wives add to the prediction of men's mortality. We will pick up these issues again in the next section in which we move beyond individual-based indicators of SES.

The evidence relating social class to health among older men and women is conflicting. Common wisdom is that SES wanes in importance as a predictor of mortality and morbidity in the elderly. While some studies support this notion, many other studies continue to show that SES is a critical predictor of health outcomes throughout the life course (Berkman, 1988). Of particular interest are the recent findings on the relationship between low educational level and risk of Alzheimer's disease and other cognitive and functional declines (White et al., 1994).

Is there a way to explain these differences in study findings? Differences among studies in the magnitude of SES effects in elderly populations may well be the result of the same set of methodological issues that besets investigators studying other subpopulations. Perhaps most relevant to the study of older men and women is the imprecision with which SES is measured. Many investigators gather data on current occupation (or most recent occupation) or current income. In the majority of cases, current income or occupation is not an accurate reflection of lifetime or usual occupation or usual income. In order to obtain information in these areas, investigators must ask questions about accumulated wealth, savings and ownership as well as usual occupation. Of course, issues of ascertainment of SES among older women or minority elders are compounded by ascertainment of SES in these groups. Furthermore, many investigators fall back on education-based indicators of SES for older populations. While this has many advantages, the most important being that it is a stable measure of SES and unlikely to be influenced by health status in old age, there are some drawbacks to using education as a marker for SES in this age group. The most important disadvantage relates to the limited number of years of schooling many older people have had who are part of the birth cohorts born between 1910 and 1930. Many Americans during this time had few years of formal education yet were occupationally quite mobile. Thus, their educational level may not be a very precise indicator of their social class. This truncated distribution and weak correlation with other indicators of SES may not be so important in future studies as more recent birth cohorts, at least in the USA, have a broader range of educational experiences.

Similar to the case for the ascertainment of SES in women and minorities, closer attention to the precision with which we measure SES using standard tools and indicators may not be sufficient to capture the true variance in risk of adverse health outcomes related to social class in older populations (Berkman et al., 1989). In order to understand SES gradients more fully in older populations, it would be helpful to have more information on social stratification on a community level – how more extended families and households contribute to the SES of older family members, how older people continue to be productive despite not being in the formal labour force, and how their status is determined based on a lifetime of experience and contribution to members of their families and communities. These areas require us to extend our thinking of social class, probing new areas and formulating new measures of SES based on more conceptually rich theories.

The assessment of SES in African-Americans and other minority ethnic groups poses a new set of issues. Racial differences in health outcomes, especially in the USA, are often attributed to underlying differences in SES rather than to genetic or inherent biological differences between races (Williams, 1990; Krieger et al., 1993). However, many studies report that health status differences between racial groups remain after 'adjustment' for SES. Investigators rarely acknowledge that the 'adjustment' for SES may not have completely accounted for SES differences. For instance, there is now a growing literature indicating that within broad categories of income, education and occupation, African-Americans earn less, have less wealth, and often have a higher cost of living than Whites in the same categories (Nam & Powers, 1983; Krieger, 1993). Such differences relate to the imprecision with which we measure SES. However, increasing the precision with which we measure SES among different ethnic groups would probably account for only a portion of the health disparities between Blacks and Whites. Additional inequalities are likely to result from racism and discrimination *per se* so that even among Blacks and Whites with exactly the same occupational and educational level disparities in SES exist. The burden that discrimination places on selected ethnic minorities limits access not only to medical care, which may influence health outcomes, but also to a broad range of life chances and opportunities. Incorporating subtle measures of social class that capture the social stratification that occurs in our society on the basis of race remains a major challenge.

Cultural differences must also be taken into account. Pearce et al. (1988) have pointed out that social class classifications based on occupation may have little meaning in traditional Maori society: 'A manual labourer performing the most menial task not infrequently turns out to be a gifted orator, or a person with exceptional prestige widely regarded by his tribe as healthy' (Durie, 1985). The linkages between education, occupation and income commonly assumed or studied in a majority culture may also differ in minority cultures, especially in newly arrived migrant groups among whom earnings may be remitted to family members remaining in the country of origin, and who may occupy particular economic or residential niches of a lower standing than their educational qualifications would normally predict in their old or new country.

What is it about SES that is so closely related to health? Conceptual clarity about the macrosocial phenomenon

Stepwise or linear gradients in morbidity and mortality risk by SES have been observed in the vast majority of studies, although policy analysts, especially in the USA, have commonly focused on the effects of poverty, using a threshold model of risk that assumes that absolute disadvantage causes poor health but that above a certain threshold there is no further effect of SES on health. As Macintyre notes, in many ways the SES gradient is not at all surprising 'since most socioeconomic classifications fit a Weberian model of a relatively continuous distribution of life chances which are likely to produce a relatively continuous distribution of health attributes' (Macintyre, in press). The gradient in risk challenges us to define what it is about social stratification *per se*, not just poverty, that is associated with poor health (Adler *et al.*, 1994). In addressing this issue many investigators have skipped over more social–structural interpretations having directly to do with class and moved on to identify mechanisms or pathways by which SES could produce poor health. Such endeavours move us from struggling with defining the macrosocial processes that are fundamental to class and stratification to what may well be the result or response of individuals to such social phenomena.

If we are to understand better how social position confers health risks it is important to focus upstream, to processes of social stratification and their implications for everyday life, as well as downstream, to the psychosocial or biological mediators of risk. Whatever the focus it is important to be clear about the underlying models with which we are operating. It is only in relation to these that the validity, reliability and utility of measures of SES can be assessed.

It is clear from the history of research on social inequalities in health that different investigators often conceptualize the same measures as being operationalizations of different underlying constructs. For example, in northern Europe occupational social class has often been seen as an indicator of direct occupational exposures (both physical and psychosocial), and its association with mortality has led investigators to study health-promoting or -damaging properties of jobs. In Great Britain occupational class has been seen as an indicator of a general style of life including residential and consumption patterns and access to a whole set of social and material resources. This may explain why in Great Britain it made sense to classify women by their husbands' occupations, as these predicted general domestic circumstances, but this makes less sense in countries in which the main focus is occupational exposures (Arber & Lahelma, 1993). The mortality risks of husbands and wives have indeed been used for some time in the United Kingdom in an effort to disentangle direct occupational risks from those associated with general life circumstances (Office of Population Censuses and Surveys, 1978).

However measured, SES gradients can be seen as expressing wealth and income differences, exposures to health-damaging circumstances, access to control over health-promoting activities and resources, or psychosocial assets such as education and coping skills. However, it is often not explicit which of these (or many other) underlying models are being assumed. For example, household ownership of homes, gardens and cars has been shown to predict mortality risk in Great Britain (Davey Smith & Egger, 1992) but the interpretation of these relationships is rarely clearly spelled out (implicitly it often seems to be that these are seen as simply indicators of wealth, it being wealth that produces health, rather than as directly health-promoting resources that wealth can help one to buy).

If we are to improve measures of SES, we must offer explicit hypotheses about the aspects of social class that we think convey health risks. For instance, if we hypothesize that relative deprivation is a more useful concept than absolute deprivation to explain SES gradients, we might refine our measures specifically to capture elements of relative deprivation. If we hypothesize that material resources are more important than psychosocial resources, we might focus increased attention on measures of wealth and on the assets or experiences that wealth produces. If we believe psychosocial resources are more critical, we might examine educational measures more carefully since they are more highly correlated with cognitive strategies and behaviours. If we hypothesize that something about status or ranking *per se*, even in the absence of resources, is the critical dimension of SES, we might examine yet other dimensions of class emphasizing prestige or hierarchy. The most important point here is that if we focus our scientific energies 'downstream' to identify consequences of SES we may miss valuable opportunities to understand the nature of the social positions we call social class.

Area- or household-based measures of SES

There are two main reasons for using household- or area-based measures of social class. The first is that individual-level data are unavailable; the second is that these more collective measures may add to or interact with individual measures and thus add explanatory power. For which of these reasons such measures are being used is not, however, always clear.

Given the prominence of occupational class analysis in Great Britain, and the way it is conceptualized as indicating broad styles of life, it is common to use household measures of class in order to get around the problem of missing individual data arising from current unemployment for whatever reason. 'Head of Household Social Class' can for example be applied to children, homemakers, and unemployed, retired or sick adults, and thus can be used for comparisons of class gradients across the life course (Ford et al., 1994). Although used to compensate for the problem of non-employment, an underlying premise of such measures is that head of household measures actually express a real form and unit of social stratification that is related to health and life expectancy.

Area-based measures of socioeconomic position, usually based on census data, have been used in a number of countries as a proxy for individual or household social class. Individuals are characterized by the aggregate socioeconomic properties of the zipcode (USA), postcode (United Kingdom), census tract, or local government area in which they live. In Great Britain there are two widely used area-based indices of deprivation based on census data: the Carstairs–Morris index [based on the percentage of unemployed people, overcrowded households, households with no car, and people in social classes IV and V (Carstairs & Morris, 1991)] and the Townsend index [based on the percentage of people with no car, in overcrowded housing, in non-owner-occupied housing, and unemployed (Phillimore et al., 1994)]. Both measures strongly predict mortality and other health measures. However, as with all ecological measures care has to be taken in interpreting these correlations. At the extremes of deprivation these indices classify fairly socially homogenous areas: Carstairs category 1 postcodes will be inhabited by people who are very well off, and category 7 postcodes by people who are very badly off. But categories in the middle will contain a mixture of better- and worse-off people and it will therefore be less valid to 'read off' personal circumstances from residence in such areas (McLoone & Boddy, 1994). It is not always clear whether investigators are using such measures simply as surrogates for individual SES ('this person lives in a high-income or middle-class area so we can infer that they have a high income or is middle class') or whether they are using them as genuinely ecological measures ('this person lives in a high-income or middle-class area so may have access to certain health-promoting local resources').

Classification of an individual's social class based on his or her personal characteristics is limited by the fact that people living together often share class position in ways not reflected by individual circumstances. Furthermore, recent evidence indicates that the socioeconomic environment or community in which one lives confers risk apart from an individual's standing in that community (Haan et al., 1987). This reasoning has led investigators to develop new techniques to assess socioeconomic position based on area-based indicators. The hypothesis is that an individual's living environment, the resources to which they have access, and the stresses to which they are exposed, are based on more than their individual characteristics. For instance, middle-class individuals living in impoverished neighbourhoods may share more experiences with their neighbours than with their middle-class counterparts in less impoverished areas. Personal and local circumstance and access to resources may interact to amplify disadvantage and health risk. For example, healthy food may be more costly and less available in poorer neighbourhoods inhabited by people with lower disposable income (Sooman et al., 1993) and 'healthy eating' has been shown to vary between neighbourhoods even after controlling for household social class, sex, income and age (Forsyth et al., 1994). It is for this reason that many public housing policies are based on the theory that poorer families will do better in middle-class suburbs than they will do if provided with straightforward subsidies to improve their individual housing.

The argument that features of the local area might amplify personal advantage or disadvantage in predicting health risk would suggest that measures should be developed that incorporate both personal and local characteristics. This has rarely been done, however, and so we are left with what may be a fake antithesis: should we be focusing on people or places? (Macintyre et al., 1993). There have been several recent attempts to examine the importance of individual and area indicators (for example, Slogget & Joshi, 1994) but by teaching these as if they are independent they may underestimate the extent to which different aspects of SES cluster and interact with each other.

The most basic aggregate data are based on household characteristics. Such data are particularly valuable for people who may not be well characterized by traditional measures. For instance, women may often be better characterized by household measures of occupation that are either averaged over working household members or in which the highest rank is given to all family members. In fact, men as well may be better characterized by this system as women increasingly are in the labour force and contributing to the economic well-being of families.

Conclusions

The measurement of SES is a serious matter that requires us to think more precisely about both conceptual issues and issues more traditionally thought of as measurement issues. Progress in this area rests on our ability to identify those aspects of SES that are most closely related to health, human development, and life expectancy. In this chapter we have reviewed measures of SES based on characteristics of the individual as well as on characteristics of the environment or more ecologically based measures. Each of these types of SES measures has strengths and weaknesses and in all likelihood taps somewhat different aspects of class. In measuring SES across diverse populations, it is also crucial to be sensitive to the ways in which measurement varies across different cultures, ethnic and demographic groups.

It is likely that as we conduct more refined research in this area we will understand more fully why SES is so profoundly related to health status. However, in order to understand this relationship, we will need to expand efforts to identify not only those psychosocial or biological processes that occur 'downstream' as a result of SES but also the nature of the social experience itself and those 'upstream' forces that place so many individuals at risk.

References

Adler, N., Boyce, T., Chesney, M., Cohen, S., Folkmen, S., Kahn, R.L. & Syme, S.L. (1994) Socioeconomic status and health; the challenge of the gradient. *Am. Psychologist*, 49, 15–24

Arber, S. & Lahelma, E. (1993) Inequalities in women's and men's ill health: Britain and Finland compared. *Soc. Sci. Med.*, 37, 1055–1068

Balarajan, R. (1989) Inequalities in health within the health sector. *Br. Med. J.*, 299, 822–825

Berkman, L. (1988) The changing and heterogeneous nature of aging and longevity: a social and biomedical perspective. *Annu. Rev. Gerontol. Geriatr.*, 8, 37–68

Berkman, L., Singer, B. & Manton, K. (1989) Black/White differences in health status and mortality among the elderly. *Demography*, 26, 661–678

Bunker, J.P., Gomby, D.S. & Kehrer, B.H. (1989) *Pathways to health; the role of social factors*. Menlo Park, CA, The Henry J. Kaiser Family Foundation

Carstairs, V. & Morris, R. (1991) *Deprivation and health in Scotland*. Aberdeen, Aberdeen University Press

Davey Smith, G. & Egger, M. (1992) Socioeconomic differences in mortality in Britain and the United States. *Am. J. Public Health*, 82, 1079–1081

Davey Smith, G., Shipley, M.J. & Rose, G. (1990) The magnitude and causes of socio-economic differentials in mortality; further evidence from the Whitehall study. *J. Epidemiol. Community Health*, 44, 265–270

Davey Smith, G., Blane, D. & Bartley, M. (1994) Explanations for socioeconomic differentials in mortality: evidence from Britain and elsewhere. *Eur. J. Public Health*, 4, 131–144

Duncan, O.D. (1961) A socioeconomic index for all occupations. In: Reiss, A.J., ed., *Occupations and social status*. New York, Free Press. pp. 109–138

Durie, M.H. (1985) A Maori perspective of health. *Soc. Sci. Med.* 20, 483–486

Featherman, D., Sobel, M. & Dickens, D. (1975) *A manual for coding occupations and industries into detailed 1970 categories and a listing of 1970-based Duncan and NORC prestige scores* (Working Paper 75-1). Madison, WI, Center for Demography and Ecology

Ford, G., Ecob, R., Hunt, K., Macintyre, S. & West, P. (1994) Patterns of class inequality throughout the lifespan; class gradients at 15, 35 and 55 in the west of Scotland. *Soc. Sci. Med.*, 39, 1037–1050

Forsyth, A., Macintyre, S. & Anderson, A. (1994) Diets for disease: extra-urban variation in reported food consumption in Glasgow. *Appetite*, 22, 259–274

Fox, A.J. (1989) *Health inequalities in European countries*. Aldershot, Gower

Fox, J. & Goldblatt, P. (1982) *Longitudinal study 1971–1975; England and Wales* (Office of Population Censuses and Surveys: Series LS No. 1). London, Her Majesty's Stationery Office

Goldberg, E.M. & Morrison, S.L. (1963) Schizophrenia and social class. *Br. J. Psychiatry*, 109, 785–802

Haan, M.N., Kaplan, G.A. & Camacho, T. (1987) Poverty and health: prospective evidence from the Alameda County study. *Am. J. Epidemiol.*, 125, 989–998

Haug, M.R. (1977) Measurement in social stratification. *Annu. Rev. Sociol.* 3, 51–77

Hollingshead, A.B. & Redlich, F.C. (1958) *Social class and mental illness: a community study*. New York, NY, John Wiley and Sons

Jones, I.G. & Cameron, D. (1984) Social class; an embarrassment to epidemiology? *Community Med.*, 6, 37–46

Kitagawa, E.M. & Hauser, P.M. (1973) *Differential mortality in the United States: a study in socioeconomic epidemiology*. Cambridge, MA, Harvard University Press

Koskinen, S. & Martelein, T. (1994) Why are socioeconomic mortality differences smaller among women than among men? *Soc. Sci. Med.*, 38, 1385–1390

Krieger, N. (1992) Overcoming the absence of socioeconomic data in medical records: validation and application of a census-based methodology. *Am. J. Public Health*, 82, 703–710

Krieger, N., Rowley, D.L., Herman, A.A., Avery, B. & Phillips, M.T. (1993) Racism, sexism and social class: implications for studies of health, disease and well being. *Am. J. Prev. Med.*, 9, 82–122

Kunst, A. & Mackenbach, J. (1994) International variations in the size of mortality differences associated with occupational status. *Int. J. Epidemiol.*, 23, 742–750

Leclerc, A., Lert, F. & Fabien, C. (1990) Differential mortality: some comparisons between England and Wales, Finland and France based on inequalities measures. *Int. J. Epidemiol.*, 19, 1001–1010

Liberatos, P., Link, B. & Kelsey, J. (1988) The measurement of social class in epidemiology. *Epidemiol. Rev.*, 10, 87–121

Lipset, S.M. (1968) Social class. *Int. Encycloped. Soc. Sci. Med.*, 15, 298–316

Lynch, P. & Oelman, B.J. (1981) Mortality from CHD in the British Army compared with the civil population. *Br. Med. J.*, 283, 405–407

Macintyre, S. (1986) The patterning of health by social position in contemporary Britain; directions for sociological research. *Soc. Sci. Med.*, 23, 393–415

Macintyre, S. The Black report and beyond; what are the issues? *Soc. Sci. Med.*, in press

Macintyre, S., MacIver, S. & Sooman, A. (1993) Area, class and health; should we be focusing on places or people? *J. Soc. Policy*, 22, 213–234

Mackenbach, J. (1994) Socioeconomic inequalities in health in the Netherlands: impact of a five year research programme. *Br. Med. J.*, 309, 1487–1491

McLoone, P. & Boddy, F.A. (1994) Deprivation and mortality in Scotland, 1981 and 1991. *Br. Med. J.*, 309, 1465–1470

Marmot, M.G., Adelstein, A. & Bulusu, L. (1984) *Immigrant mortality in England and Wales 1970–78*. London, Her Majesty's Stationery Office

Marmot, M.G., Kogevinas, M. & Elston, M.A. (1987) Social/economic status and disease. *Annu. Rev. Public Health*, 8, 111–135

Miller, D.C. (1983) *Handbook of research design and social measurement*, New York, NY, Longman

Moser, K., Pugh, H.S. & Goldblatt, P. (1988) Inequalities in women's health: looking at mortality differentials using an alternative approach. *Br. Med. J.*, 296, 1221–1224

Nam, C.B. & Powers, M.G. (1983) *The socioeconomic approach to status measurement*. Houston, TX, Cap and Gown Press

Nam, C.B. & Terrie, E.W. (1982) Measurement of socioeconomic status from United States census data. In: Powers, M.G., ed., *Measures of socioeconomic status: current issues*. Boulder, CO, Westview Press. pp. 29–42

Nam, C.B. & Terrie, E.W. (1986) *Comparing the 1980 Nam-Powers and Duncan SEI occupational scores*. Center for the Study of Population, Florida State University

Office of Population Censuses and Surveys (1978) *Occupational mortality: decennial supplement 1970–1972*. London, Her Majesty's Stationery Office

Pappas, G., Queen, S., Hadden, W. & Fisher, G. (1993) The increased disparity in mortality between socioeconomic groups in the United States 1960 and 1986. *New Engl. J. Med.*, 329, 103–109

Pearce, N., Pomare, E., Marshall, S. & Borman, B. (1993) Mortality and social class in Maori and non-Maori New Zealand men: changes between 1975–77 and 1985–87. *New Zealand Med. J.*, 106, 193–196

Phillimore, P., Beattie, A. & Townsend, P. (1994) Widening inequality of health in northern England 1981–91. *Br. Med. J.*, 308, 1125–1128

Siegel, P.M. (1971) *Prestige in the American occupational structure* [Dissertation]. University of Chicago

Slogget, A. & Joshi, H. (1994) Higher mortality in deprived areas: community or personal disadvantage? *Br. Med. J.*, 309, 1470–1474

Sooman, A., Macintyre, S. & Anderson, A. (1993) Scotland's health: a more difficult challenge for some? The price and availability of healthy food in contrasting localities in the west of Scotland. *Health Bull.*, 51, 276–284

Stevens, G. & Featherman, D.L. (1981) A revised socio-economic index of occupational status. *Soc. Sci. Res.*, 10, 364–395

Stevenson, T.H.C. (1928) The vital statistics of wealth and poverty. *J. R. Stat. Soc.*, XLI, 209–210

Syme, S.L. & Berkman, L.F. (1976) Social class, susceptibility and sickness. *Am. J. Epidemiol.*, 104, 1–8

Townsend, P., Davidson, N. & Whitehead, M., eds (1992) *Inequalities in health; the Black report and the health divide.* London, Penguin Books

Treiman, D.J. (1977) *Occupational prestige in comparative perspective.* New York, NY, Academic Press

Vågerö, D. & Lundberg, O. (1989) Health inequalities in Britain and Sweden. *Lancet*, ii, 35–36

Valkonen, T. (1989) Adult mortality and level of education; a comparison of six countries. In: Fox, J., ed., *Health inequalities in European countries.* Aldershot, Gower. pp. 142–162

Weber, M. (1946) Class, status and party. In: Gerth, H. & Mills, C.W., eds, *From Max Weber: essays in sociology.* New York, NY, Oxford University Press. pp. 180–195

White, L., Katzman, R., Losonczy, K., Salive, M., Wallace, R., Berkman, L., Taylor, J., Fillenbaum, G. & Havlik, R. (1994) Association of education with incidence of cognitive impairment in three estabished populations for epidemiologic studies of the elderly. *J. Clin. Epidemiol.*, 47, 363–374

Williams, D.R. (1990) Socioeconomic differentials in health; a review and redirection. *Soc. Psychol. Q.*, 53, 81–99

Corresponding author:
L.F. Berkman
Harvard School of Public Health, 677 Huntington Avenue, Boston, MA 02115, USA

Socioeconomic differences in cancer incidence and mortality

F. Faggiano, T. Partanen, M. Kogevinas and P. Boffetta

This chapter summarizes accumulated data on the presence, magnitude and consistency of socioeconomic differentials in mortality and incidence of all malignant neoplasms and 24 individual types of neoplasms in 37 populations in 21 countries. More or less consistent excess risks in men in lower social strata were observed for all respiratory cancers (nose, larynx and lung) and cancers of the oral cavity and pharynx, oesophagus, stomach, and, with a number of exceptions, liver, as well as for all malignancies taken together. For women, low-class excesses were consistently encountered for cancers of the oesophagus, stomach, cervix uteri and, less consistently, liver. Men in higher social strata displayed excesses of colon and brain cancers and skin melanoma. In the two Latin American populations for which data were available, lung cancer was more frequent in higher social strata. Excesses in high female socioeconomic strata were seen in most populations for cancers of the colon, breast and ovary and for skin melanoma. Longitudinal data from England and Wales suggested widening over time of social class differences in men for all cancers combined and for cancers of the lung, larynx and stomach, and in women for all cancers combined and for cervical cancer.

In this chapter we examine data on the presence, magnitude and consistency of socioeconomic differences in cancer incidence and mortality of all malignant neoplasms and 24 individual types of neoplasms in 37 populations in 21 countries. Time trends are presented for the United Kingdom, for which historical mortality data are available (Logan, 1982).

The data for this review derive from both published and unpublished sources. A 1966–1994 MEDLINE search and reference lists of the recovered sources identified the published data. In addition, a letter requesting data on socioeconomic status and cancer mortality or incidence was mailed to 77 institutions and investigators who were considered to have access to such data.

Study design
The source data derived from surveillance systems, cohort studies and case–control studies of 35 populations. Table I provides details of the studies included. With very few exceptions, ecological studies, based on geographical rates, were excluded.

Socioeconomic indicators
The concepts of social class and socioeconomic status incorporate essential economic, political and cultural components. Such a comprehensive conceptualization offers the obvious advantage that various empirical indicators of social class or socioeconomic status can be derived, as exemplified by occupational categories, education, housing and income (see the chapters by Susser and by Berkman and MacIntyre in this volume).

Occupation is historically the most commonly employed indicator of social class in health research and demography, at least in Europe. A widely used classification was developed by the Registrar General of England and Wales in 1911 (Table II). It has been modified at regular intervals.

The scales based on occupation usually do not consider inactive persons, and they classify correctly only a small proportion of women. Married women are frequently classified according to their husband's occupation. A number of job titles defy unique assignment into singular social categories because their positions in the social structure are ambiguous. An example is offered by a 14-category French scale introduced by Desplanques (1985). We used six of them on the basis of their prevalence and unambiguous social position in the site-specific cancer mortality tables for France.

Education is occasionally preferred as a social class indicator for adults over occupation, since it applies not only to employed men but also to women and inactive men. In addition, it usually does not change during adult life (Valkonen, 1989). Moreover, it permits relatively valid international comparisons based on years of attained education. Usually acquired in youth, education has an additional advantage of being unaffected by a health-based decline in social position in adults.

Housing tenure, as an indicator of wealth and income, has been mainly used in England and Wales and in Italy, in census-based record-linkage studies. An advantage of this classification is the possibility of categorizing the whole population irrespective of age and gender.

Income data are difficult to collect. Only a few studies presented in this review employed an income-based indicator of social class.

Measures of association

For the purposes of this review, all input data were converted into ratio-type summary measures of association. A commonly reported measure in longitudinal studies was the ratio of observed and expected counts of cases, indirectly standardized for age (standardized mortality or incidence ratio; SMR or SIR) or directly (comparative mortality figures; CMF).

When relative risk figures for social categories were provided, they appear as such in Tables 1–51. This implies variable reference rates for the relative risks presented. In a number of studies, the rate for the highest social category was used as the reference. In others it was the lowest one, while in still others it was the rate for the entire population.

A number of sources provided directly standardized rates. In these instances we calculated rate ratios (RRs), with the rate for the total population as reference (RR = 1) if reported. When the population rate was not available, we used the rate for the highest social category as the reference rate.

In case–control studies, odds ratios (ORs) using a high social class category as the reference (OR = 1) were used. In some studies, for reasons of statistical stability, a social category with a large number of study subjects was employed as the reference.

Four studies (Bouchardy *et al.*, 1993; Levi *et al.*, 1988; Williams & Horm, 1977; E. Regidor, pers. commun.) provided proportional measures of association, such as proportional mortality ratios (PMRs).

A number of studies provided confidence intervals for the point estimates of the rates or ratios, while others presented *P* values, usually for the social class trend. There were yet others that did not address statistical precision in any quantitative manner. No quantitative indicator of precision is therefore given in the tables. An overall impression of precision in the different studies may be obtained from the numbers of observed cases and population sizes presented in Table I.

Detailed comparisons of social class differentials between various populations and time periods are not encouraged because of different social scales employed, different cut-points within scales, different measures of association, different methods of standardization, and other definitional and operational variations.

Presentation of data

The site-specific mortality and incidence ratios are exhaustively presented in Tables 1–51 for the cancer sites listed in Table III. Risk estimates are relative risks or their approximations (SMR, SIR, OR, RR and PMR), ordered from high to low social status. When available, the absolute number of observed cases is included (N).

For the United Kingdom, comparable historical data were available decades back in time. United Kingdom trends since 1911 are presented in Figures 1–10. Results from surveillance system statistics (*Decennial supplements*) are used for comparability purposes.

International evidence of social differences

Tables 1–51 summarize social class differences in cancer mortality and incidence by site, population and period. In the following discussion, a positive social class gradient refers to excess mortality or incidence in high social strata, and a negative gradient to excess in lower strata.

All-cause mortality (Table 1)

Mortality from all causes of death offers a convenient vantage point for the scrutiny of cancer mortality and incidence. In the present data, which are restricted to populations also providing cancer data, male mortality from all causes followed a more or less consistent negative social gradient, with deprived social categories experiencing highest risks of death. This was to be expected from more comprehensive statistics. Particularly high excess fractions for the deprived classes were encountered for men in urban Canada, Great Britain 1979–1983, Finland, France, New Zealand and London (United Kingdom).

The negative gradient was reproduced by the data for women. The social class differences were, however, less pronounced in women than in men, with the exceptions of Italy 1981–1982, Scotland 1959–1963, the United States of America 1960, and the United States population survey and census cohort (Black) 1979–1985.

All neoplasms (Tables 2–3)
Despite variations in the age structures, in the proportion of cancer deaths out of all deaths (which ranged from less than 4% in subSaharan Africa to well over 20% in established market economies in 1990; Murray & Lopez, 1994), and in the more general cause-of-death structures in the different populations, mortality from all cancers correlated fairly well with mortality from all causes of death. The majority of the populations followed a negative social gradient, which was usually less steep than for all causes of death. There were, however, a number of populations for which no consistent trend was apparent: men in California (United States) 1960, Hungary, Japan and Norway; and women in Hungary, Italy, Japan and Norway.

Incidence data for all cancers were available for a smaller number of populations than data for all-cancer mortality. The negative social class trend was less obvious than for mortality. Negative trends were seen in Finland, Turin (Italy), Du Pont employees (United States), England and Wales, Spain and, to a lesser extent, Denmark. An inverted trend was suggested for men in Cali (Colombia), with the highest incidence being associated with the highest social category. There was no trend for either men or women in Sweden, women in Cali (Colombia) and women in Denmark.

Cancers of the mouth and pharynx (Tables 4–5)
In men, an excess mortality from cancers of the oral cavity and pharynx in the socially disadvantaged categories was evident in all populations for which data for these sites were available, with the exception of Japan and California (United States) (Table 4). The negative trend was particularly pronounced in Great Britain 1979–1983, France, Italy and New Zealand. The São Paolo (Brazil) data for men differentiated between mortality from cancers of the mouth and pharynx, with the social gradient for cancer of the mouth appearing weaker than that for pharynx. Women's mortality data did not reveal clear-cut social trends, except for data from the United Kingdom, which suggested a negative trend.

No general picture emerged from the incidence data (Table 5). For cancer of the mouth, there were positive [Cali (Colombia); men], and negative (Denmark, men; Sweden, women) trends, although most data were not suggestive of any trend. For pharynx cancer, data for women in Cali (Colombia), Finland and Sweden, and data for men in the United States 1969–1971 and possibly Denmark, suggested a negative social gradient. No population revealed a clear positive trend.

Cancer of the oesophagus (Tables 6–7)
With the exception of the slightly irregular trend in the earlier New Zealand data, excess mortality in men from cancer of the oesophagus concentrated on the lower social strata (Table 6). Incidence data for men followed the same pattern (Table 7) but this was less pronounced. An inverted association was seen for men in Cali (Colombia) and in the United States. Mortality and incidence data for women followed a negative trend in the majority of the populations.

Stomach cancer (Tables 8–9)
Male mortality from stomach cancer showed a highly consistent tendency towards an excess in lower socioeconomic groups (Table 8). Risks rose steeply and usually regularly from the top to the bottom of the social scale. Incidence data for men followed the mortality trend (Table 9). The negative gradient was also identified in women, except for indeterminate trends in the mortality data of Hungary, Norway and Vaud (Switzerland), and incidence data of Denmark.

Colon cancer (Tables 10–11)
In most countries a positive social class gradient was observed for colon cancer in both genders. Low risks were associated with low social strata, both for mortality (Table 10) and incidence (Table 11). The trend is well exemplified by data for mortality in both sexes in Sao Paolo (Brazil) and for incidence in men in Finland and Hong Kong. A considerable number of exceptions to the positive gradients were observed, however, particularly in North America.

Cancer of the rectum (Tables 12–13)
No consistent social trend emerged for cancer of the rectum. Mortality data for men in Sao Paolo (Brazil) and incidence data for men in Hong Kong and Istanbul (Turkey) revealed highest risks in high social groups, while the opposite trend was suggested by mortality data of the United Kingdom during the 1970s and 1980s, and incidence data of Cali (Colombia), Milan (Italy) and, to a lesser extent, Montreal (Canada). In the remaining populations, no clear trends could be identified.

Patterns for women were similar to those for men. Hungary 1980 and Sao Paolo (Brazil) mortality data, and Hong Kong incidence data, displayed highest risks among high social categories, while mortality data for the United Kingdom and incidence data for Cali (Colombia) and Milan (Italy) suggested the opposite trend.

Liver cancer (Tables 14–15)
The data suggested either a negative social gradient (excess risk concentrating in lower social classes) or no gradient. Negative trends were identified for mortality in Italy (men), New Zealand (men) and Sao Paolo (Brazil; both genders). Negative trends were identified for incidence in Cali (Colombia; both genders) and possibly Denmark (men), Milan (both genders), and the United States 1969–1971 (men). The mortality data from the United Kingdom were suggestive of a negative gradient, but not consistently. There was no clear gradient in a number of other populations.

Pancreatic cancer (Tables 16–17)
Occurrence of pancreatic cancer was not consistently associated with social class. In men, pancreas cancer mortality and incidence followed irregular patterns both between and within countries. Positive and negative, but mostly indeterminate, gradients were encountered.

In women, pancreas cancer mortality was in excess in the lowest social stratum in California (United States), England and Wales 1930–1932 and 1970–1972, Great Britain 1979–1983, Japan, Sao Paolo (Brazil) and Vaud (Switzerland). Incidence data showed excesses, deficits and irregularities, none of which was of outstanding magnitude, with the exception of a doubling of incidence from the lowest to the highest social stratum in Cali (Colombia).

Cancer of the nose and nasal cavities (Tables 18–19)
The data for nasal cancer was scanty, and it suggested excess rates for lower social strata but not consistently. The United Kingdom mortality data were not completely consistent but suggested a negative social class trend in both sexes. The same pattern was reproduced by the Finnish incidence data, particularly in men, but not in Denmark or Sweden.

Larynx cancer (Tables 20–21)
The data for larynx cancer suggested a clear negative social class gradient in men, with the exceptions of Cali (Colombia), Japan (mortality) and Sweden (incidence). Strong mortality excesses were observed in the lowest stratum in France, Great Britain, Italy and New Zealand, as contrasted with the upper social strata. The data for women were less consistent. Negative gradients were observed in the United Kingdom from 1959 (mortality), and less strongly in Denmark and Finland (incidence).

Lung cancer (Tables 22–23)
Male lung cancer risk followed a negative social class gradient in industrialized countries, particularly during recent decades. With the exceptions of the United Kingdom 1912–1912, England and Wales 1930–1932 and possibly 1949–1953, Scotland 1959–1963, Sao Paolo (Brazil) 1978–1982, and the United States population survey and census cohort (Black) 1979–1985, the 34 mortality gradients showed a negative trend for men. Some of the trends were based on the same data, using various indicators of socioeconomic status. The male mortality gradients were reproduced by 13 negative gradients for male incidence. The Latin American male populations [Cali (Colombia) and Sao Paolo (Brazil)] represented the only positive social class gradients.

In women, the patterns were less consistent. Some of the populations, such as those of Athens (Greece) 1978–1986, Canada, Denmark, Finland, Scotland, the United Kingdom 1970–1972 and 1979–1983 (mortality) and 1971–1981 (incidence), and the United States 1960 and 1969–1971, suggested a negative social class gradient, while others were inconsistent to varying degrees and still others, such as Cali (Colombia), Greater Athens (Greece) 1987–1989, Hungary 1980, Sao Paolo (Brazil) and Turin (Italy), suggested a positive trend.

Bone cancer (Tables 24–25)

Mortality data for bone cancer were available for New Zealand, Sao Paolo (Brazil), Switzerland and the United Kingdom, and incidence data were available for Denmark, Finland and Sweden. The numbers were small and the risk ratios therefore imprecise, except for the United Kingdom. Data for Finland (men), Sao Paolo (Brazil; both sexes) and Sweden (women, incidence) suggested a positive trend, while data for men in Great Britain 1979–1983 and for women in Sao Paolo (Brazil) revealed an excess for lower social strata.

Connective tissue cancer (Tables 26–27)

For cancer of the connective tissue, only the United Kingdom rates were available for mortality, and these suggested no clear association with social class in either men or women. Of populations with incidence data, Nordic countries showed either a somewhat elevated risk for more privileged social classes (Finland) or no tendencies, while the United States data 1969–1971 suggested a higher risk for lower social strata, when education was used as the social indicator.

Malignant melanoma (Tables 28–29)

Data for malignant melanoma suggested a regular pattern with the highest risk observed in the highest social strata, with very few exceptions. An excess in lower social strata was observed for Istanbul (Turkey) in men.

Female breast cancer (Tables 30–31)

The data for female breast cancer followed a consistent gradient rising from lower to higher social classes, with the exceptions of England and Wales 1971–1981 (incidence), Great Britain 1979–1983 (mortality) and Portugal. The excess fraction reached over 150% for Hong Kong, Istanbul (Turkey), Sao Paolo (Brazil) and the United States White population (population survey and census cohort 1979–1985). Northern European populations, with the exception of Sweden, suggested mortality excesses of about 100% for the highest social classes.

Cancers of the cervix and uterus (Tables 32–33)

Cervical cancer followed mortality and incidence gradients that increased, usually steeply, from the highest to lowest social category. The excess fractions were higher than 100% for Cali (Colombia), urban Canada 1971 and 1986, Sao Paolo (Brazil), Turin (Italy), the United Kingdom for all periods from 1949, and the United States 1960.

For cancer of the corpus uteri, the pattern was inconsistent between countries. In four populations, the excess concentrated on higher social classes: Cali (Colombia), Finland 1971–1985 (incidence), Sao Paolo (Brazil) and the United States 1969–1971. Excesses in lower classes were encountered for Canada, Denmark (mortality) and Italy, and possibly Finland (mortality 1969–1972). Irregular or no trends were observed for incidence data in Denmark, Istanbul (Turkey), Sweden or Turin (Italy).

Ovarian cancer (Tables 34-35)

A declining gradient from the highest to the lowest social class both for mortality and incidence of ovarian cancer was observed in populations of Mediterranean and South American countries. Excess fractions for the higher social strata exceeded 100% in Italy, Istanbul (Turkey) and Sao Paolo (Brazil). In the data for Canada, the United Kingdom, Japan, northern Europe, and the United States White population (population survey and census cohort 1979–1985), the socioeconomic trends were irregular or nonexistent.

Prostate cancer (Tables 36–37)

Prostate cancer mortality or incidence was not strongly associated with socioeconomic status. Positive trends with excesses in higher strata were encountered for Cali (Colombia), Finland (incidence 1971–1985 but not mortality 1969–1975) and possibly Istanbul (Turkey). Weak negative gradients were observed for England and Wales 1970–1972 and Spain.

Testis cancer (Table 38–39)

In Cali (Colombia) and northern European and United Kingdom populations, incidence data of testis cancer suggested, although not with compelling consistency, an excess in higher social categories. In a number of populations, however, no social class gradient was observed.

Bladder cancer (Tables 40-41)

Male bladder cancer mortality data did not suggest a consistent social class trend across populations.

A positive gradient was observed for California (United States) 1949–1951, Japan and Norway, while a negative trend was seen for Great Britain 1979–1983, Spain, and, in a somewhat irregular fashion, New Zealand 1984–1987, with the deprived social classes being at risk with excesses up to about 100%. Data for men in England and Wales 1970–1972, Italy, New Zealand 1974–1978, Sao Paolo (Brazil) and Switzerland showed an inclination towards an inverted U-shaped trend, with the peak of the risk occurring at middle social categories. This trend may be a transient phenomenon, as suggested by the disappearance of it and the emergence of a more negative gradient in the data for subsequent decades in Great Britain and New Zealand. Data for a number of male populations were not indicative of a social gradient.

The majority of the data for women were associated with irregular or no social trends. Positive social gradients in women were represented by Finland, Italy and Sao Paolo (Brazil). The data for Spain and Vaud (Switzerland) suggested a negative trend. The United Kingdom data for women were irregular across social classes until 1979–1983, when a negative gradient emerged.

Kidney cancer (Tables 42–43)
In the majority of male populations, there was no indication of a social gradient. A positive trend was suggested, however, for seven populations: Cali (Colombia), Denmark, Finland, early United Kingdom data (in subsequent United Kingdom data, this trend disappeared and a slight tendency towards an inverted trend emerged), Sao Paolo (Brazil), Vaud (Switzerland) and possibly Japan. The Sao Paolo gradient was particularly steep, ranging from RR = 4.9 for men with more than 12 years of education down to the reference level of RR = 1 for those with less than one year of education. No evidence for a negative trend was observed in any population except a weak inclination in Montreal (Canada).

The data for women showed a positive trend in England and Wales 1949–1953 and possibly Finland, and no trend for the remainder of the populations.

Brain cancer (Tables 44–45)
The majority of populations showed no association of socioeconomic status and brain cancer mortality or incidence for men. Rates appeared to be high among higher socioeconomic groups in London (United Kingdom) 1967–1987, Sao Paolo (Brazil) and Vaud (Switzerland). The same pattern was seen in earlier years in England and Wales 1930–1932 (traces of this trend were still seen in the United Kingdom data in 1949–1953 but disappeared subsequently). No evidence was found for excess risks concentrating on lower social categories.

There was also no evidence for a negative social trend in women. A positive trend was suggested by data of Cali (Colombia), Finland (a weak trend), and mortality figures for England and Wales 1949–1953 and 1970–1972. In addition, there was a weak positive incidence trend in England and Wales 1971–1981.

Cancer of the thyroid gland (Tables 46–47)
The majority of the rates for cancer of the thyroid gland did not follow a socioeconomic pattern. A negative mortality trend was suggested by the data of Vaud (Switzerland) in both sexes, and England and Wales 1959–1963 and 1970–1972 in women. Data from Cali (Colombia; women) and the United States 1969–1971 (both genders) suggested a positive trend.

Lymphomas (Tables 48–49)
The majority of mortality and incidence data for lymphomas showed no association with social class. An excess risk, particularly for Hodgkin's disease, was observed in high social strata in some populations [Brazil, Finland, Hungary and Vaud (Switzerland)].

Leukaemia (Tables 50–51)
The leukaemia data did not suggest social class trends, with the exceptions of an excess in advantaged social strata encountered in Cali (Colombia; male incidence, but the female incidence suggested an inverted gradient), England and Wales (male and female mortality 1930–1932; the gradient disappeared subsequently), Finland (male mortality), Sao Paolo (Brazil; male and female mortality) and Turin (Italy; male and female incidence; not completely consistent between social indicators used). The United States data for the White population (population survey and census cohort 1979–1985) suggested the highest male leukaemia rate in the lowest educational stratum (RR = 1.4 for those with less than four years of elementary schooling), and the highest female rate in the highest educational stratum (RR = 2.4 for those with more than four years of college education).

Time trends

Time trends for mortality by social class were available for England and Wales from 1910. There were two sources of mortality data presented in this review for England and Wales. The 1971–1975 and 1976–1981 data came from a longitudinal study, in which the occupational information, which is the basis for the social classification, was derived uniformly from the 1971 census records. In the data for 1911, 1931, 1971 and 1981, the mortality numerators were derived from occupations recorded on death certificates, while the denominators were based on census data. For reasons of possible incomparability, we did not include the longitudinal data in the trend figures (Figures 1–10) and summarized only the *Decennial supplement* data for 1911, 1931, 1971 and 1981. Social class trends were considered for mortality from all cancers and stomach and colon cancer for both sexes, as well as from lung cancer for men and breast and cervix cancers for women.

In men, the data suggested widening over time of social class differentials for all cancers pooled and cancers of the lung, larynx and stomach. For colon cancer, higher social classes displayed a somewhat increased mortality in 1911 and 1951, but this differential seems to have largely disappeared by 1981.

In women, the social differentials seem to have widened over decades in disfavour of lower social strata in mortality from all cancer sites combined and from cervical cancer. For stomach cancer, the relative excess mortality of social class V increased to some extent, while the deficit for class I decreased. Breast cancer mortality was in excess in higher classes until 1951. The differences started to level off thereafter, and no clear social gradient was observed in 1981. For colon cancer, the trends were inconsistent.

Conclusions

In men, a number of cancers revealed a consistent social class gradient across populations, with the risk being higher in more disadvantaged categories: respiratory cancers (nose, larynx and lung) and cancers of the mouth and pharynx, oesophagus, stomach, and, with a number of populations showing no or irregular trends, liver. Figures from Latin America, which were available for Cali (Colombia) and Sao Paolo (Brazil), represented an exception for respiratory tract, where the excesses among higher social categories suggested a higher prevalence of the main risk factor, tobacco smoking, among socially advantaged strata.

Excesses in lower social strata were suggested among women for cancers of the oesophagus, stomach and, less consistently, liver, but not for respiratory sites. Incidence and mortality from cervix uteri revealed a worldwide steep tendency to be more frequent in lower social strata.

The data for colon cancer and malignant melanoma suggested a positive social gradient in men, the rate being high in high social categories. In women the cancers for which higher social classes were at higher risk were malignant neoplasms of the colon, breast and ovary, and malignant melanoma.

For a number of cancers, social class trends were inconsistent or nonexisting: cancers of the rectum, pancreas, bone, connective tissues, prostate, testis, bladder, kidney and thyroid gland, and malignant lymphomas and leukaemias, in men; and cancers of the rectum, pancreas, nose, larynx, lung, bone, connective tissues, body of the uterus, bladder and kidney, and malignant lymphomas and leukaemias, in women.

Longitudinal data from England and Wales suggested widening over time of social class differences in men for all cancers combined and for cancers of the lung, larynx and stomach, and in women for all cancers combined and for cervical cancer.

References

Bouchardy, C., Parkin, D.M., Khlat, M., Mirra, A.P., Kogevinas, M., De Lima, F.D. & De Cravalho Ferreira, C.E. (1993) Education and mortality from cancer in São Paulo, Brazil. *Ann. Epidemiol.*, 3, 64–70

Buell, P., Dunn, J.E., Jr & Breslow, L. (1960) The occupational-social class risks of cancer mortality in men. *J. Chronic Dis.*, 12, 600–621

Central Bureau of Statistics, Norway (1976) Yrke of Dødelighet 1970–1973 [Occupational mortality] (Statistiske Analyser, No. 21). Oslo

Crowther, J.S., Drasar, B.S., Hill, M.J., Maclennan, R., Magnin, D., Peach, S. & Teoh-Chan, C.H. (1976) Faecal steroids and bacteria and large bowel cancer in Hong Kong by socio-economic groups. *Br. J. Cancer*, 34, 191–198

Cuello, C., Correa, P. & Haenszel, W. (1982) Socio-economic class differences in cancer incidence in Cali, Colombia. *Int. J. Cancer*, 29, 637–643

Danmarks Statistik (1979) Dødelighed og erhverv 1970–75 (Statistiske undersøgelser No. 37). Copenhagen, Ministry of Health

Davey Smith, G. & Marmot, M.G. (1991) Trends in mortality in Britain: 1920–1986. *Ann. Nutr. Metab.*, 35, 53–63

Desplanques, G. (1973) *La mortalité des adultes suivant le milieu social (1955–1971)*. Paris, INSEE

Desplanques, G. (1985) *La mortalité des adultes*. Paris, INSEE

Doornbos, G. & Kromhout, D. (1990) Educational level and mortality in a 32-year follow-up study of 18-year-old men in the Netherlands. *Int. J. Epidemiol.*, 19, 374–379

Dosemeci, M., Hayes, R.B., Vetter, R., Hoover, R.N., Tucker, M., Engin, K., Unsal, M. & Blair, A. (1993) Occupational physical activity, socioeconomic status, and risks of 15 cancer sites in Turkey. *Cancer Causes Control*, 4, 313–321

Faggiano, F., Zanetti, R. & Costa, G. (1994) Cancer risk and social inequalities in Italy. *J. Epidemiol. Community Health*, 48, 447–452

Faggiano F., Lemma, P., Costa, G., Gnavi, R. & Pagnanelli, F. (1995) Cancer mortality by educational level in Italy. *Cancer Causes Control*, 6, 311–320

Ferraroni, M., Negri, E., La Vecchia, C., D'Avanzo, B. & Franceschi, S. (1989) Socioeconomic indicators, tobacco and alcohol in the aetiology of digestive tract neoplasms. *Int. J. Epidemiol.*, 18, 556–562

Franceschi, S., Parazzini, F., Negri, E., Booth, M., LaVecchia, C., Beral, V., Tzonou, A. & Trichopoulos, D. (1991) Pooled analysis of 3 European case-control studies of epithelial ovarian cancer: III. Oral contraceptive use. *Int. J. Cancer*, 49, 61–65

Graham, S. & Gibson, R.W. (1972) Social epidemiology of cancer of the testis. *Cancer*, 29, 1242–1249

Greenberg, R.S., Haber, M.J., Clark, W.S., Brockman, J.E., Liff, J.M., Schoenberg, J.B., Austin, D.F., Preston-Martin, S., Stemhagen, A., Winn, D.M., McLaughlin, J.K. & Blot, W.J. (1991) The relation of socioeconomic status to oral and pharyngeal cancer. *Epidemiology*, 2, 194–200

Hein, H.O., Suadicani, P. & Gyntelberg, F. (1992) Lung cancer risk and social class. The Copenhagen male study: 17-year follow up. *Dan. Med. Bull.*, 39, 173–176

Hirayama, T. (1990) *Life-style and mortality: a large scale census-based cohort study in Japan*. Basel, Karger

Józan, P. (1986) Some preliminary results of the study on cancer mortality differentials by socio-economic status. In: *4th meeting of the UN/WHO/CICRED network on socio-economic differential mortality in industrialized societies*. Hungarian Central Statistical Office; United Nations Fund for Population Activities; United Nations Population Division; World Health Organization; Committee for International Cooperation in National Research in Demography. pp. 156–166

Katsoyanni, K., Trichopoulos, D., Kalandidi, A., Tomos, P. & Riboli, E. (1991) A case-control study of air pollution and tobacco smoking in lung cancer among women in Athens. *Prev. Med.*, 20, 271–278

Kitagawa, E.M. & Hauser, P.M. (1973) *Differential mortality in the United States: a study in socioeconomic epidemiology*. Cambridge, MA, Harvard University Press

Kogevinas, M. (1990) *Longitudinal study: socio-demographic differences in cancer survival*. Office of Population Censuses and Surveys, London, Her Majesty's Stationery Office

Leclerc, A., Goldberg, P., Ricard, E., Luce, D. & Brugère, J. (1993) Laryngeal cancer and alcohol, tobacco, diet and occupation: first results from a French case-control study. XV World Congress of Otorhinolaryngology, Head and Neck Surgery, Istanbul

Lehman, P., Mamboury, C. & Minder, C.E. (1990) Health and social inequalities in Switzerland. *Soc. Sci. Med.*, 31, 365–386

Levi, F., Negri, E., La Vecchia, C. & Te, V.C. (1988) Socioeconomic groups and cancer risk at death in the Swiss canton of Vaud. *Int. J. Epidemiol.*, 17, 711–717

Logan, W.P.D. (1982) *Cancer mortality by occupation and social class* (Studies on medical and population subjects No. 44). London, Her Majesty's Stationery Office; Lyon, International Agency for Research on Cancer

Lynge, E. & Thygesen, L. (1990) *Occupational cancer in Denmark*. Copenhagen

Marmot, M.G. & McDowall, M.E. (1986) Mortality decline and widening social inequities. *Lancet*, 394, 274–276

Morris, J.N. (1979) Social inequalities undiminished. *Lancet*, i, 87–90

Murray, C.J.L. & Lopez, A.D., eds (1994) *Global comparative assessments in the health sector. Disease burden, expenditures and intervention strategies*. Geneva, World Health Organization

Näyhä, S. (1977) Social group and mortality in Finland. *Br. J. Prev. Soc. Med.*, 31, 231–237

Office of Population Censuses and Surveys (1919) *Supplement to 75th Annual Report*. London, Her Majesty's Stationery Office

Office of Population Censuses and Surveys (1927) *Decennial supplement*. London, Her Majesty's Stationery Office

Office of Population Censuses and Surveys (1938) *Decennial supplement* (Part IIa). London, Her Majesty's Stationery Office

Office of Population Censuses and Surveys (1958) *Decennial supplement* (Part II). London, Her Majesty's Stationery Office

Office of Population Censuses and Surveys (1971) *Decennial supplement*. London, Her Majesty's Stationery Office

Office of Population Censuses and Surveys (1977) *Decennial supplement*. London, Her Majesty's Stationery Office

Office of Population Censuses and Surveys (1986) *Occupational mortality 1979–80, 1982–83 (Decennial supplement)*. London, Her Majesty's Stationery Office

Office of Population Censuses and Surveys (1990) *1971–1981 Longitudinal study: mortality and social organization*. London, Her Majesty's Stationery Office

Papadimitriou, C., Day, N., Tzonou, A., Gerovassilis, F., Manousos, O. & Trichopoulos, D. (1984) Biosocial correlates of colorectal cancer in Greece. *Int. J. Epidemiol.*, 13, 155–159

Pearce, N.E. & Howard, J.K. (1986) Occupation, social class and male cancer mortality in New Zealand, 1974–78. *Int. J. Epidemiol.*, 15, 456–462

Pell, S. & D'Alonzo, C.A. (1970) Chronic disease morbidity and income level in an employed population. *Am. J. Public Health*, 60, 116–129

Pukkala, E. (1995) *Cancer risk by social class and occupation. A survey of 109,000 cancer cases among Finns of working age* (Contributions to Epidemiology and Biostatistics, Vol. 7). Basel, Karger

Registrar General for Scotland (1956) *Annual Report, 1955, No. 101, Appendix IX. Occupational mortality*. Edinburgh, Her Majesty's Stationery Office

Registrar General for Scotland (1970) *Occupational mortality 1959–1963. Second supplement to the 114th Annual Report, 1968*. Edinburgh, Her Majesty's Stationery Office

Rogot, E., Sorile, P.D., Johnson, N.J. & Schmitt, C. (1992) *A mortality study of 1.3 million persons by demographic, social, and economic factors: 1979–1985 follow-up. U.S. National Longitudinal Mortality Study* (NIH Publication No. 92-3297). National Institutes of Health, National Heart, Lung, and Blood Institute

Swerdlow, A.J., Douglas, A.J., Huttly, S.R.A. & Smith, P.B. (1991) Cancer of the testis, socioeconomic status, and occupation. *Br. J. Ind. Med.*, 48, 670–674

Trichopoulos, D., Kalandidi, A., Sparos, L. & MacMahon, B. (1981) Lung cancer and passive smoking. *Int. J. Cancer*, 27, 1–4

Vågerö, D. & Persson, G. (1986). Occurrence of cancer in socioeconomic groups in Sweden. An analysis based on the Swedish Cancer Environment Registry. *Scand. J. Soc. Med.*, 14, 151–160

Valkonen, T. (1989) Adult mortality and level of education: A comparison of six countries. In: Fox, J., ed., *Health inequalities in European countries*. Aldershot, UK, Gower

Valkonen, T., Martelin, T. & Rimpelä, A. (1990) *Socio-economic mortality differences in Finland 1971–85*. Central Statistical Office of Finland, Helsinki

Van Reek, J. (1986) Mortality by social class among adults in the Netherlands since the nineteenth century. In: *4th meeting of the UN/WHO/CICRED network on socio-economic differential mortality in industrialized societies*. Hungarian Central Statistical Office; United Nations Fund for Population Activities; United Nations Population Division; World Health Organization; Committee for International Cooperation in National Research in Demography. pp. 76–81

Williams, R.R. & Horm, J.W. (1977) Association of cancer sites with tobacco and alcohol consumption and socioeconomic status of patients: interview study from the Third National Cancer Survey. *J. Natl. Cancer Inst.*, 58, 525–547

Corresponding author:

F. Faggiano
Unit of Environmental Cancer Epidemiology,
International Agency for Research on Cancer,
150 cours Albert-Thomas, F-69372 Lyon cedex 08,
France;
and Department of Public Health, University of Turin,
Via Santena 5b, I-10126 Turin, Italy

Table I. Summary of studies included in this review[a]

Country	W/S	Period	Sites	M/F	M/I	Design	Observed	Population size	Notes	Reference
Brazil (São Paolo)	W	1978–82	20	MF	M	CCS	85 868[b]			Bouchardy et al., 1993
Canada (urban area)	W	1971	T+11	MF	M	SSS	23 957[c]	5 346 550	1	Wilkins, pers. commun.
	W	1986	T+11	MF	M	SSS	25 653[c]	8 017 860	1	Wilkins, pers. commun.
Canada (Montreal)	S	1979–85	10	M	I	PCC	4576	740[b]	2	Bourbonnais & Siemiatycki, in press
Colombia (Cali)	W	1971–75	22	MF	I	SSS	~8000[d]	903 888		Cuello et al., 1982
Denmark	W	1970–75	T+7	MF	M	SSS				Danmarks Statistik, 1979
	W	1970–80	25	MF	I	RLS	73 095			Lynge & Thygesen, 1990
Denmark (Copenhagen)	S	1971–88	1	M	I	COH	144	5249	2	Hein, 1992
Finland	W	1969–72	T+7	MF	M	SSS	179 919	2 219 985[e]		Näyhä, 1977
	W	1971–85	T+6	MF	M	RLS		~1 600 000[f]		Valkonen et al, 1990
	W	1971–85	25	MF	I	RLS		~1 600 000[f]		Pukkala, 1995
France	S	1966–71	T+2	M	M	RLS		~800 000[f]		Desplanques, 1976
	S	1975–82	T+6	M	M	RLS		~1 000 000[f]		Desplanques, 1985
France (Paris)	S	1989–91	1		I	HCC	528	305[g]	3	Leclerc et al., 1993
Greece (Greater Athens)	W	1980–81	1	F	I	HCC	971	2250[g]		Franceschi et al., 1991
	W	1987–89	1	F	I	HCC	101			Katsouyanni et al., 1991
Greece (Athens)	W	1978–86	1	F	I	HCC	51			Trichopoulos et al., 1981
Greece (Athens)	W	1979–80	1	F	I	HCC	100			Papadimitriou et al., 1984
Hong Kong	S	1971	4	MF	I	SSS	815			Crowther et al., 1976
Hungary	W	1970	8	MF	M	SSS				Jozan, 1986
	W	1980	8	MF	M	SSS				Jozan, 1986
Italy	W	1981–82	T+19	MF	M	RLS	94 163	36 690 846[f]	4	Faggiano et al., 1995
Italy (Torino)	W/S	1985–87	12	MF	I	RLS	7666	30 751[g]	4,5	Faggiano et al., 1994
Italy (Milano)	S	1983–88	10	MF[n]	I	HCC	1771	1944[g]	2	Ferraroni, 1989
Japan	S	1965–82	T+16	MF	M	COH		265 118		Hirayama, 1990

Country	W/S	Period	Sites	M/F	M/I	Design	Observed	Population size	Notes	Reference
The Netherlands	W	1959–61	1	M	M	SSS				Van Reek, 1986
	S	1951–81	T+1	M	M	COH	3456	78 505	6	Doornbos & Kromhout, 1990
New Zealand	W	1975–78	T+18	M	M	SSS	5356			Pearce & Howard, 1986
	W	1985–87	T+18	M	M	SSS				N. Pearce, pers. commun.
Norway	W	1970–73	T+8	MF	M	SSS				Central Bureau of Statistics, 1976
Portugal	W	1980–82	T+2	MF	M	CCS		3 524 432		M. Giraldes, pers. commun.
Spain	W	1980–82	14	M	M	CCS				E. Regidor, pers. commun.
Sweden	W	1961–70	24	MF	I	SSS	223 215	2 809 974[f]		Vågerö & Persson, 1986
Switzerland	W	1979–82	T+17	M	M	SSS	45 565	1 617 432[f]		C.E. Minder, pers. commun.
Switzerland (Vaud)	W	1977–84	14	MF	M	CCS	4461[b]			Levi et al., 1988
Turkey (Istanbul)	S	1979–84	12	MF	I	HCC	3865	2371	3	Dosemeci et al., 1993
UK – England & Wales	W	1910–12	T+10	M	M	SSS				OPCS, 1919
	W	1921–23	3	M	M	SSS				OPCS, 1927
	W	1930–32	T+13	MF	M	SSS				OPCS, 1938
	W	1949–53	T+16	M	M	SSS				OPCS, 1958
	W	1959–63	10	M	M	SSS				OPCS, 1971
	W	1970–72	T+23	MF	M	SSS				OPCS, 1977
	S	1971–81	T+2	M	M	RLS	8488[i]	~496 000[j]		OPCS, 1990
	S	1971–81	22	MF	I	RLS	17 402	~496 000		Kogevinas, 1990
UK – Great Britain	W	1979–83	T+23	MF	M	SSS	427 812			OPCS, 1986
UK – Scotland	W	1949–53	T+2	M	M	SSS				Registrar General of Scotland, 1956
	W	1959–63	T+7	MF	M	SSS				Registrar General of Scotland, 1970
UK – 6 cities	S	1977–81	1	M	I	HCC	259	489[g]		Swerdlow et al., 1991
UK (London)	S	1967–87	T+10	M	M	COH	1237	17 530		Davey Smith & Marmot, 1991
USA (Buffalo)	S	1945–65	1	M	I	PCC	247	2504		Graham & Gibson, 1972
USA (California)	W	1949–51	13	M	M	SSS	10 401	2 984 867		Buell et al., 1960

Table I. (Contd) Summary of studies included in this review[a]

Country	W/S	Period	Sites	M/F	M/I	Design	Observed	Population size	Notes	Reference
USA (Du Pont Co)	W	1959–67	2	M	I	COH	1274	115 000		Pell & D'Alonzo, 1970
USA (San Francisco)	S	1984–85	1	M	I	PCC	762	837[g]	2	Greenberg et al., 1991
USA	S	1969–71	20	MF	I	CCS	7518[b]		2	Williams & Horm, 1977
USA	S	1960	T+7	MF	M	RLS	62 400			Kitagawa & Hauser, 1973
USA (12 census samples)	S	1979–85	T+10	MF	M	RLS	1 281 475			Rogot et al., 1992

[a]Whole or sample populations (W/S); time period of observation (period); number of cancer sites presented (sites: T = total mortality); gender (M/F); type of occurrence measure (M/I, mortality/incidence); study design (design: CCS, case–case proportional mortality study; COH, cohort study (interview or medical examination at the time of enrolment of participants); HCC, hospital-based case–control study; PCC, population-based case–control study; RLS, record linkage study; SSS, statistics from a surveillance system;); number of observed cases; general or control population size; notes and references.
[b]Cancer patients as controls.
[c]All causes.
[d]Estimated from Parkin et al., 1992.
[e]Whole population.
[f]At-risk population.
[g]Controls.
[h]Adjusted by sex.
[i]Age 15–64.
[j]All ages.
Notes. (1) Ecological indicator of social class. (2) Adjustments for other risk factors are available in the paper. (3) Adjusted for tobacco and/or alcohol use. (4) Adjusted for geographic area of birth or residences. (5) Analysed with a case–control design. (6) Adjustment for height and health score are available in the paper.

Table II. UK Registrar General's classification as of 1971 and the prevalence of class categories among active and retired persons

Class	Description (examples)	Prevalence, %
I	Professional (e.g., accountant, doctor, lawyer)	5
II	Intermediate (e.g., manager, nurse, schoolteacher)	18
III-NM	Skilled non-manual (e.g., clerical worker, secretary, shop assistant)	12
III-M	Skilled manual (e.g., bus driver, butcher, carpenter, coal-face worker)	38
IV	Partly skilled (e.g., agricultural worker, bus conductor, postman)	18
V	Unskilled (e.g., cleaner, dock worker, labourer)	9

Table III. List of cancer sites considered in the site-specific tables in this review, and correspondent rubric of the International Classification of Diseases (9th revision)

Table	Cause of death	ICD–9	Other ICD groups occasionally included
1	All causes of death	000–999	
2–3	All cancer sites	140–139	
4–5	Cancer of the buccal cavity and pharynx	140–150	Mouth (ICD 141–145); pharynx (ICD 146,148–149); upper digestive–respiratory tracts (ICD 140–150,161); hypopharynx (ICD 148)
6–7	Cancer of the oesophagus	150	
8–9	Stomach cancer	151	
10–11	Colon cancer	153	Intestine (ICD 152–154); colorectal (ICD 153–154)
12–13	Cancer of the rectum	154	
14–15	Liver and gallbladder cancer	155–156	Liver (ICD 155)
16–17	Pancreatic cancer	157	
18–19	Cancer of the nose and nasal cavity	160	
20–21	Larynx cancer	161	
22–23	Lung cancer	162	
24–25	Bone cancer	170	
26–27	Connective tissue cancer	171	
28–29	Malignant melanoma	173	Skin (ICD 172–173)
30–31	Female breast cancer	174	
32–33	Cancer of the uterus	179–180,182	Cervix (ICD 180); corpus (ICD 182); other than cervix (ICD 179, 182)
34–35	Ovarian cancer	183	
36–37	Prostate cancer	185	
38–39	Testis cancer	186	
40–41	Bladder cancer	188	
42–43	Kidney cancer	189	
44–45	Brain cancer	191–192	
46–47	Cancer of the thyroid gland	193	
48–49	Lymphoma	200–203	Hodgkin's disease (ICD 201); non–Hodgkin lymphoma (ICD 202); multiple myeloma (ICD 203); other combinations
50–51	Leukaemia	204–208	Leukemias and lymphomas (ICD 200–208); lymphoid leukaemia (ICD 204); acute, chronic lymphocytic leukaemia (ICD 204 with different morphology)

Table 1. All causes mortality

Study base	Indicators	Social scale	N	Male RR	N	Female RR	Study design
Canada (urban area) 1971 all ages	Income CMF	Q1		0.79		0.84	Surveillance system statistics using 1971 census data as denominator. Neighbourhood income quintiles as social indicator [R. Wilkins, pers commun.]
		Q2		0.82		0.84	
		Q3		0.95		0.99	
		Q4		0.99		1.01	
		Q5		1.41		1.28	
Canada (urban area) 1986 all ages	Income CMF	Q1		0.72		0.84	Surveillance system statistics using 1986 census data as denominator. Neighbourhood income quintiles as social indicator [R. Wilkins, pers commun.]
		Q2		0.80		0.83	
		Q3		0.89		0.95	
		Q4		1.08		1.03	
		Q5		1.50		1.28	
Denmark 1970–1975 age: 20–64	Occupational group SMR	Employees: I		0.79		0.98	Record-linkage study using 1970 census and 1970–1975 mortality data. Employees classified according to educational level [Danmarks Statistik, 1979]
		Employees: II		0.83		0.95	
		Employees: III		0.96		0.96	
		Employees: IV		1.15		1.96	
		Skilled workers		1.08		1.00	
		Unskilled workers		1.10		1.08	
Finland 1969–1972 age: 15–64 (married women)	Social class CMF	Upper white-collar		0.78		0.95	Surveillance system statistics using 1970 census data as denominator. Social class indicator based on occupation [Näyhä, 1977]
		Lower white-collar		0.95		1.00	
		Skilled workers		0.92		1.02	
		Unskilled workers		1.48		1.08	
		Farmers		0.87		0.96	
Finland 1971–1985 age: 35–64	Social class RR	Upper white-collar		1		1	Record-linkage study using 1970, 1975 and 1980 census data and 1971–1985 mortality [Valkonen et al., 1990]
		Lower white-collar		1.38		1.15	
		Skilled workers		1.67		1.38	
		Unskilled workers		2.30		–	
		Farmers		1.42		1.19	
France 1966–1971 age: 45–54	Occupational group RR	Managers		0.52			A sample of about 800 000 of 1955 censused population followed-up until 1971. The scale shown represents a choice of the total scale [Desplanques, 1973]
		Intermediate		0.70			
		Self-employed		0.85			
		Clerks		0.86			
		Skilled workers		1.09			
		Unskilled workers		1.58			
France 1966–71 age: 55–64	Occupational group CMF	Managers		0.63			A sample of about 800 000 of 1955 censused population followed-up until 1971. The scale shown represents a choice of the total scale [Desplanques, 1973]
		Intermediate		0.74			
		Self-employed		0.87			
		Clerks		0.90			
		Skilled workers		1.09			
		Unskilled workers		1.43			
France 1975–82 age: 45–64	Occupational group CMF	Managers		0.59			A sample of about 1000 000 of 1975 censused population followed-up until 1982. The scale shown represents a choice of the total scale [Desplanques, 1985]
		Intermediate		0.82			
		Self-employed		0.88			
		Clerks		1.05			
		Skilled workers		1.28			
		Unskilled workers		1.81			

Table 1. (Contd) All causes mortality

Study base	Indicators	Social scale	N	Male RR	N	Female RR	Study design
France 1975–1982 age: 55–64	Occupational group CMF	Managers Intermediate Self-employed Clerks Skilled workers Unskilled workers		0.65 0.92 0.93 1.20 1.21 1.40			A sample of about 1 000 000 of 1975 censused population followed-up until 1982. The scale shown represents a choice of the total scale [Desplanques, 1985]
Italy 1981–1982 age: 18–74	Educational level RR	University High school Middle school Primary school Literate Illiterate	1759 3380 8139 32855 12171 3237	1 1.05 1.26 1.37 1.38 1.64	290 1251 3017 16096 8802 3166	1 1.09 1.24 1.35 1.44 1.81	Record-linkage between 1981 census and mortality in the following six months [Faggiano et al., 1995]
Japan 1965–1982 age: 40+	Social class SMR	I, II III IV V		0.81 0.96 0.99 1		0.81 1.00 0.88 1	265 000 Japanese interviewed in 1965 and followed-up until 1982. Social class based on occupation. Reference category is farmers and miners [Hirayama, 1990]
The Netherlands 1951–1981	Education level RR	4 (high) 3 2 1 (low)		0.67 0.77 0.82 1			1951–1981 follow-up of 78 505 Dutch men medically examined in 1950–1951 for military service [Doornbos & Kronhout, 1990]
New Zealand 1974–1978 age: 15–64	Social class RR	I II III-NM III-M IV V		1 1.03 1.11 1.20 1.47 1.97			Surveillance system statistics using 1976 census data as denominators. UK Registrar General's social class classification [Pearce, 1986]
New Zealand 1984–1987 age: 15–64	Social class RR	I II III-NM III-M IV V		1 1.35 1.27 1.67 2.06 2.03			Surveillance system statistics using 1986 census data as denominators. UK Registrar General's social class classification [Pearce & Bethwaite, pers. commun.]
Norway 1970–1973 age: 20–69	Social class CMF	A B C D E (farmers)		0.91 1.11 1.02 1.12 0.81		0.98 1.00 1.00 1.09 0.94	Surveillance system statistics using 1970 census data as denominators. Social class based on occupation [Central Bureau of Statistics, 1976]

Table 1. (Contd) All causes mortality

Study base	Indicators	Social scale	N	Male RR	N	Female RR	Study design
Switzerland 1979–1982 age: 15–74	Social class SMR	I II III-NM III-M IV-V		0.64 0.77 1.04 1.2 1.01			Surveillance system statistics using 1980 census as denominator. UK Registrar General's social-class classification [Minder, pers. commun.]
UK – England and Wales 1910–1912 age: 15–64	Social class SMR	I II III IV V		0.88 0.94 0.96 0.53 1.42			Surveillance system statistics using 1910 census data as denominator. For social classification see Introduction [OPCS, 1919]
UK – England and Wales 1930–1932 age: 15–64	Social class SMR	I II III IV V		0.90 0.94 0.97 1.02 1.11			Surveillance system statistics using 1930 census data as denominator. For social classification see Introduction. Women classified according to husband's occupation [OPCS, 1938]
UK – England and Wales 1949–1953 age: 15–64	Social class SMR	I II III IV V		0.98 0.86 1.01 0.94 1.18			Surveillance system statistics using 1950 census data as denominator. For social classification see Introduction [OPCS, 1958]
UK – England and Wales age: 15–64 (married women)	Social class SMR	I II III-NM III-M I V		0.77 0.81 0.99 1.06 1.14 1.37		0.82 0.87 0.92 1.15 1.19 1.35	Surveillance system statistics using 1970 census data as denominator. For social classification 1970–1972 see Introduction. Women classified according to husband's occupation [OPCS, 1977]
UK – England and Wales 1971–1975 age: 15–64	Social class SMR	I II III-NM III-M IV V		0.80 0.80 0.92 0.90 0.97 1.15			Record-linkage study (Longitudinal Study) between 1971 census and 1971-75 mortality data for a 1% sample of the total population. UK Registrar General's social class classification [OPCS, 1990]
UK – England and Wales 1976–1981 age: 15–64	Social class SMR	I II III-NM III-M IV V		0.67 0.77 1.05 0.96 1.09 1.25			Record-linkage study (longitudinal study) between 1971 census and 1971–1975 mortality data for a 1% sample of the total population. UK Registrar General's social class classification [OPCS, 1990]

Study base	Indicators	Social scale	N	Male RR	N	Female RR	Study design
UK – Great Britain 1979–1980, 1982–1983 age: 20-64 (married women, 20–59)	Social class SMR	I II III-NM III-M IV V	10 808 56 535 33 370 116 218 69 415 36 574	0.66 0.76 0.94 1.06 1.16 1.65	3532 17 518 8420 32 609 17 958 7194	0.75 0.83 0.93 1.11 1.25 1.60	Surveillance system statistics using 1980 census data as denominator. For social classification see Introduction. Women classified according to husband's occupation [OPCS, 1986]
UK (London) 1967–1987	Employment grade RR	Administrators Professionals Clerical Other	141 2322 905 663	1 1.30 1.82 2.10			17 530 London civil servants, medically examined 1967–1969, followed-up until 1987 [Davey Smith & Marmot, 1991]
UK – Scotland 1949–1953 age: 20–64	Social class SMR	I II III IV V		1.08 0.86 1.03 0.90 1.13			Surveillance system statistics using 1950 census data as denominator. UK Registrar General's social class classification [Registrar General for Scotland, 1956]
UK – Scotland 1959–1963 20–64 (married women)	Social class SMR	I II III IV V		0.83 0.87 0.97 0.99 1.42		0.66 0.84 0.96 1.05 1.49	Surveillance system statistics using 1960 census data as denominators. UK Registrar General's classification. Women classified according to husband's occupation [Registrar General for Scotland, 1970]
USA 1960 age: 25–64 (white)	Educational level SMR	College High school Elementary school <8 years of school		0.77 0.79 1.07 1.14		0.80 0.89 1.08 1.31	Record-linkage study using 1960 mortality data and census [Kitagawa & Hauser, 1973]
USA (12 census samples) White population 1979–1985 age: 25+	Education SMR	College: 5+ y 4 y 1-3 y High school: 4 y 1-3 y Elementary school: 8 y 5-7 y 0-4 y		0.65 0.77 0.93 0.96 1.13 1.11 1.12 1.08		0.79 0.84 0.93 0.98 1.02 1.06 1.07 1.14	Census linkage [Rogot et al., 1992]
USA Black population 1979–1985 age: 25+	Education SMR	College: 5+ y 4 y 1-3 y High school: 4 y 1-3 y Elementary school: 8 y 5-7 y 0-4 y		0.59 0.60 0.81 0.92 1.08 1.14 1.06 1.02		0.56 0.65 0.73 0.88 1.11 1.07 1.13 1.00	Census linkage [Rogot et al., 1992]

Table 2. All cancer sites mortality

Study base	Indicators	Social scale	N	Male RR	N	Female RR	Study design
Canada (urban area) 1971 all ages	Income CMF	Q1 Q2 Q3 Q4 Q5		0.87 0.89 0.99 0.98 1.22		0.94 0.89 1.00 1.05 1.11	Surveillance system statistics using 1971 census data as denominator. Neighbourhood income quintiles as social indicator [R. Wilkins, pers commun.]
Canada (urban area) 1986 all ages	Income CMF	Q1 Q2 Q3 Q4 Q5		0.83 0.88 0.91 1.06 1.32		0.95 0.96 1.00 1.00 1.09	Surveillance system statistics using 1986 census data as denominator. Neighbourhood income quintiles as social indicator [R. Wilkins, pers commun.]
Denmark 1970–1975 age: 20–64	Occupational group SMR	Employees: I Employees: II Employees: III Employees: IV Skilled workers Unskilled workers		0.72 0.83 1.04 1.19 1.17 1.06		0.97 1.07 0.97 0.99 1.20 1.03	Record-linkage study using 1970 census and 1970–1975 mortality data. Employees classified according to the educational level [Danmarks Statistik, 1979]
Finland 1971–1985 age 35–64	Social class RR	Upper white-collar Lower white-collar Skilled workers Unskilled workers Farmers		1 1.20 1.54 1.78 1.20		1 1.02 1.05 – 0.90	Record-linkage study using 1970, 1975 and 1980 censuses data and 1971–85 mortality [Valkonen et al., 1990]
France 1966–1971 age: 45–64	Occupational group RR	Managers Intermediate Self-employed Clerks Skilled workers Unskilled workers		0.57 0.77 0.88 1.01 1.24 1.32			A sample of about 800 000 of 1955 censused population followed-up until 1971. The scale shown represents a choice of the total scale [Desplanques, 1973]
France 1966–1971 age: 55–64	Occupational group RR	Managers Intermediate Self-employed Clerks Skilled workers Unskilled workers		0.67 0.80 0.91 1.04 1.27 1.39			A sample of about 800 000 of 1955 censused population followed-up until 1971. The scale shown represents a choice of the total scale [Desplanques, 1973]
France 1975–1982 age: 45–54	Occupational group RR	Managers Intermediate Self-employed Clerks Skilled workers Unskilled workers		0.43 0.60 0.61 0.81 0.96 1.11			A sample of about 1 000 000 of 1975 censused population followed-up until 1982. The scale shown represent a choice of the total scale [Desplanques, 1985]

Table 2. (Contd) All cancer sites mortality

Study base	Indicators	Social scale	N	Male RR	N	Female RR	Study design
France 1975–1982 age: 45–54	Occupational group RR	Managers Intermediate Self-employed Clerks Skilled workers Unskilled workers		0.68 0.96 0.94 1.09 1.25 1.25			A sample of about 1 000 000 of 1975 censused population followed-up until 1982. The scale shown represents a choice of the total scale [Desplanques, 1985]
Hungary 1970 age: 25–64	Years of education SMR	15+ 12–14 8–11 0–7		0.88 0.98 1.56 0.87		1.19 1.33 1.17 0.93	Surveillance system statistics using 1970 census data as denominator [Jozan, 1986]
Hungary 1980 age: 25–64	Years of education SMR	15+ 12–14 8–11 0–7		0.95 0.92 1.06 0.99		1.15 1.28 1.07 0.88	Surveillance system statistics using 1980 census data as denominator [Jozan, 1986]
Italy 1981–1982 age: 18–74	Education level RR	University High school Middle school Primary school Literate Illiterate	607 1134 2726 11 688 3703 751	1 1.10 1.29 1.39 1.27 1.24	160 612 1289 6092 2651 657	1 1.03 1.03 1.00 0.94 0.90	Record-linkage between 1981 census and mortality in the following six months [Faggiano et al., 1995]
Japan 1965–1982 age: 40+	Social class SMR	I, II III IV V		0.91 1.06 1.09 1		0.90 1.10 0.92 1	265 000 Japanese interviewed in 1965 and followed-up until 1982. Social class based on occupation. Reference category is farmers and miners [Hirayama, 1990]
The Netherlands 1951–1981 age: 18	Educational level RR	4 (high) 3 2 1 (low)		0.75 0.85 0.87 1			1951–1981 follow-up of 78 505 Dutch men undergoing medical examination in 1950–1951 for military service [Doornbos & Kromhout, 1990]
New Zealand 1974–1978 age: 15–64	Social class RR	I II III-MN III-M IV V		1 1.09 1.16 1.33 1.11 1.74			Surveillance system statistics using 1976 census data as denominator. UK Registrar General's social class classification [Pearce & Howard, 1986]
New Zealand 1984–1987 age: 15–64	Social class RR	I II III-MN III-M IV V		1 1.50 1.39 1.60 2.02 1.89			Surveillance system statistics using 1986 census data as denominator. UK Registrar General social-class classification [Pearce & Bethwaite, in press]

Table 2. (Contd) All cancer sites mortality

Study base	Indicators	Social scale	N	Male RR	N	Female RR	Study design
Norway 1970–1973 age: 20–69	Social class CMF	A		0.89		1.14	Surveillance system statistics using 1970 census data as denominator. Social class indicator based on occupation [Central Bureau of Statistics, 1976]
		B		1.11		0.99	
		C		1.07		0.99	
		D		1.07		1.01	
		E (farmers)		0.79		0.91	
Portugal 1980–82 age: 20–64	Occupational group RR	Managers		1.00		1.00	Surveillance system statistics using 1980 census data as denominator
		Professionals		1.87		2.30	
		Clerks		1.87		1.86	
		Sales workers		2.13		1.08	
		Service workers		1.57		0.44	
		Agriculture, forestry and fishery		2.34		0.32	
		Other manual workers		2.37		1.81	[M. Giraldes, pers. commun., SMRs calculated by authors]
Spain 1980–1982 age: 30–64	Occupational group PMR	Professionals and managers		0.94			Proportional analysis on death certificates
		Manual workers		1.15			
		Agricultural workers		0.90			[E. Regidor, pers commun.]
Switzerland 1979–1982 age: 30–49	Social class SMR	I		0.60			Surveillance system statistics using 1980 census data as denominator. UK Registrar General's social class classification [Lehmann, 1990]
		II		0.67			
		III-MN		1.05			
		III-M		1.49			
		IV-V		1.05			
UK – England and Wales 1910–1912 age: 15–64	Social class SMR	I		0.93			Surveillance system statistics using 1910 census data as denominator. For social classification see Introduction [OPCS, 1919]
		II		0.91			
		III		1.01			
		IV		0.92			
		V		1.21			
UK – England and Wales 1930–1932 age: 15–64 (married women)	Social class SMR	I		0.83		0.97	Surveillance system statistics using 1930 census data as denominator. For social classification see Introduction. Women classified according to husband's occupation [OPCS, 1938]
		II		0.92		0.97	
		III		0.99		1.02	
		IV		1.02		0.95	
		V		1.14		1.02	
UK – England and Wales 1970–1972 age: 15–64 (married women)	Social class SMR	I		0.75		0.99	Surveillance system statistics using 1970 census data as denominator. For social classification see Introduction. Women classified according to husband's occupation [OPCS, 1977]
		II		0.80		0.97	
		III-NM		0.91		0.99	
		III-M		1.13		1.13	
		IV		1.16		1.13	
		V		1.31		1.16	

Table 2. (Contd) All cancer sites mortality

Study base	Indicators	Social scale	N	Male RR	N	Female RR	Study design
UK – England and Wales 1971–1975 age: 15–64	Social class SMR	I II III-NM III-M IV V	33 165 91 402 218 111	0.70 0.75 0.77 1.02 1.04 1.19			Record-linkage study (longitudinal study) between 1971 census and 1971–1975 mortality data for a 1% sample of the total population. UK Registrar General's social class classification [OPCS, 1990]
UK – England and Wales 1976–1981 age: 15–64	Social class SMR	I II III-NM III-M IV V	35 208 121 501 254 103	0.58 0.81 0.91 1.02 1.07 1.13			Record-linkage study (longitudinal study) between 1971 census and 1976–1981 mortality data for a 1% sample of the total population. UK Registrar General's social class classification [OPCS, 1990]
UK – Great Britain 1979–1980, 1982–1983 age: 20–64 (married women, 20–59)	Social class SMR	I II III-NM III-M IV V	3143 16392 8936 34909 20094 9771	0.69 0.77 0.89 1.13 1.17 1.54	2087 9938 4533 16014 8309 2933	0.89 0.95 1.01 1.10 1.17 1.32	Surveillance system statistics using 1980 census data as denominator. For social classification see Introduction. Women classified according to husband's occupation [OPCS, 1986]
UK (London) 1967–1987	Employment grade RR	Administrators Professionals Clerical Other	47 713 265 212	1 1.25 1.69 1.99			17 530 London civil servants, medically examined 1967–1969, and followed-up until 1987 [Davey Smith & Marmot, 1991]
UK – Scotland 1949–1953 age: 20–64	Social class SMR	I II III IV V		1.04 0.93 1.09 0.96 1.06			Surveillance system statistics using 1950 census data as denominators. UK Registrar General social class classification [Registrar General for Scotland, 1956]
UK – Scotland 1959–1963 age: 20–64 (married women)	Social class SMR	I II III IV V		0.77 0.82 1.00 0.99 1.43		0.76 0.98 0.95 1.00 1.24	Surveillance system statistics using 1960 census data as denominator. UK Registrar General's classification. Women classified according to husband's occupation [Registrar General for Scotland, 1970]

Table 2. (Contd) All cancer sites mortality

Study base	Indicators	Social scale	N	Male RR	N	Female RR	Study design
US – California 1949–1951 age: 25–64	Social class SMR	I II III IV V		0.97 0.90 1.03 0.97 1.12			Surveillance system statistics using 1950 census data as denominator. Social class indicator based on occupation [Buell et al., 1960]
USA 1960 age: 25–64 (white)	Educational level SMR	College High school Elementary school <8 years of school		0.83 0.94 1.12 1.09		0.92 0.94 1.09 1.13	Record-linkage study using 1960 mortality and census data [Kitagawa & Hauser, 1973]
USA (12 census samples) White population 1979–1985 age: 25+	Education SMR	College: 5+ y 4 y 1-3 y High school: 4 y 1-3 y Elementary school: 8 y 5-7 y 0-4 y		0.34 0.48 0.88 1.01 1.17 1.26 0.90 0.97		1.08 0.92 1.07 0.99 1.02 0.97 1.03 0.89	Census linkage. [Rogot et al.,1992]
USA (12 census samples) Black population 1979–1985 age: 25+	Education SMR	College: 5+ y 4 y 1-3 y High school: 4 y 1-3 y Elementary school: 8 y 5-7 y 0-4 y		0.34 0.48 0.88 1.01 1.17 1.26 0.90 0.97		0.86 0.73 0.95 0.86 1.17 1.05 1.02 0.92	Census linkage. [Rogot et al.,1992]

Table 3. All cancer sites incidence

Study base	Indicators	Social scale	N	Male RR	N	Female RR	Study design
Colombia (Cali) all ages	Social class RR	I II III		1 0.92 0.76		1 1.14 0.96	Data from 1973 census were used for rate denominators. Social class based on the area of residence [Cuello et al., 1982]
Denmark 1970–1980 all ages	Occupational group RR	Self-employed Employees: I Employees: II Employees: III Employees: IV Skilled workers Unskilled workers	12 893 1315 2378 5949 2795 7114 15 054	0.89 0.92 0.97 1.08 1.12 1.12 1.02	1549 334 1410 2475 6787 190 9162	1.05 1.06 1.00 1.04 1.04 1.04 1.00	Record-linkage study using 1970 census and 1970–1980 incidence data. Employees classified according to educational level [Lynge & Thygesen, 1990]
Finland 1971–1985 birth cohort: 1906–45	Social class SIR	Upper white-collar Lower white-collar Skilled workers Unskilled workers		0.84 0.91 1.03 1.10		1.13 1.08 0.96 0.95	Record-linkage study using 1970 census and 1971–1985 incidence data. Social class based on occupation [Pukkala, 1993]
Italy (Torino) 1985–1987 age: 20–69	Educational level OR	University High school Middle school Primary school	262 599 1026 2328	1 1.03 1.04 1.15	129 436 942 1944	1 0.87 0.90 0.76	Record-linkage study between 1971 and 1981 censuses and 1985–1987 incidence data [Faggiano et al., 1994]
Italy (Torino) 1985–1987 age: 20–69	Occupational group OR	Managers Clerks Self-employed Manual workers	478 793 496 2031	1 0.97 1.05 1.14	112 489 216 451	1 0.93 0.80 0.86	Record-linkage study between 1971 and 1981 censuses and 1985–1987 incidence data [Faggiano et al., 1994]
Italy (Torino) 1985–1987 age: 20–69	Housing tenure OR	Owners Tenants	1810 2305	1 1.14	1520 1844	1 1.06	Record-linkage study between 1971 and 1981 censuses and 1985–1987 incidence data [Faggiano et al., 1994]
Sweden 1961–1970 all ages	Social class SIR	Employees: I Self-employees: II Indep. farmers: III White-collars: IV Blue-collars: V	14 056 13 147 14 853 34 758 82 175	1.03 1.02 0.87 1.05 1.00	– 2612 – 31 353 28 035	– 1.01 – 1.02 0.97	Record-linkage study between 1961 census and incidence data. Social class indicator based on occupation [Vågerö & Persson, 1986]
UK – England and Wales 1971–1981 all ages	Social class SIR	I II III-MN III-M IV V	274 1501 925 2880 1761 936	0.81 0.90 0.94 1.01 1.05 1.12			Record-linkage study between 1971 census and 1971–1981 incidence data (1% sample). UK Registrar General's social class classification [Kogevinas, 1990]
UK – England and Wales 1971–1981 all ages	Housing tenure SIR	Owner occupier Private rented Council tenant	4284 1805 2882	0.89 1.06 1.16	4320 1607 2504	0.97 0.97 1.05	Record-linkage study between 1971 census and 1971–1981 incidence data (1% sample). UK Registrar General's social class classification [Kogevinas, 1990]

Table 3. (Contd) All cancer sites incidence

Study base	Indicators	Social scale	N	Male RR	N	Female RR	Study design
USA (Du Pont) 1959–1967 all ages	Income level SIR	1 2 3 4 5	25 206 103 771	0.68 0.97 0.94 1.11 1.03	169		1959–1967 follow-up of 115 000 employees of the Du Pont Co. [Pell, 1970]

Table 4. Mouth and pharynx cancer mortality

Study base	Indicators	Social scale	N	Male RR	N	Female RR	Study design
Brazil (São Paulo) 1978–1982 age: 35–74	Years of education OR	12+ 9–11 1–8 <1		0.7 1.1 1.5 1			Case–control study using deaths from other causes as controls ICD-9: 141-5; Mouth [Bouchardy et al., 1993]
Brazil (São Paulo) 1978–1982 age: 35–74	Years of education OR	12+ 9–11 1–8 <1		0.4 0.6 1.0 1			Case–control study using deaths from other causes as controls ICD-9: 146, 148, 149; Pharynx [Bouchardy et al., 1993]
Canada (urban area) 1971 all ages	Income CMF	Q1 Q2 Q3 Q4 Q5		0.48 0.80 0.92 1.16 1.48			Surveillance system statistics using 1971 census data as denominator. Neighbourhood income quintiles as social indicator [R. Wilkins, pers. commun.]
Canada (urban area) 1986 all ages	Income CMF	Q1 Q2 Q3 Q4 Q5		0.54 0.56 0.63 1.32 1.92		0.73 0.73 1.00 0.87 1.60	Surveillance system statistics using 1986 census data as denominators. Neighbourhood income quintiles as social indicator Bucal cavity [R. Wilkins, pers. commun.]
France 1975–1982 age: 45–54	Occupational group RR	Managers Intermediate Self-employed Clerks Skilled workers Unskilled workers		0.23 0.43 0.60 1.17 1.49 2.49			A sample of about 1 000 000 of 1975 censused population followed-up until 1982. The scale shown represents a choice of the total scale [Desplanques, 1985]
France 1975–1982 age: 55–64	Occupational group RR	Managers Intermediate Self-employed Clerks Skilled workers Unskilled workers		0.31 0.65 0.70 1.17 1.57 1.91			A sample of about 1 000 000 of 1975 censused population followed up until 1982. The scale shown represents a choice of the total scale [Desplanques, 1985]

Table 4. (Contd) Mouth and pharynx cancer mortality

Study base	Indicators	Social scale	N	Male RR	N	Female RR	Study design
Italy 1981–1982 age: 18–74	Educational level RR	University	10	1	–	–	Record-linkage between 1981 census and mortality in the following six months
		High school	19	0.88	11	1	
		Middle school	81	1.82	8	0.46	
		Primary school	433	2.58	56	0.71	
		Literate	155	3.42	24	0.68	
		Illiterate	25	3.10	8	0.84	[Faggiano et al., 1995]
Japan 1965–1982 age: 40+	Social class SMR	I, II		0.91		–	265 000 Japanese interviewed in 1965 and followed-up until 1982. Social class based on occupation. Reference category is farmers and miners [Hirayama, 1990]
		III		1.41		0.57	
		IV		1.05		–	
		V		1		1	
New Zealand 1974–1978 age: 15–64	Social class RR	I		1			Surveillance system statistics using 1976 census data as denominator. UK Registrar General's social class classification [Pearce & Howard, 1986]
		II		2.33			
		III-NM		3.58			
		III-M		3.58			
		IV		3.42			
		V		3.75			
New Zealand 1984–1987 age: 15–64	Social class RR	I	2	1			Surveillance system statistics using 1986 census data as denominators. UK Registrar General's social class classification [Pearce & Bethwaite, in press]
		II	12	3.22			
		III-NM	14	1.89			
		III-M	37	5.56			
		IV	31	6.11			
		V	19	8.89			
Switzerland (Vaud) 1977–1984 all ages	Social class PMR	I, II		0.59		0.78	Proportional mortality study UK Registrar General's social class classification ICD-9: 140-150, 161. (No. of males = 403, females = 63) [Levi, 1988]
		III		1.29		1.25	
		IV, V		1.24		1.07	
Switzerland 1979–1982 15–74	Social class SMR	I		0.46			Surveillance system statistics using 1980 census data as denominators. UK Registrar General's social class classification [C. E. Minder, pers commun.]
		I		0.62			
		III-NM		0.80			
		III-M		1.40			
		IV-V		1.26			
UK – England and Wales 1930–1932 age: 15–64 (married women)	Social class SMR	I		0.72		–	Surveillance system statistics using 1930 census data as denominators. For social classification see Introduction. Women classified according to husband's occupation [OPCS, 1938]
		II		0.68		0.80	
		III		0.63		1.04	
		IV		1.10		1.19	
		V		1.46		1.00	
UK – England and Wales 1970–1972 age: 15–64 (married women)	Social class SMR	I		1.16		0.90	Surveillance system statistics using 1970 census data as denominators. For social classification see Introduction. Women classified according to husband's occupation [OPCS, 1977]
		II		0.87		0.88	
		III-NM		1.04		0.89	
		III-M		0.94		1.15	
		IV		1.04		1.03	
		V		1.63		1.66	

Table 4. (Contd) Mouth and pharynx cancer mortality

Study base	Indicators	Social scale	N	Male RR	N	Female RR	Study design
UK – Great Britain 1979–1980, 1982–1983 age: 20–64 (married women, 20–59)	Social class SMR	I	53	0.61	10	0.50	Surveillance system statistics using 1980 census data as denominator. For social classification see Introduction. Women classified according to husband's occupation [OPCS, 1986]
		II	295	0.73	87	0.96	
		III-NM	161	0.86	29	0.74	
		III-M	597	1.02	138	1.10	
		IV	383	1.20	72	1.16	
		V	241	2.04	37	1.91	
USA – California 1949–1951 age: 25–64	Social class SMR	I		0.67			Surveillance system statistics using 1950 census data as denominator. Social class indicator based on occupation ICD-9: 140-148 [Buell et al., 1960]
		II		0.80			
		III		1.02			
		IV		1.12			
		V		1.00			

Table 5. Mouth and pharynx cancer incidence

Study base	Indicators	Social scale	N	Male RR	N	Female RR	Study design
Colombia (Cali) 1971–1975 all ages	Social class RR	I		1		1	Data from 1973 census were used for rate denominators. Social class based on area of residence Mouth [Cuello et al., 1982]
		II		0.60		1.09	
		III		0.53		1.00	
Colombia (Cali) 1971–1975 all ages	Social class RR	I		1		1	Data from 1973 census were used for rate denominators. Social class based on area of residence Pharynx [Cuello et al, 1982]
		II		1.33		3.50	
		III		0.67		5.00	
Denmark 1970–1980 all ages	Occupational group RR	Self-employed	61	0.83	5	0.94	Record-linkage study using 1970 census and 1970–1980 incidence data. Employees classified according to educational level Mouth [Lynge & Thygesen, 1990]
		Employees: I	5	0.67	1	0.97	
		Employees: II	9	0.71	3	0.70	
		Employees: III	35	1.20	9	1.21	
		Employees: IV	15	1.22	14	0.72	
		Skilled workers	46	1.48	0	–	
		Unskilled workers	70	0.94	34	1.19	
Denmark 1970–1980 all ages	Occupational group RR	Self-employed	83	0.83	1	0.22	Record-linkage study using 1970 census and 1970-80 incidence data. Employees classified according to educational level Pharynx [Lynge & Thygesen, 1990]
		Employees: I	6	0.56	0	–	
		Employees: II	18	0.96	5	1.31	
		Employees: III	43	1.03	7	1.00	
		Employees: IV	20	1.12	13	0.68	
		Skilled workers	61	1.35	1	1.88	
		Unskilled workers	108	1.04	36	1.27	

Table 5. (Contd) Mouth and pharynx cancer incidence

Study base	Indicators	Social scale	N	Male RR	N	Female RR	Study design
Finland 1971–1985 birth cohort: 1906–1945	Social class SIR	Upper white-collar Lower white-collar Skilled workers Unskilled workers		1.37 0.70 0.98 1.32		1.37 1.13 0.88 1.00	Record-linkage study using 1970 census and 1971–1985 incidence data. Social class based on occupation Mouth [Pukkala, 1993]
Finland 1971–1985 birth cohort: 1906–1945	Social class SIR	Upper white-collar Lower white-collar Skilled workers Unskilled workers		1.23 1.05 0.94 1.00		0.71 0.91 0.96 1.31	Record-linkage study using 1970 census and 1971–1985 incidence data. Social class based on occupation Pharynx [Pukkala, 1993]
Italy (Milano) 1983–1988 age: <75	Years of education OR	12+ 7–11 <7		3[a] 7[a] 40[a]		0.16[a] 0.29[a] 1[a]	Hospital-based case–control study. Adjusted for sex [Ferraroni et al., 1989]
Italy (Milano) 1983–1988 age: <75	Social class RR	I, II III IV, V		3[a] 9[a] 27[a]		0.50[a] 0.43[a] 1[a]	Hospital-based case–control study. UK Registrar General's social class classification. Adjusted for sex [Ferraroni et al., 1989]
Italy (Torino) 1985–1987 age: 20–69	Educational level OR	University High school Middle school Primary school	20 52 108 298	1 0.92 1.16 1.71	7 21 39	1 1.19 0.82	Record-linkage study between 1971 and 1981 censuses and 1985–1987 incidence data ICD-9: 140-150, 161 [Faggiano et al., 1994]
Italy (Torino) 1985–1987 age: 20-69	Occupational group OR	Managers Clerks Self-employed Manual workers	30 77 56 261	1 1.37 1.53 2.51			Record-linkage study between 1971 and 1981 censuses and 1985–1987 incidence data ICD-9: 140-150, 161 [Faggiano et al., 1994]
Italy (Torino) 1985–1987 age: 20-69	Housing tenure OR	Owners Tenants	173 289	1 1.64	20 46	1 1.99	Record-linkage study between 1971 and 1981 censuses and 1985–1987 incidence data ICD-9: 140-150, 161 [Faggiano et al., 1994]
Sweden 1961–1970 all ages	Social class SIR	Employees: I Self-employed: II Ind. farmers: III White-collar: IV Blue-collar: V	65 61 39 173 314	1.17 1.19 0.61 1.16 0.95	– 6 – 80 84	– 0.76 – 0.94 1.02	Record-linkage study between 1961 census and 1961–1970 incidence data. Social class indicator based on occupation Mouth [Vågerö & Persson, 1986]
Sweden 1961–1970 all ages	Social class SIR	Employees: I Self-employed: II Ind. farmers: III White-collar: IV Blue-collar: V	36 65 19 144 260	0.81 0.52 0.41 1.11 1.00	– 1 – 23 46	– 0.33 – 0.70 1.39	Record-linkage study between 1961 census and 1961–1970 incidence data. Social class indicator based on occupation Hypopharynx [Vågerö & Persson, 1986]

Table 5. (Contd) Mouth and pharynx cancer incidence

Study base	Indicators	Social scale	N	Male RR	N	Female RR	Study design
UK – England and Wales 1971–1981 all ages	Housing tenure SIR	Owner occupier Private rented Council tenant	71 41 54	0.77 1.29 1.14			Record-linkage study between 1971 census and 1971–1981 incidence data (1% sample). UK Registrar General's social class classification Mouth [Kogevinas, 1990]
USA 1969–1971 all ages	Educational level OR	College Less		0.96 1		1.89 1	Case–control study based on US Third National Cancer Survey, using deaths for other causes as controls. Mouth. [Williams & Horm, 1977]
USA 1969–1971 all ages	Educational level OR	College Less		0.65 1		1.38 1	Case–control study based on US Third National Cancer Survey, using deaths from other causes as controls Pharynx [Williams & Horm, 1977]
USA 1969–1971 all ages	Family income level OR	>US$ 10 000 Less		0.78 1		1.13 1	Case–control study based on US Third National Cancer Survey, using deaths for other causes as controls Mouth [Williams & Horm, 1977]
USA 1969–1971 all ages	Family income level OR	>US$ 10 000 Less		0.63 1		0.82 1	Case–control study based on US Third National Cancer Survey, using deaths for other causes as controls Pharynx [Williams & Horm, 1977]

[a]Data not stratified by sex.

Table 6. Oesophagus cancer mortality

Study base	Indicators	Social scale	N	Male RR	N	Female RR	Study design
Brazil (São Paulo) 1978–1982 age: 35–74	Years of education OR	12+ 9–11 1–8 <1		0.3 0.4 0.6 1		0.4 0.2 0.4 1	Case–control study using deaths from other causes as controls [Bouchardy et al., 1993]
France 1975–1982 age: 45–54	Occupational group RR	Managers Intermediate Self-employed Clerks Skilled workers Unskilled workers		0.25 0.46 0.67 1.17 1.62 2.25			A sample of about 1 000 000 of 1975 census population followed up until 1982. The scale shown represents a choice of the total scale [Desplanques, 1985]
France 1975–1982 55–64	Occupational group RR	Managers Intermediate Self-employed Clerks Skilled workers Unskilled workers		0.31 0.61 0.86 1.08 1.45 1.88			A sample of about 1 000 000 of 1975 census population age: followed up until 1982. The scale shown represents a choice of the total scale [Desplanques, 1985]
Italy 1981–1982 age: 18–74	Educational level RR	University High school Middle school Primary school Literate Illiterate	10 9 33 346 95 19	1 0.53 0.92 2.39 2.27 3.00			Record-linkage between 1981 census and mortality in the following six months [Faggiano et al., 1995]
Japan 1965–1982 age: 40+	Social class SMR	I, II III IV V		0.86 0.95 1.07 1		– 1.00 0.42 1	265 000 Japanese interviewed in 1965 and followed-up until 1982. Social class based on occupation. Reference category is farmers and miners [Hirayama, 1990]
New Zealand 1974–1978 age: 15–64	Social class RR	I II III-NM III-M IV V		1 0.69 1.27 0.94 0.86 1.35			Surveillance system statistics using 1976 census data as denominators. UK Registrar General's social class classification [Pearce & Howard, 1986]
New Zealand 1984–1987 age: 15–64	Social class RR	I II III-NM III-M IV V	4 13 18 32 37 13	1 1.58 1.21 2.32 3.58 2.95			Surveillance system statistics using 1976 census data as denominators. UK Registrar General social-class classification [Pearce & Bethwaite, in press]
Spain 1980–1982 age: 30–64	Occupational group PMR	Professionals and managers Manual workers Agricultural workers		0.84 1.19 0.81			Proportional analysis on death certificates [E. Regidor, pers. commun.]

Table 6 (Contd) Oesophagus cancer mortality

Study base	Indicators	Social scale	N	Male RR	N	Female RR	Study design
Switzerland 1979–1982 age: 15–74	Social class SMR	I II III-NM III-M IV-V		0.44 0.62 0.80 1.32 1.17			Surveillance system statistics using 1980 census data as denominator. UK Registrar General's social class classification [C.E. Minder, pers. commun.]
UK – England and Wales 1930–1932 age: 15–64 (married women)	Social class SMR	I II III IV V		0.74 0.87 0.98 0.94 1.30		0.95 0.85 1.01 0.95 1.16	Surveillance system statistics using 1930 census data as denominator. For social classification see Introduction. Women classified according to husband's occupation [OPCS, 1938]
UK – England and Wales 1970–1972 age: 15–64 (married women)	Social class SMR	I II III-NM III-M IV V		0.81 0.86 0.85 1.08 1.13 1.39		0.76 0.72 1.03 1.20 1.19 1.42	Surveillance system statistics using 1970 census data as denominator. For social classification see Introduction. Women classified according to husband's occupation [OPCS, 1977]
UK – Great Britain 1979–1980, 1982–1983 age: 20–64 (married women, 20-59)	Social class SMR	I II IIIN III-M IV V	132 602 340 1267 722 356	0.80 0.77 0.93 1.12 1.14 1.51	22 125 69 231 160 55	0.65 0.80 1.01 1.07 1.45 1.58	Surveillance system statistics using 1980 census data as denominator. For social classification see Introduction. Women classified according to husband's occupation [OPCS, 1986]
UK (London) 1967–1987	Employment grade RR	Administrators Professionals Clerical Other	1 21 6 8	1 2.2 2.6 3.8			17 530 London civil servants, medically examined 1967–1969 and, followed-up until 1987 [Davey Smith & Marmot, 1991]
USA – California 1949–1951 age: 25–64	Social class SMR	I II III IV V		0.54 0.67 0.86 1.15 1.45			Surveillance system statistics using 1950 census data as denominator. Social class indicator based on occupation [Buell et al., 1960]

Table 7. Oesophagus cancer incidence

Study base	Indicators	Social scale	N	Male RR	N	Female RR	Study design
Canada (Montreal) 1979–1985 age: 35–70 (French)	Income level OR	High Middle Low		1 2.1 1.9			Population-based case–control study. Tertiles of total family income [Bourbonnais & Siemiatycki, in press]
Canada (Montreal) 1979–1985 age: 35–70 (French)	Education OR	High Middle Low		1 1.8 1.9			Population-based case–control study. Tertiles of years of education [Bourbonnais & Siemiatycki, in press]
Canada (Montreal) 1979–1985 age: 35–70 (French)	Occupational prestige scale OR	High Middle Low		1 1.6 2.1			Population-based case–control study. Tertiles of the occupational prestige scale [Bourbonnais & Siemiatycki, in press]
Colombia (Cali) 1971–1975 all ages	Social class RR	I II III		1 0.65 0.75		1 1.25 1.75	Data from 1973 census were used for rate denominator. Social class based on area of residence [Cuello et al., 1982]
Denmark 1970–1980 all ages	Occupational group RR	Self-employed Employees: I Employees: II Employees: III Employees: IV Skilled workers Unskilled workers	146 11 22 54 24 89 220	0.82 0.65 0.79 0.83 0.85 1.24 1.24	8 1 5 4 17 0 35	1.56 1.05 1.36 0.59 0.97 – 1.26	Record-linkage study using 1970–80 incidence data and 1970 census. Employees classified according to educational level [Lynge & Thygessen 1990]
Finland 1971–1985 birth cohort: 1906–1945	Social class SIR	Upper white-collar Lower white-collar Skilled workers Unskilled workers		0.61 0.80 1.03 1.35		0.29 0.77 1.08 1.26	Record-linkage study using 1970 census and 1971–1985 incidence data. Social class based on occupation [Pukkala, 1993]
Italy (Milano) 1983–1988 age: <75	Years of education RR	12+ 7–11 <7		22[a] 39[a] 148[a]		0.36[a] 0.50[a] 1[a]	Hospital-based case–control study. Adjusted for sex [Ferraroni et al., 1989]
Italy (Milano) 1983–1988 age: <75	Social class RR	I, II III IV, V		9[a] 55[a] 115[a]		0.38[a] 0.60[a] 1[a]	Hospital-based case–control study. Adjusted for sex. [Ferraroni et al., 1989]
Sweden 1961–1970 all ages	Social class SIR	Employees: I Self-employees: II Indep. farmers: III White-collars: IV Blue-collars: V	166 195 134 461 1106	0.94 1.15 0.69 0.96 1.06	8 92 141	– 0.65 – 0.82 1.18	Record-linkage study between 1961 census and incidence data. Social class indicator based on occupation [Vågerö & Persson, 1986]

96

Socioeconomic differences in cancer incidence and mortality

Table 7. (Contd) Oesophagus cancer incidence

Study base	Indicators	Social scale	N	Male RR	N	Female RR	Study design
UK – England and Wales 1971–1981 all ages	Housing tenure SIR	Owner occupiers Private rented Council tenant	90 27 65	0.92 0.79 1.31	67 33 45	0.90 1.11 1.14	Record-linkage study between 1971 census and 1971–1981 incidence data (1% sample). UK Registrar General's social class classification. [Kogevinas, 1990]
USA 1969–1971 all ages	Educational level OR	College Less		0.59 1		1.23 1	Case–control study based on US Third National Cancer Survey, using deaths from other causes as controls. [Williams & Horm, 1977]
USA 1969–1971 all ages	Family income level OR	>US$ 10 000 Less		1.23 1		0.51 1	Case–control study based on US Third National Cancer Survey, using deaths from other causes as controls. [Williams & Horm, 1977]

aData not stratified by sex.

Table 8. Stomach cancer mortality

Study base	Indicators	Social scale	N	Male RR	N	Female RR	Study design
Brazil (São Paulo) 1978–1982 age: 35–74	Years of education OR	12+ 9–11 1–8 <1		0.3 0.4 0.6 1		0.3 0.3 0.7 1	Case–control study using deaths from other causes as controls [Bouchardy et al., 1992]
Canada (urban area) 1971 all ages	Income CMF	Q1 Q2 Q3 Q4 Q5		0.85 1.09 0.91 0.99 1.13		0.60 0.64 1.09 1.13 1.38	Surveillance system statistics using 1971 census data as denominators. Neighbourhood income quintiles as social indicator [R. Wilkins, pers. commun.]
Canada (urban area) 1986 all ages	Income CMF	Q1 Q2 Q3 Q4 Q5		1.00 0.87 0.87 1.08 1.16		0.78 0.97 1.09 1.09 1.09	Surveillance system statistics using 1986 census data as denominators. Neighbourhood income quintiles as social indicator [R. Wilkins, pers. commun.]
Denmark 1970–1975 age: 20–64	Occupational group SMR	Employees: I Employees: II Employees: III Employees: IV Skilled workers Unskilled workers		0.77 0.88 1.02 1.22 1.09 1.06		– 0.99 0.89 0.92 – 1.08	Record–linkage study using 1970 census and 1970–1975 mortality data. Employees classified according to educational level [Danmarks Statistik, 1979]

Table 8. (Contd) Stomach cancer mortality

Study base	Indicators	Social scale	N	Male RR	N	Female RR	Study design
Finland 1969–1972 age: 15–64 (married women)	Social class CMF	Upper white-collar Lower white-collar Skilled workers Unskilled workers Farmers		0.79 0.84 0.88 1.33 1.17		0.58 1.00 1.16 1.08 1.19	Surveillance system statistics using 1970 census data as denominators. Social class indicator based on occupation [Näyhä, 1977]
Finland 1971–1985 age: 15–64	Social class RR	Upper white-collar Lower white-collar Skilled workers Unskilled workers Farmers	>20 >20 >20 >20 >20	1 1.33 1.67 1.92 1.67	>20 >20 >20 >20 >20	1 1.40 1.38 – 1.82	Record-linkage study using 1970, 1975 and 1980 census data and 1971–1985 mortality [Valkonen et al., 1990]
France 1975–1982 45–54	Occupational group RR	Managers Intermediate Self-employed Clerks Skilled workers Unskilled workers		0.45 0.64 0.91 1.00 1.36 1.27			A sample of about 1 000 000 of 1975 census population age: followed up until 1982. The scale shown represents a choice of the total scale [Desplanques, 1985]
France 1975–1982 55–64	Occupational groups RR	Managers Intermediate Self-employed Clerks Skilled workers Unskilled workers		0.42 0.69 0.89 1.92 1.22 1.00			A sample of about 1 000 000 of 1975 census population age: followed up until 1982. The scale shown represents a choice of the total scale [Desplanques, 1985]
Hungary 1970 age: 25–64	Years of education SMR	15+ 12–14 8–11 0–7		0.46 0.57 1.33 1.01		0.70 0.93 0.96 0.95	Surveillance system statistics using 1970 census data as denominator [Jozan, 1986]
Hungary 1980 age: 25–64	Years of education SMR	15+ 12–14 8–11 0–7		0.58 0.79 1.00 1.12		0.98 0.96 1.11 0.95	Surveillance system statistics using 1980 census data as denominator [Jozan, 1986]
Italy 1981–1982 age: 18–74	Educational level RR	University High school Middle school Primary school Literate Illiterate	36 85 242 1377 499 114	1 1.42 1.98 2.77 2.85 3.43	7 33 78 596 327 80	1 1.23 1.31 1.96 2.20 2.25	Record-linkage between 1981 census and mortality in the following six months [Faggiano et al., 1995]
Japan 1965–1982 age: 40+	Social class SMR	I, II III IV V		0.78 1.04 1.09 1		0.76 0.95 0.94 1	265 000 Japanese interviewed in 1965 and followed up until 1982. Social class based on occupation. Reference category is farmers and miners [Hirayama, 1990]

Table 8. (Contd) Stomach cancer mortality

Study base	Indicators	Social scale	N	Male RR	N	Female RR	Study design
New Zealand 1974–1978 age: 15–64	Social class RR	I II III-NM III-M IV V		1 1.80 2.23 3.03 2.20 4.23			Surveillance system statistics using 1976 census data as denominator. UK Registrar General's social class classification [Pearce & Howard, 1986]
New Zealand 1984–1987 age: 15–64	Social class RR	I II III-NM III-M IV V	6 35 48 81 54 26	1 2.85 2.08 4.27 1.85 4.23			Surveillance system statistics using 1986 census data as denominators. UK Registrar General social-class classification [Pearce & Bethwaite, in press]
Norway 1970–1973 age: 20–69	Social class CMF	A B C D E (farmers)		0.78 1.11 1.06 1.08 0.85		1.05 0.99 0.94 1.01 1.06	Surveillance system statistics using 1970 census data as denominators. Social class indicator based on occupation ICD: 150-154 [Central Bureau of Statistics, 1976]
Spain 1980–1982 age: 30–64	Occupational group PMR	Professionals and managers Manual workers Agricultural workers		0.74 1.16 1.14			Proportional analysis on death certificates [E. Regidor, pers commun.]
Switzerland (Vaud) 1977–1984 all ages	Social class PMR	I, II III IV, V		0.97 0.92 1.17		0.70 1.35 0.94	Proportional mortality study. UK Registrar General's social class classification (No. of males = 159; females = 36) [Levi, 1988]
Switzerland 1979–1982 age: 15–74	Social class SMR	I II III-NM III-M IV-V		0.52 0.63 0.97 1.25 1.07			Surveillance system statistics using 1980 census data as denominators. UK Registrar General's social-class classification [C.E. Minder, pers. commun.]
UK – England and Wales 1910–1912 age: 15–64	Social class SMR	I II III IV V		0.75 0.96 1.02 0.91 1.29			Surveillance system statistics using 1910 census data as denominator. For social classification see Introduction [OPCS, 1919]
UK – England and Wales 1930–1932 age: 15–64 (married women)	Social class SMR	I II III IV V		0.55 0.83 0.98 1.12 1.22		0.49 0.77 1.05 1.06 1.21	Surveillance system statistics using 1930 census data as denominator. For social classification see Introduction. Women classified according to husband's occupation [OPCS, 1938]

Table 8. (Contd) Stomach cancer mortality

Study base	Indicators	Social scale	N	Male RR	N	Female RR	Study design
UK – England and Wales 1970–1972 age: 15–64 (married women)	Social class SMR	I II III-NM III-M IV V		0.50 0.66 0.79 1.18 1.25 1.47		0.60 0.84 0.76 1.22 1.23 1.45	Surveillance system statistics using 1970 census data as denominators. For social classification see Introduction. Women classified according to husband's occupation [OPCS, 1977]
UK – Great Britain 1979–1980, 1982–1983 age: 20–64 (married women, 20–59)	Social class SMR	I II III-NM III-M IV V	181 1132 664 2926 1776 817	0.50 0.67 0.83 1.19 1.27 1.58	57 266 125 551 299 118	0.77 0.79 0.86 1.18 1.28 1.61	Surveillance system statistics using 1980 census data as denominator. For social classification see Introduction. Women classified according to husband's occupation [OPCS, 1986]
UK (London) 1967–1987	Employment grade RR	Administrators Professionals Clerical Other	2 54 24 20	1 1.81 2.88 3.56			17 530 Londoner civil servants, medically examined 1967–69 and, followed-up until 1987 [Davey Smith & Marmot, 1991]
UK – Scotland 1959–1963 20–64 (married women)	Social class SMR	I II III IV V		0.49 0.77 0.58 1.05 1.66		1.11 0.56 0.90 2.53 1.04	Surveillance system statistics using 1960 census data as denominator. UK Registrar General's classification. Women classified according to husband's occupation [Registrar General for Scotland, 1970]
US – California 1949–1951 age: 25–64	Social class SMR	I II III IV V		0.51 0.72 0.93 0.99 1.65			Surveillance system statistics using 1950 census data as denominator. Social class indicator based on occupation [Buell et al., 1960]
USA 1960 age: 25–64 (white)	Educational level SMR	College High school Elementary school <8 years of school	0.56 0.97 1.07 1.25		0.45 0.94 1.03 1.22		Record-linkage study using 1960 mortality and census data [Kitagawa & Hauser, 1973]
USA (12 census samples) 1979–1985 age: 25+	Education SMR	College High school Elementary school	5+ y 4 y 1-3 y 4 y 1-3 y 8 y 5-7 y 0-4 y	0.47 0.31 0.92 1.02 1.06 1.07 1.21 1.73			Census linkage [Rogot et al., 1992]

Table 9. Stomach cancer incidence

Study base	Indicators	Social scale	N	Male RR	N	Female RR	Study design
Canada (Montreal) 1979–1985 age: 35–70 (French)	Income level OR	High Middle Low		1 2.2 2.3			Population-based case–control study. Tertiles of total family income [Bourbonnais & Siemiatycki, in press]
Canada (Montreal) 1979–1985 age: 35–70 (French)	Education OR	High Middle Low		1 1.4 1.6			Population-based case–control study. Tertiles of years of education [Bourbonnais & Siemiatycki, in press]
Canada (Montreal) 1979–1985 age: 35–70 (French)	Occupational prestige scale OR	High Middle Low		1 1.2 1.3			Population-based case–control study. Tertiles of the occupational prestige scale [Bourbonnais & Siemiatycki, in press]
Colombia (Cali) 1971–1975 all ages	Social class RR	I II III		1 1.62 1.56		1 1.57 1.48	Data from 1973 census were used for rate denominator. Social class based on area of residence [Cuello et al., 1982]
Denmark 1970–1980 all ages	Occupational group RR	Self-employed Employees: I Employees: II Employees: III Employees: IV Skilled workers Unskilled workers	666 42 78 235 129 299 875	0.91 0.61 0.68 0.89 1.10 1.01 1.20	40 10 24 42 121 5 209	1.15 1.48 0.85 0.85 0.93 1.38 1.09	Record-linkage study using 1970 census and 1970–1980 incidence data. Employees classified according to educational level [Lynge & Thygesen, 1990]
Finland 1971–1985 birth cohort: 1906–1945	Social class SIR	Upper white-collar Lower white-collar Skilled workers Unskilled workers		0.64 0.85 1.06 1.18		0.76 0.95 1.04 1.03	Record-linkage study using 1970 census and 1971–1985 incidence data. Social class based on occupation [Pukkala, 1993]
Hong Kong 1971 age: 35–64 (Chinese)	Income level RR	Higher Medium Lower	2 35 95	1 2.61 2.85	2 19 45	1 1.51 1.35	Surveillance system statistics using 1971 census data as denominators. Income levels based on residence [Crowther et al., 1976]
Italy (Milano) 1983–1988 age: <75	Social class RR	I, II III IV, V	10[a] 121[a] 194[a]	0.24[a] 0.79[a] 1[a]			Hospital-based case–control study. UK Registar General's social class classification. Adjusted by sex [Ferraroni et al., 1989]
Italy (Milano) 1983–1988 age: <75	Years of education RR	12+ 7–11 <7	37[a] 88[a] 272[a]	0.35[a] 0.63[a] 1[a]			Hospital-based case–control study. Adjusted by sex [Ferraroni et al., 1989]
Italy (Torino) 1985–1987 age: 20–69	Educational level OR	University High school Middle school Primary school	11 28 48 152	1 0.83 1.02 1.48	9 22 79	– 1 2.47 2.84	Record-linkage study between 1971 and 1981 censuses and 1985–1987 incidence data [Faggiano et al., 1994]

Table 9. (Contd) Stomach cancer incidence

Study base	Indicators	Social scale	N	Male RR	N	Female RR	Study design
Italy (Torino) 1985–1987 age: 20–69	Occupational group OR	Managers Clerks Self employed Manual workers	24 40 24 123	1 0.99 1.19 1.30	– 13 4 22	– 1 0.45 2.21	Record-linkage study between 1971 and 1981 censuses and 1985–1987 incidence data [Faggiano et al., 1994]
Italy (Torino) 1985–1987 age: 20–69	Housing tenure OR	Owners Tenants	91 138	1 1.38	43 64	1 1.09	Record-linkage study between 1971 and 1981 censuses and 1985–1987 incidence data [Faggiano et al., 1994]
Sweden 1961–1970 all ages	Social class SIR	Employees: I Self-employees: II Indep. farmers: III White-collars: IV Blue-collars: V	1058 1125 1768 2115 7546	0.92 1.00 1.09 0.78 1.08	– 107 – 1015 1335	– 0.91 – 0.88 1.12	Record-linkage study between 1961 census and incidence data. Social class indicator based on occupation [Vågerö & Persson, 1986]
Turkey (Istanbul) 1979–1984 all ages	Social class OR	Higher Medium Lower	8 61 155	1 1.5 1.4			Hospital-based case–control study. Social class indicator based on occupation [Dosemeci, 1993]
UK – England and Wales 1971–1981 all ages	Housing tenure SIR	Owner occupation Private rented Council tenant	357 156 235	0.89 1.09 1.14	223 93 157	0.92 0.94 1.24	Record-linkage study between 1971 census and 1971–81 incidence data (1% sample). UK Registrar General's social class classification. [Kogevinas, 1990]
USA 1969–1971 all ages	Educational level OR	College Less		0.42 1		0.60 1	Case–control study based on US Third National Cancer Survey, using deaths from other causes as controls. [Williams & Horm, 1977]
USA 1969–1971 all ages	Family income level OR	>US$ 10 000 Less		1.06 1		0.88 1	Case–control study based on US Third National Cancer Survey, using deaths from other causes as controls. [Williams, & Horm 1977]

[a]Data not stratified by sex.

Table 10. Colon cancer mortality

Study base	Indicators	Social scale	N	Male RR	N	Female RR	Study design
Brazil (São Paulo) 1978–1982 age: 35–1974	Years of education OR	12+ 9–11 1–8 <1		3.0 2.0 1.6 1		2.2 2.1 1.4 1	Case–control study using deaths from other causes as controls [Bouchardy et al., 1993]
Canada (urban area) 1971 all ages	Income CMF	Q1 Q2 Q3 Q4 Q5		0.84 0.92 0.98 1.00 1.20		0.99 1.05 0.95 1.00 1.02	Surveillance system statistics using 1971 census data as denominator. Neighbourhood income quintiles as social indicator [R. Wilkins, pers. commun.]
Canada (urban area) 1986 all ages	Income CMF	Q1 Q2 Q3 Q4 Q5		0.84 1.06 0.87 1.08 1.16		1.03 1.00 1.15 1.01 0.85	Surveillance system statistics using 1986 census data as denominator. Neighbourhood income quintiles as social indicator [R. Wilkins, pers. commun.]
Finland 1969–1972 age: 15–64	Social class CMF	Upper white-collar Lower white-collar Skilled workers Unskilled workers Farmers		2.04 0.96 0.58 0.87 0.54		1.24 1.07 1.03 0.87 0.78	Surveillance system statistics using 1970 census data as denominator. Social class indicator based on occupation ICD-9: 152-154 [Näyhä, 1977]
Finland 1971–1985 age: 35–64	Social class RR	Upper white-collar Lower white-collar Skilled workers Unskilled workers Farmers		1 0.92 0.79 0.70 0.61		1 0.88 0.92 – 0.77	Record-linkage study using 1970, 1975 and 1980 censuses and 1971–85 mortality data. Social class based on occupation ICD-9: 153-154 [Valkonen, 1990]
Hungary 1970 age: 25–64	Years of education SMR	15+ 12–14 8–11 0–7		1.20 2.12 1.51 0.73		0.34 1.05 1.43 0.76	Surveillance system statistics using 1970 census data as denominator [Jozan, 1986]
Hungary 1980 age: 25–64	Years of education SMR	15+ 12–14 8–11 0–7		1.66 1.46 1.06 0.71		1.34 1.57 1.27 0.76	Surveillance system statistics using 1980 census data as denominator [Jozan, 1986]
Italy 1981–1982 age: 18–74	Educational level RR	University High school Middle school Primary school Literate Illiterate	44 56 131 491 143 27	1 0.77 0.87 0.79 0.63 0.62	16 30 90 376 157 31	1 0.49 0.65 0.51 0.45 0.37	Record-linkage between 1981 census and mortality in the following six months [Faggiano et al., 1995]
Japan 1965–1982 age: 40+	Social class SMR	I, II III IV V		1.29 1.61 1.48 1		1.31 1.17 0.91 1	265 000 Japanese interviewed in 1965 and followed-up until 1982. Social class based on occupation. Reference category is farmers and miners. ICD-9: 152-154 [Hirayama, 1990]

Table 10. (Contd) Colon cancer mortality

Study base	Indicators	Social scale	N	Male RR	N	Female RR	Study design
New Zealand 1974–1978 age: 15–64	Social class RR	I		1			Surveillance system statistics using census data as denominator. UK Registrar General's social class classification [Pearce & Howard, 1986]
		II		1.00			
		III-NM		1.05			
		III-M		0.96			
		IV		0.61			
		V		0.92			
New Zealand 1984–1987 age: 15–64	Social class RR	I	30	1			Surveillance system statistics using census data as denominators. UK Registrar General's social class classification [Pearce & Bethwaite, in press]
		II	70	1.18			
		III-NM	98	0.95			
		III-M	99	1.03			
		IV	67	0.95			
		V	21	0.69			
Spain 1980–1982	Occupational group PMR	Professionals and managers		1.34			Proportional analysis on death certificates
		Manual workers		1.12			
		Agricultural workers		0.76			[E. Regidor, pers. commun.]
Switzerland (Vaud) 1977–1984 all ages	Social class PMR	I, II		1.29		1.11	Proportional mortality study. UK Registrar General's social class classification ICD-9: 152-154 (No. of males = 283; females = 159) [Levi et al., 1988]
		III		0.89		0.89	
		IV, V		0.82		1.04	
Switzerland 1979–1982 age: 15–74	Social class SMR	I		0.71			Surveillance system statistics using 1980 census data as denominator. UK Registrar General's social class classification [C.E. Minder, pers. commun.]
		II		0.94			
		III-NM		1.17			
		III-M		1.17			
		IV-V		0.81			
UK – England and Wales 1910–1912 age: 15–64	Social class SMR	I		1.27			Surveillance system statistics using 1910 census data as denominator. For social classification see Introduction ICD-9: 152-153 [OPCS, 1919]
		II		1.01			
		III		0.97			
		IV		0.85			
		V		0.98			
UK – England and Wales 1930–1932 age: 15–64 (married women)	Social class SMR	I		1.10		1.19	Surveillance system statistics using 1930 census data as denominator. For social classification see Introduction. Women classified according to husband's occupation [OPCS, 1938]
		II		1.04		0.99	
		III		1.02		1.02	
		IV		0.99		0.89	
		V		0.94		1.02	
UK – England and Wales 1949–1953 age: 15–64 (married women)	Social class SMR	I		1.21		1.15	Surveillance system statistics using 1950 census data as denominators. For social classification see Introduction [OPCS, 1958]
		II		1.01		1.06	
		III		1.02		0.99	
		IV		0.92		1.01	
		V		0.99		0.95	

Table 10. (Contd) Colon cancer mortality

Study base	Indicators	Social scale	N	Male RR	N	Female RR	Study design
UK – England and Wales 1970–1972 age: 15–64 (married women)	Social class SMR	I II III-NM III-M IV V		1.04 1.00 1.06 1.06 1.00 1.11		1.18 0.93 0.96 1.17 1.12 1.10	Surveillance system statistics using 1970 census data as denominator. For social classification see Introduction. Women classified according to husband's occupation [OPCS, 1977]
UK – Great Britain 1979–1980, 1982–1983 age: 20–64 (married women, 20-59)	Social class SMR	I II III-NM III-M IV V	300 1214 601 1834 899 421	1.14 0.99 1.05 1.03 1.01 1.16	143 629 304 906 445 129	1.07 1.04 1.16 1.08 1.05 0.98	Surveillance system statistics using 1980 census data as denominator. For social classification see Introduction. Women classified according to husband's occupation [OPCS, 1986]
UK (London) 1967–1987	Employment grade RR	Administrators Professionals Clerical Other	6 74 21 13	1 1.16 1.08 1.19			17 530 London civil servants, medically examined 1967–1969 and, followed-up until 1987 [Davey Smith & Marmot, 1991]
UK – Scotland 1959–1963 age: 20–64 (married women)	Social class SMR	I II III IV V		0.60 1.07 0.94 1.05 1.28		0.72 0.79 1.09 1.45 1.20	Surveillance system statistics using 1960 census data as denominator. UK Registrar General's classification. Women classified according to husband's occupation. ICD-9: 153-154 [Registrar General for Scotland, 1970]
USA – California 1949–1951 age: 25–64	Social class SMR	I II III IV V		1.11 1.12 1.07 0.86 0.92			Surveillance system statistics using 1950 census data as denominator. Social class indicator based on occupation. ICD-9:152-154 [Buell et al., 1960]
USA 1960 age: 25–64 (White)	Educational level SMR	College High school Elementary school <8 years of school		0.98 0.90 0.95 1.19		0.74 0.91 1.11 1.23	Record-linkage study using 1960 mortality and census data ICD-9: 152-153 [Kitagawa & Hauser, 1973]
USA (12 census samples) 1979–1985 age: 25+	Education SMR	College: 5+ y 4 y 1-3 y High school: 4 y 1-3 y Elementary school: 8 y 5-7 y 0-4 y		0.89 0.80 0.90 1.04 1.11 1.17 0.86 0.80		0.97 0.67 1.05 1.11 1.06 0.75 1.32 0.66	Census linkage [Rogot et al., 1992]

Table 11. Colon cancer incidence							
Study base	Indicators	Social scale	N	Male RR	N	Female RR	Study design
Canada (Montreal) 1979–1985 age: 35–70 (French)	Income level OR	High Middle Low		1 1.1 1.0			Population-based case–control study. Tertiles of total family income [Bourbonnais & Siemiatycki, in press]
Canada (Montreal) 1979–1985 age: 35–70 (French)	Education OR	High Middle Low		1.1 1.2 1.2			Population-based case–control study. Tertiles of years of education [Bourbonnais & Siemiatycki, in press]
Canada (Montreal) 1979–1985 age: 35–70 (French)	Occupational prestige scale OR	High Middle Low		1 1.2 1.2			Population-based case–control study. Tertiles of the occupational prestige scale [Bourbonnais & Siemiatycki, in press]
Colombia (Cali) 1971–1975 all ages	Social class RR	I II III		1 0.73 0.60		1 1.11 0.33	Data from 1973 census were used for rate denominator. Social class based on area of residence [Cuello et al., 1982]
Denmark 1970–1980 all ages	Occupational group RR	Self-employed Employees: I Employees: II Employees: III Employees: IV Skilled workers Unskilled workers	928 120 181 391 184 423 840	0.97 1.31 1.17 1.11 1.17 1.06 0.88	103 18 91 169 410 10 567	0.94 0.86 1.48 1.11 1.04 0.92 0.95	Record-linkage study using 1970 census and 1970–1980 incidence data. Employees classified according to educational level [Lynge & Thysesen, 1990]
Finland 1971–1985 birth cohort: 1906–1945	Social class SIR	Upper white-collar Lower white-collar Skilled workers Unskilled workers		1.42 1.15 0.97 0.65		1.10 1.17 0.96 0.86	Record-linkage study using 1970 census and 1971–1985 mortality-data. Social class based on occupation [Pukkala, 1993]
Greece (Athens) 1979–1980 all ages	Years of education	0 1-5 6-11 12+	12[a] 26[a] 49[a] 12[a]	1[a] 0.9[a] 1.0[a] 1.3[a]			Case–control study matched by sex [Papadimitriou et al., 1984]
Hong Kong 1971 age: 35–64 (Chinese)	Income level RR	Higher Medium Lower	3 16 27	1 0.84 0.54	– 13 28	– 1 0.81	Surveillance system statistics using 1971 census data as denominator. Income levels based on residence [Crowther et al., 1976]
Italy (Milano) 1983–1988 age: <75	Social class RR	I, II III IV, V	45[a] 155[a] 170[a]	1.34[a] 1.15[a] 1[a]			Hospital-based case–control study. UK Registrar General's social class classification. Adjusted by sex [Ferraroni et al., 1989]

Table 11. (Contd) Colon cancer incidence

Study base	Indicators	Social scale	N	Male RR	N	Female RR	Study design
Italy (Milano) 1983–1988 age: <75	Years of education RR	12+ 7–11 <7	96[a] 120[a] 239[a]	1.20[a] 1.05[a] 1[a]			Hospital-based case–control study. Adjusted by sex [Ferraroni et al., 1989]
Italy (Torino) 1985–1987 age: 20–69	Educational level OR	University High school Middle school Primary school	49 64 114 200	1 0.54 0.59 0.48	15 40 84 214	1 0.78 0.75 0.71	Record-linkage study between 1971 and 1981 censuses and 1985–1987 incidence data. ICD-9: 153-154 [Faggiano et al., 1994]
Italy (Torino) 1985–1987 age: 20–69	Occupational group OR	Managers Clerks Self-employed Manual workers	70 86 49 173	1 0.69 0.72 0.63	12 43 20 35	1 0.84 0.67 0.68	Record-linkage study between 1971 and 1981 censuses and 1985–1987 incidence data ICD-9: 153-154 [Faggiano et al., 1994]
Italy (Torino) 1985–1987 age: 20–69	Housing tenure OR	Owners Tenants	207 208	1 0.92	162 181	1 1.01	Record-linkage study between 1971 and 1981 censuses and 1985–1987 incidence data ICD-9:153-154 [Faggiano et al., 1994]
Sweden 1961–1970 all ages	Social class SIR	Employees: I Self-employed: II Indep. farmers: III White-collar: IV Blue-collar: V	1041 975 979 2903 5542	1.07 1.06 0.78 1.20 0.94	191 – 1970 1968	– 0.98 – 1.02 0.98	Record-linkage study between 1961 census and 1961–1970 incidence data. Social class indicator based on occupation [Vågerö & Persson, 1986]
Turkey (Istanbul) 1979–1984	Social class OR	Higher Medium Lower	7 21 65	1 0.5 0.7			Hospital-based case–control study. Social class indicator based on occupation [Dosemeci et al., 1989]
UK – England & Wales 1971–1981 all ages	Housing tenure SIR	Owner occupier Private rented Council tenant	289 108 134	1.01 1.06 0.93	387 140 197	1.02 0.96 0.98	Record-linkage study between 1971 census and 1971–1981 incidence data (1% sample). UK Registrar General's social class classification [Kogevinas, 1990]
USA (Du Pont) 1959–1967 all ages	Income level SIR	1 2 3 4 5	5 42 29 22 131	0.72 1.08 0.90 1.33 0.98			1959–1967 follow-up of 115 000 employees of the DuPont Co. ICD-7: 152-154 [Pell & D'Alonzo, 1970]
USA 1969–1971 all ages	Educational level OR	College Less		1.08 1		0.73 1	Case–control study based on US Third National Cancer Survey, using deaths from other causes as controls. [Williams & Horm, 1977]

Table 11. (Contd) Colon cancer incidence

Study base	Indicators	Social scale	N	Male RR	N	Female RR	Study design
USA 1969–1971 all ages	Family income level OR	>US$ 10 000 Less		0.98 1		0.78 1	Case–control study based on US Third National Cancer Survey, using deaths from other causes as controls. [Williams & Horm, 1977]

aData not stratified by sex.

Table 12. Rectum cancer mortality

Study base	Indicators	Social scale	N	Male RR	N	Female RR	Study design
Brazil (São Paulo) 1978–1982 age: 35–74	Years of education OR	12+ 9–11 1–8 <1		4.3 2.1 2.2 1		1.5 0.9 1.2 1	Case–control study using deaths from other causes as controls [Bouchardy et al., 1992]
Hungary 1970 age: 25–64	Years of education SMR	5+ 12–14 8–11 0–7		0.54 1.18 1.86 0.81		1.11 1.08 1.01 0.96	Surveillance system statistics using 1970 census data as denominator [Jozan, 1986]
Hungary 1980 age: 25–64	Years of education SMR	15+ 12–14 8–11 0–7		0.96 1.01 1.08 1.11		1.66 1.09 1.06 0.92	Surveillance system statistics using 1980 census data as denominator [Jozan, 1986]
Italy 1981–1982 age: 18–74	Education level RR	University High school Middle school Primary school Literate Illiterate	21 38 75 293 122 11	1 1.08 1.03 0.96 1.10 0.52	– 12 40 210 89 23	– 1 1.93 2.02 1.75 1.69	Record-linkage between 1981 census and mortality in the following six months [Faggiano et al., 1995]
Japan 1965–1982 age: 40+	Social class SMR	I, II III IV V		1.03 0.78 0.83 1		0.53 1.64 0.36 1	265 000 Japanese interviewed in 1965 and followed-up until 1982. Social class based on occupation. Reference category is farmers and miners [Hirayama, 1990]
New Zealand 1974–1978 age: 15–64	Social class RR	I II III-NM III-M IV V		1 0.61 0.89 0.84 0.64 1.09			Surveillance system statistics using 1976 census data as denominator. UK Registrar General's social class classification [Pearce & Howard, 1986]

Table 12. (Contd) Rectum cancer mortality

Study base	Indicators	Social scale	N	Male RR	N	Female RR	Study design
New Zealand 1984–1987 age: 15–64	Social class RR	I II III-NM III-M IV V	19 50 80 75 48 22	1 1.42 1.27 1.30 1.13 1.18			Surveillance system statistics using 1986 census data as denominator. UK Registrar General's social class classification [Pearce & Bethwaite, in press]
Spain 1980–1982	Occupational group PMR	Professionals and managers Manual workers Agricultural workers		1.20 1.11 0.84			Proportional analysis on death certificates [E. Regidor, pers. commun.]
Switzerland 1979–1982 age: 15–74	Social class SMR	I II III-NM III-M IV-V		0.78 0.86 1.31 0.95 0.91			Surveillance system statistics using 1980 census data as denominator. UK Registrar General's social class classification [C.E. Minder, pers. commun.]
UK – England and Wales 1910–1912 age: 15–64	Social class SMR	I II III IV V		0.99 0.95 1.07 0.98 1.00			Surveillance system statistics using 1910 census data as denominator. For social classification see Introduction [OPCS, 1919]
UK – England and Wales 1930–1932 age: 15–64	Social class SMR	I II III IV V				1.00 0.97 1.05 0.86 1.06	Surveillance system statistics using 1930 census data as denominator. For social classification see Introduction. (married women) Women classified according to husband's occupation [OPCS, 1938]
UK – England and Wales 1959–1963 age: 15–64 (married women)	Social class SMR	I II III IV V		0.79 0.89 1.06 0.98 1.20		0.69 0.81 1.07 1.06 1.32	Surveillance system statistics using 1960 census data as denominator. For social classification see Introduction [OPCS, 1971]
UK – England and Wales 1970–1972 age: 15–64 (married women)	Social class SMR	I II III-NM III-M IV V		0.84 0.90 1.03 1.14 1.06 1.08		0.99 0.98 0.98 1.15 1.05 1.35	Surveillance system statistics using 1970 census data as denominator. For social classification see Introduction. Women classified according to husband's occupation [OPCS, 1977]

Table 12. (Contd) Rectum cancer mortality

Study base	Indicators	Social scale	N	Male RR	N	Female RR	Study design
UK – Great Britain 1979–1980, 1982–1983 age: 20–64 (married women, 16–74)	Social class SMR	I II III-NM III-M IV V	174 838 405 1446 844 386	0.88 0.90 0.93 1.07 1.12 1.39	49 267 109 422 238 91	0.80 0.96 0.90 1.09 1.23 1.50	Surveillance system statistics using 1980 census data as denominator. For social classification see Introduction. Women classified according to husband's occupation [OPCS, 1986]
UK (London) 1967–1987	Employment grade RR	Administrators Professionals Clerical Other	4 26 6 6	1 0.50 0.50 0.88			17 530 London civil servants, medically examined 1967–1969 and, followed up until 1987. [Davey Smith & Marmot, 1991]

Table 13. Rectum cancer incidence

Study base	Indicators RR	Social scale	N	Male RR	N	Female RR	Study design
Canada (Montreal) 1979–1985 age: 35–70 (French)	Income level OR	High Middle Low		1 0.9 1.2			Population-based case–control study. Tertiles of total family income [Bourbonnais & Siemiatycki, in press]
Canada (Montreal) 1979–1985 age: 35–70 (French)	Education OR	High Middle Low		1 1.0 1.5			Population-based case–control study. Tertiles of the years of education [Bourbonnais & Siemiatycki, in press]
Canada (Montreal) 1979–1985 age: 35–70 (French)	Occupational prestige scale OR	High Middle Low		1 0.8 1.3			Population-based case–control study. Tertiles of the occupational prestige scale [Bourbonnais & Siemiatycki, in press]
Colombia (Cali) 1971–1975 all ages	Social class RR	I II III		1 1.64 1.55		1 3.75 2.25	Data from 1973 census were used for rate denominators. Social class based on area of residence [Cuello et al., 1982]
Denmark 1970–1980 all ages	Occupational group RR	Self-employed Employees: I Employees: II Employees: III Employees: IV Skilled workers Unskilled workers	885 81 129 344 165 365 869	0.99 0.96 0.93 1.06 1.16 1.01 0.97	57 15 41 98 222 5 354	0.91 1.28 0.88 1.15 1.01 0.84 1.04	Record-linkage study using 1970 census and 1970–1980 incidence data. Employees classified according to educational level [Lynge & Thygesen, 1990]

Table 13. Rectum cancer incidence

Study base	Indicators	Social scale	N	Male RR	N	Female RR	Study design
Finland 1971–1985 birth cohort: 1906–1945	Social class SIR	Upper white-collar Lower white-collar Skilled workers Unskilled workers		1.04 1.14 0.97 0.89		1.19 1.01 0.99 0.96	Record-linkage study using 1970 census and 1971–1985 incidence data. Social class based on occupation [Pukkala, 1993]
Hong Kong 1971 age: 35–64 (Chinese)	Income level RR	Higher Medium Lower	3 9 17	1 0.47 0.34	9 12	1 0.51	Surveillance system statistics using 1971 census data as denominator. Income levels based on residence [Crowther et al., 1976]
Italy (Milano) 1983–1988 age: <75	Social class RR	I, II III IV, V	24[a] 69[a] 146[a]	0.79[a] 0.60[a] 1[a]			Hospital-based case–control study. UK Registar General's social class classification. Adjusted by sex. [Ferraroni et al., 1989]
Italy (Milano) 1983–1988 age: <75	Years of education RR	12+ 7–11 <7	42[a] 66[a] 187[a]	0.63[a] 0.74[a] 1[a]			Hospital-based case–control study. Adjusted by sex [Ferraroni et al., 1989]
Sweden 1961–1970 all ages	Social class SIR	Employees: I Self-employed: II Indep. farmers: III White-collar: IV Blue-collar: V	737 721 876 342 4202	1.05 1.07 0.94 1.06 1.00	91 7 1047	– 0.91 – 0.84 1.02	Record-linkage study between 1961 census and 1961–1970 incidence data. Social class indicator based on occupation [Vågerö & Persson, 1986]
Turkey (Istanbul) 1979–1984	Social class OR	Higher Medium Lower	9 26 85	1 0.6 0.7			Hospital-based case–control study. Adjusted for sex. Social class based on occupation [Dosemeci et al., 1993]
UK – England and Wales 1971–1981 all ages	Housing tenure SIR	Owner occupier Private rented Council tenant	235 95 141	0.93 1.06 1.10	191 64 105	1.03 0.87 1.07	Record-linkage study between 1971 census and 1971–1981 incidence data (1% sample). UK Registrar General's social class classification [Kogevinas, 1990]
USA 1969–1971 all ages	Educational level OR	College Less		1.06 1		1.13 1	Case–control study based on US Third National Cancer Survey, using deaths from other causes as controls [Williams, 1977]
USA 1969–1971 all ages	Family income level OR	>US$ 10 000 Less		0.81 1		0.98 1	Case–control study based on US Third National Cancer Survey, using deaths from other causes as controls [Williams, 1977]

[a]Data not stratified by sex.

Table 14. Liver cancer mortality

Study base	Indicators	Social scale	N	Male RR	N	Female RR	Study design
Brazil (São Paulo) 1978–1982 age: 35–74	Years of education OR	12+ 9–11 1–8 <1		0.8 0.6 0.7 1		0.7 0.9 0.9 1	Case–control study using deaths from other causes as controls [Bouchardy et al., 1992]
Canada (urban area) 1971 all ages	Income	Q1 Q2 Q3 Q4 Q5		– 1.13 1.19 – 1.19			Surveillance system statistics using 1971 census data as denominators. Neighbourhood income quintiles as social indicator [R. Wilkins, pers. commun.]
Canada (urban area) 1986 all ages	Income	Q1 Q2 Q3 Q4 Q5		0.68 0.56 0.80 1.04 1.92			Surveillance system statistics using 1986 census data as denominators. Neighbourhood income quintiles as social indicator [R. Wilkins, pers. commun.]
Italy 1981–1982 age: 18–74	Educational level RR	University High school Middle school Primary school Literate Illiterate	32 52 143 589 219 53	1 1.03 1.36 1.34 1.26 1.24	– 34 52 292 186 61	– 1 0.72 0.93 1.08 1.09	Record-linkage between 1981 census and mortality in the following six months [Faggiano et al., 1995]
Japan 1965–1982 age: 40+	Social class SMR	I, II III IV V		1.14 1.44 1.22 1		0.89 1.10 1.08 1	265 000 Japanese interviewed in 1965 and followed-up until 1982. Social class based on occupation. Reference category is farmers and miners [Hirayama, 1990]
New Zealand 1974–1978 age: 15–64	Social class RR	I II III-NM III-M IV V		1 0.70 0.65 1.35 0.91 3.26			Surveillance system statistics using 1976 census data as denominator. UK Registrar General's social class classification [Pearce & Howard, 1986]
New Zealand 1984–1987 age: 15-64	Social class RR	I II III-NM III-M IV V		1 0.78 0.94 0.78 2.30 4.78			Surveillance system statistics using 1986 census data as denominator. UK Registrar General's social class classification [Pearce & Bethwaite, in press]
Switzerland (Vaud) 1977–1984 all ages	Social class PMR	I, II III IV, V		0.89 1.16 0.97		1.60 0.43 1.37	Proportional mortality study. UK Registrar General's social class classification (No. of males = 82; females = 18) [Levi et al., 1988]

Table 14. (Contd) Liver cancer mortality

Study base	Indicators	Social scale	N	Male RR	N	Female RR	Study design
Switzerland 1979–1982 age: 15–74	Social class SMR	I II III-NM III-M IV–V		0.61 0.94 1.18 1.05 0.89			Surveillance system statistics using 1980 census data as denominator. UK Registrar General's social class classification [C.E. Minder, pers. commun.]
UK – England and Wales 1921–1923 age: 15–64	Social class SMR	I II III IV V		0.70 1.01 0.99 1.00 1.08			Surveillance system statistics using 1920 census data as denominators. For social classification see Introduction [OPCS, 1927]
UK – England and Wales 1930–1932 age: 15–64 (married women)	Social class SMR	I II III IV V				0.76 0.95 0.97 1.10 1.15	Surveillance system statistics using 1930 census data as denominator. For social classification see Introduction. Women classified according to husband's occupation [OPCS, 1938]
UK – England and Wales 1970–1972 age: 15–64 (married women)	Social class SMR	I II III-NM III-M IV V		0.93 1.00 0.95 0.93 1.12 1.56		1.37 0.89 0.78 1.28 0.95 1.27	Surveillance system statistics using 1970 census data as denominator. For social classification see Introduction. Women classified according to husband's occupation [OPCS, 1977]
UK – Great Britain 1979–1980, 1982–1983 age: 20–64	Social class SMR	I II III-NM III-M IV V	42 177 87 368 194 120	0.87 0.78 0.81 1.12 1.05 1.76			Surveillance system statistics using 1980 census data as denominator. For social classification see Introduction. Women classified according to husband's occupation ICD: 155 [OPCS, 1986]

Table 15. Liver cancer incidence

Study base	Indicators	Social scale	N	Male RR	N	Female RR	Study design
Colombia (Cali) 1971–1975 all ages	Social class RR	I II III		1 3.67 2.33		1 1.54 0.96	Data from 1973 census were used for rate denominators. Social class based on area of residence [Cuello et al., 1982]
Denmark 1970–1980 all ages	Occupational group RR	Self-employed Employees: I Employees: II Employees: III Employees: IV Skilled workers Unskilled workers	117 13 15 71 28 76 161	0.77 0.91 0.64 1.29 1.16 1.25 1.06	4 2 7 15 29 2 52	0.49 1.24 1.03 1.23 0.89 2.09 1.11	Record-linkage study using 1970 census and 1970–1980 incidence data. Employees classified according to educational level [Lynge & Thygesen, 1990]
Finland 1971–1985 birth cohort: 1906–1945	Social class SIR	Upper white-collar Lower white-collar Skilled workers Unskilled workers		1.01 1.02 0.97 1.08		0.73 1.07 1.01 0.97	Record-linkage study using 1970 census and 1970–1980 incidence data. Social class based on occupation [Pukkala, 1993]
Italy (Milano) 1983–1988 age: <75	Social class RR	I, II III IV, V	7[a] 49[a] 70[a]	0.50[a] 0.87[a] 1[a]			Hospital-based case–control study. UK Register General's social class classification. Adjusted for sex [Ferraroni et al., 1989]
Italy (Milano) 1983–1988 age: <75	Years of education RR	12+ 7–11 <7	21[a] 36[a] 94[a]	0.54[a] 0.70[a] 1[a]			Hospital-based case–control study. Adjusted for sex [Ferraroni et al., 1989]
Sweden 1961–1970 all ages	Social class SIR	Employees: I Self-employed: II Indep. farmers: III White-collar: IV Blue-collar: V	379 410 216 1012 1949	1.12 1.25 0.55 1.13 0.97	 75 661 776	– 1.03 – 0.95 1.06	Record-linkage study between 1961–1970 census and incidence data. Social class indicator based on occupation [Vågerö & Persson, 1986]
UK – England and Wales 1971–1981 all ages	Housing tenure SIR	Owner occupier Private rented Council tenant	42 30 25	0.79 1.57 0.94			Record-linkage study between 1971 census and 1971–1981 incidence data (1% sample). UK Registrar General's social class classification [Kogevinas, 1990]
USA 1969–1971	Educational level OR	College Less		0.59 1		0.71 1	Case–control study based on US Third National Cancer Survey, using deaths from other causes as controls [Williams & Horm, 1977]
USA 1969–1971 all ages	Family income level OR	>US$ 10 000 Less		0.71 1		2.85 1	Case–control study based on US Third National Cancer Survey, using deaths from other causes as controls. [Williams & Horm, 1977]

Table 16. Pancreas cancer mortality

Study base	Indicators	Social scale	N	Male RR	N	Female RR	Study design
Brazil (São Paulo) 1978–1982 age: 35–74	Years of education OR	12+ 9–11 1–8 <1		1.1 1.0 0.9 1		0.7 0.9 0.9 1	Case–control study using deaths from other causes as controls [Bouchardy et al., 1992]
Canada (urban area) 1971 all ages	Income CMF	Q1 Q2 Q3 Q4 Q5		1.04 1.03 1.10 1.20		1.13 0.70 1.00 1.23 1.05	Surveillance system statistics using 1971 census data as denominator. Neighbourhood income quintiles as social indicator [R. Wilkins, pers. commun.]
Canada (urban area) 1986 all ages	Income CMF	Q1 Q2 Q3 Q4 Q5		1.09 0.80 0.96 0.91 1.22		1.12 0.85 1.71 1.12 1.12	Surveillance system statistics using 1986 census data as denominator. Neighbourhood income quintiles as social indicator [R. Wilkins, pers. commun.]
Italy 1981–1982 age: 18–74	Educational level RR	University High school Middle school Primary school Literate Illiterate	35 47 94 459 96 22	1 0.82 0.78 0.93 0.57 0.65	– 13 49 250 118 25	– 1 2.05 1.91 1.86 1.63	Record-linkage between 1981 census and mortality in the following six months [Faggiano et al, 1995]
Japan 1965–1982 age: 40+	Social class SMR	I, II III IV V		1.40 0.93 1.31 1		0.71 0.97 0.59 1	265 000 Japanese interviewed in 1965 and followed-up until 1982. Social class based on occupation. Reference category is farmers and miners [Hirayama, 1990]
New Zealand 1974–1978 age: 15–64	Social class RR	I II III-NM III-M IV V		1 1.13 1.39 1.16 0.95 1.59			Surveillance system statistics using 1976 census data as denominator. UK Registrar General's social class classification [Pearce, 1986]
New Zealand 1984–1987 age: 15–64	Social class RR	I II III-NM III-M IV V		1 1.24 1.32 1.32 1.54 0.93			Surveillance system statistics using 1986 census data as denominator. UK Registrar General's social class classification [Pearce & Bethwaite, in press]
Norway 1970–1973 age: 20–69	Social class CMF	A B C D E (farmers)		0.89 1.06 1.06 1.06 0.91			Surveillance system statistics using 1970 census data as denominator. Social class indicator based on occupation [Central Bureau of Statistics, 1976]

aData not stratified by sex.

Study base	Indicators	Social scale	N	Male RR	N	Female RR	Study design
Spain 1980–1982	Occupational group PMR	Professionals managers Manual workers Agricultural workers		1.23 1.07 0.90			Proportional analysis on death certificates [E. Regidor, pers. commun.]
Switzerland (Vaud) 1977–1984 all ages	Social class PMR	I, II III IV, V		1.12 0.84 1.07		1.12 0.77 1.25	Proportional mortality study. UK Registrar General's social class classification (No. of males = 113; females = 71) [Levi et al., 1988]
Switzerland 1979–1982 age: 15–74	Social class SMR	I II III-NM III-M IV-V		0.96 0.73 1.20 1.15 0.93			Surveillance system statistics using 1980 census data as denominator. UK Registrar General social-class classification [C.E. Minder, pers. commun.]
UK – England and Wales 1910–1912 age: 15–64	Social class SMR	I II III IV V		1.24 0.97 0.95 0.95 0.95			Surveillance system statistics using 1910 census data as denominator. For social classification see Introduction [OPCS, 1919]
UK – England and Wales 1930–1932 age: 15–64 (married women)	Social class SMR	I II III IV V		1.18 0.99 1.01 0.95 1.04		0.52 0.98 0.98 0.94 1.18	Surveillance system statistics using 1930 census data as denominator. For social classification see Introduction. Women classified according to husband's occupation [OPCS, 1938]
UK – England and Wales 1949–1953 age: 15–64	Social class SMR	I II III IV V		1.20 1.01 1.01 0.93 1.03			Surveillance system statistics using 1950 census data as denominator. For social classification see Introduction [OPCS, 1958]
UK – England and Wales 1970–1972 age: 15–64 (married women)	Social class SMR	I II III-NM III-M IV V		1.03 0.97 1.05 1.10 1.01 1.04		1.06 0.93 – 1.13 1.05 1.34	Surveillance system statistics using 1970 census data as denominator. For social classification see Introduction. Women classified according to husband's occupation [OPCS, 1977]
UK – Great Britain 1979–1980, 1982–1983 age: 20–64 (married women, 20–59)	Social class SMR	I II III-NM III-M IV V	163 876 437 1486 921 371	0.79 0.90 0.96 1.06 1.17 1.27	57 252 122 417 237 80	0.96 0.92 1.02 1.10 1.22 1.31	Surveillance system statistics using 1980 census data as denominator. For social classification see Introduction. Women classified according to husband's occupation [OPCS, 1986]

Table 16. (Contd) Pancreas cancer mortality

Study base	Indicators	Social scale	N	Male RR	N	Female RR	Study design
UK (London) 1967–1987	Employment grade RR	Administrators	4	1			17 530 London civil servants, medically examined 1967–1969 and followed-up until 1987 [Davey Smith & Marmot, 1991]
		Professionals	37	0.83			
		Clerical	14	1.04			
		Other	9	1.67			
USA – California 1949–1951 age: 25–64	Social class SMR	I		0.93			Surveillance system statistics using 1950 census data as denominators. Social class indicator based on occupation ICD: 162-163 [Buell, et al. 1960]
		II		0.88			
		III		1.00			
		IV		1.13			
		V		1.97			
USA (12 census samples) 1979–1985 age: 25+	Education SMR	College: 5+ y		0.78		0.73	Census linkage
		4 y		0.76		1.21	
		1-3 y		1.12		0.99	
		High school: 4 y		0.92		1.00	
		1-3 y		1.28		1.20	
		Elementary school: 8 y		0.98		0.82	
		5-7 y		1.09		1.01	
		0-4 y		0.92		0.95	[Rogot et al., 1992]

Table 17. Pancreas cancer incidence

Study base	Indicators	Social scale	N	Male RR	N	Female RR	Study design
Canada (Montreal) 1979–1985 age: 35–70 (French)	Income level OR	High		1			Population-based case–control study. Tertiles of total family income [Bourbonnais, in press]
		Middle		1.4			
		Low		1.7			
Canada (Montreal) 1979–1985 age: 35–70 (French)	Education OR	High		1			Population-based case–control study. Tertiles of years of education. [Bourbonnais, in press]
		Middle		1.3			
		Low		1.4			
Canada (Montreal) 1979–1985 age: 35–70 (French)	Occupational prestige scale OR	High		1			Population-based case–control study. Tertiles of the occupational prestige scale. [Bourbonnais, in press]
		Middle		1.1			
		Low		1.4			
Colombia (Cali) 1971–1975 all ages	Social class RR	I		1		1	Data from 1973 census were used for rate denominators. Social class based on area of residence. [Cuello, 1982]
		II		0.95		0.81	
		III		1.00		0.52	

Study base	Indicators	Social scale	N	Male RR	N	Female RR	Study design
Denmark 1970–1980 all ages	Occupational group RR	Self-employed Employees: I Employees: II Employees: III Employees: IV Skilled workers Unskilled workers	458 35 89 194 106 239 533	0.88 0.71 1.09 1.02 1.27 1.14 1.02	41 12 28 33 131 9 230	1.05 1.69 1.01 0.65 1.02 2.59 1.14	Record-linkage study using 1970 census and 1970–1980 incidence data. Employees classified according to educational level [Lynge, 1990]
Finland 1971–1985 birth cohort: 1906–1945	Social class SIR	Upper white-collar Lower white-collar Skilled workers Unskilled workers		0.95 0.95 1.00 1.08		1.14 1.09 0.92 1.06	Record-linkage study using 1970 census and 1971–1985 incidence data. Social class based on occupation [Pukkala, 1993]
Italy (Milano) 1983–1988 age: <75	Social class RR	I, II III IV, V	35[a] 61[a] 88[a]	1.87[a] 0.88[a] 1[a]			Hospital-based case–control study. UK Register General's social class classification Adjusted by sex [Ferraroni, 1989]
Italy (Milano) 1983–1988 age: <75	Years of education RR	12+ 7–11 <7	39[a] 53[a] 122[a]	0.85[a] 0.88[a] 1[a]			Hospital-based case–control study. Adjusted by sex [Ferraroni, 1989]
Sweden 1961–1970 all ages	Social class SIR	Employees: I Self employed: II Indep. farmers: III White-collar: IV Blue-collar: V	557 519 578 1407 3397	1.01 0.99 0.83 1.03 1.03	– 86 – 748 846	– 1.06 – 0.97 1.04	Record-linkage study between 1961 census and 1961–1970 incidence data. Social class indicator based on occupation [Vågerö, 1986]
UK – England and Wales 1971–1981 all ages	Housing tenure SIR	Owner occupier Private rented Council tenant	137 59 64	0.97 1.18 0.89	116 51 45	1.04 1.13 0.77	Record-linkage study between 1971 census and 1971–1981 incidence data (1% sample). UK Registrar General's social class classification [Kogevinas, 1990]
USA 1969–1971 all ages	Educational level OR	College Less		1.41 1		0.86 1	Case–control study based on US Third National Cancer Survey, using deaths from other causes as controls [Williams, 1977]
USA 1969–1971 all ages	Family income level OR	>US$ 10 000 Less		0.81 1		0.94 1	Case–control study based on US Third National Cancer Survey, using deaths from other causes as controls [Williams, 1977]

[a]Data not stratified by sex.

Table 18. Nose and nasal cavities cancer mortality

Study base	Indicators	Social scale	N	Male RR	N	Female RR	Study design
UK – England and Wales 1949–1953 age: 15–64 (married women)	Social class SMR	I II III IV V		0.90 0.91 1.04 0.86 1.17		1.20 0.88 0.91 1.14 1.37	Surveillance system statistics using 1950 census data as denominator. For social classification see Introduction. Women classified according to husband's occupation [OPCS, 1958]
UK – England and Wales 1959–1963 age: 15–64 (married women)	Social class SMR	I II III IV V		0.75 0.85 1.10 0.92 1.35		0.40 0.72 1.12 1.14 1.25	Surveillance system statistics using 1960 census data as denominator. For social classification see Introduction. [OPCS, 1971]
UK – England and Wales 1970–1972 age: 15–64 (married women)	Social class SMR	I II III-NM III-M IV V		0.71 0.85 0.90 1.01 1.00 1.97		0.84 0.78 0.89 1.14 1.54 0.82	Surveillance system statistics using 1970 census data as denominator. For social classification see Introduction. Women classified according to husband's occupation [OPCS, 1977]
UK – Great Britain 1979–1980, 1982–1983 age: 20–64 (married women, 20–59)	Social class SMR	I II III-NM III-M IV V	4 39 16 70 30 26	0.43 0.92 0.80 1.13 0.89 2.08	0 16 7 35 11 5	– 0.93 0.96 1.44 0.96 1.42	Surveillance system statistics using 1980 census data as denominator. For social classification see Introduction. Women classified according to husband's occupation [OPCS, 1986]

Table 19. Nose and nasal cavities cancer incidence

Study base	Indicators	Social scale	N	Male RR	N	Female RR	Study design
Denmark 1970–1980 all ages	Occupational group RR	Self-employed	38	0.80	2	0.98	Record-linkage study using 1970 census and 1970–1980 incidence data. Employees classified according to educational level
		Employees: I	2	0.40	0	–	
		Employees: II	9	1.03	3	1.56	
		Employees: III	17	0.88	8	2.58	
		Employees: IV	13	1.53	4	0.49	
		Skilled workers	37	1.71	0	–	
		Unskilled workers	44	0.89	13	1.16	[Lynge, 1990]
Finland 1971–1985 birth cohort: 1906–1945	Social class SIR	Upper white-collar		0.52		0.80	Record-linkage study using 1970 census and 1971–1985 incidence data. Social class based on occupation [Pukkala, 1993]
		Lower white-collar		0.87		0.82	
		Skilled workers		1.03		1.10	
		Unskilled workers		1.34		1.04	
Sweden 1961–1970 all ages	Social class SIR	Employees: I	77	0.95	–	–	Record-linkage study between 1961 census and 1961–1970 incidence data. Social class indicator based on occupation [Vågerö, 1986]
		Self-employed: II	40	0.95	7	1.60	
		Indep. farmers: III	53	0.92	–	–	
		White-collar: IV	108	0.93	37	0.78	
		Blue-collar: V	279	1.03	54	1.16	

Table 20. Larynx cancer mortality

Study base	Indicators	Social scale	N	Male RR	N	Female RR	Study design
Brazil (São Paulo) 1978–1982 age: 35–74	Years of education OR	12+		0.5			Case–control study using deaths from other causes as controls [Bouchardy, 1992]
		9–11		1.6			
		1–8		1.9			
		<1		1			
France 1975–1982 age: 45–54	Occupational group RR	Groups managers		0.21			A sample of about 1 000 000 of 1975 censused population followed-up until 1982. The scale shown represents a choice of the total scale [Desplanques, 1985]
		Intermediate		0.46			
		Self-employed		0.75			
		Clerks		1.17			
		Skilled workers		1.58			
		Unskilled workers		1.96			
France 1975–1982 age: 55–64	Occupational group RR	Managers		0.28			A sample of about 1 000 000 of 1975 censused population followed-up until 1982. The scale shown represents a choice of the total scale [Desplanques, 1985]
		Intermediate		0.65			
		Self-employed		0.81			
		Clerks		1.14			
		Skilled workers		1.49			
		Unskilled workers		1.97			
Italy 1981–1982 age: 18–74	Educational level RR	University	0	–			Record-linkage between 1981 census and mortality in the following six months [Faggiano, 1995]
		High school	33	1			
		Middle school	92	2.11			
		Primary school	503	2.80			
		Literate	153	2.69			
		Illiterate	36	3.30			

Table 20. (Contd) Larynx cancer mortality

Study base	Indicators	Social scale	N	Male RR	N	Female RR	Study design
Japan 1965–1982 age: 40+	Social class SMR	I, II III IV V		0.86 1.41 0.86 1		– 3.20 – 1	265 000 Japanese interviewed in 1965 and followed-up until 1982. Social class based on on occupation. Reference category is farmers and miners [Hirayama, 1990]
New Zealand 1974–1978 age: 15–64	Social class RR	I II III-NM III-M IV V		1 3.25 3.00 5.25 6.00 6.00			Surveillance system statistics using 1976 census data as denominator. UK Registrar General's social class classification [Pearce, 1986]
New Zealand 1984–1987 age: 15–64	Social class RR	I II III-NM III-M IV V	0 1 10 14 15 9	– 1 6.5 10.0 14.0 19.0			Surveillance system statistics using 1986 census data as denominator. UK Registrar General's social class classification [Pearce & Bethwaite, in press]
Spain 1980–1982	Occupational group PMR	Professionals managers Manual workers Agricultural workers		0.56 1.23 0.88			Proportional analysis on death certificates [E. Regidor, pers. commun.]
Switzerland 1979–1982 age: 15–74	Social class SMR	I II III-NM III-M IV-V		0.50 0.67 1.11 1.20 1.09			Surveillance system statistics using 1980 census data as denominator. UK Registrar General's social class classification [C.E. Minder, pers. commun.]
UK – England and Wales 1910–1912 age: 15–64	Social class SMR	I II III IV V		1.04 0.87 1.00 0.91 1.18			Surveillance system statistics using 1910 census data as denominator. For social classification see Introduction [OPCS, 1919]
UK – England and Wales 1930–1932 age: 15–64 (married women)	Social class SMR	I II III IV V		0.60 0.81 0.98 0.90 1.43		0.55 1.15 0.95 1.04 1.02	Surveillance system statistics using 1930 census data as denominator. For social classification see Introduction. Women classified according to husband's occupation [OPCS, 1938]

Table 20. (Contd) Larynx cancer mortality

Study base	Indicators	Social scale	N	Male RR	N	Female RR	Study design
UK – England and Wales 1959–1963 age: 15–64 (married women)	Social class SMR	I II III IV V				0.50 0.72 1.01 1.10 1.62	Surveillance system statistics using 1960 census data as denominator. For social classification see Introduction [OPCS, 1971]
UK – England and Wales 1970–1972 age: 15–64 (married women)	Social class SMR	I II III-NM III-M IV V		0.65 0.65 0.81 1.02 1.32 1.94		0.92 0.68 0.98 1.17 0.95 2.28	Surveillance system statistics using 1970 census data as denominator. For social classification see Introduction. Women classified according to husband's occupation [OPCS, 1977]
UK – Great Britain 1979–1980, 1982–1983 age: 20–64 (married women, 20–59)	Social class SMR	I II III-NM III-M IV V	18 138 50 360 251 138	0.39 0.63 0.49 1.14 1.41 2.10	2 11 4 43 26 11	0.40 0.48 0.40 1.34 1.58 2.12	Surveillance system statistics using 1980 census data as denominator. For social classification see Introduction. Women classified according to husband's occupation [OPCS, 1986]

Table 21. Larynx cancer incidence

Study base	Indicators	Social scale	N	Male RR	N	Female RR	Study design
Colombia (Cali) 1971–1975 all ages	Social class RR	I II III		1 0.62 0.40		1 0.83 0.83	Data from 1973 census were used for rate denominators. Social class based on area of residence [Cuello, 1982]
Denmark 1970–1980 all ages	Occupational group RR	Self-employed Employees: I Employees: II Employees: III Employees: IV Skilled workers Unskilled workers	241 24 29 125 55 187 364	0.76 0.76 0.56 1.03 1.06 1.43 1.14	6 0 2 10 21 1 49	1.07 – 0.40 1.18 0.90 1.60 1.46	Record-linkage study using 1970 census and 1970–1980 incidence data. Employees classified according to educational level [Lynge, 1990]
Finland 1971–1985 birth cohort: 1906–1945	Social class SIR	Upper white-collar Lower white-collar Skilled workers Unskilled workers		0.70 0.76 1.06 1.27		0.82 0.85 1.03 1.18	Record-linkage study using 1970 census and 1971–1985 incidence data. Social class based on occupation [Pukkala, 1993]

Table 21. (Contd) Larynx cancer incidence

Study base	Indicators	Social scale	N	Male RR	N	Female RR	Study design
France (Paris)	Educational level OR	Upper Medium Low		1 1.42 2.18			Hospital-based case–control study 1983–1991 [Leclerc, 1993]
Italy (Torino) 1985–1987 age: 20–69	Educational level OR	University High school Middle school Primary school	8 28 60 129	1 1.45 1.83 2.23			Record-linkage study between 1971 and 1981 censuses and 1985–1987 incidence data [Faggiano, 1994]
Italy (Torino) 1985–1987 age: 20–69	Occupational group OR	Managers Clerks Self-employed Manual workers	17 43 30 119	1 1.59 1.57 2.14			Record-linkage study between 1971 and 1981 censuses and 1985–1987 incidence data [Faggiano, 1994]
Italy (Torino) 1985–1987 age: 20–69	Housing tenure OR	Owners Tenants	89 130	1 1.48			Record-linkage study between 1971 and 1981 censuses and 1985–1987 incidence data [Faggiano, 1994]
Sweden 1961–1970 all ages	Social class SIR	Employees: I Self-employees: II Indep. farmers: III White-collars: IV Blue-collars: V	193 197 82 609 1135	1.01 1.13 0.42 1.15 1.01	– 9 – 34 46	– 2.73 – 0.72 1.21	Record-linkage study between 1961 census and incidence data. Social class indicator based on occupation [Vågerö, 1986]
Turkey (Istanbul) 1979–1984 all ages	Social class OR	Higher Medium Lower	11 170 597	1 3.2 4.1			Hospital-based case–control study. Adjusted for sex. Social class indicator based on occupation [Dosemeci, 1993]
UK – England and Wales 1971–1981 all ages	Housing tenure SIR	Owner occupier Private rented Council tenant	45 25 49	0.72 1.14 1.45			Record-linkage study between 1971 census and 1971–1981 incidence data (1% sample). UK Registrar General's social class classification [Kogevinas, 1990]
USA 1969–1971 all ages	Educational level OR	College Less		0.73 1		1.56 1	Case–control study based on US Third National Cancer Survey, using deaths from other causes as controls [Williams, 1977]
USA 1969–1971 all ages	Family income level OR	>US$ 10 000 Less		0.88 1		0.77 1	Case–control study based on US Third National Cancer Survey, using deaths from other causes as controls [Williams, 1977]

Table 22. Lung cancer mortality

Study base	Indicators	Social scale	N	Male RR	N	Female RR	Study design
Brazil (São Paulo) 1978–1982 age: 35–74	Years of education OR	12+ 9–11 1–8 <1		2.6 2.3 1.6 1		1.8 1.2 1.2 1	Case–control study using deaths from other causes as controls [Bouchardy, 1992]
Canada (urban area) 1971 all ages	Income CMF	Q1 Q2 Q3 Q4 Q5		0.82 0.67 0.98 1.02 1.41		0.84 0.88 0.71 1.32 1.20	Surveillance system statistics using 1971 census data as denominator. Neighbourhood income quintiles as social indicator [R. Wilkins, pers. commun.]
Canada (urban area) 1986 all ages	Income CMF	Q1 Q2 Q3 Q4 Q5		0.69 0.84 0.93 1.09 1.47		0.75 0.92 0.89 1.02 1.39	Surveillance system statistics unsing 1986 census data as denominator. Neighbourhood income quintiles as social indicator [R. Wilkins, pers. commun.]
Denmark 1970–1975 age: 20–64	Occupational group SMR	Employees: I Employees: II Employees: III Employees: IV Skilled workers Unskilled workers		0.51 0.68 1.07 1.16 1.35 1.15		– – 1.01 1.02 – 1.15	Record-linkage study using 1970 census and 1970–1975 mortality data. Employees classified according to the educational level [Danmarks Statistik, 1979]
Finland 1969–1972 age: 15–64 (married women)	Social class CMF	Upper white-collar Lower white-collar Skilled workers Unskilled workers Farmers		0.53 0.89 1.24 1.23 0.60		0.88 1.05 1.23 1.53 0.82	Surveillance system statistics using 1970 census data as denominator. Social class indicator based on occupation [Näyhä, 1977]
Finland 1971–1985 age: 35–64	Social class RR	Upper white-collar Lower white-collar Skilled workers Unskilled workers Farmers		1 1.70 2.84 3.52 1.89		1 1.33 1.60 – 0.68	Record-linkage study using 1970, 1975 and 1980 censuses and 1971–1985 mortality data. Social class indicator based on occupation [Valkonen, 1990]
France 1975–1982 age: 45–54	Occupational group Managers RR	Intermediate Self-employed Clerks Skilled workers Unskilled workers		0.68 1.02 0.93 1.11 1.35 1.35			A sample of about 1 000 000 of 1975 censused population was followed-up until 1982. The scale shown represents a choice of the total scale [Desplanques, 1985]
France 1975–1982 age: 55–64	Occupational group RR	Managers Intermediate Self-employed Clerks Skilled workers Unskilled workers		0.74 0.06 0.96 1.11 1.28 1.28			A sample of about 1 000 000 of 1975 censused population was followed-up until 1982. The scale shown represents a choice of the total scale [Desplanques, 1985]
Hungary 1970 age: 25–64	Years of education SMR	15+ 12–14 8–11 0–7		0.83 0.99 1.77 0.85		0.48 0.82 1.38 0.93	Surveillance system statistics using 1970 census data as denominators [Jozan, 1971]

Table 22. (Contd) Lung cancer mortality

Study base	Indicators	Social scale	N	Male RR	N	Female RR	Study design
Hungary 1980 age: 25–64	Years of education SMR	15+ 12–14 8–11 0–7		0.65 0.84 1.11 1.06		2.31 1.48 1.14 0.82	Surveillance system statistics using 1980 census data as denominators [Jozan, 1971]
Italy 1981–1982 age: 18–74	Educational level RR	University High school Middle school Primary school Literate Illiterate	386 338 898 4006 1170 201	1 1.12 1.42 1.53 1.34 1.13	0 43 94 449 185 45	– 1 1.20 1.09 0.99 1.04	Record-linkage between 1981 census and the mortality of the following six months [Faggiano, 1995]
Japan 1965–1982 age: 40+	Social class SMR	I, II III IV V		0.81 1.08 1.10 1		1.26 1.23 0.97 1	265 000 Japanese interviewed in 1965 and followed-up until 1982. Social class based on occupation. Reference category: farmers and miners [Hirayama, 1990]
The Netherlands 1959–1961 age: 40–64	Social class SMR	I IIa IIb III IV Miners		0.59 0.60 0.82 1.44 1.12 1.42			Surveillance system statistics using 1960 census data as denominator [Van Reek, 1986]
New Zealand 1974–1978 age: 15–64	Social class RR	I II III-NM III-M IV V		1 1.42 1.52 2.21 2.08 3.06			Surveillance system statistics using 1976 census data as denominator. UK Registrar General's social class classification [Pearce, 1976]
New Zealand 1984–1987 age: 15–64	Social class RR	I II III-NM III-M IV V	41 128 243 258 304 136	1 1.53 1.63 1.92 3.05 3.03			Surveillance system statistics using 1971 census data as denominator. UK Registrar General's social class classification [Pearce & Bethwaite, in press]
Norway 1970–1973 age: 20–69	Social class CMF	A B C D E (farmers)		0.81 1.13 0.74 1.22 0.40			Surveillance system statistics using 1970 census data as denominator. Social class indicator based on occupation [Central Bureau of Statistics, 1976]

Table 22. (Contd) Lung cancer mortality

Study base	Indicators	Social scale	N	Male RR	N	Female RR	Study design
Portugal 1980–1982 age: 20–64	Occupational group RR	Managers		1.00			Surveillance system statistics using 1980 census data as denominator
		Professionals		1.48			
		Clerks		1.33			
		Sales workers		1.24			
		Service workers		0.95			
		Agriculture, forestry, and fishery,		1.01			
		Other manual workers		1.61			[M. Giraldes, pers. commun.] SMRs calculated by authors
Spain 1980–1982 age: 30–64	Occupational group PMR	Professionals and managers		0.92			Proportional analysis on death certificates
		Manual workers		1.13			
		Agricultural workers		0.98			[E. Regidor, unpublished]
Switzerland (Vaud) 1977–1984 all ages	Social class PMR	I, II		0.90		1.07	Proportional mortality study. UK Registrar General's social class classification (No. of males = 907; females =103) [Levi, 1988]
		III		1.04		0.88	
		IV, V		1.07		1.13	
Switzerland 1979–1982 age: 15–74	Social class SMR	I		0.42			Surveillance system statistics using 1980 census data as denominator. UK Registrar age: General's social class classification [C.E. Minder, unpublished]
		II		0.72			
		III-NM		0.91			
		III-M		1.33			
		IV, V		1.10			
UK – England and Wales 1910–1912 age: 15–64	Social class SMR	I		0.94			Surveillance system statistics using 1910 census data as denominator. For social classification see Introduction [OPCS, 1919]
		II		1.06			
		III		1.06			
		IV		0.83			
		V		1.22			
UK – England and Wales 1930–1932 age: 15–64 (married women)	Social class SMR	I		1.07		1.00	Surveillance system statistics using 1930 census data as denominator. For social classification see Introduction. Women classified according to husband's occupation [OPCS, 1938]
		II		0.96		1.00	
		III		1.01		1.10	
		IV		0.91		0.82	
		V		1.12		0.91	
UK – England and Wales 1949–1953 age: 15–64	Social class SMR	I		0.81			Surveillance system statistics using 1950 census data as denominator. For social classification see Introduction [OPCS, 1958]
		II		0.82			
		III		1.07			
		IV		0.91			
		V		1.18			

Table 22. (Contd) Lung cancer mortality

Study base	Indicators	Social scale	N	Male RR	N	Female RR	Study design
UK – England and Wales 1970–1972 age: 15–64 (married women)	Social class SMR	I II III-NM III-M IV V		0.53 0.68 0.84 1.18 1.23 1.43		0.73 0.82 0.89 1.18 1.25 1.34	Surveillance system statistics using 1970 census data as denominator. For social classification see Introduction. Women classified according to husband's occupation [OPCS, 1977]
UK – England and Wales 1971–1975 age: 15–64	Social class SMR	I II III-NM III-M IV V		0.66 0.64 0.82 1.01 1.09 1.46			Record-linkage study (longitudinal study) between 1971 census and 1971–1975 mortality data for a 1% sample of the total population. UK Registrar General's social class classification [OPCS, 1990]
UK England and Wales 1976–1981 age: 15–64	Social class SMR	I II III-NM III-M IV V		0.42 0.68 0.83 1.08 1.31 1.24			Record-linkage study (Longitudinal Study) between 1971 census and 1976–1981 mortality data for a 1% sample of the total population. UK Registrar General's social class classification [OPCS, 1990]
UK – Great Britain 1979–1980, 1982–1983 age: 20–64 (married women, 20–59)	Social class SMR	I II III-NM III-M IV V	742 5163 3116 14266 8594 4503	0.43 0.63 0.80 1.20 1.26 1.78	147 991 485 2314 1348 524	0.50 0.73 0.81 1.22 1.38 1.70	Surveillance system statistics using 1980 census data as denominator. For social classification see Introduction. Women classified according to husband's occupation [OPCS, 1986]
UK (London) 1967–1987	Employment grade RR	Administrators Professionals Clerical Other	12 207 108 110	1 1.42 2.58 3.69			17 530 Londoner civil servants, undergoing a medical examination 1967–1969, followed-up until 1987 [Davey Smith, 1991]
UK – Scotland 1949–1953 age: 20–64	Social class SMR	I II III IV V		1.04 0.81 1.15 0.86 1.09			Surveillance system statistics using 1950 census data as denominator. UK Registrar General social class classification [Registrar General for Scotland, 1956]
UK –Scotland 1959–1963 age: 20–64 (married women)	Social class SMR	I II III IV V		0.61 0.70 1.04 0.98 1.51		0.59 0.81 0.78 0.85 1.38	Surveillance system statistics using 1960 census data as denominator. UK Registrar General classification. Women classified according to husband's occupation. [Registrar General for Scotland, 1970]

Table 22. (Contd) Lung cancer mortality

Study base	Indicators	Social scale	N	Male RR	N	Female RR	Study design
USA – California 1949–1951 age: 25–64	Social class SMR	I II III IV V		0.85 0.77 1.12 1.08 1.12			Surveillance system statistics using 1950 census data as denominator. Social class indicator based on occupation. ICD 162-163 [Buell, 1960]
USA 1960 age: 25–64 (White)	Educational level SMR	College High school Elementary school <8 years of school		0.61 0.95 1.14 1.18		0.90 0.94 0.96 1.23	Record-linkage study using 1960 mortality and census data. [Kitagawa, 1973]
USA (12 census samples) White population 1979–1985 age: 25+	Education SMR	College: 5+ y 4 y 1-3 y High school: 4 y 1-3 y Elementary school: 8 y 5-7 y 0-4 y		0.51 0.69 0.85 0.92 1.27 1.11 1.31 1.04		0.41 0.64 1.13 1.05 1.23 0.83 0.89 0.92	Census linkage [Rogot et al., 1992]
USA (12 census samples) Black population 1979–1985 age: 25+	Education SMR	College: 1-3 y High school: 4 y 1-3 y Elementary school: 8 y 5-7 y 0-4 y		1.00 0.84 1.15 1.23 0.84 1.18			Census linkage [Rogot et al., 1992]

Table 23. Lung cancer incidence

Study base	Indicators	Social scale	N	Male RR	N	Female RR	Study design
Canada (Montreal) 1979–1985 age: 35–70 (French)	Income level OR	High Middle Low		1 2.5 3.7			Population-based case–control study. Tertiles of total family income [Bourbonnais, in press]
Canada (Montreal) 1979–1985 age: 35–70 (French)	Education OR	High Middle Low		1 1.6 2.3			Population-based case–control study. Tertiles of years of education [Bourbonnais, in press]
Canada (Montreal) 1979–1985 age: 35–70 (French)	Occupational prestige scale OR	High Middle Low		1 2.2 3.8			Population-based case–control study. Tertiles of the occupational prestige scale [Bourbonnais, in press]

Table 23. (Contd) Lung cancer incidence

Study base	Indicators	Social scale	N	Male RR	N	Female RR	Study design
Colombia (Cali) 1971–1975 all ages	Social class RR	I II III		1 0.71 0.72		1 1.00 0.93	Data from 1973 census were used for rate denominators. Social class based on area of residence [Cuello, 1982]
Denmark 1970–1980 all ages	Occupational group RR	Self-employed Employees: I Employees: II Employees: III Employees: IV Skilled workers Unskilled workers	2674 191 338 1241 589 1725 3773	0.80 0.61 0.67 1.04 1.13 1.31 1.13	83 7 43 124 363 17 707	0.88 0.37 0.56 0.91 0.99 1.78 1.27	Record-linkage study using 1970 census and 1970–1980 incidence data. Employees classified according to educational level [Lynge, 1990]
Denmark (Copenhagen) 1971–1988 age: 35–74	Social class RR	I II III IV V	755 414 776 1684 469	1 1.6 2.3 2.9 3.7			5249 Male employees, aged 40–59, followed up 1971–1988. Social class indicator based on occupation [Hein et al., 1992]
Greece (Athens) 1978–1986	Education low/high					1.30	[Trichopoulos et al., 1981]
Greece (Greater Athens) 1987–1989	Education 0 y/7 y					0.56	[Katsoyanni et al., 1991]
Finland 1971–1985 birth cohort: 1906–1945	Social class SIR	Upper white-collar Lower white-collar Skilled workers Unskilled workers		0.45 0.73 1.07 1.38		1.08 1.11 0.92 1.04	Record-linkage study using 1970 census and 1971–1985 incidence data. Social class based on occupation [Pukkala, 1993]
Italy (Torino) 1985–1987 age: 20–69	Educational level OR	University High school Middle school Primary school	31 100 223 475	1 1.66 2.03 2.47	0 26 44 91	– 1 0.74 0.62	Record-linkage study between 1971 and 1981 censuses and 1985–1987 incidence data [Faggiano, 1995]
Italy (Torino) 1985–1987 age: 20–69	Occupational group OR	Managers Clerks Self-employed Manual workers	71 154 124 487	1 1.30 1.80 1.81	0 34 18 16	– 1 0.86 0.45	Record-linkage study between 1971 and 1981 censuses and 1985–1987 incidence data [Faggiano, 1995]
Italy (Torino) 1985–1987 age: 20–69	Housing tenure OR	Owners Tenants	344 573	1 1.44	57 99	1 1.44	Record-linkage study between 1971 and 1981 censuses and 1985–1987 incidence data [Faggiano, 1995]
Sweden 1961–1970 all ages	Social class SIR	Employees: I Self-employees: II Indep. farmers: III White-collars: IV Blue-collars: V	1760 1598 719 4274 10638	1.08 1.04 0.41 0.97 1.10	0 84 0 860 925	– 1.09 – 0.92 1.09	Record-linkage study between 1961 census and 1961–1970 incidence data. Social class indicator based on occupation [Vågerö, 1971]

Table 23. (Contd) Lung cancer incidence

Study base	Indicators	Social scale	N	Male RR	N	Female RR	Study design
Turkey (Istanbul) 1979–1984 all ages	Social class OR	Higher Medium Lower	64 294 790	1 1.0 0.9			Hospital-based case–control study. Social-class indicator based on occupation [Dosemeci, 1993]
UK – England and Wales 1971–1981 all ages	Social class SIR	I II III-NM III-M IV V	48 383 250 888 584 313	0.48 0.77 0.86 1.05 1.16 1.24			Record-linkage study between 1971 census and 1971–81 incidence data (1% sample). UK Registrar General's social class classification [Kogevinas, 1990]
UK – England and Wales 1971–1981 all ages	Housing tenure SIR	Owner occupier Private rented Council tenant	1062 574 1016	0.75 1.16 1.38	304 153 246	0.83 1.11 1.22	Record-linkage study between 1971 census and 1971–81 incidence data (1% sample). UK Registrar General's social class classification [Kogevinas, 1990]
USA (Du Pont) 1959–1967 all ages	Income level SIR	1 2 3 4 5	7 44 42 17 171	0.79 0.93 1.07 0.84 1.04			1959–1967 follow-up of 115 000 employees of the Du Pont Co. [Pell, 1970]
USA 1969–1971 all ages	Educational level OR	College Less		0.62 1		0.60 1	Case–control study based on US Third National Cancer Survey, using deaths from other causes as controls. [Williams, 1977]
USA 1969–1971 all ages	Family income level OR	>US$ 10 000 Less		0.89 1		0.60 1	Case–control study based on US Third National Cancer Survey, using deaths from other causes as controls. [Williams, 1977]

Table 24. Bone cancer mortality

Study base	Indicators	Social scale	N	Male RR	N	Female RR	Study design
Brazil (São Paulo) 1978–1982 age: 35–74	Years of education OR	12+ 9–11 1–8 <1		1.7 1.6 1.8 1		0.3 0.6 0.9 1	Case–control study using deaths from other causes as controls [Bouchardy, 1992]
New Zealand 1974–1978 age: 15–64	Social class RR	I II III-NM III-M IV V		1 0.83 1.0 0.83 0.83 2.0			Surveillance system statistics using 1976 census data as denominator. UK Registrar General's social class classification [Pearce, 1986]
New Zealand 1984–1987 age: 15–64	Social class RR	I II III-NM III-M IV V		– – 1 5.0 1.0 –			Surveillance system statistics using 1986 census data as denominator. UK Registrar General's social class classification [Pearce & Bethwaite, in press]
Switzerland 1979–1982 age: 15–74	Social class SMR	I II III-NM III-M IV-V		0.27 1.16 0.86 0.95 0.78			Surveillance system statistics using 1980 census data as denominator. UK Registrar General's social class classification [C.E. Minder, unpublished]
UK – England and Wales 1921–1923 age: 15–64	Social class SMR	I II III IV V		0.73 1.14 1.05 0.95 0.86			Surveillance system statistics using 1920 census data as denominator. For social classification see Introduction [OPCS, 1927]
UK – England and Wales 1949–1953 age: 15–64	Social class SMR	I II III IV V		1.30 0.89 1.08 0.86 0.96			Surveillance system statistics using 1950 census data as denominator. For social classification see Introduction. Women classified according to husband's occupation [OPCS, 1958]
UK – England and Wales 1959–1963 age: 15–64 (married women)	Social class SMR	I II III IV V		0.74 0.87 1.09 0.91 1.12		1.10 0.81 1.09 0.92 1.28	Surveillance system statistics using 1960 census data as denominator. For social classification see Introduction [OPCS, 1971]
UK – England and Wales 1970–1972 age: 15–64 (married women)	Social class SMR	I II III-NM III-M IV V		0.95 0.89 0.91 1.08 1.02 1.12		2.04 0.75 1.06 1.14 0.83 1.63	Surveillance system statistics using 1970 census data as denominator. For social classification see Introduction. Women classified according to husband's occupation [OPCS, 1977]

Table 24. (Contd) Bone cancer mortality

Study base	Indicators	Social scale	N	Male RR	N	Female RR	Study design
UK – Great Britain 1979–1980, 1982–1983 age: 20–64	Social class SMR	I II III-NM III-M IV V	13 59 28 104 53 24	0.85 0.95 0.86 1.05 1.09 1.31			Surveillance system statistics using 1980 census data as denominator. For social classification see Introduction. Women classified according to husband's occupation [OPCS, 1986]

Table 25. Bone cancer incidence

Study base	Indicators	Social scale	N	Male RR	N	Female RR	Study design
Denmark 1970–1980 all ages	Occupational group RR	Self-employed Employees: I Employees: II Employees: III Employees: IV Skilled workers Unskilled workers	31 5 14 6 22 34 	1.21 0.51 0.53 1.00 0.72 0.99 0.99	0 0 3 3 10 1 12	– – 1.37 0.92 1.09 3.05 1.21	Record-linkage study using 1970 census and 1970–1980 incidence data. Employees classified according to educational level [Lynge, 1990]
Finland 1971–1985 birth cohort: 1906–1945	Social class SIR	Upper white-collar Lower white-collar Skilled workers Unskilled workers		1.09 1.01 1.02 0.85		0.61 1.26 1.00 0.75	Record-linkage study using 1970 census and 1971–1985 incidence data. Social class based on occupation [Pukkala, 1993]
Sweden 1961–1970 all ages	Social class SIR	Employees: I Self-employed: II Indep. farmers: III White-collar: IV Blue-collar: V	21 30 35 87 195	0.74 1.21 1.04 1.03 0.99	– 5 – 51 44	– 1.39 – 1.01 1.02	Record-linkage study between 1961 census and 1961–1970 incidence data. Social class indicator based on occupation [Vågerö, 1986]

Table 26. Cancer of the connective tissue mortality

Study base	Indicators	Social scale	N	Male RR	N	Female RR	Study design
UK – England and Wales 1959–1963 age: 15–64 (married women)	Social class SMR	I II III IV V		1.08 0.95 1.09 0.93 0.97		0.71 1.12 0.98 1.11 0.87	Surveillance system statistics using 1960 census data as denominator. For social classification see Introduction [OPCS, 1971]
UK – England and Wales 1970–1972 age: 15–64 (married women)	Social class SMR	I II III-NM III-M IV V		0.80 0.88 0.89 1.05 1.17 1.00		1.48 1.12 0.90 0.95 0.94 1.01	Surveillance system statistics using 1970 census data as denominator. For social classification see Introduction. Women classified according to husband's occupation [OPCS, 1977]
UK – Great Britain 1979–1980, 1982–1983 age: 20–64	Social class SMR	I II III-NM III-M IV V	21 86 53 193 64 24	0.87 0.84 1.08 1.26 0.83 1.19			Surveillance system statistics using 1980 census data as denominator. For social classification see Introduction. Women classified according to husband's occupation [OPCS, 1986]

Table 27. Cancer of the connective tissue – incidence

Study base	Indicators	Social scale	N	Male RR	N	Female RR	Study design
Denmark 1970–1980 all ages	Occupational group RR	Self-employed Employees: I Employees: II Employees: III Employees: IV Skilled workers Unskilled workers	60 8 18 22 9 38 59	1.06 1.23 1.38 0.84 0.69 1.12 0.92	0 1 6 7 30 0 23	– 1.07 1.23 0.89 1.34 – 0.87	Record-linkage study using 1970 census and 1970–1980 incidence data. Employees classified according to educational level [Lynge, 1990]
Finland 1971–1985 birth cohort: 1906–1945	Social class SIR	Upper white-collar Lower white-collar Skilled workers Unskilled workers		1.18 0.97 0.99 0.98		1.22 1.05 0.99 0.87	Record-linkage study using 1970 census and 1971–1985 incidence data. Social class based on occupation [Pukkala, 1993]
Sweden 1961–1970 all ages	Social class SIR	Employees: I Self employed: II Indep. farmers: III White-collar: IV Blue-collar: V	97 76 115 319 648	0.95 0.83 0.98 1.12 0.98	– 15 – 203 184	– 0.99 – 0.97 1.01	Record-linkage study between 1961 census and 1961–1970 incidence data. Social class indicator based on occupation [Vågerö, 1986]

Table 27. (Contd) Cancer of the connective tissue – incidence

Study base	Indicators	Social scale	N	Male RR	N	Female RR	Study design
USA 1969–1971 all ages	Educational level OR	College Less		0.56 1		0.65 1	Case–control study based on US Third National Cancer Survey, using deaths from other causes as controls [Williams, 1977]
USA 1969–1971 all ages	Family income level OR	>US$ 10 000 Less		0.96 1		1.96 1	Case–control study based on US Third National Cancer Survey, using deaths from other causes as controls [Williams, 1977]

Table 28. Malignant melanoma mortality

Study base	Indicators	Social scale	N	Male RR	N	Female RR	Study design
Brazil (São Paulo) 1978–1982 age: 35–74	Years of education OR	12+ 9–11 1–8 <1		8.0 6.3 2.1 1		0.6 2.3 2.0 1	Case–control study using deaths from other causes as controls [Bouchardy, 1992]
Italy 1981–1982 age: 18–74	Education level RR	University High school Middle school Primary school Literate Illiterate	0 27 20 82 17 4	– 1 0.58 0.73 0.51 0.61	0 19 17 67 18 6	– 1 0.59 0.62 0.42 0.62	Record-linkage between 1981 census and the mortality in the following six months [Faggiano, 1995]
New Zealand 1974–1978 age: 15–64	Social class RR	I II III-NM III-M IV V		1 1.18 0.94 0.84 0.56 0.53			Surveillance system statistics using 1976 census data as denominator. UK Registrar General's social class classification [Pearce, 1986]
New Zealand using 1986 1984–1987 age: 15–64	Social class RR	I II III-NM III-M IV V		1 1.23 1.13 1.25 1.20 0.45			Surveillance system statistics census data as denominator. UK Registrar General's social class classification [Pearce and Bethwaite, in press]
Switzerland (Vaud) 1977–1984 all ages	Social class PMR	I, II III IV, V		1.39 0.82 0.86		1.38 0.70 0.98	Proportional mortality study. UK Registrar General's social class classification. ICD-9: 172-173 (No. of males = 180; females = 50). [Levi, 1988]

Table 28. (Contd) Malignant melanoma mortality

Study base	Indicators	Social scale	N	Male RR	N	Female RR	Study design
Switzerland 1979–1982 age: 15–74	Social class SMR	I II III-NM III-M IV-V		0.79 0.87 1.20 1.27 0.71			Surveillance system statistics using 1980 census data as denominator. UK Registrar General's social class classification. ICD-9: 172-173 [C.E. Minder, unpublished]
UK – England and Wales 1949–1953 age: 15–64 (married women)	Social class SMR	I II III IV V		1.45 0.98 1.07 0.81 0.85		1.45 1.11 1.04 0.71 0.95	Surveillance system statistics using 1950 census data as denominator. For social classification see Introduction. Women classified according to husband's occupation [OPCS, 1958]
UK – England 1959–1963 age: 15–64 (married women)	Social class SMR	I II III IV V		1.50 1.16 1.00 0.95 0.84		0.90 1.04 1.13 0.77 0.95	Surveillance system statistics and Wales using 1960 census data as denominator. For social classification see Introduction [OPCS, 1971]
UK – England and Wales 1970–1972 age: 15–64 (married women)	Social class SMR	I II III-NM III-M IV V		1.37 1.35 1.21 0.88 0.73 1.05		1.74 1.34 1.04 0.97 0.95 0.67	Surveillance system statistics using 1970 census data as denominator. For social classification see Introduction. Women classified according to husband's occupation [OPCS, 1977]
UK – Great Britain 1979–1980, 1982–1983 age: 20–64 (married women, 20–59)	Social class SMR	I II III-NM III-M IV V	79 320 153 315 163 55	1.33 1.26 1.34 0.85 0.89 0.82	48 179 70 251 102 36	1.21 1.07 1.01 1.07 0.97 1.12	Surveillance system statistics using 1980 census data as denominator. For social classification see Introduction. Women classified according to husband's occupation [OPCS, 1986]

Table 29. Malignant melanoma incidence

Study base	Indicators	Social scale	N	Male RR	N	Female RR	Study design
Colombia (Cali) 1971–1975 all ages	Social class RR	I II III		1 2.33 1.33		1 1.22 0.22	Data from 1973 census were used for rate denominators. Social class based on area of residence [Cuello, 1982]
Denmark 1970–1980 all ages	Occupational group RR	Self-employed Employees: I Employees: II Employees: III Employees: IV Skilled workers Unskilled workers	234 53 113 206 80 161 244	0.83 1.50 1.56 1.47 1.21 0.93 0.76	44 13 77 117 270 7 257	1.05 1.22 1.33 1.31 1.06 0.84 0.87	Record-linkage study using 1970 census and 1970–1980 incidence data. Employees classified according to educational level [Lynge, 1990]
Finland 1971–1985 birth cohort: 1906–1945	Social class SIR	Upper white-collar Lower white-collar Skilled workers Unskilled workers		1.63 1.11 0.92 0.69		1.29 1.18 0.88 0.91	Record-linkage study using 1970 census and 1971–1985 incidence data. Social class based on occupation [Pukkala, 1993]
Sweden 1961–1970 all ages	Social class SIR	Employees: I Self-employed: II Indep. farmers: III White-collar: IV Blue-collar: V	283 196 220 1200 1557	1.05 0.84 0.82 1.38 0.86	0 45 0 937 522	– 0.93 – 1.14 0.82	Record-linkage study between 1961 census and 1961–1970 incidence data. Social class indicator based on occupation [Vågerö, 1986]
Turkey (Istanbul) 1979–1984	Social class OR	Higher Medium Lower	2 11 39	1 1.3 1.6			Hospital-based case–control study. Social class indicator based on occupation [Dosemeci, 1993]
UK – England and Wales 1971–1981 all ages	Housing tenure SIR	Owner occupier Private rented Council tenant	49 9 32	1.02 0.57 1.21			Record-linkage study between 1971 census and 1971–1981 incidence data (1% sample). UK Registrar General's social class classification [Kogevinas, 1990]
USA 1969–1971 all ages	Educational level OR	College Less		2.27 1		0.79 1	Case–control study based on US Third National Cancer Survey, using deaths from other causes as controls [Williams, 1977]
USA 1969–1971 all ages	Family income level OR	>US$ 10 000 Less		1.88 1		0.94 1	Case–control study based on US Third National Cancer Survey, using deaths from other causes as controls [Williams, 1977]

Table 30. Female breast cancer mortality

Study base	Indicators	Social scale	N	RR	Study design
Brazil (São Paulo) 1978–1982 age: 35–74	Years of education OR	12+ 9–11 1–8 <1		2.6 2.4 1.6 1	Case–control study using deaths from causes as controls [Bouchardy, 1992]
Canada (urban area) 1971 all ages	Income CMF	Q1 Q2 Q3 Q4 Q5		1.08 1.00 1.01 0.95 0.99	Surveillance system statistics using 1971 census data as denominators. Neighbourhood income quintiles as social indicator [K. Wilkins, unpublished]
Canada (urban area) 1986 all ages	Income CMF	Q1 Q2 Q3 Q4 Q5		1.06 0.98 0.99 1.04 0.95	Surveillance system statistics using 1986 census data as denominators. Neighbourhood income quintiles as social indicator [K. Wilkins, unpublished]
Denmark 1970–1975 age: 20–64	Occupational groups SMR	Employees: I Employees: II Employees: III Employees: IV Skilled workers Unskilled workers		– 1.49 1.06 1.05 – 0.87	Record-linkage study using 1970 census and 1970–1975 mortality data. Employees classified according to educational level [Danmarks Statistik, 1979]
Finland 1969–1972 age: 15–64 (married women)	Social class CMF	Upper white-collar Lower white-collar Skilled workers Unskilled workers Farmers		1.47 1.07 0.94 0.79 0.74	Surveillance system statistics using 1970 census data as denominators. Social class indicator based on occupation [Näyhä, 1977]
Finland 1971–1985 age: 35–66	Social class RR 20	Upper white-collar Lower white-collar Skilled workers Unskilled workers Farmers	≥20 ≥20 ≥20 ≥20	1 0.85 0.69 – 0.60	Record-linkage study using 1970, 1975 and 1980 censuses and 1971–1985 mortality data [Valkonen, 1990]
Hungary 1970 age: 25–64	Years of education SMR	15+ 12–14 8–11 0–7		1.84 1.66 1.31 0.86	Surveillance system statistics using 1970 census data as denominators [Jozan, 1986]
Hungary 1980 age: 25–64	Years of education SMR	15+ 12–14 8–11 0–7		1.85 1.68 1.06 0.77	Surveillance system statistics using 1980 census data as denominators [Jozan, 1986]
Italy 1981–1982 age: 18–74	Educational level RR	University High school Middle school Primary school Literate Illiterate	46 186 338 1328 479 95	1 1.17 1.04 0.86 0.74 0.56	Record-linkage between 1981 census and mortality in the following six months. [Faggiano, 1995]

Table 30. (Contd) Female breast cancer mortality

Study base	Indicators	Social scale	N	RR	Study design
Japan 1965–1982 age: 40+	Social class SMR	I, II III IV V		2.33 1.45 1.58 1	265 000 Japanese interviewed in 1965 and followed-up until 1982. Social class based on occupation. Reference category: farmers and miners. [Hirayama, 1990]
Norway 1970–1973 age: 20–69	Social class CMF	A B C D E (farmers)		1.39 1.09 1.10 0.72 0.66	Surveillance system statistics using 1970 census data as denominator. Social class indicator based on occupation. [Central Bureau of Statistics, 1976]
Portugal 1980–1982 ages: 20–64	Occupational group RR	Mangers Professionals Clerks Sales workers Service workers Agriculture, forestry and fishery Other manual workers		1.00 2.68 1.95 1.08 0.41 0.11 1.10	Surveillance system statistics using 1980 census data as denominator [M. Giraldes, pers. commun.; SMRs calculated by authors]
Switzerland (Vaud) 1977–1984 all ages	Social class PMR	I, II III IV, V		1.01 1.11 0.80	Proportional mortality study. UK Registrar General's social class classification (No. = 275) [Levi, 1988]
UK – England and Wales 1930–1932 age: 15–64 (married women)	Social class SMR	I II III IV V		1.38 1.16 1.03 0.84 0.82	Surveillance system statistics using 1930 census data as denominator. For social classification see Introduction. Women classified according to husband's occupation [OPCS, 1938]
UK – England and Wales 1949–1953 age: 15–64	Social class SMR	I II III IV V		1.37 1.10 1.04 0.84 0.85	Surveillance system statistics using 1950 census data as denominator. For social classification see Introduction [OPCS, 1958]
UK – England and Wales 1970–1972 age: 15–64 (married women)	Social class SMR	I II III-NM III-M IV V		1.17 1.12 1.10 1.09 1.03 0.92	Surveillance system statistics using 1970 census data as denominators. For social classification see Introduction. Women classified according to husband's occupation. [OPCS, 1977]
UK – Great Britain 1979–1980, 1982–1983 age: 20–64 (married women, 20–59)	Social class SMR	I II III-NM III-M IV V	815 3498 1591 4784 2359 711	1.09 1.05 1.14 1.04 1.07 1.04	Surveillance system statistics using 1980 census data as denominators. For social classification see Introduction. Women classified according to husband's occupation [OPCS, 1986]

Table 30. (Contd) Female breast cancer mortality

Study base	Indicators	Social scale	N	RR	Study design
UK – Scotland 1959–1963 age: 20–64 (married women)	Social class SMR	I II III IV V		1.15 1.11 1.01 0.89 1.02	Surveillance system statistics using 1960 census data as denominator. UK Registrar General's classification. Women classified according to husband's occupation [Registrar General for Scotland, 1970]
USA 1960 age: 25–64 (White)	Education level SMR	College High school Elementary school <8 years of school		1.11 1.03 0.98 0.87	Record-linkage study using 1960 mortality and census data. [Kitagawa, 1973]
USA (12 census samples) White population age: 25+	Education SMR	College High school Elementary school		1.67 1.15 1.07 1.01 0.84 1.10 0.72 0.61	Census linkage [Rogot et al., 1992]

Table 31. Female breast cancer incidence

Study base	Indicators	Social scale	N	RR	Study design
Colombia (Cali) 1971–1975 all ages	Social class RR	I II III		1 0.64 0.45	Data from 1973 census were used for rate denominators. Social class based on area of residence [Cuello, 1982]
Denmark 1970–1980 all ages	Occupational group RR	Self-employed Employees: I Employees: II Employees: III Employees: IV Skilled workers Unskilled workers	459 124 507 792 2015 50 2147	1.16 1.38 1.25 1.20 1.08 1.00 0.84	Record-linkage study using 1970 census and 1970–1980 incidence data. Employees classified according to educational level [Lynge, 1990]
Finland 1971–1985 birth cohort: 1906–1945	Social class SIR	Upper white-collar Lower white-collar Skilled workers Unskilled workers		1.42 1.19 0.90 0.82	Record-linkage study using 1970 census and 1971–1985 incidence data. Social class based on occupation [Pukkala, 1993]

Table 31. (Contd) Female breast cancer incidence

Study base	Indicators	Social scale	N	RR	Study design
Hong Kong 1971 age: 35–64 (Chinese)	Income level RR	Higher Medium Lower	10 39 55	1 0.62 0.33	Surveillance system statistics using 1971 census data as denominator. Income levels based on residence [Crowther et al., 1976]
Italy (Torino) 1985–1987 age: 20–69	Educational level OR	University High school Middle school Primary school	44 161 316 589	1 0.86 0.84 0.66	Record-linkage study between 1971 and 1981 censuses and 1985–1987 incidence data. [Faggiano, 1994]
Italy (Torino) 1985–1987 age: 20–69	Occupational group OR	Managers Clerks Self-employed Manual workers	38 187 68 159	1 0.89 0.67 0.77	Record-linkage study between 1971 and 1981 censuses and 1985–1987 incidence data [Faggiano, 1994]
Italy (Torino) 1985–1987 age: 20–69	Housing tenure OR	Owners Tenants	503 580	1 0.98	Record-linkage study between 1971 and 1981 censuses and 1985–1987 incidence data [Faggiano, 1994]
Sweden 1961–1970 all ages	Social class SIR	Employees: I Self-employed: II Indep. farmers: III White-collar: IV Blue-collar: V	– 729 – 10040 6708	– 1.08 – 1.12 0.86	Record-linkage study between 1961 census and 1961–1970 incidence data. Social class indicator based on occupation [Vågerö, 1986]
Turkey (Istanbul) 1979–1984	Social class OR	Higher Medium Lower	18 86 127	1 0.4 0.4	Hospital-based case–control study. Social class indicator based on occupation [Dosemeci, 1993]
UK – England and Wales 1971–1981 all ages	Housing tenure SIR	Owner occupier Private rented Council tenant	1074 348 571	1.02 0.93 0.99	Record-linkage study between 1971 census and 1971–1981 incidence data (1% sample). UK Registrar General's social class classification [Kogevinas, 1990]
USA 1969–1971 all ages	Educational level OR	College Less		1.44 1	Case–control study based on US Third National Cancer Survey, using deaths from other causes as controls. [Williams, 1977]
USA 1969–1971 all ages	Family income level OR	>US $10 000 Less		1.30 1	Case–control study based on US Third National Cancer Survey, using deaths from other causes as controls [Williams, 1977]

Table 32. Cervical and endometrial cancer mortality

Study base	Indicators	Social scale	N	Cervix	N	Corpus	Study design
Brazil (São Paulo) 1978–1982 age: 35–74	Years of education OR	12+ 9–11 1–8 <1		0.2 0.4 0.7 1		2.3 1.9 2.0 1	Case–control study using deaths from other causes as controls [Bouchardy, 1992]
Canada (urban area) 1971 all ages	Income CMF	Q1 Q2 Q3 Q4 Q5				0.40 0.63 0.98 1.23 1.65	Surveillance system statistics using 1971 census data as denominator. Neighbourhood income quintiles as social indicator. ICD-9: 179-182 [R. Wilkins, unpublished]
Canada (urban area) 1986 all ages	Income CMF	Q1 Q2 Q3 Q4 Q5				0.52 0.96 0.94 0.92 1.60	Surveillance system statistics using 1986 census data as denominators. Neighbourhood income quintiles as social indicator. ICD-9: 179-182 [R. Wilkins, unpublished]
Denmark 1970–1975 age: 20–64	Occupational group SMR	Employees: I Employees: II Employees: III Employees: IV Skilled workers Unskilled workers				– 0.82 0.93 0.98 – 1.15	Record-linkage study using 1970 census and 1970–1975 mortality data. Employees classified according to educational level [Danmarks Statistik, 1979]
Finland 1969–1972 age: 15–64 (married women)	Social class CMF	Upper white-collar Lower white-collar Skilled workers Unskilled workers Farmers		0.68 1.06 1.35 1.33 0.57		0.84 1.03 0.99 0.97 1.17	Surveillance system statistics using 1970 census data as denominator. Social class indicator based on occupation [Näyhä, 1977]
Finland 1971–1985 age: 35–64 (married women)	Social class RR	Upper white-collar <20 Lower white-collar ≥20 Skilled workers ≥20 Unskilled workers Farmers ≥20		1 2.60 3.71 – 1.72			Record-linkage study using 1970, 1975 and 1980 censuses and 1971–1985 mortality data. Social class indicator based on occupation [Valkonen, 1990]
Italy 1981–1982 age: 18–74	Education level RR	University High school Middle school Primary school Literate Illiterate			0 52 113 512 240 102	– 1 1.35 1.23 1.20 1.76	Record-linkage between 1981 census and mortality in the following six months [Faggiano, 1995]
Japan 1965–1982 age: 40+	Social class SMR	I, II III IV V		0.70 1.21 0.99 1			265 000 Japanese interviewed in 1965 and followed up until 1982. Social class based on occupation. Reference category is farmers and miners [Hirayama, 1990]

Table 32. (Contd) Cervical and endometrial cancer mortality

Study base	Indicators	Social scale	N	Cervix	N	Corpus	Study design
UK – England and Wales 1949–1953 age: 15–64 (married women)	Social class SMR	I II III IV V		0.64 0.75 0.99 1.05 1.34		1.03 0.93 1.06 0.92 0.99	Surveillance system statistics using 1950 census data as denominator. For social classification see Introduction. Women classified according to husband's occupation [OPCS, 1958]
UK – England and Wales 1959–1963 age: 15–64 (married women)	Social class SMR	I II III IV V		0.34 0.64 1.00 1.16 1.81		1.00 0.94 1.03 0.99 1.22	Surveillance system statistics using 1960 census data as denominator. For social classification see Introduction [OPCS, 1971]
UK – England and Wales 1970–1972 age: 15–64 (married women)	Social class SMR	I II III-NM III-M IV V		0.42 0.66 0.69 1.20 1.40 1.61		0.75 0.97 1.03 1.16 1.20 1.02	Surveillance system statistics using 1970 census data as denominator. For social classification see Introduction. Women classified according to husband's occupation [OPCS, 1977]
UK – Great Britain 1979–1980, 1982–1983 age: 20–64 (married women, 20–59)	Social class SMR	I II III-NM III-M IV V	47 399 193 1073 544 268	0.33 0.65 0.75 1.25 1.37 2.20	20 128 64 185 104 39	0.73 1.01 1.15 1.05 1.15 1.37	Surveillance system statistics using 1930 census data as denominator. For social classification see Introduction. Women classified according to husband's occupation [OPCS, 1986]
UK – Scotland 1959–1963 age: 20–64 (married women)	Social class SMR	I II III IV V		0.31 0.49 0.94 1.37 2.19		1.16 1.09 0.92 1.15 1.20	Surveillance system statistics using 1960 census data as denominator. UK Registrar General's classification. Women classified according to husband's occupation [Registrar General for Scotland, 1970]
USA 1960 age: 25–64 (white)	Educational level SMR	College High school Elementary school <8 years of school		0.68 0.88 1.11 1.42			Record-linkage study using 1960 mortality and census data. Uterus and ovary [Kitagawa, 1973]

Table 33. Cervical and endometrial cancer incidence

Study base	Indicators	Social scale	N	Cervix	N	Corpus	Study design
Colombia (Cali) 1971–1975 all ages	Social class RR	I II III		1 2.44 2.95		1 1.03 0.55	Data from 1973 census were used for rate denominator. Social class based on area of residence. [Cuello, 1982]
Denmark 1970–1980 all ages	Occupational group RR	Self-employed Employees: I Employees: II Employees: III Employees: IV Skilled workers Unskilled workers	157 15 85 184 737 25 1349	1.12 0.43 0.49 0.67 0.94 1.07 1.38	108 33 71 161 465 15 574	1.02 1.56 0.86 1.07 1.15 1.43 0.92	Record-linkage study using 1970 census and 1970–1980 incidence data. Employees classified according to educational level. [Lynge, 1990]
Finland 1971–1985 birth cohort: 1906–1945	Social class SIR	Upper white-collar Lower white-collar Skilled workers Unskilled workers		0.63 0.86 1.02 0.83		1.18 1.11 0.98 0.84	Record-linkage study using 1970 census and 1971–1985 incidence data. Social class based on occupation [Pukkala, 1993]
Italy (Torino) 1985–1987 age: 20–69	Education level OR	University High school Middle school Primary school	0 10 32 92	– 1 1.77 2.33	0 28 50 118	– 1 0.93 0.81	Record-linkage study between 1971 and 1981 censuses and 1985–1987 incidence data. [Faggiano, 1995]
Italy (Torino) 1985–1987 age: 20–69	Occupational group OR	Managers Clerks Self-employed Manual workers	0 16 11 24	– 1 1.88 2.15	0 27 8 34	– 1 0.73 1.35	Record-linkage study between 1971 and 1981 censuses and 1985–1987 incidence data [Faggiano, 1995]
Italy (Torino) 1985–1987 age: 20–69	Housing tenure OR	Owners Tenants	34 97	1 2.27	100 93	1 0.98	Record-linkage study between 1971 and 1981 censuses and 1985–1987 incidence data [Faggiano, 1995]
Sweden 1961–1970 all ages	Social class SIR	Employees: I Self employed: II Indep. farmers: III White-collar: IV Blue-collar: V	– 154 – 2189 2242	– 0.98 – 0.84 1.22	– 159 – 2248 1829	– 0.94 – 1.07 0.93	Record-linkage study between 1961 census and 1961–1970 incidence data. Social class indicator based on occupation [Vågerö, 1986]
Turkey (Istanbul) 1979–1984	Social class OR	Higher Medium Lower	1 13 44	1 0.8 2.3	1 8 22	1 0.8 1.6	Hospital-based case–control study. Social class indicator based on occupation [Dosemeci, 1993]
UK – England and Wales 1971–1981 all ages	Housing tenure SIR	Owner occupier Private rented Council tenant	133 79 140	0.72 1.25 1.34	183 53 94	1.07 0.84 0.99	Record-linkage study between 1971 census and 1971–1981 incidence data (1% sample). UK Registrar General's social class classification [Kogevinas, 1990]

Table 33. (Contd) Cervical and endometrial cancer incidence

Study base	Indicators	Social scale	N	Cervix	N	Corpus	Study design
USA 1969–1971 all ages	Educational level OR	College Less		0.30 1		1.24 1	Case–control study based on US Third National Cancer Survey, using deaths from other causes as controls [Williams, 1977]
USA 1969–1971 all ages	Family income level OR	>US$ 10 000 Less		0.45 1		1.23 1	Case–control study based on US Third National Cancer Survey, using deaths from other causes as controls [Williams, 1977]

Table 34. Ovarian cancer mortality

Study base	Indicators	Social scale	N	RR	Study design
Brazil (São Paulo) 1978–1982 age: 35–74	Years of education OR	12+ 9–11 1–8 <1		2.2 2.3 1.7 1	Case–control study using deaths from other causes as controls [Bouchardy, 1992]
Canada (urban area) 1971 all ages	Income CMF	Q1 Q2 Q3 Q4 Q5		1.19 0.73 1.24 0.89 0.94	Surveillance system statistics using 1971 census data as denominator. Neighbourhood income quintiles as social indicator [K. Wilkins, unpublished]
Canada (urban area) 1986 all ages	Income CMF	Q1 Q2 Q3 Q4 Q5		0.97 0.97 1.03 0.93 1.07	Surveillance system statistics using 1986 census data as denominator. Neighbourhood income quintiles as social indicator [K. Wilkins, unpublished]
Italy 1981–1982 age: 18–74	Education level RR	University High school Middle school Primary school Literate Illiterate	13 35 71 323 113 19	1 0.72 0.70 0.67 0.58 0.45	Record-linkage between 1981 census and mortality in the following six months [Faggiano, 1994]
Japan 1965–1982 age: 40+	Social class SMR	I, II III IV V		1.16 0.77 1.34 1	265 000 Japanese interviewed in 1965 and followed up until 1982. Social class based on occupation. Reference category's farmers and miners. [Hirayama, 1990]
UK – England and Wales 1930–1932 age: 15–64 (married women)	Social class SMR	I II III IV V		1.43 1.16 1.02 0.77 0.83	Surveillance system statistics using 1930 census data as denominator. For social classification see Introduction. Women classified according to husband's occupation [OPCS, 1938]

Table 34. (Contd) Ovarian cancer mortality

Study base	Indicators	Social scale	N	RR	Study design
UK – England and Wales 1949–1953 age: 15–64 (married women)	Social class SMR	I II III IV V		1.57 1.06 1.06 0.80 0.82	Surveillance system statistics using 1950 census data as denominator. For social classification see Introduction [OPCS, 1958]
UK – England and Wales 1970–1972 age: 15–64 (married women)	Social class SMR	I II III-NM III-M IV V		1.18 1.04 1.08 1.12 1.08 0.93	Surveillance system statistics using 1970 census data as denominator. For social classification see Introduction. Women classified according to husband's occupation [OPCS, 1977]
UK – Great Britain 1979–1980, 1982–1983 age: 20–64 (married women, 20–59)	Social class SMR	I II III-NM III-M IV V	212 944 439 1327 728 224	1.04 1.02 1.11 1.03 1.15 1.13	Surveillance system statistics using 1980 census data as denominator. For social classification see Introduction. Women classified according to husband's occupation. Ovary, unoccupied: 0.13 [OPCS, 1986]
UK – Scotland 1959–1963 age: 20–64 (married women)	Social class SMR	I II III IV V		1.04 1.22 1.00 1.02 0.92	Surveillance system statistics using 1960 census data as denominator. UK Registrar General's classification. Women classified according to husband's occupation [Registrar General for Scotland, 1970]
USA (12 census samples) White population 1979–1985 age: 25+	Education SMR	College High school Elementary school 		1.15 0.84 1.48 0.88 0.93 1.14 0.82 0.71	Census linkage [Rogot et al.]

Table 35. Ovarian cancer incidence

Study base	Indicators	Social scale	N	RR	Study design
Colombia (Cali) 1971–1975 all ages	Social class RR	I II III		1 0.85 0.76	Data from 1973 census were used for rate denominators. Social class based on area of residence. [Cuello, 1982]
Denmark 1970–1980 all ages	Occupational group RR	Self-employed Employees: I Employees: II Employees: III Employees: IV Skilled workers Unskilled workers	113 21 103 169 494 7 662	1.07 0.93 1.06 1.01 1.07 0.55 1.00	Record-linkage study using 1970 census and 1970–1980 incidence data. Employees classified according to educational level [Lynge, 1990]
Finland 1971–1985 birth cohort: 1906–1945	Social class SIR	Upper white-collar Lower white-collar Skilled workers Unskilled workers		1.00 1.08 0.97 0.97	Record-linkage study using 1970 census and 1971–1985 incidence data. Social class based on occupation [Pukkala, 1993]
Greece (Greater Athens) 1980–1981	Sociocultural indicator low/high			1.31	[Franceschi, 1991]
Sweden 1961–1970 all ages	Social class SIR	Employees: I Self employed: II Indep. farmers: III White-collar: IV Blue-collar: V		– 1.06 – 1.00 0.99	Record-linkage study between 1961 census and 1961–1970 incidence data. Social class indicator based on occupation [Vågerö, 1986]
Turkey (Istanbul) 1979–1984	Social class OR	Higher Medium Lower	2 16 31	1.0 0.7 0.5	Hospital-based case–control study. Social class indicator based on occupation [Dosemeci, 1993]
UK – England and Wales 1971–1981 all ages	Housing tenure SIR	Owner occupier Private rented Council tenant	243 62 105	1.14 0.81 0.88	Record-linkage study between 1971 census and 1971–1981 incidence data (1% sample). UK Registrar General's social class classification [Kogevinas, 1990]
USA 1969–1971 all ages	Educational level OR	College Less		1.12 1	Case–control study based on US Third National Cancer Survey, using deaths from other causes as controls [Williams, 1977]
USA 1969–1971 all ages	Family income level OR	>US$ 10 000 Less		1.10 1	Case–control study based on US Third National Cancer Survey, using deaths for other causes as controls. [Williams, 1977]

Table 36. Prostate cancer mortality

Study base	Indicators	Social scale	N	RR	Study design
Brazil (São Paulo) 1978–1982 age: 35–74	Years of education OR	12+ 9–11 1–8 <1		0.9 0.7 0.8 1	Case–control study using deaths from other causes as controls [Bouchardy, 1992]
Canada (urban area) 1971 all ages	Income CMF	Q1 Q2 Q3 Q4 Q5		0.79 1.41 1.23 0.72 0.85	Surveillance system statistics using 1971 census data as denominator. Neighbourhood income quintiles as social indicator [R. Wilkins, unpublished]
Canada (urban area) 1986 all ages	Income CMF	Q1 Q2 Q3 Q4 Q5		1.00 0.88 1.08 0.92 1.16	Surveillance system statistics using 1986 census data as denominator. Neighbourhood income quintiles as social indicator [R. Wilkins, unpublished]
Denmark 1970–1975 age: 20–64	Occupational group SMR	Employees: I Employees: II Employees: III Employees: IV Skilled workers Unskilled workers		0.90 1.02 1.06 1.16 1.17 0.96	Record-linkage study using 1970 census and 1970–1975 mortality data. Employees classified according to educational level [Danmarks Statistik, 1979]
Finland 1969–1972 age: 15–64 (married women)	Social class CMF	Upper white-collar Lower white-collar Skilled workers Unskilled workers Farmers		1.29 1.00 0.81 1.03 0.87	Surveillance system statistics using 1970 census data as denominator. Social class indicator based on occupation [Näyhä, 1977]
Hungary 1970 age: 25–64	Years of education SMR	15+ 12–14 8–11 0–7		0.94 1.35 1.67 0.87	Surveillance system statistics using 1970 census data as denominator [Jozan, 1986]
Hungary 1980 age: 25–64	Years of education SMR	15+ 12–14 8–11 0–7		1.59 1.16 1.13 0.83	Surveillance system statistics using 1980 census data as denominator [Jozan, 1986]
Italy 1981–1982 age: 18–74	Educational level RR	University High school Middle school Primary school Literate Illiterate	21 47 104 441 193 38	1 1.57 1.60 1.38 1.22 1.02	Record-linkage between 1981 census and mortality in the following six months [Faggiano, 1995]
Japan 1965–1982 age: 40+	Social class SMR	I, II III IV V		0.88 0.83 0.78 1	265 000 Japanese interviewed in 1965 and followed up until 1982. Social class based on occupation. Reference category's farmers and miners. [Hirayama, 1990]

Table 36. (Contd) Prostate cancer mortality

Study base	Indicators	Social scale	N	RR	Study design
New Zealand 1974–1978 age: 15–64	Social class RR	I		1	Surveillance system statistics using 1976 census data as denominator. UK Registrar General's social class classification [Pearce, 1986]
		II		1.02	
		III-NM		0.58	
		III-M		1.08	
		IV		0.66	
		V		1.23	
New Zealand 1984–1987 age: 15–64	Social class RR	I		1	Surveillance system statistics using 1986 census data as denominator. UK Registrar General's social class classification [Pearce and Bethwaite, in press]
		II		0.82	
		III-NM		0.71	
		III-M		0.77	
		IV		1.38	
		V		0.89	
Norway 1970–1973 age: 20–69	Social class CMF	A		0.91	Surveillance system statistics using 1970 census data as denominator. Social class indicator based on occupation [Central Bureau of Statistics, 1976]
		B		1.27	
		C		0.99	
		D		0.90	
		E (farmers)		1.20	
Spain 1980–1982	Occupational group PMR	Professionals managers		0.92	Proportional analysis on death certificates
		Manual workers			
		Agricultural workers		1.13	
				0.98	[E. Regidor, unpublished]
Switzerland 1979–1982 age: 15–74	Social class SMR	I		0.65	Surveillance system statistics using 1980 census data as denominator. UK Registrar General's social class classification [C.E. Minder, unpublished]
		II		0.93	
		III-NM		1.17	
		III-M		1.02	
		IV-V		0.84	
UK – England and Wales 1910–1912 age: 15–64	Social class SMR	I		1.44	Surveillance system statistics using 1910 census data as denominator. For social classification see Introduction [OPCS, 1919]
		II		1.02	
		III		0.96	
		IV		0.90	
		V		0.75	
UK – England and Wales 1930–1932 age: 15–64	Social class SMR	I		1.11	Surveillance system statistics using 1930 census data as denominator. For social classification see Introduction. Women classified according to husband's occupation [OPCS, 1938]
		II		0.98	
		III		1.06	
		IV		0.88	
		V		1.06	
UK – England and Wales 1970–1972 age: 15–64	Social class SMR	I		0.91	Surveillance system statistics using 1970 census data as denominator. For social classification see Introduction. Women classified according to husband's occupation [OPCS, 1977]
		II		0.89	
		III-NM		0.99	
		III-M		1.15	
		IV		1.06	
		V		1.15	

Table 36. (Contd) Prostate cancer mortality

Study base	Indicators	Social scale	N	RR	Study design
UK – Great Britain 1979–1980, 1982–1983 age: 20–64	Social class SMR	I II III-NM III-M IV V	80 527 259 831 426 179	0.77 1.04 1.03 1.12 0.97 1.09	Surveillance system statistics using 1980 census data as denominator. For social classification see Introduction. Women classified according to husband's occupation Ovary, unoccupied: 0.13 [OPCS, 1986]
UK (London) 1967–191987	Employment grade RR	Administrators Professionals Clerical Other	5 62 11 10	1 0.95 0.57 0.85	17 530 London civil servants, medically examined 1967–1969, and followed up until 1987. [Davey Smith, 1991]
USA –California 1949–1951 age: 25–64	Social class SMR	I II III IV V		1.35 1.12 0.96 0.88 1.07	Surveillance system statistics using 1950 census data as denominator. Social class indicator based on occupation ICD 162-163 [Buell, 1960]
USA 1960 age: 25–64 (White)	Educational level SMR	College High school Elementary school <8 years of school		1.77 0.86 0.95 0.91	Record-linkage study using 1960 mortality and census data [Kitagawa, 1973]
USA (12 census samples) White population 1979–1985 age: 25+	Education SMR	College: 5+ y 4 y 1-3 y High school: 4 y 1-3 y Ellementary school: 8 y 5-7 y 0-4 y		0.83 1.31 1.07 0.94 0.93 0.95 0.95 1.38	Census linkage [Rogot et al., 1992]

Table 37. Prostate cancer incidence

Study base	Indicators	Social scale	N	RR	Study design
Canada (Montreal) 1979–1985 age: 35–70 (French)	Income level OR	High Middle Low		1 1.0 1.4	Population–based case–control study. Tertiles of total family income [Bourbonnais, in press]
Canada (Montreal) 1979–1985 age: 35–70 (French)	Education OR	High Middle Low		1 1.1 1.1	Population-based case–control study. Tertiles of years of education [Bourbonnais, in press]
Canada (Montreal) 1979–1985 age: 35–70 (French)	Occupational prestige scale OR	High Middle Low		1.0 0.9 1.0	Population-based case–control study. Tertiles of the occupational prestige scale. [Bourbonnais, in press]
Colombia (Cali) 1971–1975 all ages	Social class RR	I II III		1 0.84 0.64	Data from 1973 census were used for rate denominator. Social class based on area of residence [Cuello, 1982]
Denmark 1970–1980 all ages	Occupational group RR	Self-employed Employees: I Employees: II Employees: III Employees: IV Skilled workers Unskilled workers	1001 90 158 383 174 393 983	0.95 1.01 1.15 1.14 1.12 1.01 0.96	Record-linkage study using 1970 census and 1970–1980 incidence data. Employees classified according to educational level [Lynge, 1990]
Finland 1971–1985 birth cohort: 1906–1945	Social class SIR	Upper white-collar Lower white-collar Skilled workers Unskilled workers		1.24 1.10 0.98 0.83	Record-linkage study using 1970 census and 1971–1985 incidence data. Social class based on occupation [Pukkala, 1993]
Italy (Torino) 1985–1987 age: 20–69	Educational level OR	University High school Middle school Primary school	20 29 30 95	1 0.81 0.45 0.66	Record-linkage study between 1971 and 1981 censuses and 1985–1987 incidence data [Faggiano, 1994]
Italy (Torino) 1985–1987 age: 20–69	Occupational group OR	Managers Clerks Self employed Manual workers	21 32 26 76	1 0.73 1.17 0.94	Record-linkage study between 1971 and 1981 censuses and 1985–1987 incidence data [Faggiano, 1994]
Italy (Torino) 1985–1987 age: 20–69	Housing tenure OR	Owners Tenants	96 76	1 0.80	Record-linkage study between 1971 and 1981 censuses and 1985–1987 incidence data [Faggiano, 1994]
Sweden 1961–1970 all ages	Social class SIR	Employees: I Self employed: II Indep. farmers: III White-collar: IV Blue-collar: V	2521 2353 3441 5781 13920	1.04 0.99 1.01 1.06 0.97	Record-linkage study between 1961 census and 1961–1970 incidence data. Social class indicator based on occupation [Vågerö, 1986]
Turkey (Istanbul) 1979–1984	Social class OR	Higher Medium Lower	8 6 13	1 0.2 0.2	Hospital-based case–control study. Social class indicator based on occupation [Dosemeci, 1993]

Table 37. (Contd) Prostate cancer incidence

Study base	Indicators	Social scale	N	RR	Study design
UK – England and Wales 1971–1981 all ages	Housing tenure SIR	Owner occupier Private rented Council tenant	380 132 185	1.00 0.95 1.04	Record-linkage study between 1971 census and 1971–1981 incidence data (1% sample). UK Registrar General's social class classification [Kogevinas, 1990]
USA 1969–1971 all ages	Educational level OR	College Less		1.00 1	Case–control study based on US Third National Cancer Survey, using deaths from other causes as controls [Williams, 1977]
USA 1969–1971 all ages	Family income level OR	>US$ 10 000 Less		0.86 1	Case–control study based on US Third National Cancer Survey, using deaths from other causes as controls. [Williams, 1977]

Table 38. Testis cancer mortality

Study base	Indicators	Social scale	N	RR	Study design
Finland 1971–1985 age: 35–66	Social class RR	Upper white-collar	>20	1	Record-linkage study using 1970,
		Lower white-collar	>20	0.85	1975 and 1980 censuses and
		Skilled workers	>20	0.69	1971–1985 mortality data
		Unskilled workers	0	–	
		Farmers	>20	0.60	[Valkonen, 1990]
New Zealand 1974–1978 age: 15–64	Social class RR	I		1	Surveillance system statistics using
		II		2.7	1976 census data as denominator.
		III-NM		2.0	UK Registrar General's social class
		III-M		2.7	classification
		IV		1.0	
		V		4.9	[Pearce, 1986]
New Zealand 1984–1987 age: 15–64	Social class RR	I	0	–	Surveillance system statistics using
		II	6	1	1986 census data as denominator.
		III-NM	7	0.9	UK Registrar General's social class
		III-M	12	1.4	classification
		IV	6	0.9	
		V	2	0.4	[Pearce and Bethwaite, in press]
UK – England and Wales 1921–1923 age: 15–64	Social class SMR	I		0.83	Surveillance system statistics using
		II		1.67	1920 census data as denominator.
		III		0.89	For social classification see
		IV		0.89	Introduction
		V		0.78	[OPCS, 1927]
UK – England and Wales 1949–1953 age: 15–64	Social class SMR	I		1.64	Surveillance system statistics using
		II		1.21	1950 census data as denominator.
		III		0.92	For social classification see
		IV		0.98	Introduction.
		V		0.90	[OPCS, 1958]
UK – England and Wales 1970–1972 age: 15–64	Social class SMR	I		1.57	Surveillance system statistics using
		II		1.06	1970 census data as denominator.
		III-NM		1.25	For social classification see
		III-M		0.89	Introduction. Women classified
		IV		1.05	according to husband's occupation
		V		0.86	[OPCS, 1977]
UK – Great Britain 1979–1980, 1982–1983 age: 20–64	Social class SMR	I	29	0.80	Surveillance system statistics using
		II	112	0.85	1980 census data as denominator.
		III-NM	80	1.22	For social classification see
		III-M	213	1.04	Introduction. Women classified
		IV	92	1.04	according to husband's occupation
		V	36	1.12	[OPCS, 1986]

Table 39. Testis cancer incidence

Study base	Indicators	Social scale	N	RR	Study design
Colombia (Cali) 1971–1975 all ages	Social class RR	I II III		1 0.62 0.75	Data from 1973 census were used for rate denominators. Social class based on area of residence [Cuello, 1982]
Denmark 1970–1980 all ages	Occupational group RR	Self-employed Employees: I Employees: II Employees: III Employees: IV Skilled workers Unskilled workers	227 37 133 204 121 306 313	0.97 0.93 1.25 1.17 1.11 1.03 0.82	Record-linkage study using 1970 census and 1970–1980 incidence data. Employees classified according to educational level. [Lynge, 1990]
Finland 1971–1985 birth cohort: 1906–1945	Social class SIR	Upper white-collar Lower white-collar Skilled workers Unskilled workers		1.69 1.08 0.85 0.90	Record-linkage study using 1970 census and 1971–1985 incidence data. Social class based on occupation [Pukkala, 1993]
Sweden 1961–1970 all ages	Social class SIR	Employees: I Self employed: II Ind. farmers: III White-collar: IV Blue-collar: V	66 48 60 413 548	0.99 0.87 0.98 1.34 0.85	Record-linkage study between 1961 census and 1961–1970 incidence data. Social class indicator based on occupation [Vågerö, 1986]
Turkey (Istanbul) 1979–1984	Social class OR	Higher Medium Lower	7 48 136	1 1.1 1.0	Hospital-based case–control study. Social class indicator based on occupation. [Dosemeci, 1993]
UK 1977–1981 Age >10 years	Social class OR	I II III-NM III-M IV V		1.99 1.61 1.42 1 1.11 1.00	Hospital-based case–control study. Controls were cancer and no-cancer patients [Swerdlow et al. 1991]
UK – England and Wales 1971–1981 all ages	Housing tenure SIR	Owner occupier Private rented Council tenant	41 19 22	0.98 1.29 0.91	Record-linkage study between 1971 census and 1971–1981 incidence data (1% sample). UK Registrar General's social class classification [Kogevinas, 990]
USA Buffalo 1969–1971 all ages	Occupation OR	Professionals All other occupations Semi-unskilled		1 1.44 1.89	Hospital-based case–control study [Graham & Gibson, 1972]

Table 40. Bladder cancer mortality

Study base	Indicators	Social scale	N	Male RR	N	Female RR	Study design
Brazil (São Paulo) 1978–1982 age: 35–74	Years of education OR	>11 9–11 1–8 <1		1.4 2.2 1.2 1		2.1 0.6 1.5 1	Case–case study using deaths from other causes as controls [Bouchardy, 1992]
Canada (urban area) 1971 all ages	Income CMF	Q1 Q2 Q3 Q4 Q5		1.27 0.97 0.70 1.23 0.91			Surveillance system statistics using 1971 census data as denominator. Neighbourhood income quintiles as social indicator [K. Wilkins, unpublished]
Canada (urban area) 1986 all ages		Q1 Q2 Q3 Q4 Q5		0.88 0.88 1.00 1.13 1.06			Surveillance system statistics using 1986 census data as denominator. Neighbourhood income quintiles as social indicator [K. Wilkins, unpublished]
Italy 1981–1982 age: 18–74	Educational level RR	University High school Middle school Primary school Literate Illiterate	30 50 129 446 147 35	1 1.11 1.34 1.02 0.79 0.78	0 7 13 79 41 9	– 1 0.91 0.95 0.90 0.69	Record-linkage between 1981 census and the mortality in the following six months [Faggiano, 1995]
Japan 1965–1982 age: 40+	Social class SMR	I, II III IV V		1.23 1.13 0.99 1		1.00 1.24 – 1	265 000 Japanese interviewed in 1965 and followed-up until 1982. Social class based on occupation. Reference category is farmers and miners [Hirayama, 1990]
New Zealand 1974–1978 age: 15–64	Social class RR	I II III-NM III-M IV V		1 1.04 1.17 1.78 0.91 1.26			Surveillance system statistics using 1976 census data as denominators. UK Registrar General's social class classification [Pearce, 1986]
New Zealand 1984–1987 age: 15–64	Social class RR	I II III-NM III-M IV V	3 8 11 18 12 7	1 1.36 1.00 1.86 1.57 2.07			Surveillance system statistics using 1986 census data as denominator. UK Registrar General's social class classification [Pearce & Bethwaite, in press]
Norway 1970–1973 age: 20–69	Social class CMF	A B C D E (farmers)		1.32 1.01 1.14 0.88 –			Surveillance system statistics using 1970 census data as denominator. Social class indicator based on occupation [Central Bureau of Statistics, 1976]

Table 40. (Contd) Bladder cancer mortality

Study base	Indicators	Social scale	N	Male RR	N	Female RR	Study design
Spain 1980–1982	Occupational group PMR	Professionals managers		0.98			Proportional analysis on death certificates
		Manual workers					
		Agricultural workers, etc.		1.20			
				0.75			[E. Regidor, unpublished]
Switzerland (Vaud) 1977–1984 all ages	Social class PMR	I, II		0.84		0.70	Proportional mortality study. UK Registrar General's social class classification. (No. of males = 147; females = 27) [Levi, 1988]
		III		1.26		1.09	
		IV, V		0.90		1.39	
Switzerland 1979–1982 age: 15–74	Social class SMR	I		0.75			Surveillance system statistics using 1980 census data as denominator. Registrar General's social class classification [C.E. Minder, unpublished]
		II		0.84			
		III-NM		1.09			
		III-M		1.24			
		IV-V		0.98			
UK – England and Wales 1910–1912 age: 15–64	Social class SMR	I		0.96			Surveillance system statistics using 1910 census data as denominator. For social classification see Introduction [OPCS, 1919]
		II		0.95			
		III		1.00			
		IV		0.89			
		V		1.21			
UK – England and Wales 1930–1932 age: 15–64 (married women)	Social class SMR	I		0.76		0.60	Surveillance system statistics using 1930 census data as denominator. For social classification see Introduction. Women classified according to husband's occupation [OPCS, 1938]
		II		0.98		1.10	
		III		1.06		1.06	
		IV		0.94		0.80	
		V		1.06		0.93	
UK – England and Wales 1949–1953 age: 15–64 (married women)	Social class SMR	I		1.06		0.76	Surveillance system statistics using 1950 census data as denominator. For social classification see Introduction [OPCS, 1958]
		II		0.77		0.99	
		III		1.09		1.06	
		IV		0.96		1.04	
		V		1.07		0.92	
UK – England and Wales 1970–1972 age: 15–64 (married women)	Social class SMR	I		0.79		0.54	Surveillance system statistics using 1970 census data as denominator. For social classification see Introduction. Women classified according to husband's occupation [OPCS, 1977]
		II		0.83		0.89	
		III-NM		0.91		0.71	
		III-M		1.20		1.31	
		IV		1.05		1.16	
		V		1.15		1.23	

Table 40. (Contd) Bladder cancer mortality

Study base	Indicators	Social scale	N	Male RR	N	Female RR	Study design
UK – Great Britain 1979–1980, 1982–1983 age: 20–64 (married women, 20–59)	Social class SMR	I II III-NM III-M IV V	100 464 258 983 614 251	0.80 0.77 0.89 1.13 1.22 1.34	11 70 31 154 81 27	0.58 0.80 0.80 1.26 1.28 1.35	Surveillance system statistics using 1980 census data as denominator. For social classification see Introduction. Women classified according to husband's occupation [OPCS, 1986]
UK (London) 1967–1987	Employment grade RR	Administrators Professionals Clerical Other	2 29 6 8	1 1.12 0.62 1.44			17 530 London civil servants, medically examined 1967–1969 and followed-up until 1987. [Davey Smith, 1991]
USA (12 census samples) White population 1979–1985 age: 25+	Education SMR	College: 5+ y 4 y 1-3 y High school: 4 y 1-3 y Elementary school: 8 y 5-7 y 0-4 y		0.80 0.99 0.91 1.29 0.65 0.92 1.06 1.14			Census linkage [Rogot et al., 1992]
USA – California 1949–1951 age: 25–64	Social class SMR	I II III IV V		1.06 1.06 0.99 0.98 0.82			Surveillance system statistics using 1950 census data as denominator. Social class indicator based on occupation [Buell, 1960]

Table 41. Bladder cancer incidence

Study base	Indicators	Social scale	N	Male RR	N	Female RR	Study design
Canada (Montreal) 1979–1985 age: 35–70 (French)	Income level OR	High Middle Low		1 1.3 1.4			Population-based case–control study. Tertiles of total family income [Bourbonnais, in press]
Canada (Montreal) 1979–1985 age: 35–70 (French)	Education OR	High Middle Low		1 1.1 1.0			Population-based case–control study. Tertiles of years of education [Bourbonnais, in press]
Canada (Montreal) 1979–1985 age: 35–70 (French)	Occupational prestige scale OR	High Middle Low		1 1.0 1.0			Population-based case–control study. Tertiles of occupational prestige scale [Bourbonnais, in press]
Colombia (Cali) 1971–1975 all ages	Social class RR	I II III		1 0.47 0.27		1 1.59 0.53	Data from 1973 census were used for rate denominator. Social class based on area of residence. [Cuello, 1982]
Denmark 1970–1980 all ages	Occupational group RR	Self-employed Employees: I Employees: II Employees: III Employees: IV Skilled workers Unskilled workers	1103 112 218 575 272 665 1317	0.83 0.88 1.03 1.17 1.26 1.22 0.98	27 4 24 50 136 2 224	0.73 0.57 0.87 1.00 1.05 0.58 1.12	Record-linkage study using 1970 census and 1970–1980 incidence data. Employees classified according to educational level [Lynge, 1990]
Finland 1971–1985 birth cohort: 1906–1945	Social class SIR	Upper white-collar Lower white-collar Skilled workers Unskilled workers		1.03 0.98 1.00 1.03		1.29 1.12 0.95 0.90	Record-linkage study using 1970 census and 1971–1985 incidence data. Social class based on occupation [Pukkala, 1993]
Italy (Torino) 1985–1987 age: 20–69	Educational level OR	University High school Middle school Primary school	23 58 100 252	1 1.03 1.10 1.16			Record-linkage study between 1971 and 1981 censuses and 1985–1987 incidence data [Faggiano, 1994]
Italy (Torino) 1985–1987 age: 20–69	Occupational group OR	Managers Clerks Self employed Manual workers	54 75 57 214	1 0.79 0.99 0.98			Record-linkage study between 1971 and 1981 censuses and 1985–1987 incidence data [Faggiano, 1994]
Italy (Torino) 1985–1987 age: 20–69	Housing tenure OR	Owners Tenants	189 231	1 1.17			Record-linkage study between 1971 and 1981 censuses and 1985–1987 incidence data. [Faggiano, 1994]

Table 41. (Contd) Bladder cancer incidence

Study base	Indicators	Social scale	N	Male RR	N	Female RR	Study design
Sweden 1961–1970 all ages	Social class SIR	Employees: I Self employed: II Indep. farmers: III White-collar: IV Blue-collar: V	1013 889 704 2687 5448	1.10 1.02 0.82 1.14 0.99	– 60 – 700 603	– 0.95 – 1.08 0.94	Record-linkage study between 1961 census and 1961–1970 incidence data. Social class indicator based on occupation [Vågerö, 1986]
Turkey (Istanbul) 1979–1984 all ages	Social class OR	Higher Medium Lower	15 70 182	1 1.1 1.1			Hospital-based case–control study. Social-class indicator based on occupation [Dosemeci, 1993]
UK – England and Wales 1971–1981 all ages	Housing tenure SIR	Owner occupier Private rented Council tenant	278 122 185	0.89 1.10 1.16	110 43 57	1.01 0.99 0.98	Record-linkage study between 1971 census and 1971–1981 incidence data (1% sample). UK Registrar General's social class classification [Kogevinas, 1990]
USA 1969–1971 all ages	Educational level OR	College Less		0.82 1		0.95 1	Case–control study based on US Third National Cancer Survey, using deaths from other causes as controls [Williams, 1977]
USA 1969–1971 all ages	Family income level OR	>US$10 000 Less		1.09 1		0.88 1	Case–control study based on US Third National Cancer Survey, using deaths from other causes as controls. [Williams, 1977]

Table 42. Kidney cancer mortality

Study base	Indicators	Social scale	N	Male RR	N	Female RR	Study design
Brazil (São Paulo) 1978–1982 age: 35–74	Years of education OR	2+ 9–11 1–8 <1		4.9 3.0 1.4 1			Case–control study using deaths from other causes as controls [Bouchardy, 1992]
Canada (urban area) 1971 all ages	Income CMF	Q1 Q2 Q3 Q4 Q5		1.00 1.08 1.11 1.22 0.64			Surveillance system statistics unsing 1971 census data as denominator. Neighbourhood income quintiles as social indicator [K. Wilkins, unpublished]
Canada (urban area) 1986 all ages	Income CMF	Q1 Q2 Q3 Q4 Q5		0.84 0.82 1.21 0.89 1.18		0.69 1.13 1.13 0.81 1.06	Surveillance system statistics using 1986 census data as denominator. Neighbourhood income quintiles as social indicator [K. Wilkins, unpublished]
Italy 1981–1982 age: 18–74	Educational level RR	University High school Middle school Primary school Literate Illiterate	18 31 77 230 67 10	1 1.04 1.24 0.89 0.77 0.62	0 10 19 103 28 3	– 1 1.09 1.15 0.67 0.31	Record-linkage between 1981 census and mortality in the following six months [Faggiano, 1995]
Japan 1965–1982 age: 40+	Social class SMR	I, II III IV V		2.17 1.07 1.45 1		2.20 1.87 0.47 1	265 000 Japanese interviewed in 1965 and followed-up until 1982. Social class based on occupation. Reference category is farmers and miners [Hirayama, 1990]
New Zealand 1974–1978 age: 15–64	Social class RR	I II III-NM III-M IV V		1 1.06 0.97 0.97 0.89 1.69			Surveillance system statistics using 1976 census data as denominator. UK Registrar General's social class classification [Pearce, 1986]
New Zealand 1985–1987 age: 14–64	Social class RR	I II III-NM III-M IV V	7 18 42 23 20 13	1 1.79 1.86 1.07 1.29 2.00			Surveillance system statistics using 1986 census data as denominator. UK Registrar General's social class classification [Pearce & Bethwaite, in press]
Switzerland (Vaud) 1977–1984 all ages	Social class PMR	I, II III IV, V		1.19 1.16 0.65		1.27 0.78 1.02	Proportional mortality study. UK Registrar General's social class classification (No. of males = 65; females = 33). [Levi, 1988]
UK – England and Wales 1910–1912 age: 15–64	Social class SMR	I II III IV V		1.36 0.91 0.91 0.83 0.91			Surveillance system statistics using 1910 census data as denominator. For social classification see Introduction [OPCS, 1919]

Table 42. (Contd) Kidney cancer mortality

Study base	Indicators	Social scale	N	Male RR	N	Female RR	Study design
UK – England and Wales 1949–1953 age: 15–64 (married women)	Social class SMR	I II III IV V		1.34 1.00 1.06 0.96 0.82		1.58 1.03 1.00 0.95 0.86	Surveillance system statistics using 1950 census data as denominator. For social classification see Introduction. Women classified according to husband's occupation [OPCS, 1958]
UK – England and Wales 1959–1963 age: 15–64 (married women)	Social class SMR	I II III IV V		0.89 0.93 1.05 0.98 1.09		0.91 0.86 1.06 1.01 1.21	Surveillance system statistics using 1960 census data as denominator. For social classification see Introduction [OPCS, 1971]
UK – England and Wales 1970–1972 age: 15–64 (married women)	Social class SMR	I II III-NM III-M IV V		1.01 1.03 1.12 1.03 1.02 1.10		1.05 1.04 1.11 1.09 1.12 1.03	Surveillance system statistics using 1970 census data as denominator. For social classification see Introduction. Women classified according to husband's occupation [OPCS, 1977]
UK – Great Britain 1979–1980, 1982–1983 age: 20–64	Social class SMR	I II III-NM III-M IV V	101 483 256 740 401 171	0.95 0.98 1.12 1.04 1.02 1.18			Surveillance system statistics using 1980 census data as denominator. For social classification see Introduction. Women classified according to husband's occupation [OPCS, 1986]
USA – California 1949–1951 age: 25–64	Social class SMR	I II III IV V		0.92 0.89 1.16 0.98 1.07			Surveillance system statistics using 1950 census data as denominator. Social class indicator based on occupation [Buell, 1960]

Table 43. Kidney cancer incidence

Study base	Indicators	Social scale	N	Male RR	N	Female RR	Study design
Canada (Montreal) 1979–1985 age: 35–70 (French)	Income level OR	High Middle Low		1 2.0 1.4			Population-based case–control study. Tertiles of total family income [Bourbonnais, in press]
Canada (Montreal) 1979–1985 age: 35–70 (French)	Education OR	High Middle Low		1 1.7 1.7			Population-based case–control study. Tertiles of years of education [Bourbonnais, in press]
Canada (Montreal) 1979–1985 age: 35–70 (French)	Occupational prestige OR	High Middle Low		1 1.2 1.8			Population-based case–control study. Tertiles of the occupational prestige scale [Bourbonnais, in press]
Colombia (Cali) 1971–1975 all ages	Social class RR	I II III		1 0.44 0.16		1 1.89 1.00	Data from 1973 census were used for rate denominator. Social class based on area of residence. [Cuello, 1982]
Denmark 1970–1980 all ages	Occupational group RR	Self-employed Employees: I Employees: II Employees: III Employees: IV Skilled workers Unskilled workers	514 75 102 231 104 248 491	0.95 1.40 1.12 1.11 1.15 1.08 0.89	40 7 27 39 142 3 184	1.21 1.07 1.01 0.82 1.13 0.89 0.97	Record-linkage study using 1970 census and 1970–1980 incidence data. Employees classified according to educational level [Lynge, 1990]
Finland 1971–1985 birth cohort: 1906–1945	Social class SIR	Upper white-collar Lower white-collar Skilled workers Unskilled workers		1.22 1.12 1.00 0.73		1.13 1.11 0.95 0.95	Record-linkage study using 1970 census and 1971–1985 incidence data. Social class based on occupation [Pukkala, 1993]
Sweden 1961–1970 all ages	Social class SIR	Employees: I Self-employed: II Indep. farmers: III White-collars: IV Blue-collars: V	710 630 647 1938 3820	1.07 1.05 0.82 1.13 0.96	– 62 – 775 799	– 0.83 – 0.98 1.01	Record-linkage study between 1961 census and incidence data. Social class indicator based on occupation [Vågerö, 1986]
UK – England and Wales 1971–1981 all ages	Housing tenure SIR	Owner occupier Private rented Council tenant	72 24 37	1.03 0.88 0.99			Record-linkage study between 1971 census and 1971–1981 incidence data (1% sample). UK Registrar General's social class classification [Kogevinas, 1990]
USA 1969–1971 all ages	Educational level OR	College Less		0.56 1		0.76 1	Case–control study based on US Third National Cancer Survey, using deaths for other causes as controls. [Williams, 1977]

Table 43. (Contd) Kidney cancer incidence

Study base	Indicators	Social scale	N	Male RR	N	Female RR	Study design
USA 1969–1971 all ages	Family income level OR	>US$ 10 000 Less		0.96 1		1.08 1	Case–control study based on US Third National Cancer Survey, using deaths from other causes as controls [Williams, 1977]

Table 44. Brain cancer mortality

Study base	Indicators	Social scale	N	Male RR	N	Female RR	Study design
Brazil (São Paulo) 1978–1982 age: 35–74	Years of education OR	12+ 9–11 1–8 <1		2.6 2.8 1.5 1		0.8 1.3 1.1 1	Case–control study using deaths from other causes as controls ICD: 191-192 [Bouchardy, 1992]
Canada (urban area) 1971 all ages	Income CMF	Q1 Q2 Q3 Q4 Q5		0.93 0.84 1.28 0.93 1.02		1.12 0.82 1.15 1.26 0.62	Surveillance system statistics using 1971 census data as denominators. Neighbourhood income quintiles as social indicator [R. Wilkins, unpublished]
Canada (urban area) 1986 all ages	Income CMF	Q1 Q2 Q3 Q4 Q5		0.95 0.93 1.08 1.05 0.98		1.34 0.97 0.89 0.89 0.91	Surveillance system statistics using 1986 census data as denominators. Neighbourhood income quintiles as social indicator [R. Wilkins, unpublished]
Italy 1981–1982 age: 18–74	Educational level RR	University High school Middle school Primary school Literate Illiterate	38 56 115 345 92 22	1 0.76 0.82 0.75 0.66 0.71	10 34 48 230 89 24	1 0.92 0.68 0.74 0.68 0.71	Record-linkage between 1981 census and mortality in the following six months [Faggiano, 1995]
New Zealand 1974–1978 age: 15–64	Social class CMF	I II III-NM III-M IV V		1 0.79 0.82 0.56 0.79 0.81			Surveillance system statistics using 1976 census data as denominator. UK Registrar General's social class classification. ICD-191-192 [Pearce, 1986]
New Zealand 1984–1987 age: 15–64	Social class CMF	I II III-NM III-M IV V	14 40 47 39 38 10	1 1.54 1.04 0.91 1.22 1.19			Surveillance system statistics using 1986 census data as denominator. UK Registrar General's social class classification [Pearce and Bethwaite, in press]

Socioeconomic differences in cancer incidence and mortality

Table 44. (Contd) Brain cancer mortality

Study base	Indicators	Social scale	N	Male RR	N	Female RR	Study design
Spain 1980–1982	Occupational group PMR	Professionals managers Manual workers Agricultural workers		1.06 1.04 1.02			Proportional analysis on death certificates [E. Regidor, unpublished]
Switzerland (Vaud) 1977–1984 all ages	Social class PMR	I, II III IV, V		1.44 0.75 0.82		0.91 1.24 0.89	Proportional mortality study. UK Registrar General's social class classification. (No. of males = 75; females = 37) [Levi, 1988]
Switzerland 1979–1982 age: 15–74	Social class SMR	I II III-NM III-M IV-V		0.76 0.92 1.20 1.11 0.80			Surveillance system statistics using 1980 census data as denominator. UK Registrar General's social class classification [C.E. Minder, unpublished]
UK – England and Wales 1930–1932 age: 15–64	Social class SMR	I II III IV V		1.60 1.60 1.20 0.80 0.60			Surveillance system statistics using 1930 census data as denominator. For social classification see Introduction. Women classified according to husband's occupation [OPCS, 1938]
UK – England and Wales 1949–1953 age: 15–64 (married women)	Social class SMR	I II III IV V		1.33 0.96 1.04 0.88 0.92		1.27 1.04 1.02 0.91 0.82	Surveillance system statistics using 1950 census data as denominator. For social classification see Introduction. Women classified according to husband's occupation. [OPCS, 1958]
UK – England and Wales 1970–1972 age: 15–64 (married women)	Social class SMR	I II III-NM III-M IV V		1.08 1.01 1.11 1.05 1.00 0.92		1.37 1.08 0.98 1.11 1.00 1.00	Surveillance system statistics using 1970 census data as denominator. For social classification see Introduction. Women classified according to husband's occupation [OPCS, 1977]
UK – Great Britain 1979–1980, 1982–1983 age: 20–64 (married women, 20–59)	Social class SMR	I II III-NM III-M IV V	215 784 398 1200 577 262	1.19 0.98 1.09 1.03 0.96 1.19	97 351 138 520 196 76	1.26 1.06 0.98 1.12 0.90 1.13	Surveillance system statistics using 1980 census data as denominator. For social classification see Introduction. Women classified according to husband's occupation [OPCS, 1986]
UK (London) 1967–1987	Employment grade RR	Administrators Professionals Clerical Other	3 28 6 3	1 0.87 0.87 0.47			17 530 London civil servants, medically examined 1967–1969 and, followed-up until 1987 [Davey Smith, 1991]

163

Table 44. (Contd) Brain cancer mortality

Study base	Indicators	Social scale	N	Male RR	N	Female RR	Study design
USA – California 1949–1951 age: 25–64	Social class SMR	I		1.30			Surveillance system statistics using 1950 census data as denominator. Social class indicator based on occupation [Buell, 1960]
		II		1.27			
		III		1.08			
		IV		0.77			
		V		0.58			
USA (12 census samples) White population 1979–1985 age:25+	Education SMR	College: 5+y		1.17			Census linkage
		4y		1.11			
		1-3y		1.40			
		High school: 4y		0.90			
		1-3y		0.59			
		Elementary school: 8y		1.27			
		5-7y		0.92			[Rogot et al., 1972]

Table 45. Brain cancer incidence

Study base	Indicators	Social scale	N	Male RR	N	Female RR	Study design
Colombia (Cali) 1971–75 all ages	Social class RR	I		1		1	Data from 1973 census were used for rate denominator. Social class based on area of residence. [Cuello, 1982]
		II		1.55		0.51	
		III		1.05		0.37	
Denmark 1970–80 all ages	Occupational group RR	Self-employed	483	1.04	39	0.88	Record-linkage study using 1970 census and 1970–80 incidence data. Employees classified according to the educational level [Lynge, 1990]
		Employees: I	62	1.17	10	1.01	
		Employees: II	86	0.84	39	0.80	
		Employees: III	209	0.99	79	0.99	
		Employees: IV	94	0.95	240	1.09	
		Skilled workers	268	1.05	6	0.87	
		Unskilled workers	501	0.97	279	0.99	
Finland 1971–85 birth cohort: 1906–45	Social class SIR	Upper white-collar		1.06		1.10	Record-linkage study using 1970 census and 1971–85 incidence data. Social class based on occupation [Pukkala, 1993]
		Lower white-collar		1.10		1.05	
		Skilled workers		1.00		0.97	
		Unskilled workers		0.81		0.96	
Sweden 1961–70 all ages	Social class SIR	Employees: I	473	1.01	–	–	Record-linkage study between 1961 census and 1961–70 incidence data. Social class indicator based on occupation [Vågerö, 1986]
		Self-employed: II	402	0.99	95	1.11	
		Indep. farmers: III	525	1.02	–	–	
		White-collar: IV	1415	1.06	1174	1.02	
		Blue-collar: V	2910	0.97	973	0.97	
UK – England and Wales 1971–81 all ages	Housing tenure SIR	Owner occupier	65	0.99	53	1.07	Record-linkage study between 1971 census and 1971–81 incidence data (1% sample). UK Registrar General's social class classification [Kogevinas, 1990]
		Private rented	17	0.78	18	1.02	
		Council tenant	46	1.21	25	0.89	

Table 46. Thyroid gland cancer mortality

Study base	Indicators	Social scale	N	Male RR	N	Female RR	Study design
Switzerland (Vaud) 1977–84 all ages	Social class PMR	I, II III IV, V		0.45 0.85 2.19		0.27 1.94 1.47	Proportional mortality study. UK Registrar General's social class classification. (No. of males = 11; females = 9) [Levi, 1988]
UK – England and Wales 1949–53 age: 15–64 (married women)	Social class SMR	I II III IV V		1.00 1.19 0.98 0.97 0.88		0.64 0.93 1.05 1.04 1.00	Surveillance system statistics using 1950 census data as denominator. For social classification see Introduction. Women classified according to husband's occupation [OPCS, 1958]
UK – England and Wales 1959–63 age: 15–64 (married women)	Social class SMR	I II III IV V		0.88 1.12 1.00 0.80 1.38		0.60 0.83 1.01 1.10 1.54	Surveillance system statistics using 1960 census data as denominator. For social classification see Introduction [OPCS, 1971]
UK – England and Wales 1970–72 age: 15–64 (married women)	Social class SMR	I II III-NM IIIMM IV V		1.57 1.06 1.17 0.85 1.13 1.09		0.74 0.77 0.58 1.25 1.28 1.64	Surveillance system statistics using 1970 census data as denominator. For social classification see Introduction. Women classified according to husband's occupation [OPCS, 1977]
UK – Great Britain 1979–80, 1982–83 age: 20–64	Social class SMR	I II III-NM III-M IV V	9 45 17 63 37 10	0.98 1.08 0.89 1.05 1.15 0.85			Surveillance system statistics using 1980 census data as denominator. For social classification see Introduction. Women classified according to husband's occupation [OPCS, 1986]

Table 47. Thyroid gland cancer incidence

Study base	Indicators	Social scale	N	Male RR	N	Female RR	Study design
Colombia (Cali) 1971–1975 all ages	Social class RR	I II III		1 1.33 0.55		1 0.97 0.84	Data from 1973 census were used for rate denominator. Social class based on area of residence. [Cuello, 1982]
Denmark 1970–1980 all ages	Occupational group RR	Self-employed Employees: I Employees: II Employees: III Employees: IV Skilled workers Unskilled workers	42 3 14 22 12 27 56	0.89 0.56 1.32 1.02 1.14 0.99 1.05	4 2 6 22 41 0 43	0.57 1.22 0.68 1.50 0.99 – 0.90	Record-linkage study using 1970 census and 1970–1980 incidence data. Employees classified according to educational level [Lynge, 1990]
Finland 1971–1985 birth cohort: 1906–1945	Social class SIR	Upper white-collar Lower white-collar Skilled workers Unskilled workers		1.07 1.19 1.01 0.64		1.16 1.05 0.95 0.97	Record-linkage study using 1970 census and 1971–1985 incidence data. Social class based on occupation [Pukkala, 1993]
Sweden 1961–1970 all ages	Social class SIR	Employees: I Self-employed: II Indep. farmers: III White-collar: IV Blue-collar: V	76 91 88 261 493	0.92 1.25 0.96 1.12 0.93	33 489 370	– 1.08 – 1.01 0.98	Record-linkage study between 1961 census and 1961–1970 incidence data. Social class indicator based on occupation [Vågerö, 1986]
USA 1969–1971 all ages	Educational level OR	College Less		1.66 1		1.86 1	Case–control study based on US Third National Cancer Survey, using deaths from other causes as controls [Williams, 1977]
USA 1969–1971 all ages	Family income level OR	>US$ 10 000 Less		1.52 1		1.86 1	Case–control study based on US Third National Cancer Survey, using deaths from other causes as controls [Williams, 1977]

Table 48. Lymphoma mortality

Study base	Indicators	Social scale	N	Male RR	N	Female RR	Study design
Brazil (São Paulo) 1978–1982 age: 35–74	Years of education OR	12+ 9–11 1–8 <1		2.3 1.2 1.6 1		2.6 1.1 0.6 1	Case–control study using deaths from other causes as controls. ICD-9: 201 [Bouchardy, 1992]
Brazil (São Paulo) 1978–1982 age: 35–74	Years of education OR	12+ 9–11 1–8 <1		1.2 1.4 0.6 1		2.0 1.0 1.5 1	Case–control study using deaths from other causes as controls. ICD-9: 202 [Bouchardy, 1992]
Hungary 1970 age: 25–64	Years of education SMR	15+ 12–14 8–11 0–7		1.15 0.92 1.51 0.79		1.83 1.62 1.14 0.84	Surveillance system statistics using 1970 census data as denominator [Jozan, 1986]
Hungary 1980 age: 25–64	Years of education SMR	15+ 12–14 8–11 0–7		1.55 0.87 1.05 0.86		1.61 1.03 1.16 0.84	Surveillance system statistics using 1980 census data as denominator [Jozan, 1986]
Italy 1981–1982 age: 18–74	Education level RR	University High school Middle school Primary school Literate Illiterate	26 48 102 333 89 33	1 0.84 0.96 1.03 0.82 1.40	11 30 62 257 101 21	1 0.66 0.70 0.68 0.60 0.53	Record-linkage between 1981 census and mortality in the following six months [Faggiano, 1995]
New Zealand 1974–1978 age: 15–64	Social class RR	I II III-NM III-M IV V		1 0.37 0.46 0.50 0.25 0.83			Surveillance system statistics using 1976 census data as denominator. UK Registrar General's social class classification ICD-9: 201 [Pearce, 1986]
New Zealand 1984–1987 age: 15–64	Social class RR	I II III-NM III-M IV V	3 5 9 4 1	– 1 1.0 1.7 1.0 0.8			Surveillance system statistics using 1986 census data as denominator. UK Registrar General's social class classification ICD-9: 201 [Pearce & Bethwaite, in press]
New Zealand 1974–1978 age: 15–64	Social class RR	I II III-NM III-M IV V		1 0.55 0.58 0.63 0.54 0.52			Surveillance system statistics using 1976 census data as denominator. UK Registrar General's social class classification ICD-9: 202 [Pearce, 1986]
New Zealand 1985–1987 age: 14–64	Social class RR	I II IIIN IIIM IV V	4 24 26 24 24 10	1 3.19 2.06 1.81 2.44 2.44			Surveillance system statistics using 1986 census data as denominator. UK Registrar General's social class classification ICD-9: 202 [Pearce & Bethwaite, in press]

Study base	Indicators	Social scale	N	Male RR	N	Female RR	Study design
Spain 1980–1982	Occupational group PMR	Professionals managers		1.01			Proportional analysis on death certificates ICD-9: 201
		Manual workers		1.09			
		Agricultural workers		0.96			[E. Regidor, unpublished]
Spain 1980–1982	Occupational group PMR	Professionals managers		1.10			Proportional analysis on death certificates ICD-9: 201-202
		Manual workers		1.03			
		Agricultural workers		0.92			[E. Regidor, unpublished]
Switzerland (Vaud) 1977–1984 all ages	Social class PMR	I, II		1.39		0.81	Proportional mortality study. UK Registrar General's social class classification. ICD-9: 202 (No. of males = 77; females = 46) [Levi, 1988]
		III		0.94		1.45	
		IV, V		0.69		0.67	
Switzerland (Vaud) 1977–1984 all ages	Social class PMR	I, II		2.31		1.51	Proportional mortality study. UK Registrar General's social class classification. ICD-9: 203 (No. of males = 35; females = 18) [Levi, 1988]
		III		0.49		0.36	
		IV, V		0.73		1.88	
Switzerland 1979–1982 age: 15–74	Social class SMR	I		0.63			Surveillance system statistics using 1980 census data as denominator. UK Registrar General's social class classification ICD-8: 200-203, 208-209 [C.E. Minder, unpublished]
		II		0.89			
		III-NM		1.27			
		III-M		1.10			
		IV-V		0.83			
UK – England and Wales 1949–1953 age: 15–64 (married women)	Social class SMR	I		1.42		1.74	Surveillance system statistics using 1950 census data as denominator. For social classification see Introduction. Women classified according to husband's occupation. ICD-9:201 [OPCS, 1958]
		II		1.10		0.95	
		III		1.00		1.05	
		IV		0.93		0.95	
		V		0.87		0.74	
UK – England and Wales 1949–1953 age: 15–64 (married women)	Social class SMR	I		1.13		4.00	Surveillance system statistics using 1950 census data as denominator. For social classification see Introduction. Women classified according to husband's occupation. ICD-9: 202 [OPCS, 1958]
		II		1.34		1.06	
		III		1.02		0.96	
		IV		0.70		0.57	
		V		0.91		0.92	

Table 48. (Contd) Lymphoma mortality

Study base	Indicators	Social scale	N	Male RR	N	Female RR	Study design
UK – England and Wales 1959–1963 age: 15–64 (married women)	Social class SMR	I II III IV V		1.01 1.07 1.07 0.83 1.09		1.45 1.12 1.02 0.82 1.07	Surveillance system statistics using 1960 census data as denominator. For social classification see Introduction ICD-9: 201 [OPCS, 1971]
UK – England and Wales 1959–1963 age: 15–64 (married women)	Social class SMR	I II III IV V		1.11 1.00 1.06 0.93 1.24		– 0.96 1.13 0.81 0.91	Surveillance system statistics using 1960 census data as denominator. For social classification see Introduction. ICD-9: 202 [OPCS, 1971]
UK – England and Wales 1970–1972 age: 15–64 (married women)	Social class SMR	I II III-NM III-M IV V		1.13 1.03 1.07 1.03 1.03 0.91		1.23 0.94 1.17 1.03 1.14 1.17	Surveillance system statistics using 1970 census data as denominator. For social classification see Introduction. Women classified according to husband's occupation. ICD-9: 201 [OPCS, 1977]
UK – England and Wales 1970–1972 age: 15–64 (married women)	Social class	I II III-NM III-M IV V		1.08 0.81 1.11 1.17 1.13 0.63		0.73 1.26 1.63 1.00 0.93 0.39	Surveillance system statistics using 1970 census data as denominator. For social classification see Introduction. Women classified according to husband's occupation. ICD-9: 202 [OPCS, 1980]
UK – Great Britain 1979–1980, 1982–1983 age: 20–64 (married women, 30–59)	Social class SMR	I II III-NM III-M IV V	382 1526 724 2427 1257 538	1.07 0.97 0.98 1.05 1.04 1.21	136 639 258 919 466 149	0.95 1.04 0.98 1.06 1.15 1.19	Surveillance system statistics using 1980 census data as denominator. For social classification see Introduction. Women classified according to husband's occupation [OPCS, 1986]
USA – California 1949–1951 age: 25–64	Social class SMR	I II III IV V		1.64 1.00 1.01 0.88 1.07			Surveillance system statistics using 1950 census data as denominator. Social class indicator based on occupation ICD-9: 202-203, 205 [Buell, 1960]
USA (12 census samples) 1979–1985 age: 25+	Education SMR	College: 5+ y 4 y 1-3 y High school: 4 y 1-3 y Elementary school: 8 y 5-7 y 0-4 y		0.75 1.19 0.97 1.17 0.75 1.17 0.88 0.65		1.16 1.22 1.02 1.02 0.87 1.20 0.51 0.95	Census linkage [Rogot et al., 1992]

Table 49. Lymphoma incidence

Study base	Indicators	Social scale	N	Male RR	N	Female RR	Study design
Canada (Montreal) 1979–1985 age: 35–70 (French)	Income level OR	High Middle Low		1 1.2 1.5			Population-based case–control study. Tertiles of total family income. ICD-9: 202 [Bourbonnais, in press]
Canada (Montreal) 1979–1985 age: 35–70 (French)	Education OR	High Middle Low		1 1.0 0.9			Population-based case–control study. Tertiles of the years of education. ICD-9: 202 [Bourbonnais, in press]
Canada (Montreal) 1979–1985 age: 35–70 (French)	Occupational prestige scale OR	High Middle Low		1 1.0 1.3			Population-based case–control the occupational prestige scale. ICD-9: 202 [Bourbonnais, in press]
Colombia (Cali) 1971–1975 all ages	Social class RR	I II III		1 0.82 1.07		1 0.88 0.97	Data from 1973 census were used for rate denominator. Social class based on area of residence. [Cuello, 1982]
Colombia (Cali) 1971–1975 all ages	Social class RR	I II III		1 1.00 2.54		1 0.56 0.75	Data from 1973 census were used for rate denominator. Social class based on area of residence. ICD-9: 201 [Cuello, 1982]
Denmark 1970–1980 all ages	Occupational group RR	Self-employed Employees: I Employees: II Employees: III Employees: IV Skilled workers Unskilled workers	299 31 45 122 60 157 331	1.01 0.99 0.77 0.98 1.00 1.01 1.03	22 2 23 39 92 4 152	0.94 0.42 1.05 1.04 0.91 1.30 1.11	Record-linkage study using 1970 census and 1970–1980 incidence data. Employees classified according to educational level ICD-9: 202 [Lynge, 1990]
Denmark 1970–1980 all ages	Occupational group RR	Self-employed Employees: I Employees: II Employees: III Employees: IV Skilled workers Unskilled workers	112 12 31 55 43 84 160	0.99 0.80 0.92 0.88 1.25 0.92 1.09	9 4 7 16 48 3 56	1.12 2.14 0.63 0.88 0.94 1.55 1.04	Record-linkage study using 1970 census and 1970–1980 incidence data. Employees classified according to educational level. ICD-9: 201 [Lynge, 1990]
Finland 1971–1985 birth cohort: 1906–1945	Social class SIR	Upper white-collar Lower white-collar Skilled workers Unskilled workers		1.12 1.05 1.00 0.86		1.03 1.05 1.00 0.92	Record-linkage study using 1970 census and 1971–1985 incidence data. Social class based on occupation. ICD-9: 202 [Pukkala, 1993]

Table 49. (Contd) Lymphoma incidence

Study base	Indicators	Social scale	N	Male RR	N	Female RR	Study design
Finland 1971–1985 birth cohort: 1906–1945	Social class SIR	Upper white-collar Lower white-collar Skilled workers Unskilled workers		1.01 0.89 1.06 0.95		1.12 1.04 0.97 0.95	Record-linkage study using 1970 census and 1971–1985 incidence data. Social class based on occupation. ICD-9: 201 [Pukkala, 1993]
Finland 1971–1985 birth cohort: 1906–1945	Social class SIR	Upper white-collar Lower white-collar Skilled workers Unskilled workers		0.85 1.09 1.00 0.97		0.87 0.95 1.05 0.99	Record-linkage study using 1970 census and 1971–1985 incidence data. Social class based on occupation. ICD-9: 203 [Pukkala, 1993]
Sweden 1961–1970 all ages	Social class SIR	Employees: I Self-employed: II Indep. farmers: III White-collar: IV Blue-collar: V	138 115 142 385 906	1.06 1.00 0.95 0.99 1.00	– 11 – 191 166	– 0.84 – 0.96 1.04	Record-linkage study between 1961 census and 1961–1970 incidence data. Social class indicator based on occupation. ICD: 201 [Vågerö, 1986]
Sweden 1961–1970 all ages	Social class SIR	Employees: I Self-employed: II Indep. farmers: III White-collar: IV Blue-collar: V	313 278 431 932 1983	0.96 0.93 1.07 1.08 0.97	– 57 – 481 484	– 1.30 – 0.97 1.01	Record-linkage study between 1961 census and 1961–1970 incidence data. Social class indicator based on occupation. ICD-9: 202 [Vågerö, 1986]
UK – England and Wales 1971–1981 all ages	Housing tenure SIR	Owner occupier Private rented Council tenant	114 42 68	0.95 1.00 1.02	88 32 57	0.95 0.94 1.09	Record-linkage study between 1971 census and 1971–1981 incidence data (1% sample). UK Registrar General's social class classification [Kogevinas, 1990]
USA 1969–1971 all ages	Educational level OR	College Less		0.71 1		1.98 1	Case–control study based on US Third National Cancer Survey, using deaths from other causes as controls. ICD: 201 [Williams, 1977]
USA 1969–1971 all ages	Family income level OR	>US$ 10 000 Less		1.30 1		0.89 1	Case–control study based on US Third National Cancer Survey, using deaths from other causes as controls. ICD: 201 [Williams, 1977]

Table 50. Leukemia mortality

Study base	Indicators	Social scale	N	Male RR	N	Female RR	Study design
Brazil (São Paulo) 1978–1982 age: 35–74	Years of education OR	>11 9–11 1–8 <1		1.5 1.1 1.1 1		3.6 2.1 1.7 1	Case–case study using other causes as controls [Bouchardy, 1992]
Canada (urban area) 1971 all ages	Income	Q1 Q2 Q3 Q4 Q5		1.13 0.81 1.15 0.96 0.96		0.95 0.70 1.24 0.97 1.14	Surveillance system statistics using 1971 census data as denominator. Neighbourhood income quintiles as social indicator. ICD-9: 204-207 [R. Wilkins, unpublished]
Canada (urban area) 1986 all ages	Income	Q1 Q2 Q3 Q4 Q5		0.87 1.09 0.87 1.02 1.19		0.76 0.94 1.29 1.00 0.97	Surveillance system statistics using 1986 census data as denominator. Neighbourhood income quintiles as social indicator. ICD-9: 204-208 [R. Wilkins, unpublished]
Denmark 1970–1975 age: 20–64	Occupational group SMR	Employees: I Employees: III Employees: IV Skilled workers Unskilled workers		0.74 1.00 1.09 0.91 0.98		– 0.93 1.05 – 0.89	Record-linkage study using 1970–1975 mortality data and 1970 census. Employees classified according to educational level [Danmarks Statistik, 1979]
Finland 1969–1972 age: 15–64 (married women)	Social class CMF	Upper white-collar Lower white-collar Skilled workers Unskilled workers Farmers		1.26 1.14 0.80 0.82 0.98		1.03 1.02 1.02 1.00 0.92	Surveillance system statistics using 1970 census data as denominator. Social class indicator based on occupation [Näyhä, 1977]
Italy 1981–1982 age: 18–74	Educational level RR	University High school Middle school Primary school Literate Illiterate	21 45 98 289 126 21	1 0.98 1.14 1.12 1.36 0.97	0 30 40 201 104 23	– 1 0.89 1.12 1.32 1.11	Record-linkage between 1981 census and mortality in the following six months [Faggiano et al., 1995]
New Zealand 1974–1978 age: 15–64	Social class RR	I II III-NM III-M IV V		1 0.52 0.39 0.30 0.17 0			Surveillance system statistics using 1976 census data as denominator. UK Registrar General's social class classification. ICD-9: 204 [Pearce, 1986]
New Zealand 1984–1987 age: 15–64	Social class RR	I II III-NM III-M IV V	10 10 37 27 31 5	1 0.49 1.12 0.73 1.24 0.34			Surveillance system statistics using 1986 census data as denominator. UK Registrar General's social class classification. ICD-9: 204 [Pearce & Bethwaite, in press]

Table 50. (Contd) Leukemia mortality

Study base	Indicators	Social scale	N	Male RR	N	Female RR	Study design
Norway 1970–1973 age: 20–69	Social class CMF	A B C D E (farmers)		1.09 1.03 1.04 0.85 1.00			Surveillance system statistics using 1970 census data as denominator. Social class indicator based on occupation [Central Bureau of Statistics, 1976]
Spain 1980–1982	Occupational group PMR	Professionals managers Manual workers Agricultural workers		1.09 1.02 1.06			Proportional analysis on death certificates [E. Regidor, unpublished]
Switzerland (Vaud) 1977–1984 all ages	Social class PMR	I, II III IV, V		0.94 1.02 1.05		0.99 1.45 0.48	Proportional mortality study. UK Registrar General's social class classification [Levi, 1988]
Switzerland 1979–1982 age: 15–74	Social class SMR	I II III-NM III-M IV-V		0.77 0.95 1.27 1.06 0.79			Surveillance system statistics using 1980 census data as denominator. UK Registrar General's social class classification [C.E. Minder, unpublished]
UK – England and Wales 1930–1932 age: 15–64 (married women)	Social class SMR	I II III IV V		1.53 1.25 0.96 0.94 0.85		1.67 1.18 1.07 0.76 0.76	Surveillance system statistics using 1930 census data as denominator. For social classification see Introduction. Women classified according to husband's occupation [OPCS, 1938]
UK – England and Wales 1949–1953 age: 15–64 (married women)	Social class SMR	I II III IV V		1.23 0.98 1.04 0.93 0.89		1.45 0.92 1.02 1.04 0.87	Surveillance system statistics using 1950 census data as denominator. For social classification see Introduction [OPCS, 1958]
UK – England and Wales 1970–1972 age: 15–64 (married women)	Social class SMR	I II III-NM III-M IV V		1.13 1.00 1.07 1.01 1.04 0.95		0.88 1.08 0.98 1.05 1.10 1.27	Surveillance system statistics using 1910 census data as denominator. For social classification see Introduction. Women classified according to husband's occupation [OPCS, 1977]

Table 50. (Contd) Leukemia mortality

Study base	Indicators	Social scale	N	Male RR	N	Female RR	Study design
UK – Great Britain 1979–1980, 1982–1983 age: 20–64 (married women, 20–59)	Social class SMR	I II III-NM III-M IV V	148 525 275 923 471 200	1.10 0.90 0.99 1.07 1.06 1.22	59 293 109 422 184 64	0.92 1.08 0.95 1.10 1.05 1.19	Surveillance system statistics using 1980 census data as denominator. For social classification see Introduction. Women classified according to husband's occupation [OPCS, 1986]
USA – California 1949–1951 age: 25–64	Social class SMR	I II III IV V		1.04 1.16 1.01 0.86 1.04			Surveillance system statistics using 1950 census data as denominator. Social class indicator based on occupation [Buell, 1960]
USA (12 census samples) White population 1979–1985 age: 25+	Education SMR	College: 5+ y 4 y 1-3 y High school: 4 y 1-3 y Elementary school: 8 y 5-7 y 0-4 y		0.60 0.95 0.78 1.05 1.23 1.00 0.91 1.35		2.44 1.15 1.16 0.89 0.96 0.73 1.12 1.04	Census linkage [Rogot et al., 1992]

Table 51. Leukaemia incidence

Study base	Indicators	Social scale	N	Male RR	N	Female RR	Study design
Colombia (Cali) 1971–1975 all ages	Social class RR	I II III		1 0.90 0.75		1 1.22 1.41	Data from 1973 census were used for rate denominators. Social class based on area of residence. [Cuello, 1982]
Denmark 1970–1980 all ages	Occupational group RR	Self-employed Employees: I Employees: II Employees: III Employees: IV Skilled workers Unskilled workers	378 41 72 162 77 177 415	0.98 1.04 1.01 1.05 1.05 0.94 1.01	38 5 27 47 112 5 153	1.41 0.91 1.05 1.08 0.96 1.36 0.99	Record-linkage study using 1970 census and 1970–1980 incidence data. Employees classified according to educational level [Lynge, 1990]
Finland 1971–1985 birth cohort: 1906–1945	Social class SIR	Upper white-collar Lower white-collar Skilled workers Unskilled workers		1.15 0.98 1.01 0.93		0.99 0.98 1.01 1.01	Record-linkage study using 1970 census and 1971–1985 incidence data. Social class based on occupation [Pukkala, 1993]
Italy (Torino) 1985–1987 age: 20–69	Educational level OR	University High school Middle school Primary school	15 40 66 101	1 0.81 0.85 0.81	0 31 53 90	– 1 0.92 0.72	Record-linkage study between 1971 and 1981 censuses and 1985–1987 incidence data. ICD-9: 200-208 [Faggiano, 1994]
Italy (Torino) 1985–1987 age: 20–69	Occupational group OR	Managers Clerks Self-employed Manual workers	29 51 26 100	1 0.94 0.85 0.85	0 32 7 19	– 1 0.52 0.65	Record-linkage study between 1971 and 1981 census and 1985–1987 incidence data. ICD-9: 200-208 [Faggiano, 1995]
Italy (Torino) 1985–1987 age: 20–69	Housing tenure OR	Owners Tenants	88 125	1 1.08	71 100	1 1.03	Record-linkage study between 1971 and 1981 censuses and 1985–1987 incidence data. ICD-9: 200-208 [Faggiano, 1994]
Sweden 1961–1970 all ages	Social class SIR	Employees: I Self-employed: II Indep. farmers: III White-collar: IV Blue-collar: V	373 382 506 957 2279	0.98 1.08 1.01 1.01 0.99	– 48 – 530 481	– 1.06 – 1.04 0.98	Record-linkage study between 1961 census and 1961–1970 incidence data. Social class indicator based on occupation [Vagero, 1986]
UK – England and Wales 1971–1981 all ages	Housing tenure SIR	Owner occupier Private rented Council tenant	103 25 44	1.11 0.76 0.89	90 36 43	1.01 1.05 0.89	Record-linkage study between 1971 census and 1971–1981 incidence data (1% sample). UK Registrar General's social class classification [Kogevinas, 1990]
USA 1969–1971 all ages	Family income level OR	>US$ 10 000 Less		1.64 1			Case–control study based on US Third National Cancer Survey' using deaths from other causes as controls Acute lymphocytic leukaemia [Williams, 1977]

Table 51. (Contd) Leukaemia incidence

Study base	Indicators	Social scale	N	Male RR	N	Female RR	Study design
USA 1969–1971 all ages	Family income level OR	>US$ 10 000 Less		0.49 1		1.10 1	Case–control study based on US Third National Cancer Survey, using deaths from other causes as controls Chronic lymphocytic leukaemia [Williams, 1977]
USA 1969–1971 ages	Educational level OR	College Less		1.31 1		0.67 1	Case–control study based on US Third National Cancer Survey, using deaths from all other causes as controls Acute lymphocytic leukaemia [Williams, 1977]
USA 1969–1971 ages	Educational level OR	College Less		0.56 1		0.21 1	Case–control study based on US Third National Cancer Survey, using deaths from all other causes as controls Chronic lymphocytic leukaemia [Williams, 1977]

Socioeconomic differences in cancer survival: a review of the evidence

M. Kogevinas and M. Porta

In the discussion of social inequalities in health there has been much debate on the role of medical care. Large differences in cancer incidence and mortality from cancer have been consistently observed. To understand the potential importance of socioeconomic differences in prompt detection and treatment of cancer it is essential to have data on cancer survival. These have been examined less extensively than differences in cancer incidence. We have reviewed 42 studies on social class differences in cancer survival. Twenty-three studies were conducted in North America, and 15 in western European countries. Twenty-three studies were carried out through population-based cancer registries and 17 through hospitals or hospital-based registries. Seven studies examined survival differences for multiple cancer sites. Social class differences in cancer survival appear remarkably general. Patients in low social classes had consistently poorer survival than those in high social classes. The magnitude of the differences for most cancer sites was fairly narrow, with most relative risks falling between 1 and 1.5. The widest differences were observed for cancers of good prognosis and specifically cancers of the female breast, corpus uteri, bladder and colon. The pattern of the social differences in survival did not vary consistently by sex, country, or source of the study population and did not depend on the socioeconomic indicator used.

In the discussion of social inequalities in health there has been much debate on the role of medical care. Large differences in cancer incidence and mortality from cancer have been consistently observed among a variety of social groups. To understand the potential importance of socioeconomic differences in prompt detection and treatment of cancer it is essential to have data on cancer survival. These have been examined less extensively than differences in cancer incidence. Studies conducted by Cohart in 1955 detected an association between socioeconomic status and cancer survival only for breast cancer (Cohart, 1955). About three decades later, interest in survival patterns was renewed when large differences among ethnic groups in the United States of America became evident (Young et al., 1984). In this chapter the available evidence on the magnitude of socioeconomic differences in cancer survival is reviewed, and then issues of interpretation are briefly discussed. The factors determining the occurrence of social differences in survival are further discussed in the chapters in this book by Auvinen and Karjalainen and by Segnan.

The studies

Studies were identified through MEDLINE (Digital Library Systems, Inc.) and bibliographic references of published studies. Overall, 42 studies on cancer survival differences were reviewed. A few additional studies on less frequent cancers, such as soft-tissue sarcoma and multiple myeloma, have not been included (Savage et al., 1984). For each investigation, a general description of the study (place, source of the study population, time period of enrolment and follow-up, vital status ascertainment, and number of subjects), the socioeconomic indicators used, and brief comments are given in Table 1. Twenty-three studies (55%) were conducted in North America, 15 (36%) studies in western European countries, two studies in Asia and one study in Australia; and one study covered four countries (Table 1). Twenty-five studies (60%) were carried out through population-based cancer reg-

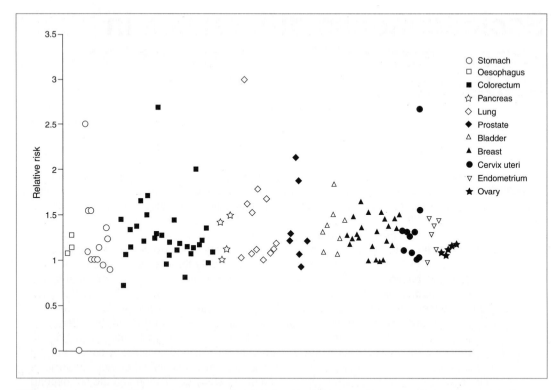

Figure 1. Socioeconomic differences in cancer survival. Relative risks for patients of low versus high socioeconomic status, as observed in 42 studies.

istries and 17 through hospitals or hospital-based registries. Seven studies examined survival differences in multiple (10 or more) cancer sites.

Studies on differences between ethnic groups were not systematically reviewed. In the United States, race is closely related to socioeconomic status; differences in cancer incidence and survival among races have been shown also to relate to socioeconomic factors (Devesa & Diamond, 1980).

Authors of the studies reviewed here used a variety of socioeconomic classifications. Residence was the most frequently used socioeconomic indicator (21 studies), mostly in studies conducted in North America. Census tracts (and less frequently, other units of residence such as census block and county) were ranked on the basis of information on sociodemographic characteristics of the population living in the tract, such as median family income, average education, percentage of working class men or composite indexes such as the Carstairs index. Occupation was used in seven studies either in the context of a social class scheme (in the United Kingdom and Finland) or for simply comparing large population categories such as blue- and white-collar workers. In four studies information was extracted from hospital records concerning type of insurance or type of hospital (for example, public versus private). Level of education was used in six studies, housing (ownership and amenities) in two studies, employment status in one study and Duncan's index of socioeconomic status in one study. In most studies results were reported for a single socioeconomic classification; occasionally, more than one classification was examined. The alternative classifications did not substantially modify the findings.

Results

A summary picture of the findings of all studies on socioeconomic differences in cancer survival is shown in Figure 1. The figure compares case fatality rates for patients of low socioeconomic status patients with those for patients of high socioeconomic status. In studies examining survival in more than two groups, differences between the two extreme groups of the socioeconomic classification are plotted – for example, high versus low income and social class I versus social class V. Approximately 130 independent relative risks, derived from the 40 studies that were reviewed, are shown, ordered by cancer site. Whenever available, relative risks are plotted separately for each sex. Estimates of the relative risk were directly available in those studies using regression or similar techniques. In most studies, however, a classical life table analysis was applied and five-year crude or relative survival rates were provided. In those studies we calculated the social class ratio for the proportion of dead subjects at five years since diagnosis (one minus the five-year survival rate), therefore deriving estimates of the relative risk.

Social class differences in cancer survival after diagnosis appear remarkably consistent across different populations (Figure 1). Nearly all relative risk estimates were above unity, suggesting that those in low social classes have poorer survival than those in high social classes. Most relative risks were low, ranging between values of 1 and 1.5. The highest relative risks were observed for cancers of fairly good, or good, prognosis, such as breast, colorectal, bladder and uterine corpus. The pattern of survival differences did not vary consistently by sex, country, socioeconomic indicator, or source of the study population.

The major findings of the studies are summarized in Tables 2–13. Each of these tables is organized in two sections: one section lists studies in which the parameter modelled was survival (in which case a high value signifies an advantage), and the other lists studies in which the parameter modelled was mortality (in which case a high value indicates a disadvantage).

Four studies examined socioeconomic differences in survival for all neoplasms (Table 2). In all studies, survival was poorest in the low socioeconomic groups.

Socioeconomic differences in survival of patients with oesophageal cancer have been examined in two studies (Table 3). In both studies survival was poorest in the low socioeconomic groups.

Socioeconomic differences in survival of patients with stomach cancer have been examined in eight studies (Table 4). In three studies survival was poorest in the low socioeconomic groups. The reverse pattern was seen in one study, while no consistent pattern was observed in four studies.

Socioeconomic differences in survival of patients with colorectal cancer have been examined in 15 studies (Table 5). Out of the 11 studies examining colon cancer, survival was poorest in the low socioeconomic groups in eight studies. In four of those studies differences were statistically significant or were wider than 10% at five years after diagnosis (one study reported survival at 10 months after diagnosis). No appreciable differences were seen in three studies. Out of the eight studies examining rectal cancer, seven found that survival was poorest in the low socioeconomic group. In three of those studies differences were statistically significant or were wider than 10% at five years after diagnosis (one study reported survival at 10 months after diagnosis). Four studies reported survival jointly for colorectal cancer. In three of those studies survival was poorest in the low socioeconomic group. In two of those studies differences were statistically significant or were wider than 10% at five years after diagnosis.

Socioeconomic differences in survival of patients with pancreatic cancer have been examined in three studies (Table 6). In two studies survival was poorest in the low socioeconomic groups, while no appreciable difference was seen in the third study.

Socioeconomic differences in survival of patients with lung cancer have been examined in 10 studies (Table 7). In eight studies survival was poorest in the low socioeconomic groups. In two of those studies differences were statistically significant or were wider than 10% at five years after diagnosis. No appreciable difference was seen in one study, and risk estimates were not presented in another study.

Socioeconomic differences in survival of patients with prostate cancer have been examined in seven studies (Table 8). In six studies, survival was poorest in the low socioeconomic groups. In three of those studies differences were wider than 10% at five years after diagnosis (one study reported survival at 10 months after diagnosis). The reverse pattern was seen in one study.

Socioeconomic differences in survival of patients with bladder cancer have been examined in five studies (Table 9). In all studies survival was poorest in the low socioeconomic groups. In four of those studies differences were statistically significant or were wider than 10% at five years after diagnosis (one study reported survival at 10 months after diagnosis).

Socioeconomic differences in survival of breast cancer patients have been examined in 24 studies (Table 10). In 19 studies survival was poorest in the low socioeconomic groups. In addition, in a multicentric study the same pattern was seen for three out of four countries. In 13 of those studies differences were statistically significant or were wider than 10% at five years after diagnosis (one study reported survival at 10 months after diagnosis). Three studies showed no association between socioeconomic status and survival and in one study survival was poorest in the high socioeconomic group.

Socioeconomic differences in survival of patients with cervical cancer have been examined in 10 studies (Table 11). In eight studies survival was poorest in the low socioeconomic groups; in five of those studies differences were statistically significant or were wider than 10% at five years after diagnosis (one study reported survival at 10 months after diagnosis). Two studies showed no association between socioeconomic status and survival.

Socioeconomic differences in survival of patients with cancer of the corpus uteri have been examined in six studies (Table 12). In five studies survival was poorest in the low socioeconomic groups; in three of those studies differences were statistically significant or were wider than 10% at five years after diagnosis (one study reported survival at 10 months after diagnosis). The reverse pattern was seen in one study.

Socioeconomic differences in survival of patients with ovarian cancer have been examined in five studies (Table 13). In four studies survival was poorest in the low socioeconomic groups, although differences were narrow. The reverse pattern was seen in one study.

Issues of interpretation

Socioeconomic differences in cancer survival, if not artifactual, may be related to differences in timing of diagnosis, in treatments applied, in the biological characteristics of the neoplasm or in host factors (Vågerö & Persson, 1987). These and other issues are discussed in depth in the chapters by Auvinen and Karjalainen and by Segnan in this volume. Here we briefly address only some factors capable of biasing comparisons among social groups: clinical lead-time bias, variations in staging practices, length bias, and the accuracy of measurements of the cause of death and of social class.

Diagnostic patterns have been shown to affect comparisons of incidence and case fatality rates. The validity of long-term comparisons of survival has also been questioned (Enstrom & Austin, 1977). A problem common to all survival studies is that case fatality rates have been shown to be less valid than incidence or mortality rates. Furthermore, the survival in high socioeconomic groups could appear to be better not because prompt diagnosis altered the natural course of the disease but simply because of lead-time bias – that is, because the diagnosis took place earlier in the natural history of the disease in one group than in the comparison group. Lead-time bias came to be understood and is often considered primarily within the context of screening programmes, when asymptomatic disease is targeted in a well-defined group of persons invited to be screened. However, only a fraction of the 'population of cancers' can be detected by screening programmes; the vast majority of cancers are diagnosed when persons with symptoms seek medical attention and, hence, the survival of most patients can be computed only from clinical diagnosis of symptomatic disease. Measuring survival from the time of onset of symptoms might be thought to overcome some of the disadvantages of measuring it from time of diagnosis, by discounting the effect of diagnostic delays due to socioeconomic barriers to medical diagnosis. However, such a possibility is not without problems of its own – mainly because the perception, assessment, recall and reporting of symptoms differ substantially among social groups. Thus, again, the 'time zero' from which survival is computed may not be similar for the different groups. Therefore, clinical lead-time bias occurs when a decrease in the time of symptomatic disease (or in the duration of symptoms, or in the interval from symptom onset to treatment onset) appears spuriously associated with longer survival (Porta et al., 1991; Maguire et al., 1994).

Is there evidence suggesting that the point in the natural history of the cancer at which diagnosis takes place is uniform or has a similar distribution across social groups? Such information as is available on stage at presentation indicates that high socioeconomic groups are frequently diagnosed earlier. Therefore, clinical (symptomatic) lead-time could be one of the factors contributing to cancer survival differences. Unfortunately, little attention has been paid to quantitative estimation of the magnitude of clinical lead-time, and of the ensuing differences in survival, among different socioeconomic groups. Studies on 'diagnostic delay' could be reoriented to bridge this gap (Maguire et al., 1994).

Analysing differences in survival across socioeconomic groups within strata of the stage of the cancer at diagnosis is one option that might also overcome some of the problems mentioned above. However, this would require that the staging effort was similar in the different social groups. Without similar staging practices, we would face a problem similar to that caused by differing access among groups to screening and diagnosis: if lower social groups had their cancers less accurately staged than higher groups (for example, tumours differentially deemed to be less disseminated in the lower than in the higher classes), stratifying or otherwise adjusting by stage would not be sufficient to produce a valid estimate of differences in survival within each stage stratum. Certainly, the intensity of the staging effort in different socioeconomic groups is difficult to ascertain, since it depends on several factors that are related to each other and to sociocultural factors: the timing in the natural course of the cancer at which patients can and choose to seek medical attention, the quality of care in every given health care setting (which may depend on workload, technology, referral options and so on), and how much value the patient, the family and the health professionals place on an accurate diagnosis and staging of the cancer (Mechanic, 1972; Eisenberg, 1980; Twaddle, 1981; Feinstein et al., 1985; Funch, 1988; Gifford, 1986; Greenberg et al., 1991; Franks & Clancy, 1993). In this context, it is worth remembering that information on stage in cancer registries may be of questionable quality. In a study in the United States, stage was wrongly coded in 20% of the cases. The major misclassifications occurred between regional and distant stages of disease; the percentage of patients presenting with local disease was fairly accurate (Feigl et al., 1982).

Data are also very scant on differences among social groups in the distribution of histological subtypes and of markers of tumour aggressiveness (Hulka et al., 1984). Yet, host–tumour interactions may differ in society, and some clinical and biological factors (such as nutritional and immunological status) may be important mediating variables of their effect upon survival. Given that the rate of tumour growth is related both to survival and to the likelihood of early clinical detection (slower-growing tumours being more amenable to the latter), the possibility of length bias should be kept in mind, too (Porta et al., 1991). It would also be meaningful to assess to what extent the causes of each type of cancer influence its prognosis, since many exposures (for example, occupational and nutritional ones) are unevenly distributed in society.

In the studies reviewed here the proportion of patients lost to follow-up did not generally differ between socioeconomic groups. Yet, diagnosis and certification of the actual cause of death may be another source of error. If valid diagnoses are less common in the disadvantaged social groups, then the case fatality rates for specific cancer could be artificially low. In addition, low social classes may be affected more by 'competing risks', since all-cause mortality is higher in low compared with high social classes. In general, this would leave fewer low socioeconomic status survivors to die of cancer. These problems were addressed in a number of studies examining survival patterns for different sets of case fatality rates, which were calculated for mortality from all causes, all cancers or only the cancer at diagnosis. Statistical models that take into account mortality from competing causes were also used. These analyses suggest that competing risks may not seriously affect the overall patterns of social differences in cancer survival.

Finally, it should be borne in mind that changes in socioeconomic status over the lifespan are seldom part of the analyses in this area. Social class and many other socioeconomic indicators are often measured inaccurately; hence, differences in cancer survival among the social groups involved may be concealed.

Concluding remarks

- A clear pattern was observed in social class differences in cancer survival. Patients in low social classes had consistently poorer survival than those in high social classes.
- The magnitude of the differences for most cancer sites is moderate, with most relative risks falling between 1 and 1.5.
- The widest differences tended to occur in cancers of good prognosis and specifically cancers of the female breast, corpus uteri, bladder and colon.

- Even if in some settings socioeconomic status per se is not considered to be a strong and independent predictor of cancer survival, it is important to analyse how biological and clinical predictors of survival correlate with socioeconomic variables.
- Further efforts should also be devoted to assessing quantitatively the magnitude of clinical lead-time, variations in diagnostic and staging practices, and length bias, and their possible influence on observed differences in survival among different socioeconomic groups.

Acknowledgements

This work was partly funded by a grant from the Ministry of Education and Science, Spain (DGICT SAB95-0189), by Fondo de Investigación Sanitaria, Spain (grant 92/0311) and by CIRIT, Spain (GRQ93-9304).

References

Auvinen, A. (1992) Social class and colon cancer survival in Finland. *Cancer*, 70, 402–409

Auvinen, A., Karjalainen, S. & Pukkala, E. (1995) Social class and cancer patient survival in Finland. *Am. J. Epidemiol.*, 142, 1089–1102

Bain, R.B., Greenberg, R.S. & Whitaker, J.P. (1986) Racial differences in survival of women with breast cancer. *J. Chronic Dis.*, 39, 631–642

Bako, G., Ferenczi, L., Hanson, J., Hill, G.H. & Dewar, R. (1985) Factors influencing the survival of patients with cancer of the stomach. *Clin. Invest. Med.*, 8, 22–28

Bassett, M.T. & Krieger, N. (1986) Social class and black-white differences in breast cancer survival. *Am. J. Public Health*, 76, 1400–1403

Berg, J.W., Ross, R. & Latourette, H.B. (1977) Economic status and survival of cancer patients. *Cancer*, 39, 467–477

Boffetta, P., Merletti, F., Winkelmann, R., Magnani, C., Cappa, A.P. & Terracini, B. (1993) Survival of breast cancer patients from Piedmont, Italy. *Cancer Causes Control*, 4, 209–215

Bonett, A., Roder, D. & Esterman, A. (1984) Determinants of case survival for cancers of the lung, colon, breast and cervix in South Australia. *Med. J. Aust.*, 141, 705–709

Brenner, H., Mielck, A., Klein, R. & Ziegler, H. (1991) The role of socioeconomic factors in the survival of patients with colorectal cancer in Saarland/Germany. *J. Clin. Epidemiol.*, 44, 807–815

Cohart, E.M. (1955) Socioeconomic distribution of cancer of the female sex organs in New Haven. *Cancer*, 8, 34–41

Chirikos, T.N. & Horner, R.D. (1985) Economic status and survivorship in digestive system cancers. *Cancer*, 56, 210–217

Chirikos, T.N., Reiches, N.A. & Moeschberger, M.L. (1984) Economic differentials in cancer survival: a multivariate analysis. *J. Chronic Dis.*, 37, 183–193

Dayal, H.H., Power, R.N. & Chiu, C. (1982) Race and socioeconomic status in survival from breast cancer. *J. Chronic Dis.*, 35, 675–683

Dayal, H.H., Polissar, L. & Dahlberg, S. (1985) Race, socioeconomic status, and other prognostic factors for survival from prostate cancer. *J. Natl. Cancer Inst.*, 74, 1001–1006

Dayal, H., Polissar, L., Yang, C.Y. & Dahlberg, S. (1987) Race, socioeconomic status, and other prognostic factors for survival from colorectal cancer. *J. Chronic Dis.*, 40, 857–864

Devesa, S.S. & Diamond, E.L. (1980) Association of breast cancer and cervical cancer incidences with income and education among whites and blacks. *J. Natl. Cancer Inst.*, 65, 515–528

Eisenberg, L. (1980) What makes persons "patients" and patients "well"? *Am. J. Med.*, 69, 277–286

Ell, K., Nishimoto, R., Mediansky, L., Mantell, J. & Hamovitch, M. (1992) Social relations, social support and survival among patients with cancer. *J. Psychosom. Res.*, 36, 531–541

Enstrom, J.E. & Austin, D.F. (1977) Interpreting cancer survival rates. *Science*, 195, 847–851

Feigl, P., Polissar, L., Lane, W.W. & Guinee, V. (1982) Reliability of basic cancer patient data. *Stat. Med.*, 1, 191–204

Feinstein, A.R., Sosin, D.M. & Wells, C.K. (1985) The Will Rogers phenomenon: stage migration and new diagnostic techniques as a source of misleading statistics for survival in cancer. *New Engl. J. Med.*, 312, 1604–1608

Franks, P. & Clancy, C.M. (1993) Physician gender bias in clinical decisionmaking: screening for cancer in primary care. *Med. Care*, 31, 213–218

Funch, D.P. (1988) Predictors and consequences of symptom reporting behaviors in colorectal cancer patients. *Med. Care*, 26, 1000–1008

Gifford, S.M. (1986) The meaning of lumps: a case study of the ambiguities of risk. In: Janes, C.R., Stall, R. & Gifford, S.M., eds, *Anthropology and epidemiology. Interdisciplinary approaches to the study of health and disease.* Dordrecht, Reidel. pp. 213–246

Gordon, N.H., Crowe, J.P., Brumberg, D.J. & Berger, N.A. (1992) Socioeconomic factors and race in breast cancer recurrence and survival. *Am. J. Epidemiol.*, 135, 609–618

Greenberg, E.R., Baron, J.A., Dain, B.J., Freeman, D.H., Jr, Yates, J.W. & Korson, R. (1991) Cancer staging may have different meanings in academic and community hospitals. *J. Clin. Epidemiol.*, 44, 505–512

Hulka, B.S., Chambless, L.E., Wilkinson, W.E., Deubner, D.C. & McCarty, K.S., Jr (1984) Hormonal and personal effects on estrogen receptors in breast cancer. *Am. J. Epidemiol.*, 119, 692–704

Karjalainen, S. & Pukkala, E. (1990) Social class as a prognostic factor in breast cancer survival. *Cancer*, 66, 819–826

Kato, I., Tominaga, S. & Ikari, A. (1992) The role of socioeconomic factors in the survival of patients with gastrointestinal cancers. *Jpn. J. Clin. Oncol.*, 22, 270–277

Keirn, W. & Metter, G. (1985) Survival of cancer patients by economic status in a free care setting. *Cancer*, 55, 1552–1555

Kogevinas, M. (1990) *OPCS Longitudinal study. Socio-demographic differences in cancer survival* (Series LS No. 5). London, Her Majesty's Stationery Office. pp. 1971–1983

Kogevinas, M., Marmot, M.G., Fox, J. & Goldblatt, P.O. (1991) Socioeconomic differences in cancer survival. *J. Epidemiol. Community Health*, 45, 216–219

Lamont, D.W., Symonds, R.P., Brodie, M.M., Nwabineli, N.J. & Gillis, C.R. (1993) Age, socio-economic status and survival from cancer of cervix in the West of Scotland 1980–87. *Br. J. Cancer*, 67, 351–357

LeMarchand, L., Kolonel, L.N. & Nomura, A.M.Y. (1984) Relationship of ethnicity and other prognostic factors to breast cancer survival patterns in Hawaii. *J. Natl. Cancer Inst.*, 73, 1259–1265

Linden, G. (1969) The influence of social class in the survival of cancer patients. *Am. J. Public Health*, 59, 267–274

Lipworth, L., Abelin, T. & Connelly, R.R. (1970) Socioeconomic-factors in the prognosis of cancer patients. *J. Chronic Dis.*, 23, 105–116

Lipworth, L., Bennett, B. & Parker, P. (1972) Prognosis of nonprivate cancer patients. *J. Natl. Cancer Inst.*, 48, 11–16

Maguire, A., Porta, M., Malats, N., Gallén, M., Piñol, J.L. & Fernandez, E. for the ISDS II Project Investigators (1994) Cancer survival and the duration of symptoms. An analysis of possible forms of the risk function. *Eur. J. Cancer*, 30A, 785–792

Mechanic, D. (1972) Social psychologic factors affecting the presentation of bodily complaints. *New Engl. J. Med.*, 286, 1132–1139

Milner, P.C. & Watts, M. (1987) Effect of socioeconomic status on survival from cervical cancer in Sheffield. *J. Epidemiol. Community Health*, 41, 200–203

Monnet, E., Boutron, M.C., Faivre, J. & Milan, C. (1993) Influence of socioeconomic status on prognosis of colorectal cancer. A population-based study in Cote D'Or, France. *Cancer*, 72, 1165–1170

Morrison, A.S., Lowe, C.R., MacMahon, B., Ravnihar, B. & Yuasa, S. (1977) Incidence risk factors and survival in breast cancer: report on five years of follow-up observation. *Eur. J. Cancer*, 13, 209–214

Murphy, M., Goldblatt, P., Thornton-Jones, H. & Silcocks, P. (1990) Survival among women with cancer of the uterine cervix: influence of marital status and social class. *J. Epidemiol. Community Health*, 44, 293–296

Nandakumar, A., Anantha, N., Venugopal, T.C., Sankaranarayanan, R., Thimmasetty, K. & Dhar, M. (1995) Survival in breast cancer: a population based study in Bangalore, India. *Int. J. Cancer*, 60, 593–596

Nomura, A., Kolonel, L., Rellahan, W., Lee, J. & Wegner, E. (1981) Racial survival patterns for lung cancer in Hawaii. *Cancer*, 48, 1265–1271

Porta, M., Gallén, M., Malats, N. & Planas, J. (1991) Influence of "diagnostic delay" upon cancer survival: an analysis of five tumour sites. *J. Epidemiol. Community Health*, 45, 225–230

Savage, D., Lindenbaum, J., Van Ryzin, J., Struening, E. & Garrett, T.J. (1984) Race, poverty, and survival in multiple myeloma. *Cancer*, 54, 3085–3094

Schrijvers, C.T., Mackenbach, J.P., Lutz, J.M., Quinn, M.J. & Coleman, M.P. (1995a) Deprivation and survival from breast cancer. *Br. J. Cancer*, 72, 738–743

Schrijvers, C.T., Coebergh, J.W., van der Heijden, L.H. & Mackenbach, J.P. (1995b) Socioeconomic variation in cancer survival in the southeastern Netherlands, 1980–1989. *Cancer*, 75, 2946–2953

Stavraky, K.M., Kincade, J.E., Stewart, M.A. & Donner, A.P. (1987) The effect of socioeconomic factors on the early prognosis of cancer. *J. Chronic Dis.*, 40, 237–244

Stavraky, K.M., Donner, A.P., Kincade, J.E. & Stewart, M.A. (1988) The effect of psychosocial factors on lung cancer mortality at one year. *J. Clin. Epidemiol.*, 41, 75–82

Steinhorn, S.C., Myers, M.H., Hankey, B.F. & Pelham, V.F. (1986) Factors associated with survival differences between black women and white women with cancer of the uterine corpus. *Am. J. Epidemiol.*, 124, 85–93

Twaddle, A.C. (1981) Sickness and the sickness career: some implications. In: Eisenberg, L. & Kleinman, A., eds, *The relevance of social science for medicine*. Dordrecht, Reidel. pp. 111–133

Vågerö, D. & Persson, G. (1987) Cancer survival and social class in Sweden. *J. Epidemiol. Community Health*, 41, 204–209

Waxler-Morrison, N., Hislop, T.G., Mears, B. & Kan, L. (1991) Effects of social relationships on survival for women with breast cancer: a prospective study. *Soc. Sci. Med.*, 33, 177–183

Wegner, E.L., Kolonel, L.N., Nomura, A.M. & Lee, J. (1982) Racial and socioeconomic status differences in survival of colorectal cancer patients in Hawaii. *Cancer*, 49, 2208–2216

Young, J.L., Jr, Gloeckler-Ries, L. & Pollack, E.S. (1984) Cancer patient survival among ethnic groups in the United States. *J. Natl. Cancer Inst.*, 73, 341–352

Corresponding author:
M. Kogevinas
Institut Municipal d'Investigació Mèdica, Universitat Autònoma de Barcelona, Carrer del Doctor Aiguader 80, E-08003 Barcelona, Spain

Table 1. Socioeconomic differences in cancer survival: description of the studies

Reference; country	Study design	Social indicator	Comments
Auvinen, 1992; Finland	Registry-based (nationwide); enrolment 1979–1982; follow-up 1987; vital status ascertainment 100%; 1951 women and 1196 men with colon cancer	Social class on the basis of own occupation in 1970 census and, for housewives, of husband's occupation: social class I, professional and administrative; class II, lower administrative and self-employed; class III, skilled workers; class IV, unskilled workers; farmers	Population overlapping with Auvinen et al., 1995
Auvinen et al., 1995; Finland	Registry-based, record-linkage study (1970 census, Finnish Cancer Registry); enrolment 1971–1985; follow-up 1990; vital status ascertainment about 100%; 106 661 subjects aged 25–64 at census; 12 cancer sites	Social class on the basis of own occupation in 1970 census and, for housewives, of husband's occupation: social class I, professional and administrative; class II, lower administrative and self-employed; class III, skilled workers; class IV, unskilled workers	
Bain et al., 1986; USA	Registry-based (Atlanta); enrolment and follow-up 1978–1982; 2858 women with breast cancer	County of residence, in two groups: Fulton county (low status) versus all other counties (high)	Definition and validity of social scale is not well specified
Bako et al., 1985; Canada	Hospital-registry-based (Edmonton); enrolment and follow-up 1969–1973; 332 males and 135 females with stomach cancer	Last occupation: agriculture; blue-collar; white-collar; not in labour force	
Bassett & Krieger, 1986; USA	Registry-based (West Washington Cancer Surveillance System); enrolment and follow-up 1973–1983; vital status ascertainment 97.8% Whites and 96% Blacks; 1506 women with breast cancer	Residence, in two groups on the basis of 1980 census block group characteristics including percentage working class (wage earners in specific occupational categories). Comparisons between blocks with <35% or >35% working class	
Berg et al., 1977; USA	Hospital-based (Tumour Registry of the University of Iowa Hospitals); enrolment 1940–1969; follow-up 1974; vital status ascertainment above 98%; 1621 subjects; 39 cancer types	Economic status, in three categories: private patients (high status); clinic pay patients (mid-to-high status); indigent patients (low status)	
Boffetta et al., 1993; Italy	Pathology department records and hospital registers (Piedmont region); enrolment 1979–1981; follow-up 1987; vital status 95%; 5265 women with breast cancer	Education: less than seven years; seven years or more	

Table 1. (Contd) Socioeconomic differences in cancer survival: description of the studies

Reference; country	Study design	Social indicator	Comments
Bonett et al., 1984; Australia	Registry-based, South Australia; enrolment 1977–1983; vital status around 90%; subjects born in Australia or Europe; four cancer sites (2676 women with breast cancer, 2227 subjects with colon cancer, 2934 subjects with lung cancer and 420 women with cervical cancer)	Residence, in two groups on the basis of median male income of postcode at 1981 census	
Brenner et al., 1991; Germany	Registry-based (Saarland, Germany); enrolment 1974–1983; 2627 colorectal cancer patients aged 45–74	Residence, in three groups on the basis of number of blue-collar workers and persons with less than nine years of schooling in the community	
Chiricos et al., 1984; USA	Hospital-based (Ohio State University Hospital); enrolment and follow-up 1977–1981; 1180 White men	Occupation, classified as blue-collar/white-collar, and economic status (mean dollar income), estimated on the basis of information from 1970 census in three groups (>US$ 13 000; US$ 6000–13 000; <US$ 6000)	
Chirikos & Horner, 1985; USA	Hospital-registry-based (Ohio State University); enrolment and follow-up 1977–1981; 84 men with colorectal cancer	Expected income derived from occupation (on the basis of information at the 1970 census), in three groups (>US$ 13 000; US$ 6000–12 999; US$ <5999)	
Dayal et al., 1982; USA	Hospital-based (Medical College of Virginia); enrolment and follow-up 1968–1972; vital status ascertainment 94%; 323 women with breast cancer	Residence, in three groups on the basis of six 1970 census tract characteristics including education and income	
Dayal et al., 1985; USA	Hospital-based (11 centres); enrolment 1977–1981; follow-up 1984; 2513 Caucasian and Black subjects with prostate cancer	Residence (zip) codes: quartiles on the basis of educational level (percentage of high-school graduates)	
Dayal et al., 1987; USA	Hospital-based (11 centres); enrolment 1977–1982; follow-up 1981; 3617 colon and 1528 rectal cancer patients	Residence (zip) codes: tertiles on the basis of percentage of high-school graduates	
Ell et al., 1992; USA	Hospital-based (Univ. S. California Comprehensive Cancer Center); three cancer sites; enrolment and follow-up dates not available; 166 women with breast cancer	Duncan's socioeconomic index, based on income, education and occupational status	

Table 1. (Contd) Socioeconomic differences in cancer survival: description of the studies

Reference; country	Study design	Social indicator	Comments
Gordon et al., 1992; USA	Hospital-based (Ohio); 1392 women with breast cancer diagnosed between 1974–1985; follow-up 1990; loss to follow-up 2.9%	Residence, using 1980 census tract indices including education, income, and percentage below poverty line	
Karjalainen & Pukkala, 1990; Finland	Registry-based, record-linkage study (Finish Cancer Registry, 1970 population census); enrolment 1971–1980; follow-up 1982; vital status ascertainment, complete; 10 181 women aged 25–69 with breast cancer	Social class on the basis of own occupation in 1970 census and, for housewives, of husband's occupation: social class I, professional and administrative; class II, lower administrative and self-employed; class III, skilled workers; class IV, unskilled workers	Population overlapping with Auvinen et al., 1995
Kato et al., 1992; Japan	Registry-based (Aichi Cancer Registry); enrolment and follow-up 1983–1988; two cancer sites (4485 subjects with stomach cancer and 2618 with colorectal cancer)	Occupation, in four groups. Men: professional; clerical; production; service. Women: professional–clerical; production; service; housewife	
Keirn & Metter, 1985; USA	Hospital-based (City of Hope Medical Center, CA); enrolment 1976–1981; three cancer sites (430 subjects with breast cancer, 265 with colon cancer and 406 with lung cancer)	Economic status, in two groups: low status patients defined as those receiving indigent insurance; high status those with non-indigent insurance	
Kogevinas et al., 1991; England and Wales	Registry-based, record-linkage study (1971 census, National Cancer Registration Scheme); enrolment 1971–1981; follow-up 1983; 6737 men and 6470 women; 18 cancer sites	Housing tenure, in two categories: owner occupiers (high status); council tenants (low status)	Results also available for Registrar General's social class (Kogevinas, 1990)
Lamont et al., 1993; Scotland	Registry-based (West of Scotland); enrolment 1980–1987; 1588 women with invasive cervical cancer	Residence (postcode), using Carstairs–Morris index of deprivation (seven categories)	
LeMarchand et al., 1984; USA	Registry-based (Hawaii Tumour Registry); enrolment 1960–1979; follow-up 1980; vital status ascertainment 93.2%; 2956 women with breast cancer	Residence, in three groups on the basis of 1960 and 1970 census tract characteristics (education and income)	
Linden, 1969; USA	Registry-based (California Tumour Registry); enrolment 1942–1962; follow-up 1966; 1662 White women aged 55–64 with localized breast cancer	Type of hospital: public/county (low status); private (high)	

Table 1. (Contd) Socioeconomic differences in cancer survival: description of the studies

Reference; country	Study design	Social indicator	Comments
Lipworth et al., 1970; USA	Hospital-based (Boston non-private hospitals and clinics); enrolment and follow-up 1957–1963; 79 men and 21 women; 10 cancer sites	Residence, classified in two categories on the basis of median family income in 1960 census tracts: >US$ 5000; <US$ 5000	
Lipworth et al., 1972; USA	Hospital-based (Boston hospitals participating in state cancer registry); enrolment and follow-up 1964–1966; 122 men and 42 women; 10 sites	Two patient groups: private patients (high status); non-private patients (low status)	
Milner & Watts, 1987; UK	Trent Cancer Registry; enrolment 1971–1984; follow-up 1986; 548 women with cervical cancer	Residence. Electoral wards ranked in five groups according to percentage of unskilled and semiskilled workers: 1 (high) to 5 (low)	
Monnet et al., 1993; France	Registry-based (Côte d'Or); enrolment 1976–1980; follow-up 1987; vital status ascertainment 98%; 771 patients with colorectal cancer	Type of housing, as registered in 1970 census in three categories: comfortable; midcomfort; no comfort	
Morrison et al., 1977; four countries	Hospital-based (Boston, MA, USA; Glamorgan, Wales; Slovenia, Yugoslavia; Tokyo, Japan); 3146 women with breast cancer; loss to follow-up less than 2%	Education, in four categories	
Murphy et al., 1990; England and Wales	Registry-based (South Thames); enrolment and follow-up 1977–1981; 1728 women (879 with social class information) with cervical cancer	Social class on the basis of occupation: social class I and II (high); III, IV and V (low)	Not defined whether social class in women is based on own or husband's occupation
Nandakumar et al., 1995; India	Registry-based (Bangalore); enrolment 1982–1989; follow-up 1993; 1514 women with breast cancer	Education, in two groups: illiterate; literate	
Nomura et al., 1981; USA	Registry-based (Hawaii Tumour Registry); enrolment 1960–1974; follow-up 1976; vital status ascertainment 98.5%; 1900 subjects with lung cancer	Residence, in three groups on the basis of census tract information (average income and average education of persons living in the tract)	
Roberts et al., 1990; Scotland	Registry-based (Edinburgh); enrolment 1979; follow-up 1986; 87 women with breast cancer	Residence (postcode sector), in two groups on the basis of census data (percentage of the population in social class IV and V)	

Table 1. (Contd) Socioeconomic differences in cancer survival: description of the studies

Reference; country	Study design	Social indicator	Comments
Rosso et al., pers. commun.; Italy	Registry-based, record-linkage study (1981 census, Piedmont Cancer Registry); enrolment 1985-87; follow-up 1993; 11 053 subjects; 21 cancer sites	Education: primary or less; middle; high; university. Relative risk also provided for housing tenure (owner occupier; council tenant)	Results also available for housing tenure
Shelton et al., 1992; USA	Registry-based (Connecticut Tumour Registry); enrolment 1984–1988; follow-up 1990; vital status ascertainment around 90%; 3711 women with in situ and invasive carcinoma of the cervix	Residence, in three groups on the basis of 1980 census tract information (percentage high-school education)	Results also available for other socioeconomic indicators (percentage living below poverty line, and median family income)
Schrijvers et al., 1995a; England	Registry-based (South Thames Regional Health Authority); 29 676 women with breast cancer diagnosed between 1980 and 1989	Residence (enumeration districts): five groups using Carstairs index (overcrowding, male unemployment, low social class and car ownership)	
Schrijvers et al., 1995b; The Netherlands	Eindhoven Cancer Registry; enrolment 1980–1989; follow-up 1991; loss to follow-up 1%; 15 016 subjects; five cancer sites	Residence (postcode): five groups on the basis of education of head of household	
Stavraky et al., 1988; Canada	Hospital-based (two hospitals, Ontario); enrolment and follow-up 1980–1982; 25–70 years at diagnosis; 224 English-speaking subjects with lung cancer	Education, in three categories: high (≥12 grade); average (grades 8–11); low (<8 grade)	
Stavraky et al., 1987; Canada	Hospital-based (two hospitals, Ontario); enrolment and follow-up 1980–1982; 25–70 years at diagnosis; 975 English-speaking subjects; all cancers combined	Education, in three categories: high (≥12 grade); average (grades 8–11); low (<8 grade)	Results also available for other socioeconomic indicators such as the seven-point Hollinshead scale of occupation
Steinhorn et al., 1986; USA	Registry-based (San Francisco-Oakland, Detroit, Atlanta); enrolment 1973–1977; 5415 women with cancer of the corpus uteri	Residence, in two groups on the basis of 1970 census tract information on median family income and mean highest education	
Vågerö & Persson, 1987; Sweden	Registry-based, record-linkage study (1960 census, Swedish Cancer Registry); enrolment and follow-up 1961–1979; 5936 men and 39 012 women aged 20–64 at diagnosis, economically active in 1960; 13 cancer sites	Occupational status at 1960 census, in two categories: blue-collar; white-collar. (In men, also self-employed agricultural workers)	

Table 1. (Contd) Socioeconomic differences in cancer survival: description of the studies

Reference; country	Study design	Social indicator	Comments
Waxler-Morrison et al., 1991; Canada	Hospital-based (AMEC, Vancouver); enrolment 1980–1981; follow-up 1985; 168 women with breast cancer	Employment status (employed; not employed) and education	
Wegner et al., 1982; USA	Registry-based (Hawaii Tumour Registry); enrolment 1960–1974; follow-up 1981; vital status ascertainment >95%; 1446 subjects with colon cancer and 881 with rectal cancer	Residence, in three groups on the basis of census tract information (average years of education and average income of persons living in the tract)	

Table 2. Socioeconomic differences in cancer survival: all neoplasms

Reference; country	Social scale	Results	Comments
Study modelling survival			
Vågerö & Persson, 1987; Sweden	Men		Results provided in figures only: approximate five-year survival rates were 42% for male white-collar workers, 38% for male blue-collar workers, 64% for female white-collar workers, and 53% for female blue-collar workers
	Blue-collar	Better	
	White-collar	Worse	
	Self-employed		
	Women		
	Blue-collar	Better	
	White-collar	Worse	
Studies modelling mortality			
Chiricos et al., 1984; USA	Blue-collar	1.0	Relative risk (CI not available). Cox regression: P values for socioeconomic status around 0.05. Relative risk not significant when adjusting for stage
	White-collar	0.80	
	>US$ 13 000	0.97	
	US$ 6000–13 000	1.0	
	<US$ 6000	1.24	
Stavraky et al., 1987; Canada	Men		Relative risk (95% CI). Logistic regression, adjusting for age and other variables. Outcome: alive at one year without disease versus alive at one year with disease or dead. Non-employed men (RR = 1.5; 95% CI = 0.9–2.4) and women (RR = 1.4; 95% CI = 0.9–2.2) had worse survival than those employed. Similar results for other socioeconomic indicators such as the seven-point Hollinshead scale of occupation
	Low education	1.0	
	High education	0.9 (0.5–1.6)	
	Women		
	Low education	1.0	
	High education	0.8 (0.9–2.4)	
Kogevinas et al., 1991; England and Wales	Men		Standardized case fatality ratio (standardized for age and period of follow-up). Crude five-year survival rates were 26% for male owner occupiers, 21% for male council tenants, 43% for female owner occupiers and 36% for female council tenants
	Owner occupier	0.92	
	Council tenant	1.10	
	Women		
	Owner occupier	0.94	
	Council tenant	1.05	

CI, confidence interval; RR, relative risk.

Table 3. Socioeconomic differences in survival from oesophageal cancer

Reference; country	Social scale	Results	Comments
Study modelling survival			
Berg et al., 1977; USA	Private	41%	Crude survival rate at six months after diagnosis
	Clinic pay	45%	
	Indigent	38%	
Study modelling mortality			
Kogevinas et al., 1991; England and Wales	Men		Standardized case fatality ratio (standardized for age and period of follow-up). Crude five-year survival rates were 5% for male owner occupiers, 3% for male council tenants, 11% for female owner occupiers and 2% for female council tenants
	Owner occupier	0.93	
	Council tenant	1.03	
	Women		
	Owner occupier	0.92	
	Council tenant	1.16	

Table 4. Socioeconomic differences in survival from stomach cancer

Reference; country	Social scale	Results	Comments
Studies modelling survival			
Berg et al., 1977; USA	Private	40%	Crude survival rate at eight months after diagnosis
	Clinic pay	34%	
	Indigent	37%	
Lipworth et al., 1970; USA	Men		Relative three-year survival rate
	>US$ 5000	27.6%	
	<US$ 5000	11%	
	Women		
	>US$ 5000	0%	
	<US$ 5000	41%	
Lipworth et al., 1972; USA	Men		Crude survival rate at 10 months after diagnosis, adjusted for stage
	Private	39%	
	Non-private	25%	
	Women		
	Private	47%	
	Non-private	30%	
Kato et al., 1992; Japan	Men		Five-year cumulative survival rate. Survival differences between occupations were also observed in a Cox regression analysis, adjusting for age, extent of disease, marital status and residence, but were not statistically significant. Relative risk for professional versus service workers was 0.83 (95% CI = 0.65–1.06) in men and 0.84 (95% CI = 0.68–1.03) in women
	Professional	59%	
	Clerical	51%	
	Production	44%	
	Service	42%	
	Women		
	Professional–clerical	47%	
	Production	40%	
	Service	42%	
	Housewives	37%	

Table 4. (Contd) Socioeconomic differences in survival from stomach cancer

Reference; country	Social scale	Results	Comments
Studies modelling survival			
Vågerö & Persson, 1987; Sweden	Men Blue-collar White-collar Self-employed	No appreciable difference	Relative five-year survival rate. Results provided in figures only for men
Studies modelling mortality			
Bako et al., 1985; Canada	Men Agriculture Blue-collar White-collar Not in labour force Women Agriculture Blue-collar White-collar Not in labour force	 1.0 0.81 (0.57–1.15) 0.82 (0.37–1.83) 0.81 (0.60–1.09) 1.0 1.21 (0.59–2.45) 1.21 (0.67–2.18) 0.93 (0.45–1.92)	Relative risk (95% CI). Cox regression
Studies modelling mortality			
Kogevinas et al., 1991; England and Wales	Men Owner occupier Council tenant Women Owner occupier Council tenant	 0.96 1.06 1.02 0.96	Standardized case fatality ratio (standardized for age and period of follow-up). Crude five-year survival rates were 5% for male owner occupiers, 3% for male council tenants, 6% for female owner occupiers and 8% for female council tenants
Schrijvers et al., 1995b; The Netherlands	High Intermediate Low	1.0 0.92 (0.71–1.20) 0.89 (0.69–1.15)	Cox regression adjusted for age

CI, confidence interval.

Table 5. Socioeconomic differences in survival from colorectal cancer

Reference; country	Social scale	Results	Comments
Studies modelling survival			
Berg et al., 1977; USA	*Colon*		Crude five-year survival rate
	Private	42%	
	Clinic pay	39%	
	Indigent	28%	
	Rectum		
	Private	41%	
	Clinic pay	33%	
	Indigent	24%	
Kato et al., 1992; Japan	*Men*		Colorectal cancer. Five-year cumulative survival rate.
	Professional	62%	Survival differences between occupations were also
	Clerical	57%	observed in a Cox regression analysis, adjusting for
	Production	58%	age, extent of disease, marital status and residence,
	Service	53%	but were not statistically significant. Relative risk for
	Women		professional versus service workers was 0.83
	Professional–clerical	72%	(95% CI = 0.65–1.06) in men and 0.84
	Production	64%	(CI = 0.68–1.03) in women
	Service	57%	
	Housewives	51%	
Keirn & Metter, 1985; USA	*Local stage*		Colon cancer. Median survival (in months) for
	Indigent	53	regional and remote stages, and 75th percentile
	Non-indigent	50	survival (in months) for local stage
	Regional		
	Indigent	41	
	Non-indigent	47	
	Remote		
	Indigent	12	
	Non-indigent	10	
Lipworth et al., 1970, USA	*Men, colon*		Relative three-year survival rate
	>US$ 5000	55%	
	<US$ 5000	38%	
	Women, colon		
	>US$ 5000	36%	
	<US$ 5000	51%	
	Men, rectum		
	>US$ 5000	26%	
	<US$ 5000	24%	
	Women, rectum		
	>US$ 5000	41%	
	<US$ 5000	31%	
Lipworth et al., 1972; USA	*Men, colon*		Crude survival rate at 10 months after diagnosis,
	Private	63%	adjusted for stage
	Non-private	55%	
	Women, colon		
	Private	70%	
	Non-private	51%	

Table 5. (Contd) Socioeconomic differences in survival from colorectal cancer

Reference; country	Social scale	Results	Comments
Studies modelling survival			
	Men, rectum		
	Private	79%	
	Non-private	48%	
	Women, rectum		
	Private	71%	
	Non-private	59%	
Vågerö & Persson, 1987; Sweden	*Men, colon*		Results provided in figures only. Approximate five-year relative survival rates for colon cancer were 47% for male white-collar workers, 44% for male blue-collar workers and 37% for self-employed farmers. Approximate five-year relative survival rates for rectal cancer were 47% for male white-collar workers, 39% for male blue-collar workers and 32% for self-employed farmers
	White-collar	Best	
	Blue-collar	Medium	
	Self-employed	Worse	
	Women, colon		
	White-collar	Better	
	Blue-collar	Worse	
	Men, rectum		
	White-collar	Best	
	Blue-collar	Medium	
	Self-employed	Worse	
	Women, rectum		
	White-collar	Better	
	Blue-collar	Worse	
Studies modelling mortality			
Auvinen, 1992; Finland	I	0.88	Colon cancer. Relative risk. Life-table regression analysis corrected by cause of death and adjusting for age and sex. Five-year survival rates were 50% for social class I and 43% for class IV
	II	0.96	
	III	1.0	
	IV	1.01	
	Farmers	1.08	
	Unknown	1.12	
Bonett et al., 1984; Australia	High	1.0	Colon cancer. Relative risk (95% CI). Cox regression analysis. Five-year survival rates were 37% for low status, 43% for medium status and 53% for high status
	Low	1.26 (1.04–1.52)	
Brenner et al., 1991; Germany	*Colon*		Cox regression adjusted for urbanity, region, year of diagnosis, sex, age and stage
	High	1.00	
	Medium	1.04 (0.87–1.25)	
	Low	1.22 (1.01–1.47)	
	Rectum		
	High	1.00	
	Medium	1.05 (0.86–1.28)	
	Low	1.32 (1.09–1.60)	
Chirikos & Horner, 1985; USA	>US$ 13 000	0.29*	Colorectal cancer. Risk ratios from Cox regression adjusted for age, disease severity and treatment. Middle income is the reference group. *$P < 0.05$
	US$ 6000–12 999	1.0	
	<US$ 5999	0.78	

Table 5. (Contd) Socioeconomic differences in survival from colorectal cancer

Reference; country	Social scale	Results	Comments
Studies modelling mortality			
Dayal et al., 1987; USA	Colon		Cox regression, adjusting for age, sex and race
	High	1.00	
	Medium	0.90	
	Low	0.97	
	Rectum		
	High	1.00	
	Medium	1.02	
	Low	1.09	
Kogevinas et al., 1991; England and Wales	Men, colon		Standardized case fatality ratio (standardized for age and period of follow-up). Crude five-year survival rates for colon cancer were 25% for male owner occupiers, 13% for male council tenants, 26% for female owner occupiers and 25% for female council tenants. Crude five-year survival rates for rectal cancer were 27% for male owner occupiers, 20% for male council tenants, 24% for female owner occupiers and 33% for female council tenants
	Owner occupier	0.89	
	Council tenant	1.28	
	Women, colon		
	Owner occupier	0.92	
	Council tenant	1.02	
	Men, rectum		
	Owner occupier	0.92	
	Council tenant	1.09	
	Women, rectum		
	Owner occupier	1.04	
	Council tenant	0.85	
Monnet et al., 1993; France	Comfortable	1.0	Colorectal cancer. Cox regression, adjusting for all prognostic variables. Five-year relative survival rates were 39% for comfortable housing, 22% for midcomfort and 12% for no comfort
	Mid-comfort	1.49 (1.20–1.72)	
	No comfort	2.01 (1.29–3.18)	
Schrijvers et al., 1995b; The Netherlands	High	1.0	Colorectal. Cox regression adjusted for age and period of follow-up
	2	1.00 (0.79–1.28)	
	3	1.06 (0.87–1.30)	
	4	1.15 (0.95–1.40)	
	Low	1.17 (0.97–1.41)	
Wegner et al., 1982; USA	Colon		Relative risk (95% CI). Cox regression. Risk ratios adjusted for age, sex, race and stage
	High	0.82 (0.66–1.02)	
	Middle	0.96 (0.80–1.17)	
	Low	1.0	
	Rectum		
	High	0.79 (0.60–1.05)	
	Middle	0.79 (0.61–1.03)	
	Low	1.0	

CI, confidence interval.

Table 6. Socioeconomic differences in survival from cancer of the pancreas

Reference; country	Social scale	Results	Comments
Studies modelling survival			
Berg et al., 1977; USA	Private Clinic pay Indigent	48% 37% 34%	Crude survival rate at four months after diagnosis
Vågero & Persson, 1987; Sweden	Blue-collar White-collar Self-employed	No appreciable differences	Results provided in figures only; approximate five-year survival rates around 5%
Study modelling mortality			
Kogevinas et al., 1991; England and Wales	Men Owner occupier Council tenant Women Owner occupier Council tenant	 0.96 1.07 0.96 1.45	Standardized case fatality ratio (standardized for age and period of follow-up). Crude one-year survival rates were 9% for male owner occupiers and male council tenants, 11% for female owner occupiers and 5% for female council tenants

Table 7. Socioeconomic differences in survival from lung cancer

Reference; country	Social scale	Results	Comments
Studies modelling survival			
Berg et al., 1977; USA	Private Clinic pay Indigent	40% 34% 26%	Crude survival rate at nine months after diagnosis
Keirn & Metter, 1985; USA	*Local stage* Indigent Non-indigent *Regional* Indigent Non-indigent *Remote* Indigent Non-indigent	 15 27 15 19 7 7	Median survival (in months) for regional and remote stages, and 75th percentile survival for local stage
Lipworth et al., 1970 USA	Men >US$ 5000 <US$ 5000 Women >US$ 5000 <US$ 5000	 10% 9.6% 15% 5%	Relative three-year survival rate
Lipworth et al., 1972 USA	Men Private Non-private Women Private Non-private	 38% 23% 35% 32%	Crude survival rate at 10 months after diagnosis, adjusted for stage
Vågerö & Persson, 1987; Sweden	Men Blue-collar White-collar Self-employed Women Blue-collar White-collar	No appreciable differences	Results provided in figures only; approximate five-year survival rates in males were 10–15%
Studies modelling mortality			
Bonett et al., 1984; Australia	High Low	NA	Cox regression analysis; results not presented; differences not statistically significant
Kogevinas et al., 1991; England and Wales	Men Owner occupier Council tenant Women Owner occupier Council tenant	 0.96 1.04 0.94 1.06	Standardized case fatality ratio (standardized for age and period of follow-up). Crude five-year survival rates (not age-adjusted) were 6% for male owner occupiers, 5% for male council tenants, 8% for female owner occupiers and 3% for female council tenants
Nomura et al., 1981 USA	High Medium Low	1.0 1.17 (1.0–1.4) 1.14 (0.97–1.3)	Relative risk (95% CI). Cox regression, adjusting for age, sex, race and stage

Table 7. Socioeconomic differences in survival from lung cancer

Reference; country	Social scale	Results	Comments
Studies modelling mortality			
Schrijvers et al., 1995b; The Netherlands	High	1.0	Cox regression, adjusting for age and period of follow-up
	2	0.97 (0.80–1.18)	
	3	1.05 (0.90–1.24)	
	4	1.14 (0.98–1.33)	
	Low	1.18 (1.02–1.36)	
Stavraky et al., 1988; Canada	Low	1.0	Relative risk (95% CI). Logistic regression, adjusting for age, sex and various psychosocial factors. Increased risk in subjects with reserved personality, extremely sober or enthusiastic personality, and persons in high need for one aspect of social support
	Average	1.2 (0.4–3.4)	
	High	0.6 (0.2–1.9)	

CI, confidence interval; NA, not available.

Table 8. Socioeconomic differences in survival from cancer of the prostate

Reference; country	Social scale	Results	Comments
Studies modelling survival			
Berg et al., 1977; USA	Private Clinic pay Indigent	44% 29% 20%	Crude five-year survival rate. 'Adjusted' five-year survival rates (only cancer deaths) were 60% for private patients, 49% for clinic pay and 43% for indigent
Lipworth et al., 1970; USA	>US$ 5000 <US$ 5000	62% 52%	Relative three-year survival rate
Lipworth et al., 1972; USA	Private Non-private	89% 71%	Crude survival rate at 10 months after diagnosis, adjusted for stage
Vågerö & Persson, 1987; Sweden	Blue-collar White-collar Self-employed	Slightly better Worse Worse	Relative five-year survival rate. Results provided in figures only; approximate survival rates were 53% for white-collar workers and 50% for blue-collar and self-employed workers
Studies modelling mortality			
Dayal et al., 1985; USA	Highest Mid-high Mid-low Lowest	1.0 1.50 1.49 1.86	Cox regression; relative risk adjusted for age and sex; no significant differences by treatment modality; differences observed within each stage of the disease
Kogevinas et al., 1991; England and Wales	Owner occupier Council tenant	1.03 0.94	Standardized case fatality ratio (standardized for age and period of follow-up). Crude five-year survival rates (not age-adjusted) were 21% for owner occupiers and 25% for council tenants
Schrijvers et al., 1995b; The Netherlands	High Intermediate Low	1.0 1.05 (0.77–1.43) 1.20 (0.95–1.59)	Cox regression adjusted for age

Table 9. Socioeconomic differences in survival from bladder cancer

Reference; country	Social scale	Results	Comments
Studies modelling survival			
Berg et al., 1977; USA	Private	42%	Crude five-year survival rate. 'Adjusted' five-year survival rates (only cancer deaths) were 58% for private patients, 53% for clinic pay and 43% for indigent
	Clinic pay	35%	
	Indigent	23%	
Lipworth et al., 1970; USA	*Men*		Relative three-year survival rate
	>US$ 5000	70%	
	<US$ 5000	54%	
	Women		
	>US$ 5000	55%	
	<US$ 5000	51%	
Lipworth et al., 1972; USA	*Men*		Crude survival rate at 10 months after diagnosis, adjusted for stage
	Private	87%	
	Non-private	63%	
	Women		
	Private	84%	
	Non-private	55%	
Vågerö & Persson, 1987; Sweden	*Men*		Results provided in figures only; approximate five-year survival rates were 73% for male white-collar workers, 70% for self-employed and 68% for male blue-collar workers
	Blue-collar	Worse	
	White-collar	Better	
	Self-employed	Medium	
	Women		
	Blue-collar	Worse	
	White-collar	Better	
Study modelling mortality			
Kogevinas et al., 1991; England and Wales	*Men*		Standardized case fatality ratio (standardized for age and period of follow-up). Crude five-year survival rates were 43% for male owner occupiers, 38% for male council tenants, 49% for female owner occupiers and 33% for female council tenants
	Owner occupier	0.91	
	Council tenant	1.11	
	Women		
	Owner occupier	0.83	
	Council tenant	1.17	

Table 10. Socioeconomic differences in survival from female breast cancer

Reference; country	Social scale	Results	Comments
Studies modelling survival			
Bain et al., 1986; USA	High	78.6%	Three-year survival rate
	Low	82.1%	
Berg et al., 1977; USA	Private	54%	Crude five-year survival rate
	Clinic pay	45%	
	Indigent	37%	
Dayal et al., 1982; USA	High	50%	Approximate five-year survival rate
	Medium	42%	
	Low	39%	
Keirn & Metter, 1985; USA	*Local stage*		Median survival (in months) for remote stage, 75th percentile for regional stage and 80th percentile for local stage
	Non-indigent	53	
	Indigent	32	
	Regional		
	Non-indigent	43	
	Indigent	32	
	Remote		
	Non-indigent	19	
	Indigent	29	
Linden, 1969; USA	Private hospital	86%	Five-year relative survival rate
	Public hospital	68%	
Lipworth et al., 1970; USA	>US$ 5000	71%	Relative three-year survival rate
	<US$ 5000	62%	
Lipworth et al., 1972; USA	Private	89%	Crude survival rate at 10 months after diagnosis, adjusted for stage
	Non-private	73%	
Morrison et al., 1977; USA, Wales, Yugoslavia, Japan	*USA*		Five-year age-adjusted survival rate
	16+	69%	
	12–15	63%	
	8–11	53%	
	<8	52%	
	Wales		
	12–15	67%	
	8–11	52%	
	Yugoslavia		
	12–15	41%	
	8–11	60%	
	<8	39%	
	Japan		
	12–15	80%	
	8–11	75%	
	<8	72%	
Roberts et al., 1990; Scotland	High	70%	Approximate five-year survival rate
	Low	70%	
Vågerö & Persson, 1987; Sweden	White-collar	72%	Approximate relative five-year survival rate
	Blue-collar	65%	

Table 10. (Contd) Socioeconomic differences in survival from female breast cancer

Reference; country	Social scale	Results	Comments
Studies modelling mortality			
Auvinen, 1995; Finland	I II III IV	0.75 (0.65–0.86) 0.85 (0.76–0.94) 0.93 (0.85–1.03) 1.0	Risk ratio adjusted for age and year of diagnosis. Five-year cumulative survival: class I, 77%; class II, 75%; class III, 73%; class IV, 72%
Bassett & Krieger, 1986; USA	High Low	1.0 1.52 (1.28–1.88)	Relative risk adjusted for race, age, stage and histology
Boffetta et al., 1993; Italy	≥ 7 years < 7 years	0.7 (0.4–1.1) 1	Relative risk adjusted for age; analysis limited to subjects living in Torino
Bonett et al., 1984; Australia	High Low	1.0 1.35 (1.0–1.7)	Cox regression analysis
Ell et al., 1992; USA	Duncan's index	0.996	Relative risk
Gordon et al., 1992; USA	High Low	1.0 1.49 (1.17–1.89)	Cox regression
Karjalainen & Pukkala, 1990; Finland	I II III IV	0.78 (0.68–0.90) 0.85 (0.77–0.93) 0.92 (0.88–0.97) 1.0	Risk ratio adjusted for age, follow-up, calendar period of diagnosis, stage and the interaction of stage and follow-up period. Overlapping with Auvinen et al., 1995
Kogevinas et al., 1991; England and Wales	Owner occupier Council tenant	0.99 0.97	Standardized case fatality ratio. Crude five-year survival rates: 50% for owner occupiers and 52% for council tenants
LeMarchand et al., 1984; USA	High Medium Low	1.0 0.96 (0.77–1.2) 1.23 (0.97–1.57)	Relative risk (95% CI), adjusting for age, stage, race, histology and marital status
Nandakumar et al., 1995; India	Illiterate Literate	1.0 0.7 (0.6–0.8)	Relative risk (95% CI), adjusting for religious group, marital status and clinical extent of the disease. Five-year survival rates were 35% (illiterate) and 46% (literate)
Rosso et al., pers. commun.; Italy	University High Middle Low	0.89 (0.54–1.49) 0.94 (0.70–1.27) 1.01 (0.99–1.51) 1.0	Relative risk adjusted for age, place of birth and housing
Schrijvers et al., 1995a; England	Affluent 2 3 4 Deprived	1.0 1.15 (1.05–1.27) 1.30 (1.18–1.44) 1.31 (1.18–1.46) 1.35 (1.16–1.579	Cox regression adjusted for follow-up period and period of diagnosis. Women aged 30–64
Schrijvers et al., 1995b; The Netherlands	High 2 3 4 Low	1.0 1.06 (0.84–1.33) 1.04 (0.86–1.26) 1.15 (0.96–1.38) 1.18 (0.99–1.42)	Cox regression adjusted for age and period of follow-up
Waxler-Morrison et al., 1991; Canada	Employed Not employed	1.0 1.52	Cox regression adjusting for nodal status, stage of disease, marital status and four other factors

CI, confidence interval.

Table 11. Socioeconomic differences in survival from cervical cancer

Reference; country	Social scale	Results	Comments
Studies modelling survival			
Berg et al., 1977; USA	Private Clinic pay Indigent	73% 67% 57%	Crude five-year survival rate
Bonett et al., 1984; Australia	High Low	73% 60%	Four-year survival rate; results not statistically significant in Cox regression analysis
Lipworth et al., 1970; USA	>US$ 5000 <US$ 5000	71% 55%	Relative three-year survival rate
Lipworth et al., 1972; USA	Private Non-private	84% 77%	Crude survival rate at 10 months after diagnosis, adjusted for stage
Murphy et al., 1990; England and Wales	I and II III IV and V	No appreciable differences	Kaplan-Meier survival curves shown in figure
Vågerö & Persson, 1987; Sweden	White-collar Blue-collar	Better Worse	Approximate relative five-year survival rates were 70% for white-collar workers and 65% for blue-collar workers (results provided in figure)
Studies modelling mortality			
Kogevinas et al., 1991; England and Wales	Owner occupier Council tenant	0.95 0.97	Standardized case fatality ratio. Crude five-year survival rates were 54% for owner occupiers and 53% for council tenants
Lamont et al., 1993; Scotland	1 Most affluent 2 Affluent 3 Above average 4 Average 5 Below average 6 Deprived 7 Most deprived	0.56 (0.4–0.7) 0.62 (0.5–0.8) 0.86 (0.8–0.97) 1.06 (0.96–1.2) 1.02 (0.9–1.1) 1.16 (1.0–1.3) 1.52 (1.3–1.7)	Age-standardized cancer morbidity ratio
Milner & Watts, 1987 UK	1 (high) 2 3 4 5 (low)	0.96 0.88 0.98 0.80 1.21	Observed over expected deaths adjusted for age (ratios calculated by Kogevinas and Porta for this review)
Shelton et al., 1992 USA	High Medium Low	0.64 (0.4–1.0) 0.96 (0.6–1.4) 1.0	Relative risk (95% CI). Logistic regression adjusting for age, race and stage of the disease

CI, confidence interval.

Table 12. Socioeconomic differences in survival from cancer of the corpus uteri

Reference; country	Social scale	Results	Comments
Studies modelling survival			
Berg et al., 1977; USA	Private Clinic pay Indigent	70% 66% 57%	Crude five-year survival rate. Adjusting for stage of the disease narrowed survival differences
Lipworth et al., 1970; USA	>US$ 5000 <US$ 5000	75% 78%	Relative three-year survival rate
Lipworth et al., 1972; USA	Private Non-private	88% 61%	Crude survival rate at 10 months after diagnosis, adjusted for stage
Vågerö & Persson, 1987; Sweden	Blue-collar White-collar	Better Worse	Relative five-year survival rate. Results provided in figures only; approximate survival rates were 88% for white-collar workers and 82% for blue-collar workers
Studies modelling mortality			
Kogevinas et al., 1991; England and Wales	Owner occupier Council tenant	0.85 1.20	Standardized case fatality ratio (standardized for age and period of follow-up). Crude five-year survival rates were 65% for owner occupiers and 54% for council tenants
Steinhorn et al., 1986; USA	High income Low income	1.0 1.33	Relative risk (95% CI not available) for adenocarcinoma ($P < 0.01$); Cox regression adjusting for race, age, stage, study centre and education. Similar results for education (RR = 1.18; $P < 0.05$). Relative risk for sarcomas not significant for income (RR = 0.83) but significant for education (RR = 1.86; $P < 0.05$)

CI, confidence interval.

Table 13. Socioeconomic differences in survival from ovarian cancer

Reference; country	Social scale	Results	Comments
Study modelling survival			
Berg et al., 1977; USA	Private Clinic pay Indigent	44% 43% 40%	Crude survival rate at 18 months after diagnosis
Lipworth et al., 1970; USA	>US$ 5000 <US$ 5000	32% 30%	Relative three-year survival rate
Lipworth et al., 1972; USA	Private Non-private	58% 55%	Crude survival rate at 10 months after diagnosis, adjusted for stage
Vågerö & Persson, 1987; Sweden	Blue-collar White-collar	Better Worse	Relative five-year survival rate. Results provided in figures only; approximate survival rates were 34% for white-collar workers and 30% for blue-collar workers
Study modelling mortality			
Kogevinas et al., 1991; England and Wales	Owner occupier Council tenant	0.94 1.07	Standardized case fatality ratio (standardized for age and period of follow-up). Crude five-year survival rates were 26% for owner occupiers and 19% for council tenants

General explanations for social inequalities in health

M. Marmot and A. Feeney

Life expectancy has always differed according to status in society, with a higher mortality among those of lower social status. Although cancer and cardiovascular diseases are more common as causes of death in rich than in poor societies, in industrialized countries the major causes of death are more common in those of lower social status. In this chapter, the magnitude of socioeconomic differences in health is examined using different measures of socioeconomic status, and methodological issues relating to these measures are discussed. Much of the discussion about social inequalities in health has been focused on the health disadvantage of those of lowest socioeconomic status. However, data from the Whitehall studies show that the social gradient in morbidity and mortality exists across employment grades in British civil servants, none of whom is poor by comparison with people in developing countries, suggesting that there are factors that operate across the whole of society. A number of potential explanations are considered here. The magnitude of socioeconomic differences in health varies between societies, and over time within societies. This suggests that identification of factors that influence socioeconomic status and health, and the pathways by which they operate, is an important public health task that could lay the basis for a reduction in inequalities in health.

Inhabitants of poor countries have a shorter life expectancy than those of rich countries. There are exceptions, but low levels of gross national product (GNP) are associated with a high toll from infectious diseases – in childbirth, in the early years of life, in adulthood and in old age. The change from predominantly communicable to non-communicable diseases as causes of death in rich countries led to the notion of diseases of affluence. We now know that this view is misleading. Although cancer and cardiovascular diseases are more common as causes of death in rich compared with poor countries, in industrialized countries the major causes of death are all more common in those of lower social status. Thus, as far as we can tell, life expectancy has always differed according to status in society, even though the causes of death that make up that higher risk have changed. In the past, the task was to explain, and take action to relieve, social inequalities in deaths from communicable disease; now the public health task is to explain, and lay the basis for action on, social differentials in cancer, cardiovascular disease and accidents.

There are at least three reasons to be concerned by inequalities in health. First, governments have set targets for improvements in health. If improvements occur at a slower rate in large portions of the population, then the remainder must show much greater improvement to keep up the average for society as a whole. Second, reduction in inequalities in health has traditionally been a goal of public health for ethical and moral reasons. Third, the level and distribution of health is an important tangible indicator of the level of well-being in society.

Geoffrey Rose taught that the proportion of 'deviants', or high-risk people, is predicted by the mean (Rose & Day, 1990). Applied to a behaviour such as alcohol intake the implication is that we should concentrate not only on the high-risk people like heavy drinkers, but on shifting the whole distribution. Might this be taken as implying that we should ignore the problem of social variations in health, but concentrate on shifting the distribution? The argument might be that there will always be socioeconomic differences within a society, and so we should not attempt to change the immutable. The response to this is twofold. First, the magnitude of socioeconomic differences in health varies between societies, and over time within societies. This might be either because the

Table 1. Mortality of men aged 15–64 by social class in England and Wales, 1930–1982[a] (SMRs)

Social class		1930–1932	1949–1953	1959–1963[b]	1970–1972	1979–1980[c]	1982–1983[c]
I	Professional	90		86	75	77	66
II	Intermediate		94	92	81	81	76
III-NM	Skilled non-manual	97		101	100	99	94
III-M	Skilled manual					106	106
IV	Partly skilled		102	104	103	114	116
V	Unskilled	111	118	127	137	165	

[a]SMRs; standardized mortality ratios, all men = 100.
[b]Adjusted figures; occupations reclassified according to 1950 classification.
[c]Men aged 20–64.

magnitude of socioeconomic differences varies or because the factors influencing the link between socioeconomic status and health vary. This leads to a second response: an aim of research in this area is to understand the reasons for the link between socioeconomic status and health, in order to inform action to change them.

Inequalities or social inequalities?

Differences between individuals in genetic endowment may well contribute to differences between individuals in life expectancy. In the unlikely event that all individuals were subject to the same set of environmental influences they would not all flourish, age and die at the same rate. As Rose has made clear, however, the determinants of individual risks of disease – why one individual gets sick and another remains healthy – may be different from the determinants of population rates of disease (Rose, 1992). We are concerned with social inequalities in health, the reasons for them, and ways of reducing their magnitude. The causes of these social inequalities are likely to be different from the causes of individual differences, where genetic factors will play a bigger role.

Le Grand and Illsley, for example, have focused on individual differences in life expectancy (Le Grand, 1989; Illsley & Le Grand, 1987). They have shown that variability (using Gini coefficients) around age at death has become less, and argue therefore that there is less inequality in society. This focus on individuals, however, answers a different question from the one we address of differences in health among social groups. If the determinants of variations in health and disease between individuals are different from the determinants of variations among social groups, it is quite possible for there to be no diminution in the relative differences in health between social groups at the same time as there has been a general improvement in health. If fewer people die prematurely, variation in age at death will be less, but this may still be distributed unequally across social groups.

There are other sources of inequality in society, among them gender and race/ethnicity. Gender differences are of great interest and importance but are treated by others. In England and Wales at least, ethnic differences in mortality cannot easily be explained on the basis of conventional social class descriptions (Marmot et al., 1984a). There must be other explanations of ethnic differences; however, these are not reviewed here.

Persistent social differences in mortality within countries

Routine health statistics in England and Wales have been used to demonstrate socioeconomic differentials in mortality since they were first collected. Chadwick reported that in 1842 the average ages at death in several occupationally defined groups were as follows: 'gentlemen and persons engaged in professions, and their families… 45 years; tradesmen and their families... 26 years; mechanics, servants and labourers, and their families... 16 years' (Chadwick, 1965). Since 1921, mortality data have been available for broad occupation-based social class groups (Pamuk, 1985), providing a unique historical series that demonstrates the persistence of socioeconomic differentials in mortality risk against the background of overall improvement in life expectancy.

Table 1 shows mortality in England and Wales according to the Registrar General's social classes, for men of working age (Blane et al., 1992, pers. commun.). This is the classification that has been standard in British statistics, and is based on classifying occupations according to status and level of responsibility. The standardized mortality ratios (SMRs) allow for comparisons of relative differences in death rates at one period. They show trends over time in relative differences, not in absolute rates. What these data do not show is that the apparent widening of mortality differentials between the 1930s

Table 2. Annual age-adjusted rate of years of potential life lost per 1000 population for deaths from principal causes in England and Wales, 1971 and 1981[a]

Social class	All causes	Ischaemic heart disease[b]	Malignant neoplasms[c]	Accidents and violence[d]
1971 (Men aged 15–64)				
I	48	14	12	10
II	52	15	12	11
III-NM	63	19	14	11
III-M	66	18	16	13
IV	77	19	17	19
V	101	21	19	28
1981 (Men aged 16–64)				
I	37	12	11	10
II	42	14	12	11
III-NM	53	18	14	11
III-M	58	20	16	16
IV	68	21	17	20
V	103	29	24	35
1971 (Women aged 15–64)				
I	29	2	13	4
II	30	2	13	5
III-NM	35	3	14	5
III-M	39	4	15	3
IV	42	4	15	5
V	53	5	17	8
1981 (Women aged 16-59)				
I	13	1	8	3
II	15	1	8	3
III-NM	18	1	9	4
III-M	18	2	9	3
IV	22	2	10	4
V	28	3	12	6

[a]Figures were standardized to total population of England and Wales, 1981.
[b]ICD (International Classification of Diseases) codes 410-414, 9th edn.
[c]ICD codes 140-209, 9th edn.
[d]ICD codes 800-999, 9th edn.

Table 3. Age-adjusted mortality in 10 years (and number of deaths) by civil service grade and cause of death

Cause of death (ICD code)	10-year mortality % (number of deaths)				Relative mortality[a]				X-test for trend
	Administrators	Professional/ executive	Clerical	Other	Administrators	Professional/ executive	Clerical	Other	
Lung cancer (162-1)	0.35 (3)	0.73 (79)	1.47 (53)	2.33 (59)	0.5	1.0	2.2	3.6	54.62
Other cancer (140-239) (162-1)	1.26 (12)	1.70 (195)	2.16 (73)	2.23 (46)	0.8	1.0	1.4	1.4	7.08
Coronary heart disease (410-414)	2.16 (17)	3.58 (399)	4.90 (160)	6.59 (128)	0.6	1.0	1.4	1.7	38.24
Cerebrovascular disease (430-435)	0.13 (1)	0.49 (51)	0.64 (23)	0.58 (14)	0.3	1.0	1.4	1.2	1.70
Other cardiovascular (404, 420-429, 440-458)	0.40 (4)	0.54 (58)	0.72 (24)	0.85 (24)	0.9	1.0	1.4	2.0	6.95
Chronic bronchitis (491-492)	0.0 (0)	0.08 (8)	0.43 (15)	0.65 (13)	0	1.0	6.0	7.3	21.01
Other respiratory (460-490, 493-519)	0.21 (2)	0.22 (24)	0.52 (18)	0.87 (15)	1.1	1.0	2.6	3.1	11.99
Gastrointestinal disease (520-577)	0.0 (0)	0.13 (15)	0.20 (7)	0.45 (15)	0	1.0	1.6	2.8	6.26
Genitourinary disease (580-607)	0.09 (1)	0.09 (10)	0.07 (2)	0.24 (6)	1.3	1.0	0.7	3.1	2.46
Accident and violence (800-949, 960-978)	0.0 (0)	0.13 (17)	0.17 (5)	0.20 (3)	0	1.0	1.4	1.5	1.36
Suicide (950-949, 980-989)	0.11 (1)	0.14 (18)	0.15 (4)	0.25 (4)	0.7	1.0	1.0	1.9	0.97
Other deaths	0.0 (0)	0.16 (18)	0.26 (9)	0.40 (6)	0	1.0	1.9	2.0	4.18
Causes not related to smoking[b]									
Cancer	0.86 (9)	1.24 (145)	1.53 (50)	1.57 (33)	0.8	1.0	1.3	1.4	4.70
Non-cancer	1.00 (10)	1.94 (216)	2.76 (93)	4.19 (82)	0.6	1.0	1.5	2.0	31.83
All causes	4.73 (41)	8.00 (892)	11.67 (393)	15.64 (326)	0.6	1.0	1.6	2.1	144.05

[a] Calculated from logistic equation adjusting for age.
[b] All causes less 140-141, 143-149, 150, 157, 160-163, 188-189, 200, 202, 410-414, 491, 492.

and the 1980s took place against a background of falling death rates.

One of the problems with the SMR as a way of controlling for age is that it weights more heavily mortality at older ages. Higher relative differences at younger ages would be masked by this technique (see below). An alternative weighting strategy is to use years of potential life lost (Blane et al., 1990). This gives more weight to deaths occurring at younger ages. Table 2 shows the striking social gradient in potential years of life lost. For men up to age 64, neoplasms and ischaemic heart disease account for about the same number of years of potential life lost. For women, neoplasms are more important.

A second implication from Table 2 is that there is a social gradient in the three major causes of death shown. This is shown in more detail with data from the Whitehall study of British civil servants (Table 3) (Marmot et al., 1984b). Grade of employment is used as the marker of socioeconomic status. There is a social gradient for all the major causes of death. The magnitude of the gradient may vary from one cause to another; it is steeper for chronic respiratory disease and lung cancer, which are strongly related to smoking, but the gradient is also present in diseases not related to smoking.

Potential problems in interpretation
Apart from the difficulty of international comparisons, there are several potential problems in interpreting the national data from England and Wales: lack of comparability between numerator and denominator; changing class composition; doubt about the appropriateness of the Registrar General's social classes; doubt about the appropriateness of the classification for women; and whether these differences in mortality are an effect of belonging to a certain class or whether degree of healthiness, or lack of it, determine both the occupational class and mortality risk – that is, there is selection on the basis of health.

The numerator/denominator argument arises because the death records are not linked to census data that supply the estimates of population at risk (Office of Population Censuses and Surveys, 1978). The recording of occupation, and hence assignment to social class, at death may differ from that at the census, leading to biased estimates of mortality ratios. While this argument may apply to data in the Registrar General's supplements on occupational mortality, it cannot apply to longitudinal studies such as the Whitehall study data shown in Table 3 (Marmot et al., 1984b), or the Office of Population Censuses and Surveys (OPCS) longitudinal study, which follows a 1% sample of the 1971 census. Comparison between the longitudinal study and the 'cross-sectional' approach to examining social inequalities showed similar relative differences between the classes (Office of Population Censuses and Surveys, 1978; Goldblatt, 1990).

A version of the numerator/denominator bias was raised as casting doubt on time trends. It was argued at the time of production of the *1979/83 Decennial Supplement* that the size of the numerator/denominator bias may have changed between 1971 and 1981 (Office of Population Censuses and Surveys, 1986). If, for example, some people whose occupation would have been classed as unskilled manual (class V) in 1971 had their occupation classified as semiskilled (class IV) in 1981, this would reduce the apparent size of the denominator in class V. If a similar change did not apply to coding of occupation at death, the numerator in class V would not be reduced in the same way. This could account for the apparently increasing relative mortality disadvantage in class V, with the SMR increasing from 137 to 165.

One way of addressing this issue was simply to combine the three non-manual classes and the three manual into two groups (Marmot & McDowall, 1986). It is quite unlikely that there would be misclassification from manual to non-manual. The results are shown in Figure 1. They confirm that for total mortality, and for lung cancer, coronary heart disease, and stroke, the social differential in mortality has widened. Unlike the SMRs shown in Table 1, these have been calculated using the 1979–1983 rates as standard. This allows comparison of absolute as well as relative changes in mortality over time. It shows that the mortality of both nonmanual and manual classes has declined over the 10-year period, at the same time as the gap between them has widened.

Similar changes have been shown in Finland (Valkonen et al., 1990). Using linked data – that is, death records linked to census information – Valkonen showed that there was a clear inverse association between an occupation-based measure of social class and mortality. Furthermore, these

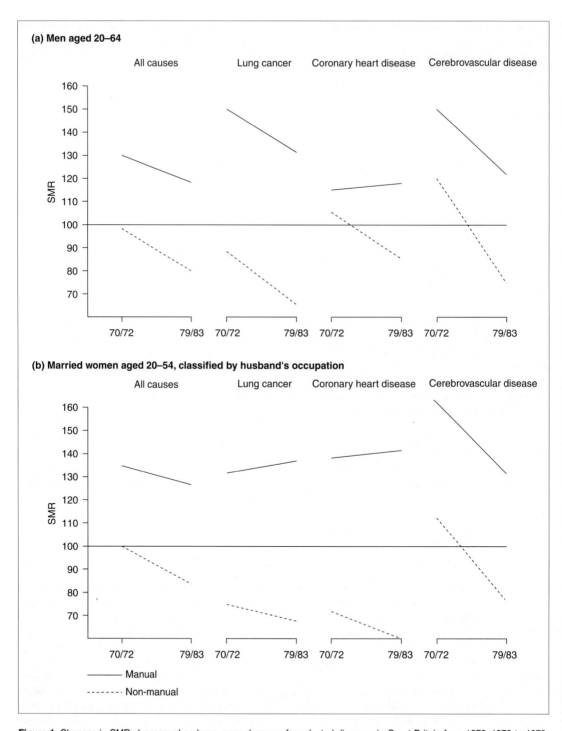

Figure 1. Changes in SMRs by manual and non-manual groups for selected diseases in Great Britain from 1970–1972 to 1979-1983 (Marmot & McDowall, 1986).

differences increased over the period 1971–1985 (Figure 2).

Figure 1 also addresses the question of changing class composition (Illsley, 1986; Strong, 1990). Illsley argues that the relatively high death rates of social class V apply to a decreasing proportion of the population and the low death rates of class I to an increasing proportion (Illsley, 1986). It is therefore difficult to state that inequalities have increased or even changed in magnitude over time. This applies less to a comparison of manual and non-manual classes. In 1971, 60% of men were in manual occupations; in 1981, 55% were. The widening gap in mortality applies therefore to groups of nearly similar size.

Pamuk dealt with changing class composition and changes in classification over a longer period by reclassifying occupations and constructing an index of inequality that takes into account both relative mortality and relative size of classes (Pamuk, 1985). Her conclusion was that class inequality in mortality in England and Wales narrowed in the 1920s and increased again during the 1950s and 1960s, so that by the early 1970s it was greater than it had been in the early part of the century both in absolute and relative terms.

The issue of the extent to which this social gradient in mortality could be produced by selective social mobility will be dealt with more fully below under 'Explanations'. The issue of the appropriateness of the classification for women is part of a more general discussion on measures of social position.

Measures of social differences in mortality
Occupation-based measures of class

The usual British approach to social class analysis, using the Registrar General's classification of occupations, has come in for criticisms additional to those discussed above (Bunker et al., 1989; Illsley, 1986; Strong, 1990). The measure is said to lack theoretical content (it is not clear what it signifies); it may not apply to people not at work; people younger than working age have to be classified by parents' occupation, and those older, or otherwise unemployed, by their previous employment; and women may not be well classified, especially housewives.

These issues have been comprehensively reviewed by the OPCS longitudinal study (Goldblatt, 1990). Goldblatt reminds us that the classification was

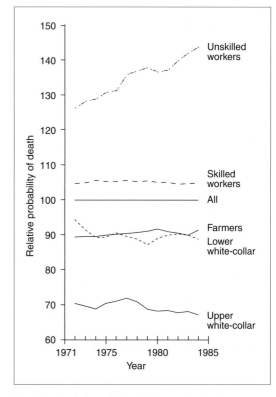

Figure 2. Relative probability of death by occupational class in Finland of men aged 35–64 (all men = 100; three-year moving averages) (Valkonen et al., 1990).

developed, by Stevenson for the *1921 Decennial Supplement on Occupational Mortality* (Stevenson, 1928), to indicate way of life. He quotes Stevenson (Goldblatt, 1990):

Classification of individuals by income was not possible under present conditions in this country, though it had been employed on a very limited scale in America. Estimation of poverty by housing conditions was very unsatisfactory, as bad housing was only one of the handicaps of poverty, so that it was impossible to determine how far the excess of mortality associated with bad housing was due to poverty and how far to the direct effects of overcrowding, etc. Even if full details of income were available, these in themselves would not provide an ideal basis for classification, as it was probably the cultural associations of wealth which promoted longevity rather than wealth itself... The method

Table 4. Mortality in 1976–1981 of men aged 15–64 by alternative social classifications

Social classifications	SMR	(%)[a]	SMR	(%)[a]	SMR	(%)[a]
Occupation-based						
Social class						
I	67	(5)	75	(24)	84	(35)
II	77	(20)				
III-NM	105	(10)				
III-M	96	(37)				
IV	109	(17)	114	(24)	103	(61)
V	125	(7)				
Other	189	(4)				
Household-based						
Private households						
Tenure						
Owner-occupied	85	(51)				
Privately rented	108	(16)	114	(47)		
Local authority	117	(31)				
Car access						
Two or more	77	(15)				
One	90	(50)	87	(65)		
None	122	(33)				
Non-private households	162	(2)				
All men aged 15–64	100	(100)				

[a]Figures in parentheses are the percentage of expected deaths attributed to each group.

advocated for meeting the conditions to be considered was that of inferring social position from occupation. By this means regard could be paid to (average) culture as well as income.

In this discussion are the seeds of arguments that continue to flourish: the extent to which we should be seeking measures that reflect material well-being and those that reflect lifestyle – 'culture' in Stevenson's terms; and the degree to which we want a measure that reflects poverty or one that is related more generally to social position that includes those not in poverty.

Social class and household measures of social circumstances

The classification based on occupation has served well the task of predicting differences in mortality across the spectrum of society, as shown in Table 1. The OPCS longitudinal study has explored the use of this and other methods of social classification based on the material conditions of households: housing tenure, and access to cars. All of these measures are strongly related to household income. The relation of these measures to mortality for men aged 15–64 is shown in Table 4 (Goldblatt, 1990). Social class, housing tenure and access to cars all predict mortality.

One of the criticisms of current analyses of social class is that the extremes of mortality apply to small groups of the population. Indeed, Table 4 shows that the SMR of 67 in social class I applies to only 5% of the expected deaths; and the SMR of 125 in class V applies to only 7%. The analysis by access to cars deals with this criticism. The SMR of 122 in those with no access to cars applies to 33% of deaths.

The longitudinal study was also able to deal with the question of applying social classification to people of different ages. Each of the three measures – social class, housing tenure, and access to cars – makes an independent contribution to the prediction of mortality at working age and each continues to predict beyond working age (Goldblatt, 1990). Lack of access to a car is less predictive at older ages, which is in accord with the decline in the percent of people with access to a car beyond age 75.

Applying these measures to women
Using the Registrar General's social classes to examine mortality of women is more problematic. Married women who are not employed outside the home are, provocatively, classed as 'unoccupied' and are therefore difficult to assign to a social class. In addition, if occupational class, in Stevenson's terms, is supposed to provide a guide to culture as well as income, a married woman's occupation will provide only part of the picture. It will more often be the case that the family circumstances are misclassified if only the woman's occupation is known than if only the man's is known. Figure 3 shows, for married women, that social class based on husband's occupation predicts mortality better than social class based on the woman's own occupation (Moser et al., 1990). As can be seen from the figures showing the proportion of expected deaths, approximately 41% of deaths are expected in women who are 'unoccupied' (mostly housewives). Among these women, husband's social class predicts mortality more strongly than among other women. For single women, own social class is a powerful predictor. For example, the SMR among women in non-manual occupations is 81, and among manual occupations it is 156.

Housing tenure and access to cars predict mortality at working ages, as well as among older women. One advantage of these measures for use among women is that they do not require the distinction between married and unmarried women (Goldblatt, 1990). They also, as with men, distinguish mortality differences among groups that make up large proportions of the total population.

Education
Another approach to measurement of the social gradient in mortality is to take education as a predictor. Valkonen (1989) has used this to compare socioeconomic differences in mortality in different European countries in the 1970s (Figure 4). For men, the slope of the relation is remarkably similar in Scandinavian countries, in England and Wales, and

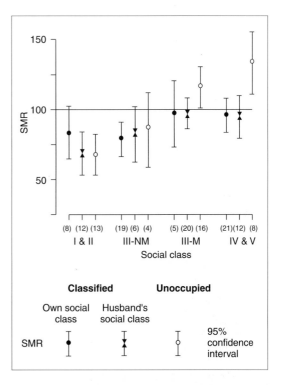

Figure 3. Mortality in 1976–1981 among married women aged 15–59; by own social class and by husband's social class for those classified to an occupation, and by husband's social class for those who were 'unoccupied' in 1971. Figures in parentheses represent expected deaths associated with each SMR (standardized mortality ratio) (as a percentage of the total for all married women). Data are from the OPCS longitudinal study (Moser et al., 1990).

in Hungary. The greater the number of years of education, the lower the mortality rate. The general relationship is similar among women, but the slope of the lines varies. As with the discussion on women's social class based on occupation, so with education; a woman's material income and wealth may be determined not only by her own characteristics but also by those of her husband. The degree to which this is so may vary from country to country and hence account to some extent for the variations in slopes.

Valkonen points out that the apparently fixed nature of the relation between education and mortality for men should not be taken as a general rule. The slope became steeper in England and Wales from 1971–1975 to 1976–1981. This is consistent with the findings for occupational social class shown in Figure 1.

Similar findings come from the United States of America. Kitagawa and Hauser (1973) showed a strong and consistent inverse association between education and mortality in the 1960s in the USA. More recently, Pappas et al. (1993) showed that the decline in mortality in the USA varied substantially by education: the greater the number of years of education, the steeper the decline.

Area-based measures of material circumstances

A quite different approach to measuring inequalities in health is to classify not individuals but areas. To some extent this has been used as a proxy for individual-based measures, where these may not be available. The use of these measures has also been justified on theoretical grounds. Townsend, one of the authors of the Black report, developed census-based measures of social deprivation precisely to examine the effect on mortality of material circumstances (Townsend et al., 1988). His index of deprivation comprised the proportion of households that had access to cars, the percentage of unemployed, the percentage of owner-occupiers, and the degree of crowding. In the Northern Region of England, this measure was strongly related to mortality – the greater the deprivation the greater the mortality (Townsend et al., 1988). A similar measure of deprivation in Scotland was strongly related to area differences in mortality (Carstairs & Morris, 1991).

We have applied the Townsend measure of deprivation to census tracts (population 7000, approximately) throughout England (Eames et al., 1993). The relation of deprivation scores to all-cause mortality is shown in Figure 5. This illustrates the continuous relation between deprivation and mortality – the more deprived tracts have higher mortality. There are alternative ways of interpreting this finding of a gradient in the relationship of deprivation to mortality. One is that deprivation is a misleading term. As with data on occupation-based social class, we are dealing not with the effects of poverty but relative position – a gradient in mortality. Most people living in wards that are classified in the second or third quintiles in Figure 5 are not deprived – yet their mortality rate is higher than that of people in the least-deprived quintile. This interpretation suggests relative deprivation rather than absolute material disadvantage.

An alternative interpretation would be of a dichotomy between deprived and non-deprived – above the threshold of deprivation, mortality is raised compared with below. This would suggest that each successive quintile contains a greater proportion of deprived households; hence the appearance of a gradient in mortality. It is not easy to distinguish between these two alternatives from ecological-based studies. The studies based on classification of individuals, reviewed above, suggest strongly that the relation between social position and mortality is graded and not a threshold effect.

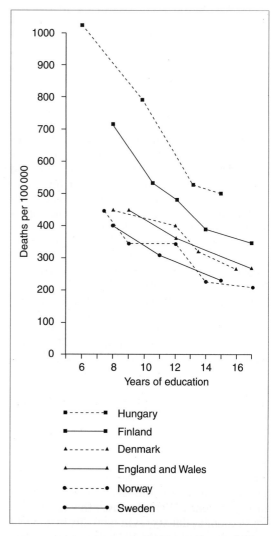

Figure 4. Age-standardized mortality (per 100 000) from all causes of death by years of education and country, for men aged 35–54 (log-scale) (Valkonen, 1989).

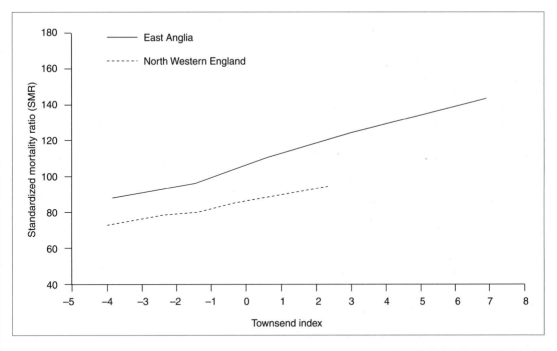

Figure 5. All-cause mortality in two regions of England: North Western England and East Anglia, with mean male mortality by quintiles of the Townsend index of deprivation. The higher the score, the greater the deprivation (Eames et al., 1993).

A second observation from Figure 5 relates to regional differences in mortality within England. The North Western Region has higher mortality than East Anglia and a greater spread of deprivation, yet at comparable levels of deprivation, the regional difference in mortality persists. In addition, the slope of the relation between deprivation and mortality appears to differ between regions. This suggests that the 'meaning' of the deprivation index differs depending on the context, or that some other factor(s) modifies the effect of deprivation on mortality within regions, and contributes to regional differences in mortality independently of deprivation.

These area-based studies may be important not only because they provide a guide to the socioeconomic status of individual residents. The Human Population Laboratory from Alameda County, California showed that people living in a poverty area experienced a higher mortality rate than people living in non-poverty areas, independent of a wide range of personal characteristics including income and health behaviours (Haan et al., 1987).

Choosing between measures
Stevenson based his social class measure on occupation because he wished it to represent status in the community: a mixture of material conditions and culture. Other methods of classifying social position have been discussed. To these we should add income as commonly used in the USA (Bunker et al., 1989). How are we to choose between them? There are two types of criteria: prediction and explanation.

We consider prediction first. The work of the OPCS longitudinal study, discussed above, shows how household measures – housing tenure and access to cars – predict mortality independently of social class and are more appropriate for married women. Education also predicts mortality. As Valkonen et al. (1990) discusses for Finland, as the majority of the middle-aged population falls into one basic education category, it distinguishes better among higher status groups than it does among lower status. A similar consideration limits the use of education in census-based studies in Great Britain (Goldblatt, 1990).

The pragmatic question of the best socioeconomic predictor will be answered differently depending on circumstances. In the special circumstances of the Whitehall study, for example, employment grade provides a precise social classification that is a powerful predictor of mortality. This particular measure is highly correlated with salary and material conditions such as housing tenure and access to cars (Davey Smith et al., 1990) and with 'culture', as well as with years of education.

This raises the more difficult question of explanation. One should be wary of the temptation of using a standard multivariate analysis as a way of determining which of a number of socioeconomic indicators is most important in the causal network (Marmot, 1989). For example, suppose mortality were analysed in a statistical model in which the predictors (independent variables) included education, social class (based on occupation), housing, cars, income, and area-deprivation score. Years of education might show the strongest prediction simply because it was measured more precisely than the others. One should be wary of concluding that education holds the best prospect for intervention to reduce social inequalities. Education may be a precise marker of social position and hence good as a predictor, but may not itself be a determinant of health status. The question of explanations will be considered at length below.

Poverty or inequality

Much of the discussion about social inequalities in health has related to the health disadvantage of those at the bottom (M'Gonigle & Kirby, 1937). This is analogous to seeing social problems as a particular problem for a disadvantaged minority. There is little doubt that poverty, or deprivation, is likely to be bad – for health among other things. The Whitehall data on mortality (Table 3) suggest that something other than absolute poverty is at

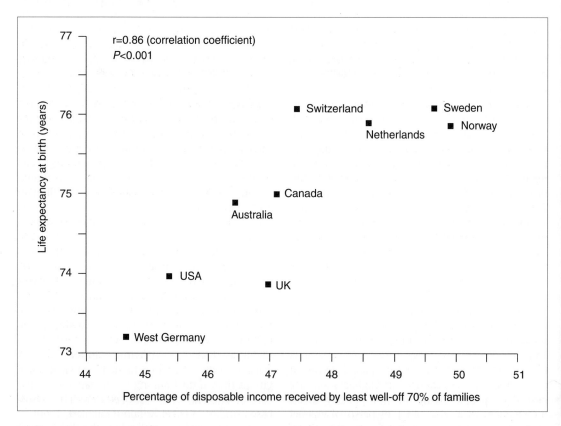

Figure 6. Relationship between life expectancy at birth (male and female combined) and percentage of post-tax and post-benefit income received by the least well-off 70% of families in a group of OECD countries in 1981 (Wilkinson, 1992).

work here. Each grade has worse health and higher mortality rates than the grade above it. Executive-grade civil servants are not poor by any absolute standard yet they have higher mortality rates than administrators. Even clerical officers who are far from well-off, with earnings at or below the national average, are not poor by comparison with people in England at an earlier period in history, or with those in developing countries.

This social gradient in mortality suggests the operation of factors across the whole of society. Whether they are relative deprivation or relative lack of access to the fruits of a wealthy society, it is clear that explanations for socioeconomic differentials in Britain in the 1990s must be broader than the notion of poverty advanced earlier in the century (M'Gonigle & Kirby, 1937).

As with the above discussion of social inequalities in mortality, international data suggest conclusions that are congruent with those from within-country comparisons. Wilkinson (1992) argues that, comparing countries, there is a relation between GNP per head and life expectancy at birth only for poor countries: in 1984, below a threshold of GNP per head of about US$ 5000, few countries had inhabitants with a life expectancy of 70 years or more. Beyond that level, however, there is little relation between GNP and life expectancy (Wilkinson, 1986). The relation of life expectancy with measures of income dispersion is much closer. Figure 6 uses as a measure of equality of income distribution the share of total post-tax household income received by the least well-off 70% of families. There is a striking correlation with life expectancy. Wilkinson (1992) tried several measures of income distribution – the share of total income received by the least well-off 10%, 20%, 30% and so on. He found that the correlation with life expectancy increased progressively until the bottom 60–70% was reached – that is, the higher the share of total income enjoyed by the bottom 60–70%, the longer the average life expectancy.

As a further test of the income inequality hypothesis, Wilkinson plotted changes in life expectancy against changes in income distribution for six countries for which these data were available. Japan had the greatest increase in equality of income distribution and the greatest increase in life expectancy; and Japan now has the most equitable distribution of income and the longest life expectancy of any country within the Organization for Economic Cooperation and Development (OECD). By contrast, in the United Kingdom the bottom 60% of households had a declining share of total income and the population had a relatively small increase in life expectancy (World Bank, 1992).

How are we to explain the relation between income inequality and life expectancy? One possibility is that greater income inequality implies a greater proportion of the population in poverty. If there were a non-linear relation between income and life expectancy – stronger at lower incomes than at high – the apparent relation between income inequality and life expectancy could possibly be explained by the differing proportions of those in poverty. Wilkinson (1992) points out that this is not the most likely explanation. If the criterion of income inequality is taken as the share of income received by the bottom 10%, the relation with life expectancy is very much weaker than the criterion of the share received by the bottom 60 or 70%. Even if increased income inequalities did result in more households in poverty, the proportion of absolute poor in OECD countries is still too small to account for the size of the relation between income inequality and life expectancy. There must be an effect of income inequality on the large bulk of the population. In other words, the main influence on life expectancy among rich countries is likely to be of relative rather than absolute deprivation.

Explanations of social inequalities in health
Medical care

A first attempt at explanation for social inequalities in health might be inequity in the distribution of medical care. Indeed, the establishment of the Black Committee in the United Kingdom was in response to the apparent failure of social inequalities in health to have disappeared 30 years after the establishment of the National Health Service. Congruent with the conclusion of McKeown (1979) on the limited role of medical care in contributing to improvements in life expectancy, the Black Committee did not attribute inequalities in health to inequity in the distribution of medical care.

One way this issue has been examined is to look at mortality from causes judged to be amenable to medical care. Figure 7 extracts data from a report by Mackenbach *et al.* (1989). It shows that the relative decline in mortality from causes amenable to medical care may have been slightly greater in social class I than in class V – consistent with inequity in medical care. Amenable causes are, however, a small proportion of the total. In absolute terms the widening gap in mortality between the classes is clearly the result of the decline in non-amenable causes in higher classes and the lack of such decline in lower classes.

Within Great Britain, an analysis of data from the OPCS longitudinal study shows that social class differences in cancer mortality are largely the result of differences in the incidence of cancer. Differentials in survival make a minor contribution to overall cancer mortality (Kogevinas et al., 1991).

Analogous to the limited role of differences in medical care in generating social inequalities in health is the limited role of variations in medical care in generating international differences in health. When comparing Japan with England and Wales, for example, we noted that Japan, like the United Kingdom, spends a relatively small proportion of GNP on medical care. The decline in mortality in Japan was observed for both amenable and non-amenable causes of death (Marmot & Davey Smith, 1989).

Access to good-quality medical care is a right that should be enjoyed equally by all members of society. It is hard to make the case that it is differential access to or provision of medical care that is responsible for inequalities in health in European countries.

Health selection

The argument here is, in essence, that health may determine social position rather than vice versa. This was one of the possible explanations considered by the Black report and rejected as a major cause of social inequalities in health (Black et al., 1988). There are several periods during the life course when selection could operate and also several potential mechanisms, with a varying degree of plausibility and evidence supporting these interpretations (Blane et al., 1993).

The most straightforward suggestion is that the sick drift down the social hierarchy, producing

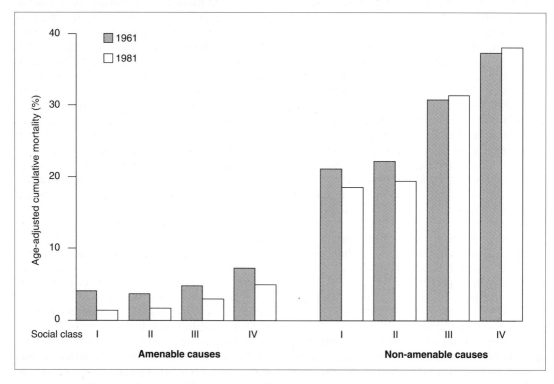

Figure 7. Mortality from amenable and non-amenable causes by social class in England and Wales in men aged 15–64 in 1961 and 1981 (Mackenbach et al., 1989).

social groups at the bottom containing a disproportionate number of individuals at high mortality risk. Three pieces of evidence bear on whether such intragenerational selection is an important contributor to mortality differentials. First, in the OPCS longitudinal study, social class mortality differentials after 1981 for subjects who were in the same social class groups in 1971 and 1981 – and could therefore not have experienced health-related social mobility – were identical to those for the whole population (Goldblatt, 1989). Second, in this study, mortality differentials according to the Registrar General's social classes persisted beyond age 75, long after retirement (Fox et al., 1985). Here social class is based on last occupation. By definition, health after retirement cannot cause downward social drift as the classification is fixed. Third, in participants free of manifest disease at entry to the Whitehall study, mortality differentials by employment grade were essentially the same as in the whole study population. As no reclassification of social position after study entry was made, mobility caused by differences in health status, at least as measured, cannot account for the mortality gradient (Marmot et al., 1978).

Recently, there has been much interest in the possibility that selection occurring at an earlier age – between early childhood and labour market entry – is an important determinant of health inequalities (Illsley, 1986; West, 1991). This could occur in two ways: if health status in childhood determines both health and social class in early adulthood; or if a common antecedent determines both adult health status and future adult social class. This second possibility is considered in the next section. There is evidence that ill health in childhood is associated with downward social mobility (Wadsworth, 1986), but the effect is of minor importance. Other studies that can relate health in childhood to health in early adulthood similarly suggest that this cannot account for class differences in health in adulthood (Power et al., 1990; Lundberg, 1991).

Factors operating early in life
A different version of the selection hypothesis suggests that while social selection based on health status is not a crucial contributor to health differentials, common background factors determine both social position and health in adulthood. This process has been termed 'indirect selection' (Wilkinson, 1986), and recognizes that people bring with them into adulthood the results of influences from their earlier days: genetic factors, biological results of early experiences, and educational, cultural, psychological and social factors.

An interesting body of work suggests that non-genetic factors operating *in utero* or in the first year of life may have long-term effects on risk of cardiovascular and other diseases (Barker, 1989). Barker's work showed that birth weight, weight at one year and thinness at birth were related to a variety of indicators of cardiovascular disease and diabetes. These associations appear to be independent of parents' social class. It is possible that both social position and health in adulthood are determined by common early life influences. The focus on the childhood origins of adult disease has been criticized precisely because influences from early life shape the lives people lead and the social environments in which they live and work (Ben-Shlomo & Davey Smith, 1991). It may be these conditions of adult life that are related to ill health, and the importance of childhood conditions may therefore be indirect. It is clearly not easy to separate the direct effects on health of early and later life experiences.

One indicator that may help is height. In his studies in Aberdeen, Illsley (Illsley, 1955) showed that women who were upwardly mobile – their husband's social class was higher than their father's – were taller than women who married within their class. More recent work, analysing data from the 1958 birth cohort in the United Kingdom, confirms that social mobility between birth and age 23 was selective with respect to height but mobility did not account for the social gradients in height (Power et al., 1990, 1991).

Whitehall data relating height to mortality provide some insight into the possible separate effects of current and past environment. Height is influenced by environment as well as by genes and is related to social status as measured by employment grade. Short height predicts adult mortality independently of grade of employment (Marmot et al., 1984b), and it is reasonable to speculate that this may in part be a reflection of a persisting influence from early life. Grade of employment, which is to some extent an index of current social influences, predicts mortality independently of height. Thus, two sets of influences may affect mortality risk:

factors from early life and current influences. As indicated, this is to oversimplify: people's current social situations are influenced by their prior experiences. Nevertheless it is important to attempt to distinguish these two sets of influences as their relative importance is crucial to determining the appropriate locus for interventions that may both improve overall adult health and reduce the socioeconomic differentials. Research is currently ongoing that may reduce the degree to which views are polarized between those who think future health status is virtually programmed in early life (Barker, 1990) and those who support the current mainstream focus on influences acting in later life (Elford et al., 1991).

It is interesting that short height is a stronger predictor of cardiovascular disease than of all-cause mortality. If short height predicts cancer mortality less strongly than it does cardiovascular disease, might this mean that cancer is less influenced by factors operating early in life?

General susceptibility or specific causes?

A striking feature of social class differences in health is the generalizability of the findings across diverse pathological conditions. In the Whitehall study, the higher risk of death among lower grades applied to deaths from lung cancer, other cancers, coronary heart disease, cerebrovascular disease, other cardiovascular disease, chronic bronchitis, other respiratory disease, gastrointestinal disease, genitourinary disease, accidents and violence (Marmot et al., 1984b). Findings such as these, suggesting that there may be common factors operating that cut across known causes of disease, have prompted speculation that there may be factors that increase general susceptibility to ill health (Berkman & Syme, 1979; Cassel, 1976).

An alternative to a general susceptibility hypothesis is that a variety of specific factors are operating to explain social class differences in mortality. Support for this view comes from the fact that some cancers, notably those of the colon, brain, prostate, haematopoietic system and breast, and melanoma, do not show the same social class variation as the causes of death listed above (Davey Smith et al., 1991).

When posed with two conflicting alternatives such as these, a reasonable working hypothesis is that they are both correct. A general susceptibility

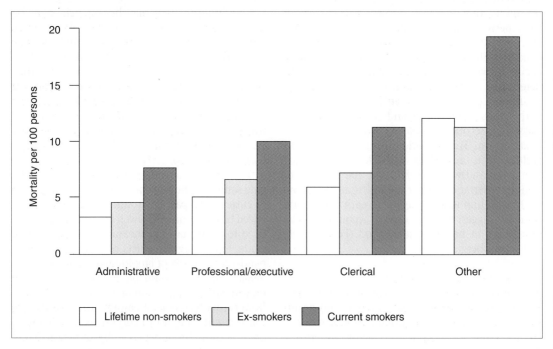

Figure 8. Ten-year mortality risk (age-standardized) by smoking behaviour and employment grade (Davey Smith & Shipley, 1991).

hypothesis implies that certain groups will be at higher risk of death whatever causes are operating. It does not deny the operation of specific causes. Diseases linked to smoking, such as chronic bronchitis and lung cancer, show a particularly strong social class gradient – stronger than cancers not linked to smoking. But the latter do show a social class gradient, as do other diseases not linked to smoking. Put another way, the general susceptibility hypothesis means that there are factors operating that cut across our current system of classifying diseases. These will increase risk of death in addition to the effect of known factors such as smoking. This can account for the fact that an administrator who smokes 20 cigarettes a day has a lower risk of lung cancer mortality than a lower-grade civil servant smoking the same amount, even after pack-years and tar content of cigarettes are taken into account; and for the gradient in mortality that occurs for coronary heart disease even among nonsmokers. Figure 8 illustrates this point (Davey Smith & Shipley, 1991).

The production of this apparently increased susceptibility may well be operating at a social level. The whole life course of people in different social locations is different, and insults to health may accumulate over the entire period from birth to death. That these influences on health cluster in such a way as to produce social groups at differing degrees of disadvantage with respect to most diseases is undeniable. Our current level of knowledge regarding this general susceptibility allows us to go little beyond this empirical observation, however.

Health-related behaviours and biological risk factors
The influential review by Doll and Peto (1981), which became the basis for the USA National Cancer Institute's goals, suggested that much of cancer occurrence could be attributed to behaviours – smoking, diet, alcohol consumption and behavioural response to screening. There are two questions here: how much of the social differential in the occurrence of cancer can be attributed to social differences in behaviours; and why are there social differences in behaviours?

The Whitehall II study established a clear social gradient, using grade of employment as an indicator of socioeconomic position, in a number of health behaviours and established biological risk factors (Marmot *et al.*, 1991). The most striking risk factor difference among grades was in smoking (see

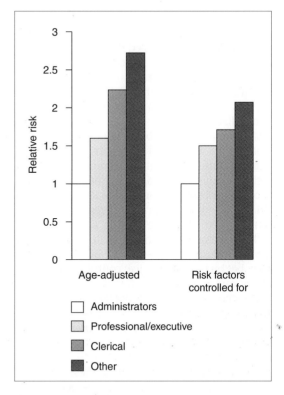

Figure 9. Relative risk of coronary heart disease death in 10 years, controlling for age and risk factors (smoking, systolic blood pressure, plasma cholesterol, height, and blood sugar) (Marmot *et al.*, 1984b).

also Figure 10). Women had a higher prevalence of smoking than men in all but the lowest (clerical and office-support) grade. As a rough indicator of dietary pattern, consumption of skimmed and semi-skimmed milk, wholemeal bread and fresh fruit and vegetables was higher in higher grades. To the extent that dietary intake of non-starch polysaccharides and of antioxidants is protective against cancer, this might provide part of the explanation for the social gradient, along with smoking.

In addition, the proportion of men and women not taking moderate or vigorous exercise in their leisure time was higher in lower grades. Possibly related, there was a significant inverse trend of mean body mass index (weight/height^2) by grade but, especially in men, the differences were small. The distribution is different, however. The prevalence of obesity (body mass index >30) was greater in lower grades, strikingly in the clerical grade.

Interestingly, average alcohol consumption was higher among the higher grades of men and, more strikingly, of women. In fact, there were more non-drinkers in the low grades, no difference by grade in the proportion of heavy drinkers, and more moderate drinkers in the high grades (Marmot et al., 1993). This is similar to results from other non-industrial populations – in the Netherlands, for example, women from higher socioeconomic groups were more likely to report higher weekly alcohol consumption than women from lower socioeconomic groups, although negligible differences were observed for men (Mackenbach, 1994). On the face of it, therefore, differences in alcohol consumption cannot be playing a major role in social differentials in cancer risk. It might be that among industrial workers, alcohol consumption may be higher and hence contribute more to higher cancer risk.

Plasma cholesterol levels did not differ by grade, and the small inverse association between grade and blood pressure level in men observed in the first Whitehall study was still present in the second study but was even smaller.

It is easier to quantify the effect these risk factor differences may have on cardiovascular disease than on cancer. The grade differences in smoking were insufficient to account for differences by grade in mortality from smoking-related diseases (Marmot et al., 1984b). There were only small differences in blood pressure between grades, and plasma cholesterol levels were higher in higher grades. The main coronary risk factors could account therefore for little of the gradient in mortality by grade of employment (Figure 9).

To the extent that behaviour does account for social gradients in health, this raises a new question: why the social gradient in behaviour? It is worth dwelling on the implications of the smoking rates in Figure 10. In the baseline examinations for the first Whitehall study, conducted between 1967 and 1969, there was a clear social gradient in smoking. Twenty years later, the prevalence of smoking has declined across the whole of society and in a new cohort of civil servants, different people from those examined 20 years earlier, the social gradient is reproduced.

Social class should not be treated as a confounder. It is reasonable to 'control' for smoking if one wishes to examine the extent to which smoking accounts for social class differences in disease. Similarly, an analysis of the health risks of smoking will be flawed if it fails to take into account that smoking is associated with social position which, in turn, is associated with adverse health for reasons in addition to smoking (Davey Smith & Shipley, 1991).

To understand the pathways by which social inequalities in health are generated, one needs to examine the links in the chain – for example, between social position and smoking – and not to control for their effects.

Material conditions

The Black report (Black et al., 1988) emphasized the importance of material conditions as an explanation for social inequalities in health. In fact, Black referred to materialist or structural explanations: emphasizing hazards to which some people have no choice but to be exposed given the present distribution of income and opportunity. These can be interpreted as broader than simply material conditions, to include psychosocial influences that are inherent in position in society.

As shown above, in the OPCS longitudinal study, measures other than social class based on occupation predicted mortality, including housing tenure (ownership) and household access to cars. These are clearly a guide to material conditions. In the Whitehall study, car ownership predicted mortality, independently of grade of employment (Davey Smith et al., 1990). In addition, men who reported they engaged in gardening had lower mortality than other men. The link between gardening and lower mortality is open to a variety of interpretations; among them is that possessing a house with a garden is a measure of wealth. These data are consistent with the link between deprivation and mortality shown in the geographic-based studies reviewed above.

The difficulty in understanding material explanations is to know how they operate. At a time when poor living conditions meant polluted water, crowded unsanitary housing with high rates of cross-infection, and appalling conditions of employment, it was not difficult to see how these could be responsible for worse health among the socially deprived. This would be additional to the effects of inadequate diet, which is part of the material conditions of life. As conditions improved, mortality of all social groups improved. But why do the social gradients persist? Are we to understand that there are residual effects of bad housing with damp and infection, as well as air pollution and other material conditions that, although they affect the lower social groups less than they used to, still affect them more than the higher social strata? If there are such residual effects it is no surprise that they do not affect all social strata equally. Can this be the whole explanation?

The Whitehall data and Wilkinson's data on income inequalities are relevant here. In both Whitehall studies, morbidity and mortality varied linearly with grade of employment. It is possible that the worse health of the second-highest grade compared with the highest could be the result of worse housing, poorer diet for children, or greater pollution, but it seems unlikely that comfortable 'middle-class' people in Great Britain are suffering from the effects of material deprivation. Similarly, comparing rich countries, Wilkinson showed that it was not differences in wealth that predicted differences in life expectancy, but differences in inequality of income distribution.

One is drawn to the view that, in addition to the multiple influences already discussed, there must be an influence of relative position in society. What may be important is not absolute deprivation but relative deprivation. This would account for a social gradient in ill health, because each group, while not necessarily suffering from greater effects of bad housing and so on, will have 'less' than the group above it. This would account for the widespread finding of social inequalities in health in societies with very different levels of health. The social gradient in ill health will vary in magnitude depending on the magnitude of the relative differences in 'deprivation'.

In a society that has met the subsistence needs of its members, what do we mean by 'less' or 'relative deprivation'? In addition to the factors explored above, we should look to psychosocial factors for part of the answer.

Psychosocial factors as potential explanations
This is not the place to review the whole field of psychosocial factors and health. There are a number of ways such factors could influence social differentials in health: characteristics of the psychosocial work environment, low control, and low

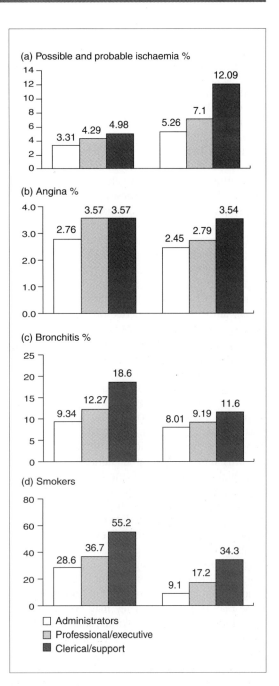

Figure 10. Prevalence of cardiorespiratory disease and smoking among men aged 40–54 in the Whitehall I (1967–1969) and Whitehall II (1985–1988) studies: age-adjusted percentages of (a) probable and possible ECG ischaemia; (b) angina pectoris; (c) chronic bronchitis; and (d) current cigarette smokers (Marmot *et al.*, 1991).

variety and low use of skills show a social gradient; there may be differences in the frequency of life events or other sources of life stress; and there may be social gradients in social supports. In the second Whitehall study, as one measure of perceived control over their health, fewer of the lower grades reported that they believed it was possible to reduce the risk of heart attack.

There is here a richness of potential explanations for social inequalities in health that tie in with other evidence on psychosocial factors and health. Three major hypotheses concern job strain, low social supports, and low control, respectively. There is a large body of evidence linking high psychological demands at work and low control to cardiovascular and other diseases (Karasek & Theorell, 1990). The relative disadvantage of lower grades with respect to control could be a factor in their higher rates of disease (Marmot & Theorell, 1988). Similarly, the evidence on low social supports and ill health suggests a further contributor to the gradient in ill health (Berkman, 1984; House et al., 1988).

The concept of 'control' may provide a common link to many of the psychosocial explanations. Perceived lack of control over health (external health locus of control) may account for social differentials in health behaviour. Perceived lack of control may also be the crucial factor in the workplace that explains why jobs of lower status are associated with higher disease risk. Syme has suggested that increasing lack of control may be a linking factor that accounts for increasing health disadvantage as the social scale is descended (Syme, 1989). Poverty, whatever else it represents, is lack of control.

The evidence linking these factors to social differentials in disease is again stronger for cardiovascular disease and mental illness than it is for cancer. It remains an open question how important these may be in cancer. They may potentially act by influencing health behaviours such as diet and smoking or by some more direct pathway such as those now being explored under the heading of psycho-neuro-immunology.

Acknowledgements

This chapter is an abridged version of a chapter in *Society and health*, edited by Amick, B., Levine, S., Tarlov, A. and Walsh, D. (Oxford University Press, in press).

References

Barker, D.J. (1989) The intrauterine and early postnatal origins of cardiovascular disease and chronic bronchitis. *J. Epidemiol. Community Health*, 43, 237–240

Barker, D.J. (1990) The fetal and infant origins of adult disease. *Br. Med. J.*, 301, 1111

Ben-Shlomo, Y. & Davey Smith, G. (1991) Deprivation in infancy or adult life: which is more important for mortality risk? *Lancet*, 337, 530–534

Berkman, L.F. (1984) Assessing the physical health effects of social networks and social support. In: Breslow, L., Fielding, J.E. & Lave, L.B., eds, *Annual Review of Public Health*. Palo Alto, USA, Annual Reviews Inc. pp. 413–432

Berkman, L.F. & Syme, S.L. (1979) Social networks, host resistance and mortality: a nine-year follow-up of Alameda County residents. *Am. J. Epidemiol.*, 109, 186–204

Black, D., Morris, J.N., Smith, C., Townsend, P. & Whitehead, M. (1988) *Inequalities in health: the Black report; the health divide*. London, Penguin Group

Blane, D., Davey Smith, G. & Bartley, M. (1990) Social class differences in years of potential life lost: size, trends and principal causes. *Br. Med. J.*, 301, 429–432

Blane, D., Davey Smith, G. & Bartley, M. (1993) Social selection: what does it contribute to social class differences in health? *Sociol. Health Illness*, 15, 1–15

Bunker, J.P., Gomby, D.S. & Kehrer, B.H., eds, (1989) *Pathways to health – the role of social factors*. Menlo Park, California, The Henry J. Kaiser Family Foundation

Carstairs, V. & Morris, R. (1991) *Deprivation and health in Scotland*. Aberdeen, Aberdeen University Press

Cassel, J.C. (1976) The contribution of the social environment to host resistance. *Am. J. Epidemiol.*, 104, 107–123

Chadwick, E. (1965) *Report on the sanitary condition of the labouring population of Great Britain, 1842*. Edinburgh, Edinburgh University Press

Davey Smith, G. & Shipley, M.J. (1991) Confounding of occupation and smoking: its magnitude and consequences. *Soc. Sci. Med.*, 32, 1297–1300

Davey Smith, G., Shipley, M.J. & Rose, G. (1990) Magnitude and causes of socioeconomic differentials in mortality: further evidence from the Whitehall study. *J. Epidemiol. Community Health*, 44, 265–270

Davey Smith, G., Leon, D., Shipley, M.J. & Rose, G. (1991) Socioeconomic differentials in cancer among men. *Int. J. Epidemiol.*, 20, 339–345

Doll, R. & Peto, R. (1981) *The causes of cancer*. New York, Oxford University Press

Eames, M., Ben-Shlomo, Y. & Marmot, M.G. (1993) Social deprivation and premature mortality: regional comparison across England. *Br. Med. J.*, 307, 1097–1102

Elford, J., Whincup, P. & Shaper, A.G. (1991) Early life experience and adult cardiovascular disease: longitudinal and case-control studies. *Int. J. Epidemiol.*, 20, 833–844

Fox, A.J., Goldblatt, P.O. & Jones, D.R. (1985) Social class mortality differentials: artefact, selection or life circumstances? *J. Epidemiol. Community Health*, 39, 1–8

Goldblatt, P. (1989) Mortality by social class, 1971–85. *Popul. Trends*, 56, 6–15

Goldblatt, P. (1990) Mortality and alternative social classifications. In: Goldblatt, P., ed., *1971–1981 Longitudinal study. Mortality and social organisation*. London, Her Majesty's Stationery Office. pp. 163–192

Haan, M., Kaplan, G.A. & Camacho, T. (1987) Poverty and health: prospective evidence from the Alameda County study. *Am. J. Epidemiol.*, 125, 989–998

House, J.S., Landis, K.R. & Umberson, D. (1988) Social relationships and health. *Science*, 241, 540–545

Illsley, R. (1955) Social class selection and class differences in relation to still-births and infant deaths. *Br. Med. J.*, 2, 1520–1524

Illsley, R. (1986) Occupational class, selection and the production of inequalities in health. *Q. J. Soc. Affairs*, 2, 151–165

Illsley, R. & Le Grand, J. (1987) The measurement of inequality in health. In: Williams, A., ed., *Economics and health*. New York, Macmillan.

Karasek, R. & Theorell, T. (1990) *Healthy work: stress, productivity, and the reconstruction of working life*. New York, Basic Books

Kitagawa, E. & Hauser, P.M. (1973) *Differential mortality in the United States. A study in socioeconomic epidemiology*. Cambridge, MA, Harvard University Press

Kogevinas, M., Marmot, M.G., Fox, A.J. & Goldblatt, P.O. (1991) Socioeconomic differences in cancer survival. *J. Epidemiol. Community Health*, 45, 216–219

Le Grand, J. (1989) An international comparison of distribution of ages-at-death. In: Fox, J., ed., *Health inequalities in European countries*. Aldershot, Gower Publishing. pp. 75–91

Lundberg, O. (1991) Childhood living conditions, health status and social mobility: a contribution to the health selection debate. *Eur. Sociol. Rev.*, 7, 149–162

M'Gonigle, G.C.M. & Kirby, J. (1937) *Poverty and public health*. London, Golancz

Mackenbach, J.P. (1994) *Ongezonde verschillen*. Assen, Van Gorcum

Mackenbach, J.P., Stronks, K. & Kunst, A.E. (1989) The contribution of medical care to inequalities in health: differences between socio-economic groups in decline of mortality from conditions amenable to medical intervention. *Soc. Sci. Med.*, 29, 369–376

McKeown, T. (1979) *The role of medicine: dream, mirage or nemesis?* Oxford, Basil Blackwell

Marmot, M.G. (1989) Future research on social inequalities. In: Bunker, J.P., Gomby, D.S. & Kehrer, B.H., eds, *Pathways to health: the role of social factors*. Menlo Park, California, The Henry J. Kaiser Family Foundation. pp. 224–250

Marmot, M.G. & Davey Smith, G. (1989) Why are the Japanese living longer? *Br. Med. J.*, 299, 1547–1551

Marmot, M.G. & McDowall, M.E. (1986) Mortality decline and widening social inequalities. *Lancet*, ii, 274–276

Marmot, M.G. & Theorell, T. (1988) Social class and cardiovascular disease: the contribution of work. *Int. J. Health Serv.*, 18, 659–674

Marmot, M.G., Rose, G., Shipley, M. & Hamilton, P.J.S. (1978) Employment grade and coronary heart disease in British civil servants. *J. Epidemiol. Community Health*, 32, 244–249

Marmot, M.G., Adelstein, A.M. & Bulusu, L. (1984a) Lessons from the study of immigrant mortality. *Lancet*, 1, 1455–1458

Marmot, M.G., Shipley, M.J. & Rose, G. (1984b) Inequalities in death – specific explanations of a general pattern. *Lancet*, 1, 1003–1006

Marmot, M.G., Davey Smith, G., Stansfeld, S., Patel, C., North, F., Head, J., White, I., Brunner, E. & Feeney, A. (1991) Health inequalities among British civil servants: the Whitehall II study. *Lancet*, 337, 1387–1393

Marmot, M.G., North, F., Feeney, A. & Head, J. (1993) Alcohol consumption and sickness absence: from the Whitehall II study. *Addiction*, 88, 369–382

Moser, K., Pugh, H. & Goldblatt, P. (1990) Mortality and the social classification of women. In: Goldblatt, P., ed., *1971–1981 Longitudinal study. Mortality and social organisation*. London, Her Majesty's Stationery Office. pp. 145–162

Office of Population Censuses and Surveys (1978) *Occupational mortality 1970–1972*. London, Her Majesty's Stationery Office

Office of Population Censuses and Surveys (1986) *Occupational mortality 1979–80, 1982–83*. London, Her Majesty's Stationery Office

Pamuk, E.R. (1985) Social class inequality in mortality from 1921 to 1972 in England and Wales. *Popul. Stud.*, 39, 17–31

Pappas, G., Queen, S., Hadden, W. & Fisher, G. (1993) The increasing disparity in mortality between socio-economic groups in the United States, 1960 and 1986. *New Engl. J. Med.*, 329, 103–109

Power, C., Manor, O., Fox, A.J. & Fogelman, K. (1990) Health in childhood and social inequalities in health in young adults. *J. R. Stat. Soc. A*, 153, 17–28

Power, C., Manor, O. & Fox, J. (1991) *Health and class: the early years*. London, Chapman & Hall

Rose, G. (1992) *The strategy of preventive medicine*. Oxford, Oxford University Press

Rose, G. & Day, S. (1990) The population mean predicts the number of deviant individuals. *Br. Med. J.*, 301, 1031–1034

Stevenson, T.H.C. (1928) The vital statistics of wealth and poverty (report of a paper to Royal Statistical Society). *Br. Med. J.*, i, 354

Strong, P.M. (1990) Black on class and mortality: theory, method and history. *J. Public Health Med.*, 12, 168–180

Syme, S.L. (1989) Control and health: a personal perspective. In: Steptoe, A. & Appels, A., eds, *Stress, personal control and health*. Chichester, Wiley. pp. 3–18

Townsend, P., Phillimore, P. & Beattie, A. (1988) *Health and deprivation: inequality in the North*. London, Croom Helm

Valkonen, T. (1989) Adult mortality and level of education: a comparison of six countries. In: Fox, J., ed., *Health inequalities in European countries*. Aldershot, Gower Publishing. pp. 142–162

Valkonen, T., Martelin, T. & Rimpela, A. (1990) *Socio-economic mortality differences in Finland 1971–85*. Helsinki, Central Statistical Office in Finland

Wadsworth, M.E.J. (1986) Serious illness in childhood and its association with later-life achievement. In: Wilkinson, R.G., ed., *Class and health*. London, Tavistock Publications. pp. 50–74

West, P. (1991) Rethinking the health selection explanation for health inequalities. *Soc. Sci. Med.*, 32, 373–384

Wilkinson, R.G. (1986) Socio-economic differences in mortality: interpreting the data on their size and trends. In: Wilkinson, R.G., ed., *Class and health*. London, Tavistock Publications. pp. 1–20

Wilkinson, R.G. (1992) Income distribution and life expectancy. *Br. Med. J.*, 304, 165–168

World Bank (1992) *World development report 1992*. Oxford, Oxford University Press

Corresponding author:
M. Marmot
Department of Epidemiology and Public Health,
University College London Medical School,
1–19 Torrington Place, London WC1E 6BT, UK

Tobacco smoking, cancer and social class

S.D. Stellman and K. Resnicow

Consumption of tobacco products, both by smoking and by other means, has long been causally connected with cancers of the lung, larynx, mouth and pharynx, oesophagus, bladder, and many other sites. Tobacco is the main specific contributor to total mortality in many developed countries and has become a major contributor in the developing countries as well. In most industrialized countries, prevalence of cigarette smoking is currently higher in low than in high social classes, although in some industrialized countries smoking was more frequent in high social classes during the first half of this century. The latter pattern of tobacco consumption is more likely to apply to developing countries. To formulate and carry out effective tobacco control activities it is essential to assess the relative incidence of tobacco-related cancers in different social strata and the prevalence of tobacco use across strata. Despite many years of data gathering the information base is far from complete, especially in developing countries where tobacco use is increasing rapidly, and where aggressive marketing by the transnational tobacco industry is occurring. A critical question is the extent to which tobacco usage can 'explain' the observed social class differences in cancer risk. Class differences in lung cancer are likely to be mostly related to the unequal distribution of tobacco smoking between social classes, and in some fairly simple situations this has been satisfactorily demonstrated. Nevertheless, there are many unresolved issues, especially with regard to the role of collateral exposures, such as hazardous occupations, poor diet, and limited access to health care. The question of whether tobacco use 'explains' socioeconomic differences in one or more of the cancers that it causes has rarely been directly addressed in epidemiological studies.

Consumption of tobacco products, both by smoking and by other means, has long been causally connected with cancers of the lung, larynx, mouth and pharynx, oesophagus, bladder, and many other sites (IARC, 1986). Tobacco is the main specific contributor to total mortality in many developed countries. Its consumption causes about 19% of all deaths in the United States of America (USA), including 11–30% of cancer deaths, in addition to substantial numbers of deaths from coronary heart disease and chronic obstructive pulmonary disease (COPD) (McGinnis & Foege, 1993). Tobacco use has become a major contributor to mortality in the developing countries as well (Boffetta & Parkin, 1994), and its role in such countries will increase in the coming years.

Tobacco use varies with social class in most countries. To formulate and carry out effective tobacco control activities it is essential to assess the relative incidence of tobacco-related cancers in different social strata and the prevalence of tobacco use across strata. Therefore, a goal of this chapter is to show how tobacco use varies with some of the social class indicators in a number of countries. Despite many years of data gathering, however, the information base is far from complete, especially in developing countries where tobacco use is increasing rapidly, and where aggressive marketing by the transnational tobacco industry is occurring. The spread of tobacco use is a dynamic process that requires systematic monitoring. This spread to new, untapped markets is in part driven by a decline in smoking in developed countries, especially among the more affluent classes, as documented by increasing cessation rates. A second goal of this chapter is to present the risk of the principal tobacco-related cancers in several countries according to a variety of social class indicators and to evaluate the extent to which tobacco usage can 'explain' the observed social class differences in cancer risk. This is related to the more general question of what proportion of these differences can be explained by identifiable lifestyle and environmental factors, including other

Table 1. Smoking prevalence (percentage) by occupational category and sex in the USA during the periods 1978–1980 and 1987–1990[a]

Occupation	Males		Females	
	1978–1980	1987–1990	1978–1980	1987–1990
White-collar	32.0	24.0	31.4	24.4
Blue-collar	45.3	40.2	36.9	34.8

[a]Nelson et al., 1994a.

risk factors that affect the same cancers caused by smoking. These other risk factors, such as occupation and diet, can act independently or they may interact with smoking to increase risk, as is the case with alcohol consumption in some upper aerodigestive tract cancers. Risk factors other than tobacco are sometimes downplayed or even overlooked in programmes of smoking cessation. We therefore thought it important to present here some data on cofactors and correlates of smoking behaviour that also play a role in the etiology of tobacco-related cancers.

Throughout this chapter we make reference to the substantial literature on cancer rates and tobacco use in African-American (Black) and White populations in the USA. The large ethnic differences in both incidence and mortality for tobacco-related cancers are obviously related to cigarette smoking, but two significant questions remain: (1) do tobacco usage patterns satisfactorily explain these differences in a quantitative manner, and (2) is there a separately discernible role of social class that in turn 'explains' or underlies the tobacco usage patterns. The first question falls within the conventional role of epidemiology, whereas the second, with its sociopolitical overtones, may not necessarily fall within the traditional domain of biomedical science. The first may eventually be simpler to answer, but the second may be the more important because, as discussed in the section 'Intervention and prevention', serious attempts to prevent these cancers in economically disadvantaged populations must be guided by its solution.

Finally, a substantial body of experience has been developed in the application of smoking cessation methods, in which sensitivity to class and culture has come to play an important role. It is useful to summarize some of the lessons learned from studies of knowledge and attitudes about smoking of members of different social strata, and to highlight successful intervention programmes that have targeted groups of individuals, especially ethnic minorities, that are often considered by public health specialists as difficult to reach.

Tobacco consumption by social class

The rate and extent to which different social classes have taken up tobacco smoking have varied markedly both within and between countries, and the distribution is constantly changing, especially in Western countries under pressure from anti-smoking organizations. Consequently, any relationship between social class and smoking that might be generally true in one country or society cannot necessarily be generalized to others. In the USA, for example, there is a strong inverse relation between social class (represented by occupational grouping or income) and smoking in men, but only a weak one in women (Stellman & Stellman, 1980). In countries where the great majority of men have been smokers at one time or other during their lives, but where widespread smoking by women may have occurred only recently, such as Japan (Wynder et al., 1992) and France (Wynder et al., 1981), even such weak generalizations may be inapplicable. Furthermore, in some Latin American countries there is a positive association between social class and tobacco use.

Prevalence of usage and cessation

Trends in tobacco consumption in the USA have been reported periodically by the National Center for Health Statistics through its annual National Health Interview Survey (NHIS) (Schoenborn &

Table 2. Distribution (percentage) of smoking habits of American women in selected occupations, 1982[a]

Occupation	Never smoked regularly	Current smoker 1–20 daily	Current smoker 21+ daily	Former smoker
Farmer	78.7	8.6	2.3	10.5
Sewer, stitcher	69.4	14.7	3.6	12.3
Factory worker	63.1	16.8	5.6	14.5
Food preparation	62.5	16.0	4.7	16.8
Teacher	60.3	11.9	3.1	24.7
Domestic service	58.8	18.2	5.0	18.0
Office clerical	54.2	16.3	6.2	23.3
Sales	53.9	17.9	6.2	22.0
Beautician	53.1	20.9	5.2	20.8
Book-keeper	53.1	16.1	7.4	21.4
Assembler	52.7	18.7	7.1	21.9
Nurse	47.8	18.4	5.7	28.1
Doctor	46.9	15.7	5.7	31.7
Social worker	45.3	15.2	6.7	32.9
Lawyer	41.7	12.8	9.3	36.2
Waitress	40.6	27.8	12.6	19.0

[a]Stellman et al., 1988.

Boyd, 1989). Trends in European tobacco usage were recently summarized by Hill (Hill, 1992). Total production and importation statistics for developing countries are sometimes available from industry sources (Mackay, 1992), but data on tobacco use within different population strata, especially with regard to income, education and other class variables, are usually available only from health surveys or epidemiological studies.

Occupation Epidemiological studies of occupational cancer have been among our most important tools for identifying environmental carcinogens, and occupational differences in cancer risks have been the subject of many investigations (Monson, 1990; Decoufle, 1982; Stellman & Stellman, 1996). Occupation itself can be used as an indicator of social class, so that clues to explain the impact of social class on cancer are often found in studies of differences in cancer rates between workers in different jobs or industries. The cancer sites most frequently affected by occupational carcinogens – those of the respiratory system and bladder – happen also to be strongly related to tobacco use. Therefore, it is

Table 3. Odds favouring current smoking in an Australian population[a]

Socioeconomic variable	Rate ratio[b] Men	Rate ratio[b] Women
Employed[c]	1.00	1.00
Unemployed	1.53*	1.43*
Not in labour force	1.40*	1.02
Professionals[c]	1.00	1.00
Managers and administrators	1.52*	1.56*
Paraprofessionals	1.44*	1.50*
Tradespersons	1.97*	2.11*
Clerks	1.62*	1.51*
Sales and service	1.74*	1.84*
Plant and machinery operators	2.19*	1.94
Labourers and related workers	2.28*	2.18*

*$P < 0.01$
[a]I. Gordon, pers. commun.
[b]Age-standardized to the 1988 total Australian population.
[c]Reference category.

essential to understand the role of tobacco in occupational carcinogenesis.

Few countries have tracked the evolution of smoking habits by occupation over time. Table 1 shows the estimated prevalence of smoking by occupational class and sex in 1978–1980 and 1987–1990 in the United States according to the NHIS (Nelson et al., 1994a). The data show that during both periods blue-collar workers smoked more than white-collar workers, and that the range was wider for men than for women. The findings also show that the class differences in smoking prevalence over time have widened in the USA.

Other American studies have also shown that men's smoking habits strongly reflect their socio-economic levels as reflected in their occupations, with higher prevalence of smoking associated with lower occupational levels, although the association is weaker for women (Stellman & Stellman, 1980). Data from the American Cancer Society prospective study (Table 2) show the distribution of smoking habits in American women in occupations in which the majority of American working women are employed. The highest prevalence of smoking was found in two economically disparate occupations – waitresses and lawyers – while teachers and domestic service workers were among the least likely to smoke.

The relationship between occupational status and smoking in male and female Australian workers is far more uniform than in America (Table 3). These smoking trends are also different from those recently reported for France by Sasco et al. (1994, 1995). There, the non-working population had among the lowest smoking rates. During 1979–1991, the smoking prevalence among mid-level employees decreased from 57% to 47% among men, but increased sharply among women, from 33% to 47% (Sasco et al., 1994).

For more than a century and a half the Registrar General for England and Wales has published periodic reports on mortality in relation to occupation. The Registrar General groups individuals into five broad social classes: I for professional occupations, II for managerial and lower professional occupations, III for skilled occupations (which may be further subdivided into manual and non-manual), IV for partly skilled occupations, and V for unskilled occupations (see, for instance, Registrar General for England and Wales, 1958). Using data from the Annual Consumer Survey of the Tobacco Advisory Council, Wald and Nicolaides-Bouman (1991) have examined the trends over time in smoking prevalence for each of the five classes for men and for women. In the late 1940s and early to mid-1950s cigarette smoking was more common among the

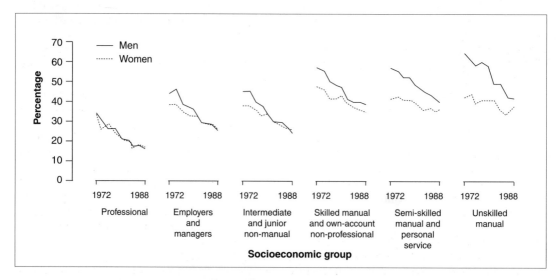

Figure 1. Percentages of men and women who smoke cigarettes (manufactured and hand-rolled) by socioeconomic group in Great Britain, 1972–1988 (men and women aged 16 and over). Data are from General Household Survey of the Office of Population Censuses and Surveys. Modified from Wald & Nicolaides-Bouman, 1991.

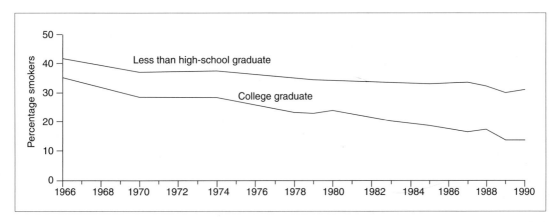

Figure 2. Prevalence of current smoking among adults by education in the United States, 1966–1990. Redrawn from Giovino et al., 1994.

higher social classes. However, by the late 1950s the prevalence of smoking manufactured cigarettes was approximately the same in all social classes, with about 58% of British men and 42% of British women current smokers in 1958. Between 1958 and 1971 the percentage in men declined from 54% to 37% in social class I, but remained about 59% in class V, after which declines in prevalence occurred in all classes.

Trends in smoking for the more recent years of rapid change are illustrated in Figure 1, which uses a slightly different socioeconomic group classification, from the General Household Survey of the Office of Population Censuses and Surveys. In both this and the Registrar General's system, it can be observed that in the higher social classes smoking rates have been about the same in men and women throughout the recent past but in lower social classes the gap between men and women has gradually narrowed, while the overall prevalence of smoking has been declining in all classes. Wald and Nicolaides-Bouman have shown that there has been a long-term reversal of the smoking habit, from professionals once being most likely to smoke to the unskilled now being the most likely smokers. The decline in overall prevalence of smoking has continued into the 1990s: results of the 1992 General Household Survey showed that 28% of adults were cigarette smokers, compared with 30% in 1990. The already wide occupational differentials continued to expand: only 14% of men in the professional group were cigarette smokers, compared with 42% of men in the unskilled manual group (Thomas et al., 1994).

Knowledge of smoking data for the occupational group of health professionals such as doctors and nurses is particularly important. An American Cancer Society study that traced smoking in doctors, nurses and dentists over several decades found that by 1982 only 17% of doctors, 14% of dentists and 23% of female nurses were still smokers (Garfinkel & Stellman, 1986). In a more recent American study of a national sample, Nelson et al. (1994b) found that smoking among physicians had fallen to 3% and among registered nurses to 18%. In the British doctors study (Doll & Hill, 1954; Doll et al., 1994) that followed doctors from 1951 to 1991, it was found that by 1991 18% of the doctors surviving were still smoking in comparison with 62% at the beginning of the study. In 1951, 85% of smokers consumed cigarettes (the remaining consuming only pipe and cigars) compared with 38% at 1991. These figures are important, because doctors and nurses serve as the only source of health information and counselling for many adults, and it is essential that they be role models if they are to function effectively.

These varying relationships between occupation and tobacco use make interpretation of epidemiological studies of occupational cohorts difficult if smoking data are not available at the individual level. Since this is frequently the case (Blair et al., 1988), numerous authors have presented detailed data on smoking prevalence by occupation and industry (Stellman & Stellman, 1980; Brackbill et al., 1988; Sterling & Weinkam, 1976, 1978; Weinkam & Sterling, 1987; Nelson et al., 1994a) to aid interpretation of analytical studies that lack smoking data. Axelson and Steenland (1988) have proposed a method for indirectly adjusting risks for smoking that utilizes prevalence data from industry-wide surveys.

Education There is considerable information available about tobacco use in relation to education for a number of countries. Data are available at frequent intervals for a probability sample of the USA (Resnicow et al., 1991) and have been presented in detail by the National Center for Health Statistics (Schoenborn & Boyd, 1989). Figure 2 shows that in 1967 the proportion of current smokers among adult Americans ranged from 35–42%, but has since fallen by nearly two-thirds to about 14% among college graduates, and has dropped only modestly among those with less than a high-school education (Giovino et al., 1994). As shown in Figure 3, there has been a rapidly increasing proportion of exsmokers among college graduates over time, compared with a much slower increase among the less educated. Out-of-school adolescents, who generally engage in more risky health behaviours than those who continue to attend, are also more likely to smoke (Centers for Disease Control, 1994a); smoking prevalence among high-school drop-outs has been reported to be as high as 70% (Pirie et al., 1988).

These findings have been replicated in many studies, utilizing different sampling techniques in a variety of American subpopulations (Novotny et al., 1988; Kabat, et al., 1991; Shea et al., 1991). In general, persons with less than high-school education are more consistently likely to start smoking cigarettes during childhood and adolescence (Escobedo et al., 1990), and among younger smokers of both sexes (age 18–30 years) those with only a high-school diploma smoke more cigarettes daily than those with at least some college education (Wagenknecht et al., 1990).

Educational gradients have been observed during the past one or two decades in many industrialized countries, including the United Kingdom (Wald & Nicolaides-Bouman, 1991), Italy (La Vecchia et al., 1992), Australia (I. Gordon, pers. commun.) and France. In the late 1970s, Wynder et al. (1981) found choice of cigarette type by French men to be strongly dependent on education, with the least educated men preferring by far (71%) nonfilter cigarettes manufactured with black tobacco (Figure 4). In 1966–1967 the lung cancer death rate for males in France was 70% that of the rate for males in the USA (Segi & Kurihara, 1972), but today the rates are nearly equal (Leclerc et al., 1990; Parkin et al., 1992). This equalization was predicted (Wynder et al., 1981) on the basis of a change towards American smoking habits in France: French smokers used to hold the lit cigarette in their mouths without inhaling; in addition, the protonated nicotine in black tobacco is absorbed through oral mucosa (Brunnemann & Hoffmann, 1974), leading smokers of black tobacco to inhale much less than smokers of blond (American-style) tobacco.

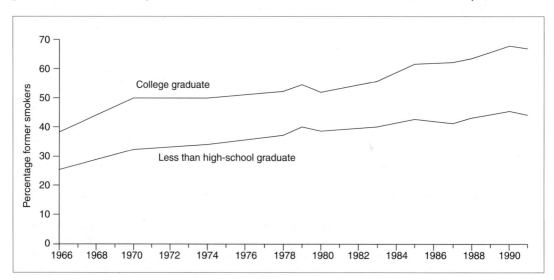

Figure 3. Prevalence of former smoking among adults by education in the United States, 1966–1990. Redrawn from Giovino et al., 1994.

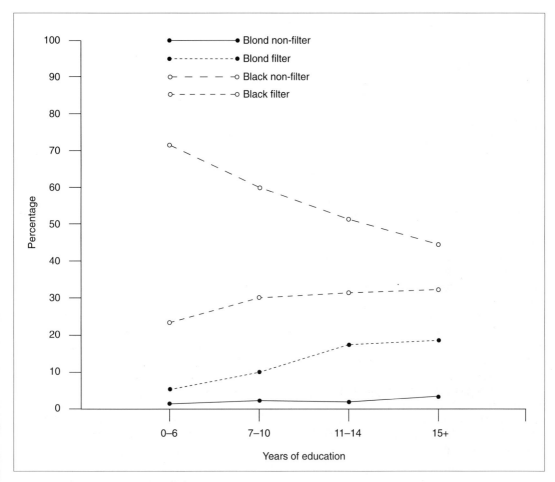

Figure 4. Tobacco type and filter preferences among French males by education. Redrawn from Wynder *et al.*, 1981.

Data from Spain have recently been published for a national random sample of the population taken in 1987 (Table 4) (Regidor *et al.*, 1994). These show that whereas smoking was more popular among college-educated men over 65 years old compared with poorly educated men of the same age, in younger cohorts the trend was reversed and more closely resembles the inverse association between cigarette smoking and education in the United Kingdom and the USA. While very few older Spanish women were cigarette smokers, the habit was very popular among young women, and showed a very strong association with higher levels of education, in sharp contrast to the data for men.

In most populations surveyed, the type of cigarette smoked is also related to education. Figure 5 shows a strong inverse relationship between education and the tar yield of cigarettes preferred by over 120 000 American men who were current smokers in 1982 (Stellman & Garfinkel, 1986). This is clearly a manifestation of the widespread belief among smokers that cigarettes with lower tar/nicotine yields carry reduced health risks.

Income, housing Cross-sectional studies of tobacco use frequently use income as a measure of social class. In Table 5, the smoking rates among Australian men and women relative to the most affluent quintile increase with decreasing income, reaching levels

43–47% greater for men and 32–53% greater for women aged 25 and older (I. Gordon, pers. commun.). Figure 6 shows the decline of never-smoking with increased income in the 1987 United States NHIS data, due principally to smoking cessation among the more affluent (Schoenborn & Boyd, 1989).

The British Office of Population Censuses and Surveys found smoking prevalence among women of ages 45–59 years was 48% for renters compared with 30% for owner-occupiers, and 50% for those without access to cars compared with 33% for those with access (Pugh et al., 1991). Owner-occupiers who were once smokers were twice as likely to quit smoking over a 17-year period than were renters.

Ethnicity Table 6 shows estimates of smoking prevalence from a variety of American sources during the years 1980–1992 by ethnic group (White, Black and Hispanic) and sex. According to Novotny et al. (1988), 40 years ago smoking prevalence in Whites and Blacks was about equal, but since that time Blacks have generally been more likely to smoke than Whites and less likely to quit.

However, Blacks who do smoke tend to consume fewer cigarettes per day than Whites do. In the NHIS, Blacks who smoked were far less likely to be heavy smokers (at least 15 per day) than Whites (odds ratio = 0.3; 95% confidence interval = 0.2–0.3); a similar odds ratio was reported by Kabat et al. (1991) among patients hospitalized for non-tobacco-related diseases. Using data from the 1987 NHIS and the National Hispanic Health and Nutrition Examination Survey (NHANES), Escobedo et al. (1990) reported that among men who started to smoke at 18 years of age or younger, Hispanic men had the highest smoking initiation rates, Whites had intermediate rates, and Blacks had the lowest rates. Furthermore, persons with less than a high-school education were consistently more likely to start smoking cigarettes during childhood and adolescence.

Population-based data on smoking among American Hispanics in San Francisco confirm their

Table 4. Prevalence of smoking (percentage) in adults according to age and education in Spain, 1987[a]

	Age							
	16–24 years		25–44 years		45–64 years		65+ years	
Education	All smokers	Smokers of 20+ per day	All smokers	Smokers of 20+ per day	All smokers	Smokers of 20+ per day	All smokers	Smokers of 20+ per day
Males								
Less than primary school	67	34	64	41	55	30	34	14
Primary school	62	26	67	40	53	30	32	11
High school	51	18	68	40	55	26	39	13
University	53	29	54	31	51	25	37	10
Females								
Less than primary school	43	12	17	5	3	1	1	0
Primary school	46	11	28	7	6	2	4	1
High school	48	10	47	12	15	10	6	1
University	62	7	52	20	12	4	5	1

[a]Regidor et al., 1994.

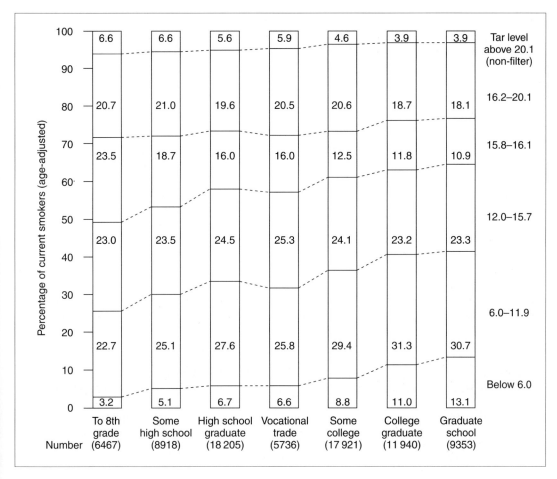

Figure 5. Age-adjusted distribution of tar yield of current cigarettes, by educational attainment, for males of age 40–89 years. Redrawn from Stellman & Garfinkel, 1986.

lower smoking prevalence especially among women. From telephone interviews conducted in 1989 with 652 randomly selected subjects, Perez-Stable et al. (1994) reported that 26% of Hispanic males and 8% of Hispanic females were current smokers, compared with 30% and 29% of White males and females, respectively. The corresponding figures for all Hispanics in the USA in 1987–1991 were 28.6% and 17.0% for males and females, respectively (Giovino et al., 1994).

Published estimates of smoking prevalence, however, are not always comparable with each other because of changes in population coverage both with respect to geographical area and ages of eligible subjects. In order to understand the degree to which smoking patterns explain ethnic differences in rates of lung cancer, much larger analytical studies will be required that account in a detailed way for individual smoking dosage histories, rather than simply recording whether or not individuals ever smoked. Few American studies have addressed this issue for lung cancer.

Cofactors and correlates of smoking behaviour
Much of the literature on tobacco and health treats cigarette smoking as if it were the only risk factor for cancer when in fact smokers are frequently exposed to other lifestyle and environmental factors that may affect cancer risk. Two of the most important of these correlated behaviours are occupation and diet. In epidemiological studies of these factors, cigarette smoking is often treated as

Table 5. Rate ratios for cigarette smoking, by sex, age and income in Australia, 1989–1990[a]

Quintile of socioeconomic disadvantage	Age					
	18–24		25–64		65+	
	Males	Females	Males	Females	Males	Females
First	1.00	1.00	1.00	1.00	1.00	1.00
Second	1.16	1.16	1.13	1.25*	1.01	0.93
Third	1.04	1.08	1.27*	1.32*	1.53	1.13
Fourth	1.02	1.26	1.30*	1.42*	1.40	1.16
Fifth	1.24	1.22	1.43*	1.53*	1.47	1.32

*$P < 0.01$
[a] I. Gordon, pers. commun.

a confounding factor to be adjusted. For instance, in one American lung cancer study, after adjusting for smoking Morabia *et al.* (1992) computed a population attributable risk of 9.2% for occupation, independently of cigarette smoking. Comparative presentation of both smoking and occupationally related cancer risks within the same study is rare.

Occupation The 1986 Surgeon General's Report on Smoking and Health considered many occupational exposures, including those to petrochemicals, aromatic amines, pesticides, asbestos, radon daughters, and cotton dust, as factors in lung cancer and chronic lung disease (US Department of Health and Human Services, 1985).

Exposure to some substances, notably asbestos, increases the risk of smoking-related disease far above the amount expected if smoking and asbestos exerted their effects independently. This effect, called synergism, has important implications for predicting future numbers of cases of implicated diseases (Selikoff, 1981).

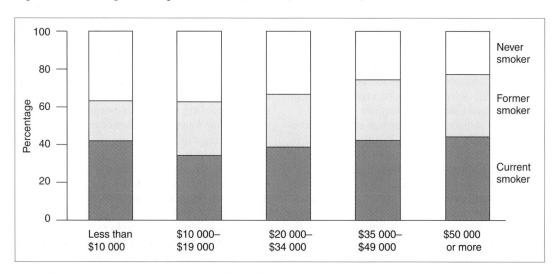

Figure 6. Distribution of smoking habits by annual income among males in the United States, 1987. Redrawn from Schoenborn & Boyd, 1989.

Table 6. Prevalence (percentage) of smoking by ethnic group and sex in the USA, 1980–1992

White, Black and Hispanic		White		Black		Hispanic		Year	Age	Reference
Males	Females	Males	Females	Males	Females	Males	Females			
–	–	35.1	30.8	45.6	36.3	–	–	1985	25–64	Novotny et al., 1988
–	–	29.7	23.8	49.1	29.5	–	–	1980–1990	20+	Kabat et al., 1991
–	–	29.3	23.7	32.5	25.1	–	–	1986	17+	Centers for Disease Control, 1987
31.2	26.5	30.6	27.5	38.9	28.2	30.0	18.0	1987	18+	American Cancer Society, 1991
31.2	26.5	30.5	26.7	39.0	28.0	30.0	18.0	1987	18+	Schoenborn & Boyd, 1989
–	–	27.4	23.8	35.1	24.4	25.2	15.5	1991	18+	Centers for Disease Control, 1993
28.6	24.6	28.6	25.9	32.3	24.1	23.6	18.0	1992	18+	Centers for Disease Control, 1994b

Synergism between asbestos-related occupations and smoking has been reported in both cohort and case–control studies. A classic example was reported for asbestos insulation workers by Hammond et al. (1979) who observed a lung death rate per 100 000 person-years of 11.3 in non-exposed nonsmokers, 122.6 in nonexposed smokers, 58.4 in nonsmoking asbestos-exposed workers, and 601.6 in smoking asbestos-exposed workers. An additive model would have predicted a rate in the latter of 170, while a multiplicative model predicted 633.6, which was close to the observed rate (Stellman, 1986).

A striking example of synergism in a case–control study is displayed in Figure 7, which shows the relative risk for lung cancer according to the number of cigarettes smoked per day and whether or not subjects worked in a shipyard, an occupation associated with heavy asbestos exposures (Blot & Fraumeni, 1981). Superimposed on the figure are theoretical risks that would be expected based upon both an additive and a multiplicative (synergistic) interaction. For heavy smokers, the multiplicative model closely predicted the observed lung cancer risk. Other examples of occupational exposures that may interact synergistically with smoking include exposure to radon daughters in underground mining in Sweden (Damber & Larsson, 1985), exposure to a variety of industrial chemicals in Italy (Pastorino et al., 1984), and working with metal materials in Japan (Hirayama, 1981).

Diet Diet has long been known to play an important role in the development of many cancers, including those that are also caused by cigarette smoking. The association of smoking with various dietary factors, including those that may affect lung cancer risk, has been studied extensively. In the American Cancer Society follow-up study, Stellman showed that men with the lowest consumption of foods rich in vitamins A and C were twice as likely to smoke cigarettes than men with high consumption, as shown in Figure 8 (Stellman, 1985). This was confirmed by Hebert and Kabat (1990) in a hospital-based study and extended to a variety of other foods. Wang and Hammond (1985) found consistently higher standardized mortality ratios for lung cancer among men who consumed little or no fruits or juices compared with those who consumed them daily. In Japan, Gao et al. (1993) reported a strong protective effect against lung cancer in smokers who consumed fresh fruits and vegetables frequently. Willett has reviewed the role of vitamin A

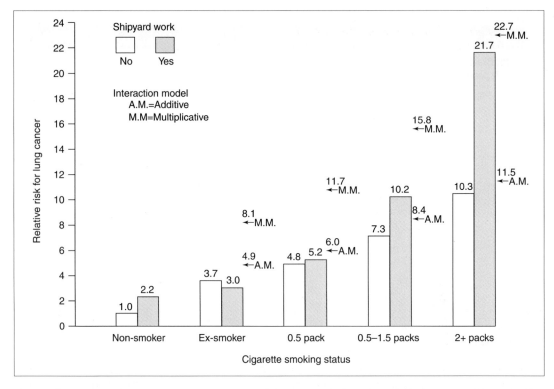

Figure 7. Relative risk for lung cancer according to number of cigarettes smoked per day, and whether or not subject worked in a shipyard. Redrawn from Stellman, 1986.

in lung cancer and concluded that dietary carotenoids are associated with reduced lung cancer risk (Willett, 1990).

However, the interplay of diet with tobacco usage patterns in different social strata and subsequent effect on cancer risk has not yet received adequate attention. For instance, as discussed above, it is difficult to explain the increased risk for lung cancer in American Blacks relative to Whites solely in terms of cigarette smoking. Hebert et al. (1991) reviewed Black–White differences in rates of lung, larynx and other cancers, and suggested that ethnic differences in diet are also important. Hebert and Kabat (1990) reported that total fat, meat consumption and cholesterol intake were positively associated with cigarette smoking, while total fruit, vitamin A and fibre consumption were negatively associated. Dorgan et al. (1993) found significant inverse smoking-adjusted associations of lung cancer with vegetables, fruit and carotenoids for White women, nonsignificant inverse associations for White men, but no inverse associations for Blacks.

In Japan, low consumption of green and yellow vegetables was associated with an increased risk of lung cancer in men and women independently of smoking (Hirayama, 1990). More generally, fruits and vegetables have been suggested to reduce the risk of lung cancer among smokers in Japan (Hirayama, 1986; Gao et al. 1993).

Alcohol plays an important etiological role alongside tobacco in cancers of the mouth, larynx, oral cavity, and oesophagus (Wynder & Stellman, 1977, 1979). It has been asserted that alcohol is not a carcinogen by itself but that it promotes the carcinogenic effects of tobacco smoke. However, the fact that most heavy drinkers are also heavy smokers makes it difficult to disentangle the two effects. In one of the few studies that addressed an indicator of socioeconomic status, Mashberg et al. (1993) found for oral cavity cancer a multiplicative effect of smoking and drinking in White United States veterans, with Blacks at lower risk than Whites. A recent Swedish case–control study of stomach cancer in relation to tobacco and alcohol use observed a significant interaction between tobacco use and fruit intake, with the latter more protective among smokers than among nonsmokers, and showed that high alcohol intake was associated with increased stomach cancer risk (Hansson et al., 1994).

Metabolic and genetic studies The development and application of novel molecular techniques to epidemiology (Shields & Harris, 1991) raises a number of new issues with respect to social factors and cancer, particularly concerning differences in cancer incidence between ethnic populations. Some population subgroups have been described as 'genetically high risk' for lung cancer (Kawajiri et al., 1990). In a Japanese population the odds ratio for lung cancer among those with a cytochrome P450 polymorphism (*Msp*1) was 3 compared with those without the polymorphism. By contrast, in an American study the same allele was more prevalent in Blacks compared with Whites by the same odds ratio of 3, but there was no association between the allele and lung cancer after stratification by race (Sheilds et al., 1993).

Metabolic and genetic studies of cancer incidence in different groups have so far been rather small and subject to wide variation, so no firm conclusions can yet be drawn. However, it is inevitable that refinements in laboratory techniques coupled with a reduction in testing costs will lead to an exponential proliferation of such studies in the near future. It is unclear whether inter-ethnic differences are derived from individuals' genetic inheritance, or whether they are acquired as a result of a long-term exposure or deficiency, which could conceivably be nutritional or occupational in origin. Findings from such studies might have unforeseeable social consequences. For instance, if it were determined that some ethnically identifiable segment of the population were inherently more 'susceptible' to lung cancer than another, this could affect availability of insurance and social services. The ethical questions associated with these possibilities are beyond the scope of this chapter.

Intervention and prevention

Cigarette smoking in most industrialized countries peaked after World War II and then gradually declined. Nevertheless, as there are still millions of smokers, and thousands of youths taking up the habit each year, concerted smoking control programmes

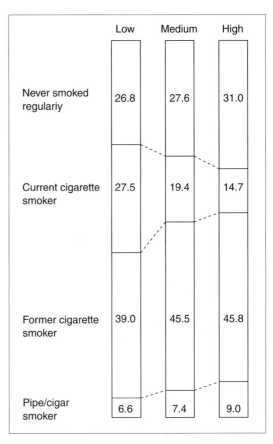

Figure 8. Distribution of smoking habits, adjusted for age, among 550 000 men surveyed in 1982 American Cancer Society follow-up study, according to whether they scored high, medium or low on a scale that measures frequency of consumption of foods rich in vitamins A and C. Redrawn from Stellman, 1986.

have become public policy in many countries. The United States in particular has invested heavily in controlled trials whose goals have included reduction in smoking prevalence in selected populations.

Two such trials recently concluded with results that have been termed disappointing by most observers. The Community Intervention Trial for Smoking Cessation (COMMIT) was a four-year randomized controlled trial in which one of each of 11 matched community pairs in the United States and Canada was randomly assigned to an intervention (COMMIT Research Group, 1995a, 1995b). There was a modest reduction in smoking levels for light-to-moderate smokers but not for heavy smokers. The Minnesota Heart Health Program was a research and demonstration project in three pairs of communites in northern–central USA and was designed to reduce risk factors for heart disease, including cigarette smoking (Lando *et al.*, 1995). The programme had no effect on smoking prevalence in men, and positive effects were seen for women in cross-sectional but not longitudinal data. Susser and others have pointed out the inherent difficulties in measuring the 'true' impact of an intervention that is conducted while secular rates are already declining, stating that 'the very same changes aimed for in such matters as smoking, diet, and exercise are those that have been progressing apace in response to the social movement...' (Susser, 1995).

Efforts to reduce smoking in the poor and lower socioeconomic minorities have been far less successful than in the affluent (Romano *et al.*, 1991). To a large extent this failure relates to our inability to deliver health-related information to the former groups through 'standard' educational channels such as those used in the community trials just described. Minorities tend also to have a lower level of knowledge of basic health risks. Several American surveys have found Blacks to have lower levels of 'cancer knowledge' than Whites, such as knowledge of smoking-related risks, as well as symptoms, screening and treatment (Robinson *et al.*, 1991). In a study of inner city Blacks in Indianapolis, 40% did not know that smoking was related to cancer (Loehrer *et al.*, 1991), and Blacks nationally were less likely than Whites to know the health risks of smoking but more likely to believe that cancer was not preventable (Jepson *et al.*, 1991).

Nevertheless, lack of knowledge is only one component of the social environment that contributes to high-risk behaviours. The social processes connected with the higher smoking rates in Blacks in the United States and the implications for developing effective cancer prevention programmes have been extensively studied. In addition to the socioeconomic factors of poverty and ethnic discrimination, higher smoking rates are associated with psychological factors such as hostility, overcrowding, and low social support (Baquet *et al.*, 1991; Romano *et al.*, 1991; Manfredi *et al.*, 1992; Schweritz *et al.*, 1992), as well as targeted cigarette advertising in minority communities (Royce *et al.*, 1993).

Therefore, it is evident that a 'one size fits all' approach to smoking cessation and other interventions is impractical, and that there is a need to develop culturally sensitive smoking cessation interventions consonant with the social and psychological characteristics of the target population. Cultural sensitivity comprises three dimensions. The first – surface structure – relates to the appearance of intervention materials. This can be achieved by using actors and settings recognized by the target audience. The second level – deep structure – may be more difficult to attain. It involves developing messages consistent with psychological, spiritual and emotional characteristics of the target audience. The third dimension – intragroup heterogeneity – recognizes that a large ethnic group, such as African-Americans, actually consists of individuals from many different cultural and social backgrounds. For example, among low-income, inner city Blacks there are varying levels of Afrocentricity (Cross, 1991). Some adopt African names and dress in Afrocentric clothing, but many do not. Moreover, a single African-American population may include immigrants from Haiti, Jamaica, and West Africa, in addition to Blacks born in the USA, and may contain groups with unique language, dress, food and norms. Failure to appreciate the significant variation within groups can result in so-called 'ethnic glossing' (Trimble, 1990–91).

Lack of culturally sensitive interventions may, at least in part, explain the lower success rate among Blacks who attempt to quit smoking. One programme tailored to a specific population in a culturally sensitive manner is the American Health Foundation's 'Harlem Health Connection', a community-based initiative to involve African-Americans in health promotion programmes with a special emphasis on reduction of cigarette smoking. Within this single community in one of the poorest neighbourhoods of New York City, smoking prevalence varied widely with the organizational 'channel' (recruitment milieu) – from 20% among those enrolled through church groups to 48% among those enrolled from (generally public) health care facilities and public housing (Resnicow *et al.*, 1996). Other intervention studies for Blacks have also worked through church groups (Stillman *et al.*, 1993).

The goal of the Harlem Health Connection was to develop and test a culturally sensitive self-help smoking cessation programme for low-income Blacks (Resnicow *et al.*, in press). The intervention comprised a video and accompanying cessation guide, a single telephone counselling call, and several quit and win contests. Intervention materials employed culturally specific cessation messages and strategies. For example, characters representing Martin Luther King, Jr, Malcolm X and Marcus Garvey – key figures from United States Black history – were used to link cessation messages to cultural and historical themes; for instance, with slogans such as 'break the chains of smoking addiction', 'be strong for your people', and 'there can be no progress without struggle'. Intervention results were mixed: there was a slight difference in self-reported point prevalence abstinence at six-month follow-up between the treatment and control groups (11.2% versus 7.9%; $P = 0.06$), but a significantly higher abstinence rate in the group that received both the materials and a booster call (16.4% versus 7.9%; $P < 0.05$).

This programme is one example of a nationwide effort, sponsored by the US National Cancer Institute (NCI) and other Federal health agencies, to develop and evaluate interventions aimed at stopping or preventing tobacco use in North America. The NCI Smoking, Tobacco, and Cancer Program has supported at least 49 intervention trials since 1984, with a total affected population of about ten million participants (Glynn *et al.*, 1993), and at least nine initiatives have been specifically targeted for reducing smoking in Blacks (Stotts *et al.*, 1991).

Risk of tobacco-related cancers in relation to indicators of social class

The relationship of social class to cancers of various sites has been reviewed comprehensively elsewhere in this book. Incidence, mortality and survival data are usually derived from official sources, such as tumour registries and vital statistics bureaus. These data sources frequently include social class indicators such as occupation, ethnic group, or education. Within many societies there is a strong gradient for each risk measure with respect to these social class variables. A large body of evidence confirms the inverse association of lung cancer with social class in many developed countries. Since this is the most common tobacco-related cancer in these countries, it is logical to ask whether risk within countries is associated with social class, and, if so, whether the association is explained by patterns of tobacco use. Unfortunately, most vital statistics sources contain no information on tobacco consumption, in which case the risk attributable to tobacco use must be inferred indirectly.

Incidence and mortality

Working with data from the Danish Cancer Registry and Central Population Register, Lynge and Thygesen (1990) reported a very strong gradient for lung and bladder cancers with respect to an occupational measure of social status in men and women (Table 7). Gradients were also observed for

Table 7. Relative risks for selected tobacco-related cancer sites for Danish men and women by occupational grouping[a]							
	Males			Females			
	Farmer	Academic	Building worker	Academic	Farmer	Clerk	Factory worker
Mouth	0.36	0.67	2.36	–	–	–	–
Pharynx	0.34	0.56	1.83	–	–	–	–
Oesophagus	0.47	0.65	1.42	–	–	–	–
Larynx	0.29	0.76	1.33	–	–	–	–
Lung	0.40	0.61	1.37	0.45	0.46	0.99	1.43
Bladder	0.51	0.88	1.15	0.61	0.83	1.05	1.19
Cervix	–	–	–	0.50	0.66	0.94	1.47

[a]Lynge and Thygesen, 1990.

several other tobacco-related sites for men but not women – mainly sites also associated with heavy alcohol consumption – and for cervical cancer in women. A similar pattern has been observed in England and Wales, indicating the predominant importance of tobacco smoking patterns for the occurrence of social class differences in cancer incidence and mortality in these societies. In an Italian case–control study, La Vecchia et al. (1992) reported a very strong inverse relation between education and several tobacco-related cancers, including oral cavity and pharynx (relative risk = 0.3 for highest versus lowest level), oesophagus (0.6), larynx (0.3) and cervix (0.7), but not lung.

An indirect estimation of the importance of smoking for the occurrence of social class differences in mortality in New Zealand (Pearce et al., 1985) indicated that smoking patterns explained much of the increased risk for social classes III and IV but not the very high mortality for class V (unskilled manual workers).

The literature on cancer mortality with respect to social class includes many studies derived from vital records, which often contain education, occupation, ethnicity, and other data such as residence from which income can be inferred through linkage with census records. Besides the classic tabulation of British mortality data by Logan (1982), there are also numerous cohort mortality studies, especially those performed in selected industries or among specific occupational groups. While occupation is dealt with elsewhere in this book, it is useful to point out the comment of Blair et al.

(1988) that many occupational cohorts, such as those constructed through linkage with official death records in Denmark (Lynge, 1990–91), Switzerland (Minder & Beer-Porizek, 1992; Minder, 1993) and New Zealand (Firth et al., 1993), lack smoking data and thus contribute only indirectly to the question of the relative roles of smoking and occupational exposures.

In the USA, Blacks constitute about one fifth of the total population but experience significantly higher cancer incidence rates than Whites for all sites combined (Baquet et al., 1991), and particularly for lung cancer (Devesa et al., 1991). Devesa et al. (1991) reported data from the National Cancer Institute's Surveillance, Epidemiology, and End Results (SEER) Program that showed drastic and disproportionate increases in lung cancer rates occuring in both Whites and Blacks during the approximately 15 years between 1969–1971, (when regular geographically based incidence reporting was established) and 1984–1986.

Baquet et al. (1991) linked 1978–1982 SEER data for three urban populations with the corresponding 1980 United States census data on socioeconomic status within individual census tracts. They found nearly equal lung cancer rates for Blacks and Whites after adjustment for either income or education and population density (Table 8). They concluded that 'the disproportionate distribution of Blacks at lower socioeconomic levels accounts for much of the excess cancer burden' (Baquet et al., 1991). These observations are extremely important because they show the existence of wide social class differences in cancer incidence in both races, but only minor race differences within each social class. They are also very important because they give rise to the testable hypothesis that patterns of tobacco use and other correlated lifestyle behaviours may 'explain' social class differences in cancer rates.

Bladder cancer is causally related to cigarette smoking, although the association is not as strong as for lung cancer (Wynder & Stellman, 1977). Anton-Culver et al. (1993) found the incidence of invasive bladder cancer in Hispanics to be half that in Whites in Orange County, California, and reported that after adjustment for smoking and occupational exposures, the risk of bladder cancer in Hispanics did not differ significantly from that

Table 8. Age-adjusted (1970-standard) rates for lung cancer by ethnic group and education in USA, 1978–1982[a]

Education	White	Black
<12 years	79.7	78.4
High school graduate	64.1	59.3
Some college education	51.0	56.9
College graduate	44.7	16.4
Adjusted	59.7	56.7
Adjusted (age only)	58.6	70.8

[a]Baquet et al., 1991.

in Whites or Asians and Pacific Islanders. This suggested that ethnic differences in bladder cancer incidence could be due largely to differences in smoking and occupational exposures.

Although literature on ethnic differences in cancer tends to be dominated by North American studies, recently population-based data on lung cancer have become available in São Paolo, Brazil; these show that, except for cancer of the oesophagus, Blacks had lower rates of most tobacco-related cancers than did Whites, and in fact their rates for lung and bladder cancer were significantly lower (odds ratios = 0.7 and 0.5, respectively) (Bouchardy et al., 1991).

Survival
The important question of social class differences in cancer survival and the reasons that underlie those differences are treated at length elsewhere in this book (se the chapters by Auvinen and Karjalainen, by Kogevinas and Porta and by Segnan). This literature, although voluminous, relates largely to differences in such prognostic factors as screening, barriers and access to medical treatment, quality of care, and host characteristics. It is possible that continued smoking might play a role as well in the occurrence of social class and race differences in cancer survival, but there have so far been relatively few studies of survival from tobacco-related cancers in which the role of tobacco use was explicitly accounted for (see, for example, Koh et al., 1984).

Conclusions and generalizations
The complex relationships between tobacco use and social class throughout the world make a facile description of tobacco–cancer linkages exceptionally challenging. One would like to propose the hypothesis that class differences in lung cancer, say, are mostly related to the unequal distribution of tobacco smoking between social classes, and in some fairly simple situations this has been satisfactorily demonstrated. Nevertheless, there are many unresolved issues, especially with regard to the role of collateral exposures, such as hazardous occupations, poor diet, and limited access to health care. This makes it essential that careful epidemiological investigations in these areas be continued.

Despite an overall decline in death rates in the USA since 1960, poor and poorly educated people have higher death rates than those with higher incomes or better educations, and this disparity increased between 1960 and 1986 (Pappas et al., 1993). Whether tobacco use 'explains' socioeconomic differences in one or more of the cancers that it causes has rarely been directly addressed in epidemiological studies. Typically, socioeconomic status is treated as a nuisance factor whose effect one hopes to render invisible by statistical adjustment in order to focus on the etiological effects of tobacco use itself.

The difference in cancer incidence and mortality between Blacks and Whites in the USA has been a subject of considerable discussion. While the medical and epidemiological literature tends to focus narrowly on specific lifestyle and environmental agents, such as tobacco use and occupation, the social processes that underlie exposure to these agents are frequently neglected (Hurowitz, 1993). The most obvious such process is racism, which is a systematic set of social inequities. Freeman has written extensively on the relationship between racism, its associated poverty, and health (Freeman, 1993a, 1993b), while Cooper has attempted to construct a theoretical framework to define pathogenic mechanisms that also incorporate social processes (Cooper, 1993). This framework is not unique to the USA; Smith et al. (1991) have also shown in the United Kingdom that no single factor, such as differences in smoking behaviour or susceptibility, can entirely account for the association between tobacco use and cancer. The need for continuing research into these relationships has not diminished.

The changing rates in the tobacco-related cancer sites with the highest incidence (lung and oral cavity), along with shifting patterns of tobacco usage due to changing preferences abetted by aggressive marketing, make it increasingly important to maintain a flow of 'reliable data on prevalence of tobacco use and morbidity and mortality rates among minorities' (Chen, 1993). This includes support for ongoing population surveys and especially uniformity in recording race and ethnicity in surveys, vital records, and census data.

But as important as it is to continue to improve the information base on tobacco use and its health consequences, it is even more urgent to interpret and channel these findings into prevention strategies. If we are only now learning how best to do this in the industrialized world, the struggle may be even more difficult in places like southern Europe, where smoking is on an upsurge (Hill, 1992), and in eastern Europe and the developing countries of South America, Asia and Africa, where the transnational tobacco companies have been aggressively expanding their markets (Connolly, 1992; Mackay, 1992). Perhaps health advocates can take a leaf from these companies and learn to export successful smoking and cessation programmes as well.

Acknowledgements

This work was supported by United States National Cancer Institute Grants CA-17613, CA-32617, CA-63021 and CA-68384 and United States Center for Substance Abuse Prevention Grant No. SPO-7250.

References

American Cancer Society (1991) *Cancer Facts and Figures for Minority Americans - 1991*, Atlanta, GA, American Cancer Society

Anton-Culver, H., Lee-Feldstein, A. & Taylor, T.H. (1993) The association of bladder cancer risk with ethnicity, gender, and smoking. *Ann. Epidemiol.*, 3, 429–433

Axelson, O. & Steenland, K. (1988) Indirect methods of assessing the effects of tobacco use in occupational studies. *Am. J. Ind. Med.*, 13, 105–118

Baquet, C.R., Horm, J.W., Gibbs, T. & Greenwald, P. (1991) Socioeconomic factors and cancer incidence among blacks and whites. *J. Natl. Cancer Inst.*, 83, 551–557

Blair, A., Steenland, K., Shy, C., O'Berg, M., Halperin, W. & Thomas, T. (1988) Control of smoking in occupational epidemiologic studies: methods and needs. *Am. J. Ind. Med.*, 13, 3–4

Blot, W.J. & Fraumeni, J.F. Jr., (1981) Cancer among shipyard workers. In: Peto, R. & Schneiderman, M., eds, *Quantification of occupational cancer*. Cold Spring Harbor, NY, Cold Spring Harbor Laboratory. pp. 37–46

Boffetta, P. & Parkin, D.M. (1994) Cancer in developing countries. *CA Cancer J. Clin.*, 44, 81–90

Bouchardy, C., Mirra, A.P., Khlat, M., Parkin, D.M., deSouza, M.J.P. & Gotlieb, S.L.D. (1991) Ethnicity and cancer risk in Sao Paolo, Brazil. *Cancer Epidemiol. Biomarkers Prev.*, 1, 21–27

Brackbill, R., Frazier, T. & Shilling, S. (1988) Smoking characteristics of US workers, 1978–1980. *Am. J. Ind. Med.*, 13, 5–41

Brunnemann, K.D. & Hoffmann, D. (1974) The pH of tobacco smoke. *Food Cosmet. Toxicol.*, 12, 115–121

Centers for Disease Control (1987) Cigarette smoking in the United States, 1986. *Morb. Mortal. Weekly Rep.*, 36, (35) 581–585

Centers for Disease Control (1993) Cigarette smoking among adults - United States, 1991. *Morb. Mortal. Weekly Rep.*, 42, (12) 230–233

Centers for Disease Control (1994a) Health risk behaviors among adolescents who do and do not attend school – United States, 1992. *Morb. Mortal. Weekly Rep.*, 43 (8), 129–132

Centers for Disease Control (1994b) Cigarette smoking among adults - United States, 1992, and changes in the definition of current cigarette smoking. *Morb. Mortal. Weekly Rep.*, 43, (19) 342–346

Chen, V.W. (1993) Smoking and the health gap in minorities. *Ann. Epidemiol.*, 3, 159–164

COMMIT Research Group (1995a) Community Intervention Trial for Smoking Cessation (COMMIT): I. Cohort results from a four-year community intervention. *Am. J. Public Health*, 85, 183–192

COMMIT Research Group (1995b) Community Intervention Trial for Smoking Cessation (COMMIT): II. Changes in adult cigarette smoking prevalence. *Am. J. Public Health*, 85, 193–200

Connolly, G.N. (1992) Worldwide expansion of transnational tobacco industry. *J. Natl. Cancer Inst. Monogr.*, 12, 29–35

Cooper, R.S. (1993) Health and the social status of Blacks in the United States. *Ann. Epidemiol.*, 3, 137–144

Covey, L.S., Zang, E.A. & Wynder, E.L. (1992) Cigarette smoking and occupational status: 1977–1990. *Am. J. Public Health*, 82, 1230–1234

Cross, W.E. (1991) *Shades of Black: diversity in African American identity*. Philadelphia, Temple University Press

Damber, L. & Larsson, L-G. (1985) Underground mining, smoking, and lung cancer: a case-control study in the iron ore municipalities in Northern Sweden. *J. Natl. Cancer Inst.*, 74, 1207–1212

Decoufle, P. (1982) Occupation. In: Schottenfeld, D. & Fraumeni, J.F., Jr, eds, *Cancer epidemiology and prevention*. Philadelphia, Saunders. pp. 318–335

Devesa, S.S., Shaw, G.L. & Blot, W.J. (1991) Changing patterns of lung cancer incidence by histological type. *Cancer Epidemiol. Biomarkers Prev.*, 1, 29–34

Doll, R. & Hill, A.B. (1954) The mortality of doctors in relation to their smoking habits. A preliminary report. *Br. Med. J.*, ii, 1525–1536

Doll, R., Peto, R., Wheatley, K., Gray, R. & Sutherland, I. (1994) Mortality in relation to smoking: 40 years' observations on male British doctors. *Br. Med. J.*, 309, 901–911

Dorgan, J.F., Ziegler, R.G., Schoenberg, J.B., Hartge, P., McAdams, M.J., Falk, R.T., Wilcox, H.B. & Shaw, G.L. (1993) Race and sex differences in associations of vegetables, fruits, and carotenoids with lung cancer risk in New Jersey (United States). *Cancer Causes Control*, 4, 273–281

Escobedo, L.G., Anda, R.F., Smith, P.F., Remington, P.L. & Mast, E.E. (1990) Sociodemographic characteristics of cigarette smoking initiation in the United States. *J. Am. Med. Assoc.*, 264, 1550–1555

Firth, H.M., Herbison, G.P., Cooke, K.R. & Fraser, J. (1993) Male cancer mortality by occupation: 1973–86. *New Zealand Med. J.*, 106, 328–330

Freeman, H.P. (1993a) The impact of clinical trial protocols on patient care systems in a large city hospital. Access for the socially disadvantaged. *Cancer*, 72 (Suppl. 9), 2834–2838

Freeman, H.P. (1993b) Poverty, race, racism, and survival. *Ann. Epidemiol.*, 3, 145–149

Gao, C.M., Tajima, K., Kuroishi, T., Hirose, K. & Inoue, M. (1993) Protective effects of raw vegetables and fruit against lung cancer among smokers and ex-smokers: a case-control study in the Tokai area of Japan. *Jap. J. Cancer Res.*, 84, 594–600

Garfinkel, L. & Stellman, S.D. (1986) Cigarette smoking among physicians, dentists, and nurses. *CA Cancer J. Clin.*, 36, 2–8

Giovino, G., Schooley, M.W., Zhu, B-P., Chrismon, J.H., Tomar, S.L., Peddicord, J.P., Merritt, R.K., Husten, C.G. & Eriksen, M.P. (1994) Surveillance for selected tobacco-use behaviors – United States, 1900–1994. *Morb. Mortal. Weekly Rep.*, 43 (SS-3), 1–43

Glynn, T.J., Manley, M.W., Mills, S.L. & Shopland, D.R. (1993) The United States National Cancer Institute and the science of tobacco control research. *Cancer Det. Prev.*, 17, 507–512

Hammond, E.C., Selikoff, I.J. & Seidman, H. (1979) Asbestos exposure, cigarette smoking and death rates. *Ann. N.Y. Acad. Sci.*, 330, 473–490

Hansson, L-E., Baron, J., Nyren, O., Bergstrom, R., Wolk, A. & Adami, H-O. (1994) Tobacco, alcohol, and the risk of gastric cancer. A population-based case-control study in Sweden. *Int. J. Cancer*, 57, 26–31

Hebert, J.R. & Kabat, G.C. (1990) Differences in dietary intake associated with smoking status. *Eur. J. Clin. Nutr.*, 44, 185–193

Hebert, J.R., Miller, D.R., Toporoff, E.D., Teas, J. & Barone, J. (1991) Black-White differences in United States cancer rates: a discussion of possible dietary factors to explain large and growing divergencies. *Cancer Prev.*, 1, 141–156

Hill, C. (1992) Trends in tobacco use in Europe. *J. Natl. Cancer Inst. Monogr.*, 12, 21–24

Hirayama, T. (1981) Proportion of cancer attributable to occupation obtained from a census, population-based, large cohort study in Japan. In: Peto, R. & Schneiderman, M., eds, *Quantification of occupational cancer*. Cold Spring Harbor, NY, Cold Spring Harbor Laboratory. pp. 631–649

Hirayama, T. (1986) Nutrition and cancer – a large scale cohort study. *Prog. Clin. Biol. Res.*, 206, 299–311

Hirayama, T. (1990) *Life-style and mortality. A large-scale census-based cohort study in Japan*. New York, Karger

Hurowitz, J.C.C. (1993) Toward a social policy for health. *New Engl. J. Med.*, 329, 130–133

International Agency for Research on Cancer (1986). *IARC Monographs on the Evaluation of Carcinogenic Risk of Chemicals to Humans*. Vol. 38, *Tobacco smoking*. Lyon, IARC

Jepson, C., Kessler, L.G., Portnoy, B. & Gibbs, T. (1991) Black-white differences in cancer prevention knowledge and behavior. *Am. J. Public Health*, 81, 501–504

Kabat, G.C., Morabia, A. & Wynder, E.L. (1991) Comparison of smoking habits of Blacks and Whites in a case-control study. *Am. J. Public Health*, 81, 1483–1486

Kawajiri, K., Nakachi, K., Imai, K., Yoshii, A., Shinoda, N. & Watanabe, N. (1990) Identification of genetically high risk individuals to lung cancer by DNA polymorphisms of the cytochrome P4501A1 gene. *FEBS Lett.*, 263, 131–133

Kogevinas, M., Marmot, M.G., Fox, A.J. & Goldblatt, P.O. (1991) Socioeconomic differences in cancer survival. *J. Epidemiol. Community Health*, 45, 216–219

Koh, H.K., Sober, A.J., Day, C.L., Lew, R.A. & Fitzpatrick, T.B. (1984) Cigarette smoking and malignant melanoma. Prognostic implications. *Cancer*, 53, 2570–2573

Lando, H., Pechacek, T.F., Pirie, P.L., Murray, D.M., Mittelmark, M.B., Lichtenstein, E., Nothwehr, F. & Gray, C. (1995) Changes in adult cigarette smoking in the Minnesota Heart Health Program. *Am. J. Public Health*, 85, 201–208

La Vecchia, C., Negri, E. & Franceschi, S. (1992) Education and cancer risk. *Cancer*, 70, 2935–2941

Leclerc, A., Lert, F. & Fabien, C. (1990) Differential mortality: some comparisons between England and Wales, Finland and France, based on inequality measures. *Int. J. Epidemiol.*, 19, 1001–1010

Loehrer, P.J., Sr, Greger, H.A., Weinberger, M., Musick, B., Miller, M., Nichols, C., Bryan, J., Higgs, D. & Brock, D. (1991) Knowledge and beliefs about cancer in a socioeconomically disadvantaged population. *Cancer*, 68, 1665–1671

Logan, W.P.D. (1982) *Cancer mortality by occupation and social class 1851–1971* (Vol. OCPS SMPS No. 43). London, Her Majesty's Stationery Office

Lynge, E. (1990–91) Occupational mortality and cancer analysis. *Public Health Rev.*, 18, 99–116

Lynge, E. & Thygesen, L. (1990) Occupational cancer in Denmark. Cancer incidence in the 1970 census population. *Scand. J. Work Environ. Health*, 16 (Suppl. 2), 3–35

Mackay, J. (1992) US tobacco export to Third World: Third world war. *J. Natl. Cancer Inst. Monogr.*, 12, 25–28

McGinnis, J.M. & Foege, W.H. (1993) Actual causes of death in the United States. *J. Am. Med. Assoc.*, 270, 2207–2212

Manfredi, C., Lacey, L., Warnecke, R. & Buis, M. (1992) Smoking-related behavior, beliefs, and social environment of young black women in subsidized public housing in Chicago. *Am. J. Public Health*, 82, 267–272

Mashberg, A., Boffetta, P., Winkelman, R. & Garfinkel, L. (1993) Tobacco smoking, alcohol drinking, and cancer of the oral cavity and oropharynx among U.S. veterans. *Cancer*, 72, 1369–1375

Minder, C.E. (1993) Socio-economic factors and mortality in Switzerland. *Soz. Präventivmed.*, 38, 313–328

Minder, C.E. & Beer-Porizek, V. (1992) Cancer mortality of Swiss men by occupation, 1979–1982. *Scand. J. Work Environ. Health*, 18 (Suppl. 3), 1–27

Monson, R.R. (1990) *Occupational epidemiology*. Boca Raton, FL, CRC Press

Morabia, A., Markowitz, S., Garibaldi, K. & Wynder, E.L. (1992) Lung cancer and occupation: results of a multicentre case-control study. *Br. J. Ind. Med.*, 49, 721–727

Nelson, D.E., Emont, S.L., Brackbill, R.M., Cameron, L.L., Peddicord, J. & Fiore, M.C. (1994a) Cigarette smoking prevalence by occupation in the United States. A comparison between 1978 to 1980 and 1987 to 1990. *J. Occup. Med.*, 36, 516–525

Nelson, D.E., Giovino, G.A., Emont, S.L., Brackbill, R., Cameron, L.L., Peddicord, J. & Mowery, P.D. (1994b) Trends in cigarette smoking among US physicians and nurses. *J. Am. Med. Assoc.*, 271, 1273–1275

Novotny, T.E., Warner, K.E., Kendrick, J.S. & Remington, P.L. (1988) Smoking by Blacks and Whites: socioeconomic and demographic differences. *Am. J. Public Health*, 78, 1187–1189

Pappas, G., Queen, S., Hadden, W. & Fisher, G. (1993) The increasing disparity in mortality between socioeconomic groups in the United States, 1960 and 1986. *New Engl. J. Med.*, 329, 103–109

Parkin, D.M., Muir, C.S., Whelan, S.L., Gao, Y-T., Ferlay, J. & Powell, J. (1992) *Cancer incidence in five continents, Vol. VI*. Lyon, International Agency for Research on Cancer

Pastorino, U., Berrino, F., Gervasio, A., Pesenti, V., Riboli, E. & Crosignani, P. (1984) Proportion of lung cancers due to occupational exposure. *Int. J. Cancer*, 33, 231–237

Pearce, N.E., Davis, P.B., Smith, A.H. & Foster, F.H. (1985) Social class, ethnic group, and male mortality in New Zealand, 1974–78. *J. Epidemiol. Commun. Health*, 39, 9–14

Perez-Stable, E.J., Marin, G. & Marin, B.V. (1994) Behavioral risk factors: a comparison of Latinos and non-Latino Whites in San Francisco. *Am. J. Public Health*, 84, 971–976

Pirie, P.L., Murray, D.M. & Luepker, R.V. (1988) Smoking prevalence in a cohort of adolescents, including absentees, dropouts, and transfers. *Am. J. Public Health*, 78, 176–178

Pugh, H., Power, C., Goldblatt, P. & Arber, S. (1991) Women, lung cancer mortality, socio-economic status, and changing smoking patterns. *Soc. Sci. Med.*, 32, 1105–1110

Regidor, E., Gutierrez-Fisac, J.L. & Rodriguez, C. (1994) *Diferencias y desigualdades en salud en Espana [Differences and inequities in health in Spain]*. Madrid, Ediciones Diaz de Santos

Registrar General for England and Wales (1958) *Decennial supplement – 1951. Occupational mortality, part II*. London, Her Majesty's Stationery Office.

Resnicow, K., Kabat, G. & Wynder, E.L. (1991) Progress in decreasing cigarette smoking. In: DeVita, V.T., Jr, Helman, S. & Rosenberg, S.A., eds, *Important Advances in Oncology*. Philadelphia, Lippincott, pp. 205–213.

Resnicow, K., Futterman, R., Weston, R.W., Royce, J., Parms, C., Freeman, H. & Orlandi, M.A. (1996) Smoking prevalence in Harlem, New York. *Am. J. Health Prom.*, 10 (3), 7–10

Resnicow, K., Vaughan, R.D., Futterman, R., Weston, R.W., Royce, J., Davis-Hearn, M., Smith, M., Parms, C., Freeman, H. & Orlandi, M.A. (1997) A self-help smoking cessation program for inner-city African Americans: results from the Harlem Health Connection Project. *Health Educ. Q.* (in press)

Robinson, R.G., Kessler, L.G. & Naughton, M.D. (1991) Cancer awareness among African Americans: a survey assessing race, social status, and occupation. *J. Natl. Med. Assoc.*, 83, 491–497

Romano, P.S., Bloom, J. & Syme, S.L. (1991) Smoking, social support, and hassles in an urban African-American community. *Am. J. Public Health*, 81, 1415–1422

Royce, J.M., Hymowitz, N., Corbett, K., Hartwell, T.D. & Orlandi, M. (1993) Smoking cessation factors among African Americans and whites. *Am. J. Public Health*, 83, 220–226

Sasco, A.J., Grizeau, D., Pobel, D., Chatard, O. & Danzon, M. (1994) Tabagisme et class sociale en France de 1974 a 1991. *Bull. Cancer*, 81, 355–359

Sasco, A.J., Grizeau, D. & Danzon, M. (1995) Is tobacco use finally decreasing in France? *Cancer Det. Prev.*, 19, 210–218

Satariano, W.A. & Swanson, G.M. (1977) Racial differences in cancer incidence: the significance of age-specific patterns. *Cancer*, 62, 2640–2653

Scherwitz, L.W., Perkins, L.L., Chesney, M.A., Hughes, G.H., Sidney, S. & Manolio, T.A. (1992) Hostility and health behaviors in young adults: The CARDIA Study. *Am. J. Epidemiol.*, 136, 136–145

Schoenborn, C.A. & Boyd, G. (1989) *Smoking and other tobacco use: United States, 1987*. Washington, DC, US Govt Printing Office

Segi, M. & Kurihara, M. (1972) *Cancer mortality for selected sites in 24 countries, No. 6, 1966–1967*. Nagoya, Japan Cancer Society

Selikoff, I.J. (1981) Constraints in estimating occupational contributions to current cancer mortality in the United States. In: Peto, R. & Schneiderman, M., eds, *Quantification of occupational cancer*. Cold Spring Harbor, NY, Cold Spring Harbor Laboratory. pp. 3–13

Shea, S., Stein, A.D., Basch, C.E., Lantigua, R., Maylahn, C., Strogatz, D.S. & Novick, L. (1991) Independent associations of educational attainment and ethnicity with behavioral risk factors for cardiovascular disease. *Am. J. Epidemiol.*, 134, 567–582

Sheilds, P.G., Caporaso, N.E., Falk, R.T., Sugimura, H., Trivers, G.E., Trump, B.F., Hoover, R.N., Weston, A. & Harris, C.C. (1993) Lung cancer, race, and a CYP1A1 genetic polymorphism. *Cancer Epidemiol. Biomarkers Prev.*, 2, 481–485

Shields, P.C. & Harris, C.C. (1991) Molecular epidemiology and the genetics of environmental cancer. *J. Am. Med. Assoc.*, 266, 681–687

Smith, G.D., Leon, D., Shipley, M.J. & Rose, G. (1991) Socioeconomic differentials in cancer among men. *Int. J. Epidemiol.*, 20, 339–345

Stellman, S.D. (1985) Chairman's remarks. *Natl. Cancer Inst. Monogr.*, 67, 145–147

Stellman, S.D. (1986) Interactions between smoking and other exposures: occupation and diet. In: Hoffmann, D. & Harris, C., eds, *Mechanisms in tobacco carcinogenesis*. Cold Spring Harbor, NY, Cold Spring Harbor Laboratory. pp. 377–393

Stellman, S.D. & Garfinkel, L. (1986) Smoking habits and tar levels in a new American Cancer Society prospective study of 1.2 million men and women. *J. Natl. Cancer Inst.*, 76, 1057–1063

Stellman, S.D. & Stellman, J.M. (1980) Women's occupations, smoking, and cancer and other diseases. *CA Cancer J. Clin.*, 31, 29–43

Stellman, S.D., Boffetta, P. & Garfinkel, L. (1988) Smoking habits of 800,000 American men and women in relation to their occupations. *Am. J. Ind. Med.*, 13, 43–58

Stellman, J.M. & Stellman, S.D. (1996) Cancer and the workplace *CA Cancer J. Clin.* 46, 70–92

Sterling, T. & Weinkam, J. (1976) Smoking characteristics by type of employment. *J. Occup. Med.*, 18, 743–754

Sterling, T.D. & Weinkam, J.J. (1978) Smoking patterns by occupation, industry, sex and race. *Arch. Environ. Health*, 33, 313–317

Stillman, F.A., Bone, L.R., Rand, C., Levine, D.M. & Becker, D.M. (1993) Heart, body, and soul: a church-based smoking-cessation program for urban African-Americans. *Prev. Med.*, 22, 335–349

Stotts, R.C., Glynn, T.J. & Baquet, C.R. (1991) Smoking cessation among Blacks. *J. Health Care Poor Underserved*, 2, 307–319

Susser, M. (1995) Editorial: The tribulations of trials – intervention in communities. *Am. J. Public Health*, 85, 156–158

Thomas, M., Goddard, E., Hickman, M. & Hunter, P. (1994) *General household survey 1992. An inter-departmental survey carried out by OPCS between April 1992 and March 1993* (Office of Population Censuses and Surveys. Social Survey Division). London, Her Majesty's Stationery Office

Trimble, J.E. (1990–91) Ethnic specification, validation prospects, and the future of drug use research. *Int. J. Addict.*, 25 (2A), 149–170

U.S. Department of Health and Human Services (1985) *The health consequences of smoking. Cancer and chronic lung disease in the workplace. A report of the Surgeon-General.* Washington, DC, US DHHS, Public Health Service, Office of Smoking and Health, Centers for Disease Control

Wagenknecht, L.E., Perkins, L.L., Cutter, G.R., Sidney, S., Burke, G.L., Manolio, T.A., Jacobs, D.R., Liu, K., Friedman, G.D., Hughes, G.H. & Hulley, S.B. (1990) Cigarette smoking is strongly related to educational status: the CARDIA study. *Prev. Med.*, 19, 158–169

Wald, N. & Nicolaides-Bouman, A. (1991) *UK smoking statistics*. London, Oxford University Press

Wang, L-D. & Hammond, E.C. (1985) Lung cancer, fruit, green salad, and vitamin pills. *Chin. Med. J.*, 98, 206–210

Weinkam, J.J. & Sterling, T.D. (1987) Changes in smoking characteristics by type of employment from 1970 to 1979/80. *Am. J. Ind. Med.*, 11, 539–561

Willett, W.C. (1990) Vitamin A and lung cancer. *Nutr. Rev.*, 48, 201–211

Williams, J., Clifford, C., Hopper, J. & Giles, G. (1991) Socioeconomic status and cancer mortality and incidence in Melbourne. *Eur. J. Cancer*, 27, 917–921

Wynder, E.L. & Stellman, S.D. (1977) Comparative epidemiology of tobacco-related cancers. *Cancer Res.*, 37, 4608–4622

Wynder, E.L. & Stellman, S.D. (1979) The impact of long-term filter use on lung and larynx cancer: a case-control study. *J. Natl Cancer Inst.*, 62, 471–477

Wynder, E.L., Mushinski, M., Stellman, S.D. & Choay, P. (1981) Tobacco usage in France: an epidemiologic survey. *Prev. Med.*, 10, 301–315

Wynder, E.L., Taioli, E. & Fujita, Y. (1992) Ecologic study of lung cancer risk factors in the US and Japan, with special reference to smoking and diet. *Jpn. J. Cancer Res.*, 83, 418–423

Corresponding author:
S.D. Stellman
American Health Foundation, 320 E. 43rd Street, New York, NY 10017, USA

Alcohol drinking, social class and cancer

H. Møller and H. Tønnesen

This chapter reviews the data on occurrence of cancers that are potentially caused by alcohol drinking (cancers of the upper gastrointestinal and respiratory tracts, and liver cancer) in relation to social class. In order to assess the role of alcohol drinking in the observed social class gradients of these cancers, we have particularly looked for consistency in the gradients of different alcohol-related cancers, and used lung cancer occurrence to judge the role of tobacco smoking, which is the major other determinant of these diseases. Additional data on levels of alcohol drinking and on the occurrence of other alcohol-related morbidity are brought into the discussion where available. A role of alcohol drinking in the observed negative social class gradients for alcohol-related cancers is very likely in men in France, Italy and New Zealand. Evidence that is less strong, but is suggestive of a role of alcohol drinking, is seen for men in Brazil, Switzerland, the United Kingdom and Denmark. Although a role of alcohol drinking is likely or possible in certain populations, other factors may contribute as well, most notably tobacco smoking and dietary habits. Additional data on the frequency of complications after surgical procedures in alcohol drinkers are reviewed briefly.

Alcohol drinking is causally associated with cancers of the mouth and pharynx, oesophagus, larynx and liver (IARC, 1988). To the extent that alcohol drinking varies between different social classes, it may therefore contribute to the observed associations between social class and the risk of these particular cancers. The purpose of this chapter is to analyse the data that are presented in detail in the chapter by Faggiano et al. from this perspective, and review additional information from published sources regarding the association between social class and alcohol drinking in different populations. A general discussion of social determinants of alcohol use or alcohol abuse falls outside the scope of the chapter. Finally, data are reviewed on the association between alcohol drinking and the frequency of complications after surgical procedures. This may be relevant to the understanding of social class differences in survival after cancer diagnosis.

Alcohol drinking and cancer risk

Epidemiological studies have established with certainty that alcohol drinking is a strong risk factor for cancers of the upper gastrointestinal and respiratory tracts and for cancer of the liver (IARC, 1988). For example, a recent cohort study of alcohol abusers in Copenhagen, Denmark showed highly elevated risks of cancers of the mouth and pharynx, oesophagus, larynx and liver (Tønnesen et al., 1994; Table 1). Cancer of the lung also occurred more frequently than expected in this cohort but this excess was thought to reflect confounding by tobacco smoking rather than a causal effect of alcohol drinking on lung cancer occurrence. The mechanisms by which the consumption of alcoholic beverages increases the risk of cancer are not known (Seitz et al., 1992). Some cancers of organs other than those of the upper gastrointestinal and respiratory tracts have also been associated with alcohol drinking, but these associations are not established conclusively at the present time, and the possible relative risks involved are lower than those for cancers of the mouth and pharynx, oesophagus, larynx and liver.

Cancers of the upper gastrointestinal and respiratory tracts

Both alcohol drinking and tobacco smoking contribute to the risk of cancers of the upper gastrointestinal and respiratory tracts. Studies of alcohol drinking in non-smokers and of tobacco smoking in non-drinkers have confirmed that each habit is truly a risk factor for cancer, even in the absence of the other (La Vecchia & Negri, 1989; Talamini et al., 1990). For these diseases, it has been found consistently

Table 1. Numbers of cases (n), relative risks (RR) and 95% confidence intervals (CI) for selected cancers in a cohort of alcohol abusers[a]

Cancer site	Men			Women		
	n	RR	95% CI	n	RR	95% CI
Mouth and pharynx	112	3.6	3.0–4.3	22	17.2	10.8–26.0
Oesophagus	57	5.3	4.0–6.9	2	4.9	0.6–17.7
Larynx	65	3.7	2.8–4.7	1	2.2	0.6–12.2
Liver	38	4.1	2.9–5.6	1	1.6	0.0–8.9
Lung	456	2.5	2.3–2.7	29	3.7	2.5–5.4

[a]Data from Tønnesen et al., 1994.

that the combination of alcohol drinking and tobacco smoking adds more to the absolute risk than the sum of the two factors separately (Tuyns et al., 1977; Blot et al., 1988; Tuyns et al., 1988). For example, in one case–control study in France, the relative risk of oesophageal cancer in persons who were in the highest category of both alcohol drinking and tobacco smoking was 44.4, compared with persons in the lowest category for both factors; however, the relative risks associated with high alcohol drinking alone and with high tobacco smoking alone were 18.0 and 5.1, respectively (Tuyns et al., 1977). When such an interaction between two factors is present, a large proportion of the excess cases of cancer is attributable to the combination of the two factors.

The high relative risks of cancer of the mouth and pharynx, oesophagus and larynx (as seen in the Danish cohort study; Table 1) therefore reflect not only the pure effect of alcohol drinking but also the interaction with tobacco smoking. The separate effects of alcohol and tobacco can be distinguished only if both factors are recorded accurately for each individual in the study and if a sufficient proportion of individuals in the study population have only one of the two habits. It is well known that smoking and drinking tend to occur together in individuals, and that most heavy drinkers smoke tobacco as well (Johnson & Jennison, 1992). In addition, other determinants of cancer risk – for example, dietary habits – may be suspected to be associated with educational level, tobacco smoking and alcohol drinking (La Vecchia et al., 1992).

An important finding is that the relative risk of cancers of the upper gastrointestinal and respiratory tracts is particularly increased at very high levels of alcohol consumption. In the case–control study from France, for example, the relative risk of oesophageal cancer appeared to increase only above a daily consumption of 40 g of ethanol per day and five- to 10-fold increases in risk were seen only above 60 g of ethanol per day (Tuyns et al., 1977). This suggests that the behaviour of main relevance to cancer risk, and to the possible influence of alcohol drinking on the observed associations between social class and cancer occurrence, is a high level of alcohol consumption. The frequency of heavy drinking, alcohol abuse or alcoholism may thus be more relevant than variation within the low range of alcohol drinking.

The association of cancers of the upper gastrointestinal and respiratory tracts with alcohol drinking and tobacco smoking may vary for different subtypes of these diseases. Cancers of the lip, salivary glands and nasopharynx, which are often tabulated as part of cancer of the mouth and pharynx, are probably associated with alcohol to a lesser extent (if at all) than other parts of the mouth and pharynx are (Tønnesen et al., 1994). Within the oesophagus, the association with alcohol and tobacco is certain for squamous-cell carcinoma, which most often occurs in the middle or upper part of the organ, but adenocarcinoma occurring at the junction between the oesophagus and the stomach is probably related to smoking and drinking to a smaller extent, if at all (Levi et al., 1990). Within the larynx, cancers of the supraglottis are etiologically similar to cancers of the mouth and pharynx and are associated with alcohol drinking and tobacco smoking. Cancers of the glottis are etiologically more similar to lung cancer and highly associated with tobacco smoking but less strongly with alcohol drinking (Tuyns et al., 1988).

Cancer of the liver

The association between alcohol drinking and liver cancer is probably associated mainly with a relatively high intake of alcoholic beverages (Corrao et al., 1993) and may be mediated by the effect of alcohol drinking on the occurrence of liver cirrhosis (Adami et al., 1992). Like the other alcohol-related cancers discussed above, liver cancer may be associated with tobacco smoking (Yu et al., 1988), but the role of smoking is smaller for liver cancer than for cancers of the upper gastrointestinal and respiratory tracts.

Causes other than drinking and smoking

It should be emphasized that all the diseases that are associated with alcohol drinking, either alone or in combination with tobacco smoking, have other causes as well. Although alcohol and tobacco are strong determinants of the risk of these diseases in most if not all human populations, the pattern of occurrence of these diseases cannot be understood solely in terms of these two risk factors. Dietary factors in particular are suspected of playing a role in the etiology of these diseases as well, with a diet rich in vegetables and fruit probably having a protective effect (Negri et al., 1991). In some populations, particularly in developing countries, hepatocellular carcinoma is strongly associated with hepatitis caused by the hepatitis B and C viruses (IARC, 1994), and it is suspected that dietary intake of some mycotoxins (for example, aflatoxin) also increases the risk (IARC, 1993). Dietary habits and viral infections may themselves be associated with social class, and can therefore contribute to the observed social class differences in cancer occurrence. For discussion of the role of these factors, see other chapters of this book.

International patterns of social class differences in alcoholic-beverage consumption and occurrence of alcohol-related cancers

The social class gradients for cancers known to be associated causally with alcohol drinking (mouth, pharynx, oesophagus, larynx and liver) are shown in Table 2 (mortality) and Table 3 (incidence). In addition, lung cancer is included in the tables because this disease serves as an indicator of the frequency of tobacco smoking in different social classes in the populations studied. The occurrence of alcohol-related cancers is clearly associated with low social class, although the pattern is not totally uniform. The direction of the social class gradient is indicated with plus or minus signs: a minus sign indicates a negative association between social class and cancer risk – that is, low social class associated with high risk; a plus sign indicates a positive association. The number of signs indicates approximately the strength of the association as indicated by the relative difference in risk between the extreme groups of social class.

Brazil

In Brazil, social class was negatively associated with mortality from alcohol-related cancers, including cancer of the liver. The gradient for lung cancer, however, was in the opposite direction. This suggests that alcohol drinking, but not tobacco smoking, may explain a part of the observed social class gradient for cancers of the pharynx, oesophagus, larynx and liver in Brazil. Bouchardy et al. (1993) suggested that the gradient in oesophageal cancer could be explained by the higher use of sugar-cane spirit, black tobacco and mate in the lower social classes. This explanation may apply to cancers of the mouth, pharynx and larynx as well.

Colombia

In contrast to most other populations, the pattern of incidence of alcohol-related cancers in Colombia tended to show positive social class gradients. The only strong, negative association was for pharyngeal cancer in women, but no gradient was seen in women for oral cancer and the gradient for oesophageal cancer was less strong than that for pharyngeal cancer. The positive associations for lung cancer, laryngeal cancer and cancer of the mouth suggest a role of tobacco smoking in the positive social class gradients for these cancers. Cuello et al. (1982) suggested that the gradients in men could be due to differences in the type of tobacco smoked (predominantly black tobacco in the lower classes), and that the type of alcohol used by the lower classes (aguardiente) is a relatively pure substance,

Table 2. Social class gradients of the mortality from alcohol-related cancers and lung cancer

Country	Sex	Mouth	Pharynx	Oesophagus	Larynx	Liver	Lung
Brazil (Bouchardy et al., 1993)	M	0	--	--	--	-	++
	F					-	++
		Mouth & pharynx					
Canada (R. Wilkins, pers. commun.)	M[a]	--				--	--
	F[a]	--					-
	M[b]	--				0	--
	F[b]						-
France (Desplanques, 1985)	M[c]	---		---	---		--
	M[d]	---		---	---		-
Italy (Faggiano et al., 1995)	M	--		--	--	-	0
	F	0				-	0
Japan (Hirayama, 1990)	M	0	0	0	0	0	0
	F	0		0	0	0	+
New Zealand (Pearce & Howard, 1986)	M	--		0	---	--	--
	M[e]	---		--	---	---	--
Switzerland (Levi et al., 1988; C.E. Minder, pers. commun.)	M[f]	--		--	--	0	-
	F[f]	0		0	0	0	0
	M	--		--	--	0	--
United Kingdom (England & Wales) (OPCS, 1977)	M	0	-		--	-	--
	F	-			--	0	--
United Kingdom (Great Britain) (OPCS, 1990)	M	--		--	---	--	--
	F	--		--	---		--
United Kingdom (London) (Davey Smith, 1991)	M			--			--
USA (California) (Buell et al., 1960)	M	-		--			0

-, Negative association; +, positive association: 0, no association; -/+, gradients less than two-fold; - -/++, gradients two- to five-fold; - - -/+++, gradients five-fold or higher. For details of the studies, see the chapter in this volume by Faggiano et al.
[a]1986 census data. [b]1971 census data. [c]Age 45–54 years. [d]Age 55–64 years. [e]N.E. Pearce, pers. commun. [f]Combined analysis of cancers of the mouth, pharynx, oesophagus and larynx.

free of tannins or other contaminants introduced by fermentation processes.

Japan
The mortality data from Japan did not suggest any association between alcohol-related cancers and social class. Similarly, the cohort study of Hirayama (1990) showed no association between social class and mortality from liver cirrhosis in men, but women in the highest social classes had a slightly elevated mortality from liver cirrhosis.

France
Mortality data for French men show strong negative associations between social class and the risk of alcohol-related cancers. The gradient for lung cancer was in the same direction, but less strong.

Desplanques (1985) showed strong negative social class gradients for mortality from liver cirrhosis and from alcoholism in French men. A similar, but less strong, pattern was seen in women.

Based on data from a survey in Lorraine, d'Houtaud et al. (1989) concluded that a greater proportion of men of high socioeconomic status than of low status regularly consumes alcohol, but those of higher socioeconomic status consume smaller amounts of alcohol per drinking occurrence. Among those of lower socioeconomic status, there were both more abstainers and more heavy drinkers. A greater proportion of women in higher social classes than of lower socioeconomic status were regular consumers, but the amounts consumed by women were considerably lower than those consumed by men.

The social class pattern of cancer mortality in men in France suggests a strong involvement of both alcohol drinking and tobacco smoking.

Italy

In Italy, negative associations were seen between social class and the incidence of and mortality from alcohol-related cancers, particularly in men, but the gradients were less strong than in France. The mortality data showed no social class effect on lung cancer, but in the incidence data on lung cancer

Table 3. Social class gradients of the incidence of alcohol-related cancers and lung cancer

Country	Sex	Mouth	Pharynx	Oesophagus	Larynx	Liver	Lung
Colombia (Cuello et al., 1982)	M	+	0	0	++	0	+
	F	0	---	-	+	0	0
Canada (Burbonnais & Siemiatycki, in press)	M[a]		-				--
	M[b]	--				--	
	M[c]				--		--
Denmark (Lynge & Thygesen, 1990)	M	--	-	-	--	-	-
	F	0	0	+	--	--	--
Finland (Pukkala, 1995)	M	0	0	--	--	0	--
	F	0	-	---	-	0	0
		Mouth & pharynx					
Italy (Milan) (Ferraroni et al., 1989)	M+F[b]	---		--		--	
	M+F[d]	--		--		-	
Italy (Torino) (Faggiano et al., 1994)	M[b]	-			--		--
	F[b]	0					++
	M[e]	--			--		-
	F[e]						++
	M[f]	-					-
	F[f]	--					-
Sweden (Vågerö & Persson, 1986)	M	0	0	0	0	0	0
	F	0	--	0	0	0	0
United Kingdom (England & Wales) (Kogevinas, 1990)	M	0		0	--	0	--
	F			-			
USA (Williams & Horm, 1977)	M[b]	0	-	-	-	-	-
	F[b]	+	+	0		+	0	-
	M[a]	0	-	0	0	0	-
	F[a]	0	0	-		0	+	-

-, Negative association; +, positive association: 0, no association; –/+, gradients less than two-fold; – –/++, gradients two- to five-fold; – – –/+++, gradients five-fold or higher. For details of the studies, see the chapter in this volume by Faggiano et al.
[a]Income level. [b]Education. [c]Occupational prestige scale. [d]Social class. [e]Occupational group. [f]Housing tenure.

there were tendencies towards negative associations in men but in women the associations were positive. The incidence of liver cancer was associated with social class about as strongly as the other alcohol-related cancers. Faggiano et al. (1994) attributed the negative associations with mouth and pharynx cancer in men partly to the known negative association between alcohol drinking and low educational level in Italian men. The negative gradients for laryngeal cancer in men probably reflect both alcohol drinking and tobacco smoking.

A strong association between low educational level and mortality from liver cirrhosis has been observed in Italy (F. Faggiano, pers. commun.). Both men and women who had not completed primary-school education had a three- to fivefold higher mortality from cirrhosis than those with a university education.

In a case–control study on digestive tract neoplasms, the observed social class gradients for cancers of the mouth and pharynx and of the oesophagus were attenuated by adjustment for smoking and alcohol drinking (Ferraroni et al., 1989). This provides direct evidence that the social class gradients are partly due to smoking and drinking.

Overall, the data suggest a strong role of alcohol drinking in the social class gradient of cancers of the mouth and pharynx, oesophagus, larynx and liver in Italy. In men in particular, tobacco smoking probably contributes as well.

Switzerland

In Switzerland, the two available analyses of mortality data suggested a negative social class gradient in men for cancers of the mouth and pharynx, oesophagus and larynx. However, the gradient for lung cancer was of similar magnitude and no association was seen between social class and liver cancer in this population. This pattern suggests a stronger role of tobacco smoking than alcohol drinking in the social class gradient of cancers of the mouth and pharynx, oesophagus and larynx in this population.

A survey performed among patients at two hospitals in the French-speaking part of Switzerland, however, reported an association between alcoholism and low socioeconomic class (Trisconi et al., 1989). The prevalence of cigarette smoking was 60% in alcoholics and 29% in other patients.

In a combined analysis of population samples from Italy, Spain, Switzerland and France, the average daily consumption of alcohol ranged from 21.5 g in professionals to 40.1 g in manual workers (Péquignot et al., 1988). The average consumption in women was lower, around 6–8 g per day, and did not vary between occupational groups. A similar analysis of tobacco smoking showed little association with occupational group in men, but white-collar and professional women smoked more than women in lower occupational groups (Berrino et al., 1988).

Sweden

Cancer incidence data from the Nordic countries did not consistently show any social class gradient in the occurrence of alcohol-related cancers. In Sweden, no association was seen except for pharyngeal cancer which, in women only, was negatively associated with social class. This may be a chance occurrence or a phenomenon unrelated to alcohol drinking.

Data from the cross-sectional Stockholm Health of the Population Study showed small differences between mean alcohol consumption and the prevalence of high consumers in various socioeconomic and educational groups in both men and women (Romelsjö, 1989). Self-employed men had a higher consumption than other groups (14.8% consuming more than 35 g of ethanol per day compared with 9.3% in the survey overall). A review of other studies in the same geographical area indicated a change over time in the social pattern of alcohol consumption (Halldin, 1985; Romelsjö, 1989). In the 1960s and 1970s, the higher social classes had a higher proportion of high alcohol consumers than the lower classes for both men and women. In the 1980s, a different pattern has emerged. In young people, high alcohol consumption is associated with low social class and low educational level. In older people, high alcohol consumption remains associated with high social class and high educational level.

A prospective study from Lundby in the south of Sweden showed a strong association between both low social class and low educational level and the probability of developing alcoholism in a 15-year period of follow-up (Öjesjö et al., 1983).

Finland

In men in Finland, negative associations with social class were seen for oesophageal and laryngeal cancers. The absence of a similar effect on cancers of the mouth, pharynx and liver, and a negative association between social class and lung cancer, suggest a stronger role of tobacco smoking than alcohol drinking in the social class gradient for oesophageal and laryngeal cancer in Finnish men. In women, however, negative gradients were seen for cancers of the pharynx, oesophagus and larynx, but the data on liver cancer and lung cancer showed no dependence on social class. This is somewhat similar to the situation in Swedish women, and it is possible that factors other than alcohol drinking and tobacco smoking are responsible for the social class trends of cancers of the pharynx, oesophagus and larynx in women in these countries.

Aro et al. (1986) studied health-related habits in a sample of white-collar and blue-collar workers in Finland. In men, the frequency of heavy alcohol intoxication was highest in the blue-collar workers, but the white-collar workers generally consumed alcohol more frequently than the blue-collar workers. Female white-collar workers generally consumed alcohol more often than blue-collar workers, and they also had a higher frequency of heavy alcohol intoxication. Particularly among women, tobacco smoking was highest in the blue-collar workers. A study of hospital admissions in Finland showed associations between low social class and admission for alcohol poisoning, alcoholism and alcoholic psychosis (Poikolainen, 1982, 1983). The associations were stronger for men than for women. No association was seen between liver cirrhosis and low social class. It was speculated that liver cirrhosis depends most strongly on daily heavy drinking, which is more common in the upper social classes. The lower classes tend to concentrate their drinking in episodes of intoxication interspersed with periods with abstinence; this pattern may carry a higher risk of alcohol poisoning, alcoholism and alcoholic psychosis.

Denmark
In Denmark, the data in men consistently showed a negative association between social class and incidence of alcohol-related cancers and lung cancer, suggesting that both alcohol drinking and tobacco smoking contribute to the social class gradients of alcohol-related cancers. In Danish women, negative social class gradients were seen for cancers of the larynx, liver and lung. The gradients in laryngeal cancer and lung cancer are probably due to differences in tobacco smoking, but the negative gradient in liver cancer, with no accompanying negative trend in cancers of the mouth, pharynx and oesophagus, is more difficult to explain. Chronic infections with hepatitis B and C viruses are rare in the Danish population.

In the Glostrup Population Studies, alcohol consumption was described by social class and employment status by Sælan et al. (1992). In both men and women, the average number of drinks per week was highest in the lower social classes. Analysis by employment status showed that, among men, self-employed persons had the highest consumption and salaried employees the lowest. In women, salaried employees had the highest consumption and workers the lowest.

United Kingdom
The recent mortality data from the United Kingdom showed a fairly consistent pattern with negative associations in both men and women between social class and the alcohol-related cancers and lung cancer. The incidence data, however, were less consistent and the negative associations were largely confined to laryngeal cancer and lung cancer.

Kogevinas (1990) noted that alcohol consumption varies between social classes in the United Kingdom, and that about a quarter of men in manual occupations were classified as heavy drinkers compared with around 10% in non-manual occupations. He noted also, however, that social class variation in tobacco smoking is pronounced, with 33% of both male and female professionals and 64% of male and 42% of female unskilled workers being smokers.

Other studies of social class and alcohol drinking in the United Kingdom have shown no or weak associations. One study in Northern Ireland showed a higher prevalence of problem-drinking in skilled manual workers than in professionals and managers (Murray & McMillan, 1993); another study in Wales showed a tendency of higher prevalence of consuming 22 drinks or more per week in the higher social classes (Farrow et al., 1988). In a study of three regions of the United Kingdom, the weekly alcohol consumption increased with household income in both men and women, and, among men, manual workers drank more alcohol than nonmanual workers (Crawford, 1988). In a sample of attendees at a London health centre, no association was seen between social class and high alcohol consumption (King, 1986).

If it is assumed that the gradients with social class suggested by the mortality data are correct, then the roughly similar associations of social class with alcohol-related cancers and lung cancer, and the lack of consistent association between social class and drinking in the United Kingdom, suggest that a major part of the social class gradients in the United Kingdom may be due to differences in tobacco smoking.

Canada
The available data from Canada showed a negative association between social class and cancers of the mouth and pharynx, oesophagus, liver and lung. This pattern may suggest a role of both tobacco smoking and alcohol drinking in the social class gradients for cancers of the mouth and pharynx, oesophagus and liver in Canada.

New Zealand
The data on cancer mortality in men in New Zealand showed very strong negative associations with social class – about as strong as those seen in France. Since both liver cancer and lung cancer are strongly negatively associated with social class, it is possible that the social class gradient in cancers of the mouth and pharynx, oesophagus and larynx are due to strong social class differences in both alcohol drinking and tobacco smoking.

Data published by Pearce et al. (1983) showed negative social class gradients for most broad disease groupings. The category 'mental disorders', which includes predominantly deaths from alcoholism, alcoholic psychosis and drug dependence, was much over-represented in the lowest classes. This, together with the strong social gradient for liver cancer, suggests a strong role of alcohol drinking in the observed gradients for cancers of the upper gastrointestinal and respiratory tracts in New Zealand.

Pearce and Howard (1986) presented crude relative risks for different social classes, and expected relative risks, the latter taking into account the known social class differences in tobacco smoking, based on census information. The authors considered that the smoking pattern may explain much of the elevated risks of cancers of the mouth and pharynx, oesophagus and larynx in social classes III and IV, but not the very high risks of these cancers in social class V. Alcohol drinking may in particular contribute to the high risks in this social class.

In addition, Pearce and Howard considered that chronic infection with hepatitis B virus may contribute to mortality from liver cancer in some parts of New Zealand.

Casswell and Gordon (1984) studied self-reported alcohol consumption in various occupational and social groups in New Zealand. The pattern of high frequency of drinking low quantities of alcohol was most common in the higher social classes; the pattern of drinking high quantities less often was more common in the lower social classes. An average daily consumption of more than 100 ml of ethanol was most frequent in the lowest social classes. Additional data on liver cirrhosis mortality showed a low mortality from this disease in the highest social class (standardized mortality ratio = 0.23) and a very high mortality in the lowest class (standardized mortality ratio = 3.59).

United States of America
In men in the United States, cancers of the mouth and pharynx, oesophagus, larynx, liver and lung tended to show negative associations with social class. This suggests that both alcohol drinking and tobacco smoking may contribute to the gradients. In women in the United States, the association between social class and alcohol-related cancers tended to be positive.

The paper by Williams and Horm (1977) confirmed the suspected associations between social class and smoking and drinking. In men, years of education was negatively associated with cigarette smoking and total alcohol consumption. In women, years of education and family income were positively associated with both drinking and smoking. Tobacco smoking and alcohol drinking were positively associated in both men and women.

Surveys were conducted of alcohol drinking patterns in 48 states in the United States in 1984 and 1990 (Midanik & Clark, 1994). In men and women combined, weekly drinking was more frequent in households with above-median income and in persons with a high educational level. However, for more heavy consumption, indicated by the consumption of five or more drinks on one occasion, the opposite trends were seen and the highest responses occurred in households with below-median income and in persons with a low level of education. The survey indicated large variations in alcohol drinking in different regions of the USA.

In the Healthy Worker Project, conducted among employees in Minneapolis-St Paul, the frequency of alcohol drinking was associated with high socioeconomic status in women, but no association was seen in men (Jeffery et al., 1991). In both men and women, tobacco smoking was strongly associated with low socioeconomic status.

A combined analysis of 10 surveys of alcohol drinking in a total sample of 9900 persons was reported by Knupfer (1989). The analysis attempted to distinguish different patterns of drinking in nine categories ranging from lifelong abstention to frequent drunkenness. Six social class groups were constructed by a combination of educational and income levels. In men, the frequency of abstention decreased from 35% in the lowest social class to 5% in the highest. The categories from frequent light drinking to moderately heavy drinking were associated with high social class; the frequency increased from 34% in the lowest class to 69% in the highest. Finally, the category indicating frequent drunkenness was weakly inversely associated with social class and decreased from 19% in the lowest social class to 15% in the highest. In particular, the highest category of drinking decreased from 6% to 2%. In women, the patterns were qualitatively similar to those in men, but the proportion of abstainers was higher and the proportion of drinkers lower than in men in all categories of social class.

Alcohol drinking and the frequency of complications after surgery

Treatment of many cancers involves a surgical resection of the tumour. An effect of alcohol drinking on the probability of postoperative complications may, therefore, contribute to the observed social class differences in survival after cancer diagnosis in populations where alcohol drinking is associated with social class. A series of studies by Tønnesen and colleagues in Denmark have shown that persons with a high consumption of alcohol at the time of operation have an increased risk of postoperative complications. Persons with a high alcohol consumption constitute 7–20% of those undergoing surgery, but carry the burden of more than half of the total postoperative morbidity. An influence of social class per se on postoperative morbidity has not been investigated except in one small uncontrolled study of eight men undergoing amputation (Hunter & Middleton, 1984).

In a retrospective study of 73 alcoholics and 73 matched controls, the postoperative morbidity after transurethral resection of the prostate was 62% and 20%, respectively (Tønnesen et al., 1988). The most common complications were infection and bleeding episodes. Bacterial infections occurred more often in the group of alcohol abusers, but the types of bacteria found in infected persons were similar in abusers and controls.

In a retrospective study of 90 alcohol abusers and 90 matched controls, the postoperative morbidity after osteosynthesis of malleolar fractures was increased in the alcohol abusers (33% versus 9%) (Tønnesen et al., 1991). The excess was particularly due to infections. The long-term outcome was also poorer in the group of alcohol abusers, who required more reoperations.

The postoperative course after evacuation of subdural haematoma was studied in a group of 106 patients (Sonne & Tønnesen, 1992). In the one-third of the patients who drank more than 60 g of ethanol per day, an increased postoperative morbidity and mortality was seen.

The postoperative morbidity after hysterectomy was studied prospectively in 229 consecutive patients; the frequency of complications was 80% in alcohol abusers (more than 60 g of alcohol per day), 27% in social drinkers (25–60 g per day) and 13% in the group of women who consumed less than 25 g of alcohol per day (Felding et al., 1992).

A prospective study was reported of 30 persons undergoing colorectal resection (15 noncirrhotic alcohol misusers who had consumed at least 60 g of alcohol per day for several years and a matched control group of 15 persons who consumed less than 25 g of alcohol per day) (Tønnesen et al., 1992). The group of alcohol abusers developed complications more often than controls (10 persons versus three) and stayed longer in hospital after surgery (median 20 days versus 12). Detailed data were collected that indicated that the effects could be due to alcohol-induced dysfunction of the heart, suppression of the cellular immune system and haemostatic imbalance.

Other studies of alcohol abusers have shown an effect on postoperative morbidity, but these studies did not evaluate alcohol consumption at the time of surgery (Kleeman & Zöller, 1986; Nguyen et al., 1990). Liver transplantation for alcoholic cirrhosis has been performed without development of more complications in selected alcoholics who had abstained from alcohol for several months before surgery (Knechtle *et al.*, 1992; Lucey *et al.*, 1992; McCurry *et al.*, 1992).

Conclusions

Clear differences have been demonstrated in disease occurrence between social classes and between groups with different levels of education. Most often we would not think of class or educational level as the cause of disease per se, but rather seek to identify the correlates of these variables that are more directly involved in the causation of the diseases. In this context, we would not think of alcohol drinking as a potential confounder of the association between social class and cancer, but merely as a mechanism underlying an association between social class and cancer risk, or as an aspect of social class that associates class with cancer risk.

The theoretical line of argument of this chapter responds to the following question: can the observed variation in cancer incidence and mortality by social class be attributed to alcohol drinking? To answer this question, attention can be restricted to the cancers with which alcohol drinking is causally associated (cancers of the mouth, pharynx, oesophagus, larynx and liver), and to the populations where social class variation in such cancers has been observed. We have put the main emphasis on the consistency of the social class pattern of the different alcohol-related cancers, because, if the social class variation in alcohol drinking is sufficiently strong to lead to an increase or decrease in one of these cancers a similar pattern would be expected to follow in the other alcohol-related cancers. For this reason we have used as the point of departure a tabulation of the social class gradients of alcohol-related cancers, and discussed, as secondary data, levels of alcohol drinking as assessed by different methods, occurrence of other alcohol-related diseases and occurrence of lung cancer (a good indicator of tobacco smoking).

The clearest evidence of a role of alcohol drinking in social gradients in cancer risk is found in men in France and Italy. In these populations, wine has traditionally been consumed with meals and particularly so among those of lower social classes such as agricultural and manual workers. These countries rank among the highest in the world in per capita levels of alcohol consumption. The consistency of the social class effect across the alcohol-related cancers and the stronger gradient for these cancers than for lung cancer increase our confidence in alcohol drinking as the main or at least a major contributory factor in the social class gradients. This interpretation is also supported by the finding of similar social class gradients in liver cirrhosis in both men and women in France and Italy.

The only other population where a very strong social class gradient can be clearly linked to alcohol drinking is that of New Zealand. As in France, the gradients of alcohol-related cancers are even stronger than for lung cancer, and additional studies of alcohol drinking, alcohol-related psychiatric morbidity and liver cirrhosis in relation to social class has confirmed a high level of these parameters in the lowest social classes of men in New Zealand. As for France and Italy, the data on women are inadequate for an assessment of the role of alcohol drinking.

There are few other populations where a social class gradient is seen consistently for the cancers known to be caused by alcohol drinking, but patterns suggestive of a role of alcohol drinking are seen in Brazil, Switzerland, United Kingdom and Denmark. The situation in Brazil is interesting, with opposite social class trends for alcohol-related cancers and lung cancer; thus, tobacco smoking can be effectively ruled out as an explanation for the observed negative social class gradients. In Switzerland, the situation may be similar to France and Italy, with habitual wine drinking among lower social classes being a factor of importance. In the United Kingdom, the mortality data were suggestive of social class gradients but no such pattern was seen in the incidence data. Finally, in Denmark, fairly consistent but not very strong negative social class gradients were seen. A role of alcohol drinking is possible and supported by some evidence of a higher level of alcohol drinking in the lower social classes in Denmark.

It is emphasized that although a role of alcohol drinking in the social class gradients of the relevant cancers is very likely in men in France, Italy and New Zealand, and probable in some other populations, other factors may contribute as well. In particular, a role of tobacco smoking is likely, as indicated by the parallel gradients in lung cancer. Dietary factors may be another main contributing factor in the social class gradients of tobacco-related cancers.

Survival after cancer diagnosis is related to surgical intervention and to other parameters. The influence of social class on postoperative morbidity has not been investigated properly, but there is plenty of evidence of association between social class and survival after cancer diagnosis (see the chapter by Kogevinas and Porta in this book). An effect of alcohol on postoperative morbidity has been observed for a range of different surgical procedures, and consists of increased frequency of infections, bleeding episodes and cardiopulmonary insufficiency. These complications are probably due to preoperative disturbances of the cellular immune system, haemostasis and heart function. Since most of the quoted studies compared alcohol abusers with people who did not abuse alcohol, it is not clearly established what level of alcohol consumption is required for the adverse effect on postoperative morbidity. However, at present it may be concluded that in populations where there is an association between alcohol abuse and low social class, the effect of alcohol abuse on postoperative morbidity may contribute to the observed negative social class gradients in survival after cancer diagnosis.

References

Adami, H.O., Hsing, A.W., McLaughlin, J.K., Trichopoulos, D., Hacker, D., Ekbom, A. & Persson, I. (1992) Alcoholism and liver cirrhosis in the aetiology of primary liver cancer. *Int. J. Cancer*, 51, 898–902

Aro, S., Rasanen, L. & Telama, R. (1986) Social class and changes in health-related habits in Finland in 1973–1983. *Scand. J. Soc. Med.*, 14, 39–47

Berrino, F., Merletti, F., Zubiri, A., del Moral, A., Raymond, L., Estève, J. & Tuyns, A.J. (1988) A comparative study of smoking, drinking and dietary habits in population samples in France, Italy, Spain and Switzerland. II. Tobacco smoking. *Rev. Epidém. Santé Publ.*, 36, 166–176

Blot, W.J., McLaughlin, J.K., Winn, D.M., Austin, D.F., Greenberg, R.S., Preston Martin, S., Bernstein, L., Schoenberg, J.B., Stemhagen, A. & Fraumeni, J.F. (1988) Smoking and drinking in relation to oral and pharyngeal cancer. *Cancer Res.*, 48, 3282–3287

Bouchardy, C., Parkin, D.M., Khlat, M., Mirra, A.P., Kogevinas, M., De Lima, F.D. & Ferreira, C.E. (1992) Education and mortality from cancer in Sao Paulo, Brazil. *Ann. Epidemiol.*, 3, 64–70

Bourbonnais, R. & Siemiatycki, J. Social class and risk of ten types of cancer in Montreal, Canada. (in press)

Buell, P., Dunn, J.E., Jr & Breslow, L. (1960) The occupational-social class risks of cancer mortality in men. *J. Chron. Dis.*, 12, 600–621

Casswell, S. & Gordon, A. (1984) Drinking and occupational status in New Zealand men. *J. Stud. Alcohol*, 45, 144–148

Corrao, G., Arico, S., Lepore, R., Valenti, M., Torchio, P., Galatola, G., Tabone, M. & Di Orio, F. (1993) Amount and duration of alcohol intake as risk factors of symptomatic liver cirrhosis: a case-control study. *J. Clin. Epidemiol.*, 46, 601–607

Crawford, A. (1988) Self-reported alcohol consumption among population sub-groups in three areas of Britain. *Drug Alcohol Depend.*, 21, 161–167

Cuello, C., Correa, P. & Haenszel, W. (1982) Socio-economic class differences in cancer incidence in Cali, Colombia. *Int. J. Cancer*, 29, 637–643

Davey Smith, G. (1991) Socioeconomic differentials in cancer among men. *Int. J. Epidemiol.*, 20, 339–345

Desplanques, G. (1985) *La mortalite des adultes*. Paris, INSEE

d'Houtaud, A., Adriaanse, H. & Field, M.G. (1989) Alcohol consumption in France: production, consumption, morbidity and mortality, prevention and education in the last three decades. *Adv. Alcohol Substance Abuse*, 8, 19–44

Faggiano, F., Zanetti, R. & Costa, G. (1994) Social inequalities in cancer risk in Italy. *J. Epidemiol. Community Health*, 48, 447–452

Faggiano, F., Lemma, P., Costa, G., Goavi, R. & Paganelli, F. Cancer mortality by educational level in Italy. *Cancer Causes Control* (in press)

Farrow, S.C., Charny, M.C. & Lewis, P.C. (1988) A community survey of alcohol consumption. *Alcohol Alcohol.*, 23, 315–322

Felding, C., Jensen, L.M. & Tønnesen, H. (1992) Influence of alcohol intake on postoperative morbidity after hysterectomy. *Am. J. Obstet. Gynecol.*, 166, 667–670

Ferraroni, M., Negri, E., La Vecchia, C., D'Avanzo, B. & Franceschi, S. (1989) Socioeconomic indicators, tobacco and alcohol in the aetiology of digestive tract neoplasms. *Int. J. Epidemiol.*, 18, 556–562

Halldin, J. (1985) Alcohol consumption and alcoholism in an urban population in central Sweden. *Acta Psychiatr. Scand.*, 71, 128–140

Hirayama, T. (1990) *Life-style and mortality: a large-scale census-based cohort study in Japan*. Basel, Karger

Hunter, J. & Middleton, F.R.I. (1984) Cold injury amputees – a psychosocial problem? *Prosthet. Orthotics Int.*, 8, 143–146

IARC (1988) *IARC monographs on the evaluation of carcinogenic risks to humans* (Vol. 44): *alcohol drinking*. Lyon, International Agency for Research on Cancer

IARC (1993) *IARC monographs on the evaluation of carcinogenic risks to humans* (Vol. 56): *some naturally occurring substances*. Lyon, International Agency for Research on Cancer

IARC (1994) *IARC monographs on the evaluation of carcinogenic risks to humans*. (Vol. 59): *hepatitis viruses*. Lyon, International Agency for Research on Cancer

Jeffery, R.W., French, S.A., Forster, J.L. & Spry, V.M. (1991) Socioeconomic status differences in health behaviors related to obesity: the Healthy Worker Project. *Int. J. Obes.*, 15, 689–696

Johnson, K.A. & Jennison, K.M. (1992) The drinking-smoking syndrome and social context. *Int. J. Addict.*, 27, 749–792

King, M. (1986) At risk drinking among general practice attenders: prevalence, characteristics and alcohol-related problems. *Br. J. Psychiatry*, 148, 533–540

Kleeman, P.P. & Zöller, B. (1986) Intra- und postoperatives Risiko béi patientien mit chronischem Alkholabusus. *Dtsch Zahnärztl. Z.*, 41, 452–456

Knechtle, S.J., Fleming, M.F., Barry, K.L., Steen, D., Pirsch, J.D., Hafez, G.R., D'Alessandro, A., Reed, A., Sollinger, H.W., Kalayoglu, M. & Belzer, F.O. (1992) Liver transplantation for alcoholic liver disease. *Surgery*, 112, 694–703

Knupfer, G. (1989) The prevalence in various social groups of eight different drinking patterns, from abstaining to frequent drunkenness: analysis of 10 U.S. surveys combined. *Br. J. Addict.*, 84, 1305–1318

Kogevinas, M. (1990) *Longitudinal study: socio-demographic differences in cancer survival*. London, Her Majesty's Stationery Office

La Vecchia, C. & Negri, E. (1989) The role of alcohol in oesophageal cancer in non-smokers, and of tobacco in non-drinkers. *Int. J. Cancer*, 43, 784–785

La Vecchia, C., Negri, E., Franceschi, S., Parazzini, F. & Decarli, A. (1992) Differences in dietary intake with smoking, alcohol and education. *Nutr. Cancer*, 17, 297–304

Levi, F., Negri, E., La Vecchia, C. & Te, V.C. (1988) Socioeconomic groups and cancer risk at death in the Swiss Canton of Vaud. *Int. J. Epidemiol.*, 17, 711–717

Levi, F., Ollyo, J.B., La Vecchia, C., Boyle, P., Monnier, P. & Savary, M. (1990) The consumption of tobacco, alcohol and the risk of adenocarcinoma in Barrett's oesophagus. *Int. J. Cancer*, 45, 852–854

Lucey, M.R., Merion, R.M., Henley, K.S., Campbell, D.A., Turcotte, J.G., Nostrant, T.T., Blow, F.C. & Beresford, T.P. (1992) Selection for and outcome of liver transplantation in alcoholic liver disease. *Gastroenterology*, 102, 1736–1741

Lynge, E. & Thygesen, L. (1990) Occupational cancer in Denmark. *Scand. J. Work Environ. Health*, 16 (Suppl. 2), 1–35

McCurry, K.R., Baliga, P., Merion, R.M., Ham, J.M., Lucey, M.R., Beresford, T.P., Turcotte, J.G. & Campbell, D.A. (1992) Resource utilization and outcome of liver transplantation for alcoholic cirrhosis. A case-control study. *Arch. Surg.*, 127, 772–777

Midanik, L.T. & Clark, W.B. (1994) The demographic distribution of US drinking patterns in 1990: description and trends from 1984. *Am. J. Public Health*, 84, 1218–1222

Murray, M. & McMillan, C. (1993) Problem drinking in Northern Ireland: results of a community survey using the CAGE questionnaire. *Alcohol Alcohol.*, 28, 477–483

Negri, E., La Vecchia, C., Franceschi, S., D'Avanzo, B. & Parazzini, F. (1991) Vegetable and fruit consumption and cancer risk. *Int. J. Cancer*, 48, 350–354

Nguyen, B.T., Thompson, J.S., Edney, J.A. & Rikkers, L.F. (1990) Comparison of ulcer surgery in a veterans administration and university hospital. *Am. Surg.*, 56, 606–609

Office of Population Censuses and Surveys (1977) *Decennial supplement, 1977*. London, Her Majesty's Stationery Office

Office of Population Censuses and Surveys (1990) *Longitudinal study: mortality and social organization*. London, Her Majesty's Stationery Office

Öjesjö, L., Hagnell, O. & Lanke, J. (1983) Class variations in the incidence of alcoholism in the Lundby study, Sweden. *Soc. Psychiatry*, 18, 123–128

Pearce, N.E. & Howard, J.K. (1986) Occupation, social class and male cancer mortality in New Zealand, 1974–78. *Int. J. Epidemiol.*, 15, 456–462

Pearce, N.E., Davis, P.B., Smith, A.H. & Foster, F.H. (1983) Mortality and social class in New Zealand II: male mortality by major disease groupings. *New Zealand Med. J.*, 96, 711–716

Péquignot, G., Crosignani, P., Terracini, B., Ascunce, N., Zubiri, A., Raymond, L., Estève, J. & Tuyns, A.J. (1988) A comparative study of smoking, drinking and dietary habits in population samples in France, Italy, Spain and Switzerland. III. Consumption of alcohol. *Rev. Epidém. Santé Publ.*, 36, 177–185

Poikolainen, K. (1982) Risk of alcohol-related hospital admissions by marital status and social class among females. *Drug Alcohol Depend.*, 10, 159–164

Poikolainen, K. (1983) Risk of alcohol-related hospital admission in men as predicted by marital status and social class. *J. Stud. Alcohol*, 44, 986–995

Pukkala, E. (1993) *Cancer risk by social class and occupation*. Basel, Karger

Romelsjö, A. (1989) The relationship between alcohol consumption and social status in Stockholm. Has the social pattern of alcohol consumption changed? *Int. J. Epidemiol.*, 18, 842–851

Sælan, H., Møller, L. & Køster, A. (1992) Alcohol consumption in a Danish cohort during 11 years. *Scand. J. Soc. Med.*, 20, 87–93

Seitz, H.K., Simanowski, U.A. & Osswald, B.R. (1992) Epidemiology and pathophysiology of ethanol-associated gastrointestinal cancer. *Pharmacogenetics*, 2, 278–287

Sonne, N.M. & Tønnesen, H. (1992) The influence of alcoholism on outcome after evacuation of subdural haematoma. *Br. J. Neurosurg.*, 6, 125–130

Talamini, R., Franceschi, S., Barra, S. & La Vecchia, C. (1990) The role of alcohol in oral and pharyngeal cancer in non-smokers, and of tobacco in non-drinkers. *Int. J. Cancer*, 46, 391–393

Tønnesen, H., Schutten, B.T., Tollund, L., Hasselqvist, P. & Klintorp, S. (1988) Influence of alcoholism on morbidity after transurethral prostatectomy. *Scand. J. Urol. Nephrol.*, 22, 175–177

Tønnesen, H., Pedersen, A., Jensen, M.R., Møller, A. & Madsen, J.C. (1991) Ankle fractures and alcoholism. The influence of alcoholism on morbidity after malleolar fractures. *J. Bone Joint Surg. Br.*, 73, 511–513

Tønnesen, H., Petersen, K.R., Højgaard, L., Stokholm, K.H., Nielsen, H.J., Knigge, U. & Kehlet, H. (1992) Postoperative morbidity among symptom-free alcohol misusers. *Lancet*, 340, 334–337

Tønnesen, H., Møller, H., Andersen, J.R., Juel, K. & Jensen, E. (1994) Cancer morbidity in alcohol abusers. *Br. J. Cancer*, 69, 327–332

Trisconi, Y., Marini, M., de Werra, P., Paccaud, F., Magnenat, P. & Yersin, B. (1989) Medicosocial characteristics of hospitalized alcoholic patients in two internal medicine departments of hospitals in French-speaking Switzerland. *Schweiz. Med. Wochenschr.*, 119, 1907–1912

Tuyns, A.J., Péquignot, G. & Jensen, O.M. (1977) Oesophageal cancer in Ille-et-Vilaine in relation to levels of alcohol and tobacco consumption. Risks are multiplying. *Bull. Cancer Paris*, 64, 45–60

Tuyns, A.J., Estève, J., Raymond, L., Berrino, F., Benhamou, E., Blanchet, F., Boffetta, P., Crosignani, P., del Moral, A., Lehmann, W., Merletti, F., Péquignot, G., Riboli, E., Sancho-Garnier, H., Terracini, B., Zubiri, A. & Zubiri, L. (1988) Cancer of the larynx/hypopharynx, tobacco and alcohol: IARC international case-control study in Turin and Varese (Italy), Zaragoza and Navarra (Spain), Geneva (Switzerland) and Calvados (France). *Int. J. Cancer*, 41, 483–491

Vågerö, D. & Persson, G. (1986) Occurrence of cancer in socioeconomic groups in Sweden. An analysis based on the Swedish Cancer Environment Registry. *Scand. J. Soc. Med.*, 14, 151–160

Williams, R.R. & Horm, J.W. (1977) Association of cancer sites with tobacco and alcohol consumption and socioeconomic status of patients: interview study from the Third National Cancer Survey. *J. Natl Cancer Inst.*, 58, 525–547

Yu, H., Harris, R.E., Kabat, G.C. & Wynder, E.L. (1988) Cigarette smoking, alcohol consumption and primary liver cancer: a case-control study in the USA. *Int. J. Cancer*, 42, 325–328

Corresponding author:
H. Møller
Center for Research in Health and Social Statistics,
The Danish National Research Foundation,
Sejrøgade 11, DK-2100 Copenhagen Ø, Denmark

Diet and cancer: possible explanations for the higher risk of cancer in the poor

J.D. Potter

Humans have always had to eat; diets have always contained the same nutrients and bioactive constituents. Therefore, some have argued, the present pattern of diseases and changes in that pattern cannot be causally linked to dietary intake. This argument, its naivety notwithstanding, raises some important issues for the way we think about the epidemiology of nutrition and disease. Current research on diet and specific diseases is based, obviously, on the premise that this argument is false. This chapter uses a broad brush to present the evidence for a significant and causal association between eating patterns and cancer. It shows that, far from being an implausible link, the relationship between dietary patterns and cancer is largely explained by the dependence of humans on their food supply – dependence not merely in the sense of providing energy to sustain life, but more related to evolutionarily adaptive patterns of food intake and contemporary aberrations in those patterns. The chapter also shows that it is plausible that at least part of the explanation for the higher risk of cancer among the poor in both rich countries and poor countries relates to the extent of the aberrations in food supply and eating patterns.

The starting point for this chapter is a diet to which humans are adapted. I note, especially, intakes of substances for which we are dependent on the environment and intakes of substances to which we have low or infrequent exposure. It is a diet with seasonal variability in total food intake and the availability of specific foods.

There are four types of aberrations in dietary patterns that could produce cancer and, perhaps, disease in general: an imbalance between energy intake and output; an alteration in the pattern of intake of either macro- or micronutrients or both; specific deficiencies – of nutrients and bioactive compounds; and the presence in the food supply, from time to time, of substances to which the organism has been almost never exposed and, therefore, to which there may not be the relevant metabolic responses.

Is there a diet to which we are well adapted?
Evoking a lost Golden Age to compare with current miseries has a long history: from Eden (where fruit-eating seems to have been part of the problem) through Rousseau's noble savage to discussions of the nature of palaeolithic diets (Eaton & Konner, 1985). We cannot know exactly the nature of diets to which humans are well adapted (although dentition, metabolic enzyme profiles, and the length and morphology of the gastrointestinal tract provide some clues). Nevertheless, is there some plausible description that can be presented of our early diet? If there is, we should also acknowledge that there must have been considerable variability in the details in the same way that there is extensive geographic variability in contemporary diets. Common features of our early diet must have included: a wide variety of foods – roots, nuts, seeds, leaves and fruit (grains have become a staple only in the last 10 to 15 thousand years but would have been gathered regularly in season); sporadic intake (despite the Tarzanist fantasies of many) of lean meat low in saturated fat (with a more secure and regular supply of fish and seafood for coastal dwellers); some intake of insects, larvae, bone marrow, and organ meats; very low intake of alcohol; low and irregular intake of eggs, milk, and milk products; little refining or fractionation of food into parts; and variability, by season, both of total amount of food available (and therefore body weight), and of kinds of foods available. Thus, there would also

have been variability in the availability of particular nutrients. Differences in this overall intake pattern (for example, diets higher in fat in extreme northern populations) would have been the result of climate and geography (as it varied over time and from place to place). However, in general, until very recently, saturated fat and alcohol intake would have been low, vegetable food (but not grain) intake high, and the kinds of plant food eaten highly varied.

An important argument in favour of human adaptation to specific eating patterns and food sources is that there are some nutrients for which we are known to be dependent upon the environment. This concept has some important implications for cancer etiology. The consequences of a variety of nutrient deficiencies are well described. But deficiency disorders can arise only if the organism is incapable of endogenous synthesis. This argues that there has been no selection pressure to develop (or maintain) such a capacity in the species and, therefore, that the essential substances are widely available in the environment. Perversely, then, because essential nutrients are widely available in naturally occurring human food, deficiencies are possible. Essential amino acids, essential fatty acids, microelements and vitamins are examples of substances that must be obtained from food. The fact that what is essential varies across species underlines the importance of the adaptation process.

The adaptation argument is as follows: essential nutrients – energy-bearing, micronutrient, and bioactive compounds – are widely available in nature; they have important functions in growth, development and reproduction; the organism is adapted to their ubiquity; and deficiencies impair growth, development and reproduction.

A plausible analogy exists in relation to substances that are necessary for the maintenance of the organism and this applies especially to substances that reduce the risk of carcinogenesis. This extension of the argument is that the normal function of cells is dependent on the presence of a variety of widespread dietary constituents probably including, but – importantly – not confined to, those necessary for growth and development. Without these substances, cells malfunction: the cells may become more susceptible to exposure to carcinogens; they may lose some specific protective mechanisms such as timely enzyme induction; or there may be an increase in replication rates as somatic cells seek to adapt to the new – deprived – conditions. Maintenance is a continuous function from birth almost to death, whereas growth, development and reproduction are confined to specific periods of life.

The converse of this argument applies to dietary constituents that are rare in nature. If the organism is exposed rarely (or not at all) to specific substances, then high intakes are likely to have untoward consequences. This is relevant both to rare exposures that result in acute toxicity and to hitherto unaccustomed levels of exposure that overwhelm the metabolic processes that normally handle lower levels. Plant, fungal and bacterial toxins are members of the first class of exposures; a Western-style high-fat/high-calorie intake that influences cholesterol and insulin metabolism, adipose storage, and sex steroid hormone production and transport is an example of the other type of exposure. Dietary patterns that are high in grains (common in agricultural communities) are often associated with a reduced intake of other plant foods; further, these diets include significant amounts of abrasive material that may result in tissue damage and reactive epithelial hyperplasia, particularly in the oesophagus. It is useful to consider that, in the same way that diets vary with geography, there may be differing degrees of adaptation in long-exposed versus relatively recently exposed populations.

A potent objection to any adaptation argument is that natural selection will be an influence only up to the age of reproduction; therefore, because chronic diseases, particularly cancers, are almost exclusively diseases of postreproductive years, dietary adaptation is an unnecessary postulate. There are four responses to this. The first is to argue that humans have a long period of juvenile dependence and that survival of parents in a healthy state is therefore likely to be selected for. The second response is based on consideration of the unit of selection. If the relevant issue is the survival of tribes, then those groups of protohumans and humans that had sufficient elders who knew how to respond to infrequent hazards (epidemic disease, food and water shortage, and natural hazards such as fire, earthquake or extreme weather) would have had a better chance of survival. Tribal wisdom maintained by the old would have facilitated survival of the whole group. Without elders and without knowledge, tribes would be more likely to succumb to the vagaries of their habitat. Tribes in which longevity was selected for would survive to pass on their wisdom, their knowledge of the local ecology, their eating habits, their adapted metabolisms, and their genes in that fascinating blend of the heritability of both culture and biology that may mark the true distinction between humans and other animals.

Third, to argue that chronic diseases are a phenomenon of older age and therefore that resistance to them cannot have been selected for is to ignore the fact that these diseases do not occur at younger ages and that, therefore, some resistance (at least to the point of postponing them to older ages) has been selected for. Fourth, a diet that reduces risk of cancer and other chronic diseases may also improve reproductive success. A wide variety of substances are both teratogenic and carcinogenic and other substances, it is becoming increasingly clear, reduce risk of both teratogenesis and cancer (for example, folate); selection for improved reproductive success via an interaction between diet and metabolism could directly select for reduced cancer risk.

This chapter explores the evidence for the existence of unaccustomed exposures and protective dietary constituents, and for risks associated with both energy and nutrient imbalance. Possible biological mechanisms are considered briefly. In the absence of unequivocal evidence for a link between poverty and cancer via the quality/quantity of food consumed, what is known about dietary patterns and poverty is considered in the light of diet and cancer links.

Dietary exposure and cancer risk

I use the adaptation argument outlined above as a framework in the following discussion and show that some of the empirical relations that have been established in the epidemiological literature can be explained by four types of aberrations in dietary patterns: energy imbalance, nutrient imbalance, specific deficiency, and specific exposure. These are illustrated in relation to certain cancers, particularly breast, colon and pancreas.

Energy imbalance

Epidemiological evidence Energy imbalance is a massive topic and still not well understood in relation to cancer; only a few aspects are touched upon here. Three measures of energy balance (Pariza & Simopoulos, 1986) have been examined in etiological studies of cancer: total intake (Potter & McMichael, 1986; Lyon et al., 1987; Willett & Stampfer, 1986), energy output (Garabrant et al., 1984) and a variety of measures of growth (Micozzi, 1985) and obesity (Helmrich et al., 1983; Paffenbarger et al., 1980). There is no consistent relationship between these measures and all cancers; even for some specific cancers, the data are not clear. In addition, there are some paradoxes.

The present evidence suggests that higher physical activity is related to a lower risk of colon cancer (Garabrant et al., 1984; Vena et al., 1985; Gerhardsson et al., 1986; Wu et al., 1987; Paffenbarger et al., 1987; Slattery et al., 1988a; Potter et al., 1993) but that obesity is probably not a risk factor (Potter & McMichael, 1986). For endometrial cancer (Elwood et al., 1977; LaVecchia et al., 1984; Folsom et al., 1989) and postmenopausal breast cancer (Lew & Garfinkel, 1979; Helmrich et al., 1983), however, obesity is a risk factor. For premenopausal breast cancer, obesity is associated with a reduced risk (Helmrich et al., 1983; Paffenbarger et al., 1980). Physical activity bears an uncertain but perhaps inverse relation to risk of breast cancer (Frisch et al., 1985; Bernstein et al., 1994). Total energy intake is an inconsistent risk factor for all three cancers. There are no established relationships of energy imbalance with other cancers but there is a general association between obesity and overall cancer risk (Lew & Garfinkel, 1979). Peripheral versus central adipose distribution as measured by ratios of waist and hip circumferences or fat folds has long been known to be related to risk of diabetes mellitus and coronary heart disease (Vague, 1956; Feldman et al., 1969; Larsson et al., 1984; Donahue et al., 1987; Selby et al., 1989). There is some evidence of a relationship with breast cancer that may be restricted to those with a family history of that cancer (Sellers et al., 1992). Fat distribution does not appear to be related to risk of endometrial cancer (Folsom et al., 1989).

Plausible mechanisms At least three mechanisms to explain the link between energy imbalance and cancer risk are plausible: mechanisms involving hormonal or mechanical processes or cellular workload. Peripheral adipose is the principal source of estrogens postmenopausally (Grodin et al., 1973) via the conversion of adrenal androstenedione.

This provides a plausible explanation for the association with endometrial and postmenopausal breast cancer; both are associated with higher, cumulative lifetime estrogen exposures. The reason for the association between obesity and a lower risk of premenopausal breast cancer is yet to be established; it is not simply a matter of failing to detect cancerous lesions in large breasts (Willett et al., 1985) but is possibly related to differences in the steroid receptor status of pre- and postmenopausal breast cancer (Potter et al., 1994).

Physical activity may exert a protective effect against colon cancer by a mechanical effect – higher activity results in a shorter mouth-to-anus transit time, although this characteristic does not generally show an inverse association with colon cancer. Because obesity is not associated with an increased risk of colon cancer (while high calorie intake and low physical activity are), there may be metabolic differences between those who get colon cancer and those who do not. Additionally, the total amount of food passing through the large bowel may represent a measure of cellular work or epithelial damage and thereby influence rates of cell replication (Potter 1989, 1992a).

The complex relationships between dietary intake, obesity and physical activity, on the one hand, and cancer risk, on the other, will become clear only when the relevant intermediate metabolic steps are understood; these include effects on gut function, on cell turnover, and on hormone production. What remains, however, is the empirical observation that aspects of energy imbalance are related to risk of all cancers and to risk of cancers at specific sites.

Adaptation It was argued above that the human organism is adapted to extensive variability in food intake. It is able to make rapid use of increases in food supply in order to survive through lean times. This is the 'thrifty gene' hypothesis originally proposed to explain the survival advantage of the predisposition to diabetes and obesity (Neel, 1969). However, a high intake of food as a regular, rather than occasional, phenomenon will 'jam open' more than insulin responses; it will also increase adipose production of estrogen, a factor perhaps originally associated with reproductive success – sufficient body fat, as well as hormonal support, to carry a child to term. Is this related to the fact that obesity is associated with a reduced risk of premenopausal breast cancer and a liability only later in life? Intriguingly, higher abdominal fat in prepregnancy (as measured by waist-to-hip ratio) is associated with significantly larger infants at delivery (Brown et al., 1996).

In fasting animals, the structure of the intestinal epithelium is simple with a low cell replication rate (Stragand & Hagemann, 1977). Following refeeding, the replication rate increases, as does the complexity and the total area of the epithelial surface. This appears to be a highly adaptive response to variable food availability – a rapid cell turnover and a large absorptive surface during feasting, but low activity during fasting, which thus conserves energy. With high intake as a regular phenomenon, however, rapid cell turnover and maximal epithelial surface provide an environment that increases the probability of cancer. Our original observation that increased meal frequency increased the risk of colon but not rectal cancer (Potter & McMichael, 1986) has now been confirmed in several other studies (LaVecchia et al., 1988a; Young & Wolf, 1988; Potter et al., 1993) – additional evidence that there is a cost for more frequent food intake.

Therefore, although prolonged obesity was probably uncommon in our ancestors, the ability to assimilate and store energy rapidly when it was available has probably been selected for. Inheriting this kind of metabolism in societies where food is widely available and consumed *ad libitum* appears to have consequences for cancer risk. It remains to be seen if the tendency for variation in body fat distribution (as measured for example by waist-to-hip ratio) – an established risk factor for several chronic diseases – is a marker for particular metabolic differences.

Imbalance of food/nutritional intake

Epidemiological evidence Currently, dietary epidemiological studies of cancer are frequently focused on the role of intakes of macronutrients, particularly fat and alcohol.

Fat is of uncertain relevance in the etiology of breast cancer but the association with alcohol is surprisingly consistent. Subpopulations with intakes of animal fat and protein that are lower than those of the general community, such as vegetarian nuns (Kinlen, 1982) and Seventh-day Adventists (Phillips et al., 1980), show little evidence of a lower risk of breast cancer. In both these groups, the reproductive rate is lower than in comparison populations but these studies certainly do not provide strong evidence for a role for ingested fat in breast carcinogenesis.

There have been at least 11 case–control studies on the relationship between meat or fat consumption and breast cancer (Phillips, 1975; Miller et al., 1978; Lubin et al., 1981; Graham et al., 1982; Talamini et al., 1984; Howe, 1985; Hirohata et al., 1985; Le et al., 1986; Lubin et al., 1986; Katsouyanni et al., 1986; Shun-Zhang et al., 1990). The study of Lubin et al. (1981) – the only study showing a significant increase in risk in association with consumption of fat – was based on frequency of consumption of just eight food items. Two of three studies (Talamini et al., 1984 and Le et al., 1986 versus Katsouyanni et al., 1986) reporting on dairy product consumption found positive associations with risk.

Of the cohort studies (Willett et al., 1987, 1992; Hirayama, 1978; Phillips & Snowdon, 1983; Kushi et al., 1992; van den Brandt, 1993), only that of Hirayama (1978) found an association with daily consumption of meat and risk of breast cancer after age 54 years. This finding was based on just 14 cases in this category.

In contrast, meat, protein and fat intake are consistently, almost universally, positively related to risk of colon cancer. Of the 16 studies that have reported on the association of colon cancer with fat and protein, 13 have shown an increased risk. Only the studies of Macquart-Moulin et al. (1986) in France and Tuyns et al. (1988) in Belgium failed to find an association; sugar was the only nutrient associated with increased risk in this latter study. [One study (Stemmermann et al., 1984), using a 24-hour recall, found an inverse association with fat in a Hawaiian Japanese cohort.] Sixteen of 27 studies have reported an increase in risk associated with higher meat consumption (see Potter et al., 1993); only Hirayama (1981), using a three-item questionnaire in a very large cohort study in Japan, found an inverse association. Of the 10 studies, both case–control and cohort, with at least 100 cases, with a food frequency questionnaire of at least 50 items, and a response rate among cases (case–control studies only) of at least 60%, eight showed a positive association with meat intake. Five studies have noted a positive association with eggs; four have shown an inverse association with fish or seafood (for a detailed review, see Potter et al., 1993).

Pancreas cancer shows an even more consistent relationship with meat and fat intake. Almost all the reported studies show an increased risk in association with higher consumption of meat, fat or fried foods (for a detailed review, see Anderson et al., 1994)

Alcohol consumption has been shown in ecological studies to be related to rectal and colon cancer (Potter et al., 1982) and to cancers of the oesophagus and larynx (McMichael, 1979). In analytical studies, there is a very consistent but not strong relationship with breast cancer and a somewhat less consistent association with cancers of the colon and rectum. Alcohol may interact with estrogen replacement therapy to increase further the risk of breast cancer (Gapstur et al., 1992; Colditz et al., 1990). The evidence for a causal association with pancreas cancer is weak (Velema et al., 1986). However, alcohol is extensively implicated in cancers of the upper digestive and respiratory tracts; it is regarded by the International Agency for Research on Cancer as an established human carcinogen for this association in particular (IARC, 1988).

High-grain-consuming areas are at higher risk of oesophageal and stomach cancer. van Rensberg (1981) has shown that there is a consistently higher risk for oesophageal cancer among populations with high corn and wheat consumption compared with those where sorghum, millet, cassava, yams or peanuts are staples. It is worth noting that a lower risk of colon cancer in high-risk areas (for example, the United States of America (USA), western Europe and Australia) is much more consistently found with high vegetable rather than high cereal intake, although cereal-eating communities are, in general, at lower risk of cancer of the bowel.

There are some data to suggest that a high intake of simple carbohydrate is associated with increased risk of colorectal cancer (Bristol et al., 1985; Tuyns et al., 1988; Bostick et al., 1994).

Plausible mechanisms A variety of mechanisms have been proposed to account for an association between cancer and dietary fat. Intriguingly, the explanations have been most prolific for breast cancer, for which the empirical evidence from analytical studies is weakest. Both direct and indirect mechanisms have been proposed.

The proposed direct mechanisms are, first, via effects of unsaturated fatty acids on cell membrane structure and function (Welsch & Aylsworth, 1983), epithelial proliferation (Kidwell et al., 1982), immune responsiveness (Vitale & Broitman, 1981) and cell–cell communication (Welsch & Aylsworth, 1983); and, second, via effects of fatty acid and cholesterol metabolites (epoxides and peroxides) on promotion of transformed cells (Petrakis et al., 1980; Gruenke et al., 1987).

The postulated indirect mechanisms are, first, via the effects of fat on hormone receptors (Welsch & Aylsworth, 1983), on prolactin production (Hill et al., 1980) or on bowel flora (Hill et al., 1971), which may either alter the bioavailability of estrogens through effects on steroid deconjugation (Gorbach, 1984), or alter the bacterial production of specific anticarcinogenic agents (Adlercreutz et al., 1982; Adlercreutz, 1984, 1991); and, second, via the effects of higher food intake on age at menarche (Frisch & McArthur, 1974), age at menopause (de Waard et al., 1964), and the accumulation of adipose tissue, already noted as the site in the body where adrenal androstenedione is converted to estrone (Grodin et al., 1973).

Although many of these proposed mechanisms may be conceptually attractive, their large number is itself a problem. Breast cancer, like all other cancers, is undoubtedly of multifactorial origin, so that a sizeable list of potential etiological pathways is not an insurmountable problem. However, there is currently no clear understanding of human breast carcinogenesis, no unequivocal precursor lesion, and no known biochemical marker. Accordingly, it is inappropriate to search for a mechanism in the absence both of an understanding of the intermediate steps and of data establishing that a high-fat diet is indeed a risk factor for mammary tumorigenesis.

For colon cancer, the dominant hypothesis has long been derived from the relationship between dietary fat intake and bile acid metabolism. Fat intake, it is proposed, increases the amount or concentration of bile acids secreted into the small bowel; bacteria present in the large bowel metabolize the primary acids to secondary acids and these, it is suggested, have greater toxicity, cocarcinogenic or promotional activity, and trophic effects. There is a considerable amount of corroborative human metabolic and animal experimental evidence for this hypothesis; the body of evidence is, however, not totally coherent (McMichael & Potter, 1985, 1986). Other roles for fat have been proposed, including direct toxic action on the bowel wall (Bruce, 1987).

Arylamines, produced when meat is cooked, have been proposed as specific colon carcinogens (Sugimura & Sato, 1983) and there is a growing literature on the role of these compounds (Sugimura, 1985) and their metabolism, including, especially, genetically variable acetylator (NAT2) status (Weber, 1987; Turesky et al., 1991; Kadlubar et al., 1992).

McMichael (1981) has argued that gastrointestinal hormones have trophic and hyperplastic effects on the exocrine pancreas and could act as mediators of known or suspected dietary (and other) risk factors. Gastrin and cholecystokinin are potent stimulators of pancreatic hyperplasia (Johnson, 1981). Cholecystokinin has been shown, in animal models, to be a significant promotor of pancreatic neoplasia (Howatson & Carter, 1985). More recently, Anderson et al. (1992) have shown that the pancreas is capable of metabolizing arylamines (found in both cooked meat and tobacco smoke, the major risk factors for pancreas cancer), thus suggesting another potential pathway from diet to cancer.

For the relationship between alcohol and breast cancer, there have been no major advances in relation to biological mechanisms beyond the mechanism originally postulated by Williams (1976) – namely, stimulation of prolactin secretion. It is also possible, as a number of workers have pointed out, that because alcohol is a significant energy source, the mechanism could be related to those postulated for obesity or caloric intake in general. Finally, it appears plausible that DNA damage could result following formation of acetaldehyde adducts; acetaldehyde is a highly reactive compound formed as the first step in the oxidative metabolism of alcohol. For colon and rectal cancer, the evidence suggests that the effect of alcohol on bile acid metabolism is rather like that of fat (McMichael & Potter, 1985, 1986). Several mechanisms have been postulated to account for the causal association of alcohol with cancers of the larynx and oesophagus (IARC, 1988). These include acting as a chronic irritant and inducing excess cell replication, acting as a solvent for direct-acting carcinogens (particularly those in cigarette smoke), being associated with specific nutrient deficiencies, and being a vehicle for other compounds present in alcoholic drinks.

Possible mechanisms for the association between grain consumption and higher risks of upper digestive tract cancer include traumatic effects of silicaceous fibres and resultant high epithelial cellular turnover, and reduced intake of other plant foods, which results in deficiencies either of micronutrients (van Rensberg, 1981) or specific bioactive substances (see below).

Adaptation The argument for adaptation in relation to these exposures is more than a general affirmation of a human incapacity to handle a high-fat or a high-alcohol intake but, nonetheless, remains rather speculative and possibly circular. The primary premise is that high-fat, high-alcohol or high-grain intakes were not part of the regular dietary patterns of early humans. An intermittent high intake of food (a feast–fast economy) produced rapid short-term adaptive responses – increased gut epithelial cell proliferation, increased secretion of appropriate hormones (both trophic and secretory-control hormones) and bile acids, and so on. These, as an energy-conserving mechanism, then subsided when food became scarce; high metabolic and cellular activity is a cost to the organism that is not a good investment in the presence of reduced food availability. This capacity for rapid response (an extension of the thrifty gene hypothesis) becomes non-adaptive in the presence of consistent high intake, leading not only to alcoholism and obesity but also to chronic high gut hormone levels and elevated epithelial proliferation rates. High intakes of abrasive fibres result in higher upper digestive tract proliferation rates and an elevated risk of carcinogenesis, particularly in the oesophagus.

There is an additional aspect to the adaptation argument that is particularly related to the internal ecology of the large gut. The large bowel can be regarded as a complex ecosystem in which the colonic contents act as a culture medium for both bacterial and upper-crypt colonic cells (McMichael & Potter, 1986; Potter, 1989, 1992a). The culture medium in turn is influenced extensively by both host conditions (including a variety of hormones) and ingested foods and alcohol. This complex ecosystem may be one of the most flexible parts of the human–environment interaction. It is argued, however, that its flexibility is finite and that sufficient disturbance of its homeostasis has consequences for carcinogenesis (McMichael & Potter, 1986; Potter, 1989, 1992a).

Specific deficiencies – nutrients and bioactive compounds

Epidemiological evidence The most obvious specific deficiencies are those of micronutrients. It is important to note that, in relation to β-carotene, retinol and ascorbate, higher risks of particular cancers have been reported in individuals with lower intakes or blood levels but these have largely been within the normal range (Wald *et al.*, 1980; Kark *et al.*, 1981; Peto *et al.*, 1981). There are several cancers that have a probable relationship with micronutrient deficiencies – notably lung cancer and cervical cancer (Peto *et al.*, 1981; Ziegler, 1989) with reduced intakes or lower levels of vitamin A. Other squamous epithelial cancers (for example, skin) may be related to lower β-carotene or retinol levels. Lower dietary ascorbate levels have been associated with a higher risk of rectal cancer (Bjelke, 1973; Potter & McMichael, 1986). Minerals, such as calcium, and trace elements, such as selenium, have also been examined for their possible role in cancer etiology (Bruce, 1987; Willett *et al.*, 1983; Slattery *et al.*, 1988b). It is worth noting, however, that supplementation with β-carotene does not reduce risk of lung cancer (the risk may even be elevated) (The Alpha-Tocopherol, Beta-Carotene Cancer Prevention Study Group, 1994) or metachronous polyp formation (Greenberg *et al.*, 1994).

Of most interest, and, until recently, not clearly identified as the most consistent finding in the dietary etiology of cancer, is the more general relationship between a higher intake of vegetables and a lower risk of cancer at a wide variety of sites including mouth and pharynx, lung, stomach, pancreas, colon and rectum (for comprehensive reviews of data and mechanisms see: Potter, 1990; Steinmetz & Potter 1991a, 1991b).

Of the 28 studies of colon cancer that have discussed findings for vegetables, 23 reported an inverse association. Inverse associations with fruit are much less common. Of the nine studies, both case–control and cohort, that have reported on vegetables, with at least 100 cases, with a dietary questionnaire of at least 50 items, and a response rate among cases (case–control studies only) of at least 60%, only the study of Peters *et al.* (1992) failed to find a reduced risk in association with higher intake of one or more measures of vegetable intake. Four other similar studies, lacking only data on response rates, showed comparable findings.

The related finding that foods high in fibre (often a measure of vegetable as well as grain intake) are protective also has been noted in 10 of 16 studies and 6 of 11 of the larger, better-conducted studies as defined above. Three of four other studies, again lacking only data on response rates, reported similar findings (for more detail, see Potter et al., 1993). Howe et al. (1992) have recently completed a formal meta-analysis of 13 case–control studies using the original data and found that there is a consistently lower risk in association with higher overall fibre intake, with odds ratios of 1.0, 0.79, 0.69, 0.63 and 0.53 for each quintile of consumption from lowest to highest (P for trend < 0.0001). However, in those populations where cereal consumption is high – southern Europe and Asia, particularly – there is a puzzling observation that risk is higher in association with higher consumption of rice (Japanese) (Wynder et al., 1969) or pasta and rice (southern Europeans) (Macquart-Moulin et al., 1986; LaVecchia et al., 1988a). Of the studies of pancreas cancer, all have reported a lower risk in association with a higher intake of vegetables or fruit or both (see Anderson et al., 1994 for more detail). Recent studies of prostate (Oishi et al., 1988), mouth and pharynx (McLaughlin et al., 1988), lung (Byers et al., 1987; Koo, 1988), cervix (LaVecchia et al., 1988b; Brock et al., 1988) and stomach (You et al., 1988; LaVecchia et al., 1987) cancer show similar findings for vegetables or fruit or both.

The obvious question is whether these data are merely providing less specific support for the association between high intakes of defined micronutrients such as β-carotene and ascorbate and lower risk of cancers or whether additional factors are at work. The failure of some of the specific supplement trials to reduce cancer risk, and data from animal and in vitro studies, suggest that this is a broader and more interesting phenomenon related to more than the commonly cited specific vitamins (but, importantly, not excluding them).

Vegetables contain a wide variety of substances that have been shown to have anticarcinogenic properties, such as phenols, isothiocyanates, flavonoids, indoles and lignans (Wattenberg, 1985; Steinmetz & Potter, 1991b), fermentable fibre, and, of course, vitamins and trace elements. A recent nested case–control study found serum lycopene levels to be markedly different between cases of pancreas cancer and controls (relative risk for low versus high tertile = 5.40). Lycopene is a carotenoid without retinoid activity (Burney et al., 1989). Higher folate intake is associated with lower risk of colon adenomatous polyps (Giovanucci et al., 1993) and colon cancer (W.C. Willett, pers. commun.).

At present, we are not able to provide a summary estimate of the intakes of most of the bioactive substances – food tables do not provide the data, most of the relevant food analysis has not been done, and it is probable that whole classes of these constituents, and certainly individual constituents, remain to be identified.

Plausible mechanisms Several roles have been identified (Wattenberg, 1985) for the known bioactive components of plant foods in reducing cancer risk, and some of these are summarized in the following list.

(1) Inhibition of formation of direct-acting carcinogens. Ascorbate is effective in blocking the formation of N-nitrosamines in vivo from precursor nitrates and amines (Mirvish, 1981; Bartsch et al., 1988).
(2) Prevention of reaction of agents with target tissues. There are a variety of ways in which this may occur (see Wattenberg, 1985). Non-nutrient compounds, such as phenols, found in plant foods (Wood et al., 1982) can react with active carcinogens.
(3) Induction of enzymes that detoxify or conjugate carcinogenic compounds. A variety of plant-related substances have been shown to have this effect (Wattenberg, 1977, 1983).
(4) Inhibition of carcinogenesis even when delivered after known carcinogen exposure in animals. Carotenoids and selenium are included in this category but the mechanisms are unclear (Wattenberg, 1985).
(5) Reduction or prevention of hyperplasia in a variety of epithelial tissues.

More generally, the steps from procarcinogen exposure to cell transformation can be considered as follows: the procarcinogen is activated to the ultimate carcinogen (each of these may be solubilized and excreted); the carcinogen passes through membranes; the carcinogen interacts with DNA – perhaps forming adducts and/or producing mutations; DNA synthesis and replication (or DNA repair) occur; repair may have varying degrees of fidelity; and cell replication with abnormal DNA and subsequent abnormal protein synthesis result (or cell differentiation occurs). At almost every one of these steps, specific known phytochemicals can alter the likelihood of carcinogenesis, occasionally in a way that enhances risk, but usually in a favourable direction. For example, substances such as glucosinolates and indoles, isothiocyanates and thiocyanates, phenols and coumarins can induce a multiplicity of solubilizing and (usually) inactivating enzymes; ascorbate and phenols block the formation of carcinogens such as nitrosamines; flavonoids and carotenoids can act as antioxidants; lipid-soluble compounds such as carotenoids and sterols may alter membrane integrity; some sulphur-containing compounds can suppress DNA and protein synthesis; and carotenoids suppress DNA synthesis and enhance differentiation (Steinmetz & Potter, 1991b; Wattenberg, 1992).

Adaptation It is here that the adaptation argument has its most interesting implications (and perhaps significant testability). There are known to be substances, including vitamins and trace elements, without which the organism cannot grow, be maintained, or reproduce optimally. The consequences of low levels (dietary or tissue) of these substances may include carcinogenesis. It is argued here that we are equally dependent on the environment to provide other substances that have specific anti-carcinogenic properties. In the absence of these substances, humans (perhaps all vertebrates) are at higher risk of cancers at a number of sites, particularly those where epithelial surfaces are more exposed to the environment – lung, digestive tract, and cervix. It is argued that these compounds supplied by the diet act to induce detoxifying enzymes, to block activation, and so on, and that the organism is reliant on them to do so.

As is clear from this whole chapter, the adaptation hypothesis can be related to each of the four dietary phenomena associated with increased cancer risk. However, there are more problems in testing the notion that humans are exposed to 'unaccustomed levels' of total energy or specific nutrients than in testing whether specific deficiencies or specific exposures are carcinogenic. There is no way to modify unaccustomed levels of one factor in a controlled trial that does not also modify others: for example, any attempt to decrease fat will result in decreased calories or increased levels of other nutrients; similarly, weight modification changes nutrient intakes, energy intake, energy expenditure, or all three. In contrast, addition of either specific compounds to the diet or modification of vegetable intake provides tests both of the protective hypothesis (and perhaps of their 'essential nutrient' status) and of possible public health strategies. There are also ways to test the adaptation argument with *in vitro* studies.

Specific exposures

Epidemiological evidence The mostly widely accepted theory of carcinogenesis implicates specific damage (either physical or chemical) to cellular DNA and now, more specifically, to proto-oncogenes and tumour suppressor genes. The diet contains a number of naturally occurring substances that have been shown to be carcinogenic – for example, aflatoxins and N-nitroso compounds – but there are very few human cancers for which a specific dietary carcinogen has been identified unequivocally [primary hepatocellular cancer, where aflatoxins are strongly implicated, is the major exception (IARC, 1976; Peers *et al.*, 1987)]. While DNA-interacting carcinogens are a major focus of animal experiments, most human studies to date have identified promoters or cocarcinogens, such as alcohol, or host phenomena, such as obesity, as discussed above. There is also, however, evidence for the importance of arylamines in colon cancer (and perhaps N-nitroso compounds in upper digestive tract cancers) and it may still be the case that specific dietary carcinogens will be identified for other epithelial cancers.

Plausible mechanisms The mechanisms of action of DNA-damaging carcinogens in general (Pitot, 1986), and aflatoxins (IARC, 1976), N-nitroso compounds (Bartsch *et al.*, 1982) and arylamines (Sugimura, 1985) in particular, have been well reviewed elsewhere.

Adaptation The adaptation argument in relation to specific exposures has four facets. First, we appear not to have developed specific mechanisms to detoxify certain carcinogens. (It is equally clear that there are mechanisms to detoxify some (Chasseaud, 1979) and to activate others.) Second, there are differences in the population distribution of specific detoxifying enzymes (for example, approximately 40% of the population lacks one component subset (M1) of the glutathione S-transferase enzymes). Third, the mechanisms may become overloaded at high exposures. Fourth, the specific enzymes may not be lacking but exposure to the agents themselves is relatively uncommon and induction of the detoxifying enzyme(s) is normally achieved by other ubiquitous substances; it seems likely that this is the nature of the relationship that we have with a wide variety of the bioactive compounds. There are areas of the world where cancer has been attributed both to reduced intakes of specific nutrients and to exposure to specific dietary carcinogens: for example, China, where oesophageal cancer may be associated with N-nitroso compounds (Yang, 1980) and low intakes and blood levels of a variety of vitamins and trace elements (Thurnham et al., 1985). To this point, supplementation with the missing nutrients has been disappointingly ineffective in reducing risk (Muñoz et al., 1985; Wahrendorf et al., 1988; Li et al., 1993). This may suggest that what are missing are not the obvious micronutrients (these may be just markers for the real deficiencies) but specific compounds that keep the detoxifying enzymes 'tuned'. The experiment that follows from this hypothesis is obvious – either add a variety of fruits and vegetables to the diet or add some specific non-nutrient enzyme inducers, blocking agents, and so on. The advantage of studying oesophageal cancer is that there is an identified precursor lesion. Assessing precursor lesions allows more rapid tests of a variety of strategies on relatively small populations over shorter time periods than studies of the cancers themselves. Similar arguments apply to testing the role of vegetables in the prevention of recurrence of adenomatous polyps in addition to existing studies of single likely preventive agents (Bertram et al., 1987). Such studies are now being undertaken in a number of settings.

Diet and social class

Clear evidence that diet contributes to the higher risk of cancer associated with lower social class is lacking but, based on the above view of dietary carcinogenesis, some important circumstantial evidence exists. This evidence particularly shows that there are unequal distributions of dietary and related risk factors across the social class spectrum in the developed world and that such a pattern may be even more pronounced in the developing world. The most obvious specific risk factors with evidence of differences by social class are fat, meat and alcohol intake and intake of vegetables and fruit. However, not all of these differences are in a direction consistent with these as agents that explain the social class gradient for cancer.

Specific evidence of a poorer-quality diet in lower social classes comes from a variety of studies in different parts of the world. In the USA, Davis et al. (1990) have shown, using data from the Nationwide Food Consumption Survey, that living alone, a lower income, a reduced expenditure on food, and unemployment are statistically independent (although clearly sociologically interrelated) predictors of a poorer-quality diet among those over 55 years of age. In this study, quality of diet was measured on the basis of the intake of a number of vitamins and minerals, some of which – for example, vitamin C, calcium and vitamin A – are known to be related to cancer risk.

Data from the Second National Health and Nutrition Examination Survey (NHANES II), and particularly from the Continuing Survey of Food Intake of Individuals (CSFII), were used by Block and Abrams (1993) to show that income has a major influence on the dietary intake and nutritional status of women between 15 and 44 years. They showed that poverty is associated with, among other things, lower intakes of folate, 'carotene', vitamins C and E, and calcium. They further showed that among poorer women (the criterion used, somewhat obscurely, was earning ≤131% of the official USA poverty level) only 53.7% reported consuming vegetables at least once in four nonconsecutive 24-hour periods. In contrast, among women with incomes over 300% of the USA poverty level, 82.0% reported such consumption. For fruit and juice, the proportions were 67.4% and 87.4%, respectively. Murphy et al. (1992) have shown that among young males (19–24 years) in the USA poverty is associated with a poorer diet, again as assessed by intakes of specific micronutrients.

In common with the poor in the developing world, there is sometimes an overall shortage of food among those dependent on government assistance programs (Taren et al., 1990) and other very poor members of USA society. This applies particularly to the elderly, single-parent families, and children (Food and Research Action Committee, 1984, 1985; Physician Task Force on Hunger in America, 1985, 1986, 1987).

The 1989 baseline survey of the New York State Health Heart Project showed that beef and whole-milk (and possibly egg) intakes were inversely related to social class as measured by educational attainment (Shea et al., 1993) – although somewhat paradoxically, this is probably explained in part by the relatively inexpensive nature and abundance of animal food in the USA, and the constant marketing of these products via the media. The opposite was true for vegetable and fruit consumption (Shea et al., 1993). Subar et al. (1992) also reported that vegetable intake was lower in the less educated and lower in Blacks and Hispanics than in Whites, but that intakes of fruit and juices were not markedly different by ethnic status. Education was inversely related to fat consumption in the Minnesota Heart Survey in the early 1980s (Kushi et al., 1988). These data could suggest that poverty and lower educational achievement in postindustrial societies, at least, may lead to a complex mix of poverty of information and skills, a distorted view of 'status foods', and an approach to dietary priorities typical of the 1940s, which was then much more focused on protein and calories.

Similar data on the quality of diets among the poor and individuals of lower socioeconomic status are available for other developed countries. A number of surveys in the United Kingdom show lower intakes of fibre, vitamin C, calcium and, among some groups, total calories in individuals of lower socioeconomic status than among the population at large (Braddon et al., 1988; Gregory et al., 1990; Cade et al., 1992). The homeless (Cade et al., 1992) and unemployed (Braddon et al., 1988) seem particularly at risk. Cade et al. (1992) also reported on the very high rates (around 70%) of smoking that prevail among those living in shared public housing (a group that is essentially a subset of the homeless).

In Australia, there is important evidence that cost of food and social status (as defined by occupational status, education and income) are determinants of diet quality. Individuals of higher socioeconomic status have diets lower in fat and refined sugar and higher in fibre (this again suggests that access to information on diet and health shows significant social class differences), although such individuals also consume more alcohol (Smith & Baghurst, 1992). When food groups rather than nutrients are considered, groups of higher socioeconomic status consume more whole-grain cereals, more low-fat milk, and more fruit. Finally, these Australian data, however, also show that individuals of higher socioeconomic status consume more meat and cheese. Very similar data were reported for Australia by Steele et al. (1991) and Baghurst et al. (1990).

Data for Canada in the 1970s showed that folate and vitamin C were strongly related to income (Myers & Kroetsch, 1978). They also showed that intakes of calcium and, in pregnant women, vitamin A were low in the poorest groups. Vegetable and fruit intake were consistently inversely related to income in most age groups and both sexes (Myers & Kroetsch, 1978).

Finnish data showed that social class was a determinant of a variety of health-related habits in the 1970s and 1980s. Blue-collar males smoked more, drank more heavily though less frequently, were less physically active, and consumed diets less well matched to official dietary recommendations than white-collar males. Women showed similar but less marked social class differences (Aro et al., 1986). The reported dietary differences particularly focused on intakes of butter and whole milk, both of which were higher among males of lower socioeconomic status, particularly (Aro et al., 1986).

A Danish nutritional survey, undertaken in 1985 with the specific purpose of understanding the social distribution of diet-related disease risk, showed that intake of a number of foods varied extensively by social class (Haraldsdottir et al., 1987). Fat intake was much higher in the lowest social class than in the highest social class among men, although the variability among women was much less marked. Intake of potatoes showed a similar pattern, but the gradient for intake of other vegetables was reversed and particularly so among women. The pattern of fruit intake was similar to that for vegetables and, again, the social class gradient was more marked among women. Bread intake showed a positive social class gradient in women but an inverse pattern among men. Beer also showed this difference by sex: intake among men in the highest social class was twice that of men in the lowest social class, but the opposite pattern was seen among women. Wine consumption showed a positive social class gradient in both sexes.

A crucial additional social class difference involves tobacco consumption – considered in detail in the chapter in this book by Stellman and Resnicow. One important observation that is relevant here is that in the developed world there is a higher prevalence of smoking among individuals of lower socioeconomic status, and that, generally, individuals who smoke have diets poorer in quality, particularly characterized by lower intake of vegetables and fruits. Perversely, smokers have a greater need (given the intake of toxic and carcinogenic compounds) of the micronurients that such foods provide. It may be, therefore, that the poorer members of society are at significantly elevated risk of cancer – both because of the interactive nature of the poor-diet–smoking combination and because of the large numbers of cancers that are related to both exposures.

Dietary differences around the world are much greater than those seen within countries. The developing world, as a broad generalization, shows patterns of intake that are much lower in meat and fat and much higher in abrasive cereals than the industrialized world but intakes of specific micronutrients are often marginal among the poor in these countries. Many of the studies that have been undertaken in the developing world have focused on nutrition among mothers, children and pregnant women, largely because this is perceived as the group at most risk of nutritional deprivation. It is likely, however, that the poor of both sexes and all ages are generally malnourished and that this is a significant contributor to cancer risk.

Intakes of a number of micronutrients – including folate, calcium, and vitamins D and C – are lower in Asian populations, for instance; but not all differences appear to be in a deleterious direction (Newman et al., 1991). In some populations, intakes of fruit (Zeitlin et al., 1992) or vegetables and fruit (Mele et al., 1991) are lower among poor women and children.

The relationship between specific nutritional and food deficiencies has been discussed extensively in the first part of this chapter. More generalized malnutrition, however, is likely to have a complex relation to carcinogenesis with some steps in the process being enhanced and others inhibited (Deo, 1981).

In only a few studies have diet and cancer data been examined to establish whether socioeconomic indicators might explain dietary associations. For instance, LaVecchia et al. (1987) noted that including socioeconomic status in models reduced the strength of the positive association between pasta and rice and risk of stomach cancer in a northern Italian population but did not modify the relationship with other dietary risk factors. In an analysis of all cancer incidence in the Iowa Women's Health Study, however, socioeconomic status, as measured by educational status, remained predictive in models that also contain vegetable intake (Potter et al., unpublished). Smith and Baghurst (1992) made the point that although there are social class differences in the Australian diet, these do not appear to be large enough to explain the social class gradient observed, particularly for coronary heart disease. What proportion of the socioeconomic-status-related variation in cancer is explained by diet remains unknown. Given the extent of misclassification of both diet and social class, there could be a considerable degree of confounding between the two kinds of variables that remains uncontrolled when both kinds of variables are used in analytical models.

In his preface to the second edition of *Poverty and Health* (Kosa & Zola, 1975), Zola notes that the 'social sciences discovered illness in the 1950s and rediscovered poverty almost a decade later'. Epidemiology, from its earliest days, clearly knew about poverty; John Snow's discussion of cholera (Snow, 1855) includes the following potent reminder to those who remember only the removal of the Broad Street pump handle:

> It is in the families of the poor that cholera is often observed to pass from one individual to another, while in cleanly dwellings, where the hand-basin and towel are in constant use, and where the rooms for cooking, eating, and sleeping are distinct from each other, the communication of cholera from person to person is rarely observed. In the houses of the poor also, the disease is hardly ever contracted by medical, clerical, and other visitors, who do not eat or drink in the sick room, while it often fares differently with the social visitor, who comes either to see the patient or attend his funeral.

Perhaps in the 1990s, nutrition and epidemiology, too, can rediscover poverty and appreciate this as a fundamental and preventable determinant of cancer risk.

Summary

There are a variety of ways in which diet may influence the development of human cancers. What is proposed here (and elsewhere – see Potter, 1992b) is a theoretical framework and an argument, with the features summarized below.

(1) There is a dietary pattern to which humans are well adapted – an 'original diet'.
(2) This original dietary pattern had specific features, which included regular exposure to a variety of substances that are required for human metabolism but that are not usually explicitly labelled as 'essential nutrients'.
(3) The original dietary pattern was low in highly abrasive cereal products (consumption of large amounts of grains is a relatively recent phenomenon), with less resultant damage and frequent cell repair, particularly to the upper gastrointestinal tract.
(4) The original dietary pattern involved variability in intake, which resulted in variability in cell replication rates particularly in the gastrointestinal tract, and little risk of obesity.
(5) The original dietary pattern involved almost no intake of alcohol and therefore little capacity for its solvent and chronic cell damage capacities.
(6) Abandonment of each of these aspects of dietary adaptation has consequences for carcinogenesis. Most notable is the reduction of intake of vegetables and fruit with subsequent loss of appropriate enzyme 'tuning' and so on, and a generally increased susceptibility to cancer at a number of sites. A high intake of fat, of grains, and of alcohol, and increased obesity, are each associated with recognizable patterns of cancers.
(7) The higher risk of cancer that exists among the poor, in both the developed and developing world, is to some, as yet unknown, degree related to the fact that the amount of variation from the diet to which we are well adapted is greater in that portion of the population who have less access to the world's goods and services. This is particularly true regarding the intake of fresh vegetables and fruit, almost universally consumed in smaller quantities among the poor in most parts of the world. Some diet-related cancers, particularly breast cancer, run counter to the general trend towards higher risks in poorer people; it is probable that social class differences in other risk factors, particularly reproductive history, explains this discrepancy, at least in part.

References

Adlercreutz, H. (1991) Diet and sex hormone metabolism. In: Rowland, I.R., ed., *Nutrition, toxicity, and cancer*. Boca Raton, CRC Press. pp. 137–195

Adlercreutz, H. (1984) Does fiber-rich food containing animal lignan precursors protect against both colon and breast cancer? An extension of the "fiber hypothesis". *Gastroenterology*, 86, 761–766

Adlercreutz, H., Fotsis, T., Heikkinen, R., Dwyer, J., Woods M., Goldin, B. & Gorbach, S. (1982) Excretion of the lignans enterolactone and enterodiol and of equol in omnivorous and vegetarian postmenopausal women and in women with breast cancer. *Lancet*, 2, 1295–1299

The Alpha-Tocopherol, Beta Carotene Cancer Prevention Study Group (1994) The effect of vitamin E and beta carotene on the incidence of lung cancer and other cancers in male smokers. *New Engl. J. Med.*, 330, 1029–1035

Anderson, K.E., Potter, J.D., Hammons, G.J., Teitel, C.H., Guengerich, F.P., Chou, H-C., Lang, N.P. & Kadlubar, F.F. (1992) Metabolic activation of aromatic amines by human pancreas. *Proc. Am. Assoc. Cancer Res.*, 33, 153

Anderson, K.E., Potter, J.D. & Mack, T.M. (1994) Pancreas. In: Schottenfeld, D., ed., *Cancer epidemiology and prevention* (2nd edn). New York, Oxford University Press

Aro, S., Räsänen, & Telama, R. (1986) Social class and changes in health-related habits in Finland in 1973–1983. *Scand. J. Soc. Med.*, 14, 39–47

Baghurst, K.I., Record, S.J., Baghurst, P.A., Syrette, J.A., Crawford, D. & Worsley, A. (1990) Sociodemographic determinants in Australia of the intake of food and nutrients implicated in cancer aetiology. *Med. J. Aust.*, 153, 444–452

Bartsch, H., Castegnaro, M., O'Neill, I.K. & Okada, M. (1982) *N-nitroso compounds: occurrence and biological effects* (IARC Scientific Publications No. 41). Lyon, International Agency for Research on Cancer

Bartsch, H., Ohshima, H. & Pignatelli, B. (1988) Inhibitors of endogenous nitrosation. Mechanisms and implications in human cancer prevention. *Mutat. Res.*, 202, 307–324

Bernstein, L., Henderson, B.E., Hanisch, R., Sullivan-Halley, J. & Ross, R.K. (1994) Physical exercise and reduced risk of breast cancer in young women. *J. Natl Cancer Inst.*, 86, 1403–1408

Bertram, J.S., Kolonel, L.N. & Meyskens, F.L. (1987) Rationale and strategies for chemoprevention of cancer in humans. *Cancer Res.*, 47, 3012–3031

Bjelke, E. (1973) *Epidemiological studies of cancer of the stomach, colon and rectum. Vol. III, Case-control study of gastrointestinal cancer in Norway. Vol. IV, Case-control study of the digestive tract cancers in Minnesota*. Ann Arbor University Microfilms

Block, G. & Abrams, B. (1993) Vitamin and mineral status of women of childbearing potential. *Ann. NY Acad. Sci.*, 678, 244–254

Bostick, R.M., Potter, J.D., Kushi, L.H., Sellars, T., Steinmetz, K., McKenzie, D., Gapstur, S. & Folsom, A. (1994) Sugar, meat, and fat intake, and non-dietary risk factors for colon cancer incidence in women. *Cancer Causes Control*, 5, 38–52

Braddon, F.E.M., Wadsworth, M.E.J., Davies, J.M.C. & Cripps, H.A. (1988) Social and regional differences in food and alcohol consumption and their measurement in a national birth cohort. *J. Epidemiol. Commun. Health*, 42, 341–349

Bristol, J.B., Emmett, P.M., Heaton, K.W. & Williamson, R.C.N. (1985) Sugar, fat, and the risk of colorectal cancer. *Br. Med. J.*, 291, 1467–1470

Brock, K.E., Berry, G., Mock, P.A., MacLennan, R., Truswell, A. & Brinton, L. (1988) Nutrients in diet and plasma and risk of in situ cervical cancer. *J. Natl Cancer Inst.*, 80, 580–585

Brown, J.E., Potter, J.D., Jacobs, D.R. Kopher, R.A., Rourke, M.J., Barosso, G., Abrams, B., Hannan, P. & Schmid, L. (1996) Maternal waist-to-hip ratio as a predictor of newborn size: results of the DIANA project. *Epidemiology*, 7, 62–66

Bruce, W.R. (1987) Recent hypotheses for the origin of colon cancer. *Cancer Res.*, 97, 4237–4242

Burney, P.G.J., Comstock, G.W. & Morris, J.S. (1989) Serologic precursors of cancer: serum micronutrients and the subsequent risk of pancreatic cancer. *Am. J. Clin. Nutr.*, 49, 895–900

Byers, T.E., Graham, S., Maughey, B.P. Marshall, J. & Swanson, M. (1987) Diet and lung cancer risk: findings from the western New York diet study. *Am. J. Epidemiol.*, 125, 351–363

Cade, J. (1992) Diet of adults living in houses in multiple occupation. *Eur. J. Clin. Nutr.*, 46, 795–801

Chasseaud, L.F. (1979) The role of glutathione and glutathione S-transferases in the metabolism of chemical carcinogens and other electrophilic agents. *Adv. Cancer Res.*, 29, 175–274

Colditz, G.A., Stampfer, M.J., Willett, W.C., Hennekens, C., Rosner, B. & Speizer, F. (1990) Prospective study of estrogen replacement therapy and risk of breast cancer in postmenopausal women. *J. Am. Med. Assoc.*, 264, 2648–2653

Davis, M.A., Murphy, S.P., Neuhaus, J.M. & Lein, D. (1990) Living arrangements and dietary quality of older U.S. adults. *J. Am. Diet. Assoc.*, 90, 1667–1672

de Waard, F., Baanders-van Halewijn, E.A. & Huizinga, J. (1964) The bimodal age distribution of patients with mammary carcinoma: evidence for the existence of 2 types of human breast cancer. *Cancer*, 17, 141–151

Deo, M.G. (1981) Implications of malnutrition in chemical carcinogenesis. *J. Cancer Res. Clin. Oncol.*, 99, 77–86

Donahue, R.B., Abbott, R.D., Bloom, B., Reed, D. & Yano, K. (1987) Central obesity and coronary heart disease in men. *Lancet*, 1, 821–824

Eaton, B.S. & Konner, M. (1985) Paleolithic nutrition. *New Engl. J. Med.*, 312, 283–289

Elwood, J.M., Cole, P., Rothman, K.J. & Kaplan, S.D. (1977) Epidemiology of endometrial cancer. *J. Natl Cancer Inst.*, 59, 1055–1060

Feldman, R., Sender, A.J. & Siegelaub, A.B. (1969) Difference in diabetic and non-diabetic fat distribution patterns by skinfold measurements. *Diabetes*, 18, 478–486

Folsom, A.R., Kaye, S.A., Potter, J.D. & Prineas, R.J. (1989) Association of incident carcinoma of the endometrium with body weight and fat distribution in older women. Early findings of the Iowa Women's Study. *Cancer Res.*, 49, 6828–6831

Food and Research Action Committee (1984) *Bitter harvest: a status report on the need for emergency food assistance in America*. Washington, DC, Food and Research Action Committee

Food and Research Action Committee (1985) *Bitter harvest II: a status report on the need for emergency food assistance in America*. Washington, DC, Food and Research Action Committee

Frisch, R.E. & McArthur, J.W. (1974) Menstrual cycles: fatness as a determinant of the minimum weight for height necessary for their maintenance or onset. *Science*, 185, 949–951

Frisch, R.E., Wyshak, G., Albright, N.L., Albright, T., Schiff, I., Jones, K., Witschi, J., Shiang, E., Koff, E. & Margulio, M. (1985) Lower prevalence of breast cancer and cancers of the reproductive system among former college athletes compared to non-athletes. *Br. J. Cancer*, 52, 885–891

Gapstur, S.M., Potter, J.D. & Sellers, T.A. (1992) Increased risk of breast cancer with alcohol consumption in postmenopausal women. *Am. J. Epidemiol.*, 136, 1221–1231

Garabrant, D.H., Peters, J.M., Mack, T.M. & Bernstein, L. (1984) Job activity and colon cancer risk. *Am. J. Epidemiol.*, 119, 1005–1014

Gerhardsson, M., Norell, S.E., Kiviranta, H., Pedersen, N.L. & Ahlbom, A. (1986) Sedentary jobs and colon cancer. *Am. J. Epidemiol.*, 123, 775–780

Giovanucci, E., Stampfer, M.J., Colditz, G.A. & Rimm, E.B. (1993) Folate, methionine, and alcohol intake and risk of colorectal adenoma. *J. Natl Cancer Inst.*, 85, 875–884

Gorbach, S.L. (1984) Estrogens, breast cancer, and intestinal flora. *Rev. Infect. Dis.*, 6 (Suppl.), 85–90

Graham, S., Marshall, J., Mettlin, C., Rzepka, T., Nemoto, T. & Byers, T. (1982) Diet in the epidemiology of breast cancer. *Am. J. Epidemiol.*, 116, 68–75

Greenberg, E.R., Baron, J.A., Tosteson, T.D. Freeman, D., Beck, G., Bond, J., Colacchio, T., Collier, J., Frankl, H., Haile, R., Mandel, J., Nierenberg, D., Rothstein, R., Snover, D., Stevens, M., Summers, R. & van Stock, R. (1994) A clinical trial of antioxidant vitamins to prevent colorectal adenoma. *New Engl. J. Med.*, 331, 141–147

Gregory, J., Foster, K., Tyler, H. & Wiseman, M. (1990) The dietary and nutritional survey of British adults: Office of Population Censuses and Surveys. London, Her Majesty's Stationery Office

Gruenke, L.D., Wrensch, M.R., Petrakis, N.H. Miike, R., Ernster, V. & Craig, J. (1987) Breast fluid cholesterol and cholesterol epoxides: relationship to breast cancer risk factors and other characteristics. *Cancer Res.*, 47, 5483–5487

Grodin, J.M., Siiteri, P.K. & MacDonald, P.C. (1973) Source of estrogen production in postmenopausal women. *J. Clin. Endocrinol. Metab.*, 36, 207–214

Haraldsdottir, J., Holm, L., Jensen, J.H. & Moller, A. (1987) *Danskernes kostvaner 1985, 2. Hvem spiser hvad?* (Publication No. 154). Levnedsmiddlestyrelsen, Soborg

Helmrich, S.P., Shapiro, S., Rosenberg, L. Kaufman, D., Slone, D., Bain, C., Miettinen, D., Stolley, P., Rosenshein, N., Knapp, R., Leavitt, T., Schottenfeld, D., Engle, R. & Levy, M. (1983) Risk factors for breast cancer. *Am. J. Epidemiol.*, 117, 35–45

Hill, M.J., Goddard, P. & Williams, R.E.O. (1971) Gut bacteria and aetiology of cancer of the breast. *Lancet*, 2, 472–473

Hill, P., Garbaczewski, L., Helman, P. Huskisson, J., Sporangisa, E. & Wynder, E. (1980) Diet, lifestyle, and menstrual activity. *Am. J. Clin. Nutr.*, 33, 1192–1198

Hirayama, T. (1978) Epidemiology of breast cancer with special reference to the role of diet. *Prev. Med.*, 7, 173–195

Hirayama, T. (1988) A large-scale cohort study on the relationship between diet and selected cancers of digestive organs. In: Bruce, W.R., Correa, P., Lipkin, M., Tannenbaum, S. & Wilkins, T., eds, *Gastrointestinal cancer: endogenous factors* (Banbury Report 7). New York, Cold Spring Harbor Laboratory. pp. 409–426

Hirohata, T., Shigematsu, T., Nomura, A.M.Y., Nomura, Y., Horie, A. & Hirohata, I. (1985) Occurrence of breast cancer in relation to diet and reproductive history: a case-control study in Fukuoka, Japan. *Natl Cancer Inst. Monogr.*, 67, 187–190

Howatson, A.G. & Carter, D.C. (1985) Pancreatic carcinogenesis – enhancement by cholecystokinin in the hamster-nitrosamine model. *Br. J. Cancer*, 51, 107–114

Howe, G.R. (1985) The use of polytomous dual response data to increase power in case-control studies: an application to the association between dietary fat and breast cancer. *J. Chronic Dis.*, 38, 663–670

Howe, G.R., Benito, E., Castellato, R., Cornée, J., Estève, J., Gallagher, R., Iscovich, J., Deng-ao, J., Kaaks, R., Kune, S., L'Abbé, K., Lee, H., Miller, A., Peters, R., Potter J., Riboli, E., Slattery, M., Trichopoulos, D., Tuyns, A., Tzounou, A., Whittemore, A., Wu-Williams, A. & Shu, Z. (1992) Dietary intake of fiber and decreased risk of cancers of the colon and rectum: evidence from the combined analysis of 13 case-control studies. *J. Natl Cancer Inst.*, 84, 1887–1896

IARC (1988) *Alcohol drinking* (IARC Monographs Vol. 44). Lyon, International Agency for Research on Cancer

IARC (1976) *Some naturally occurring substances* (IARC Monographs Vol. 10). Lyon, International Agency for Research on Cancer

Johnson, L.R. (1981) Effects of gastrointestinal hormones on pancreatic growth. *Cancer*, 47, 1640–1645

Kadlubar, F.F., Butler, M.A., Kaderlik, K.R., Chou, H.C. & Lang, N.P. (1992) Polymorphisms for aromatic amine metabolism in humans: relevance for human carcinogenesis. *Environ. Health Persp.*, 98, 69–74

Kark, J.D., Smith, A.H., Switzer, B.R. & Hames C.G. (1981) Serum vitamin A (retinol) and cancer incidence in Evans County, Georgia. *J. Natl Cancer Inst.*, 66, 7–16

Katsouyanni, K., Trichopoulos, D., Boyle, P., Xirouchaki, E., Trichopoulou, A., Lissios, B., Vasilaros, S. & McMahon, B. (1986) Diet and breast cancer: a case-control study in Greece. *Int. J. Cancer*, 38, 815–820

Kidwell, W.R., Knazek, R.A., Vonderhaar, B.K. & Losonczy, I. (1982) Effects of unsaturated fatty acids on the development and proliferation of normal and neoplastic breast epithelium. In: Arndt, M.S., van Eys, J. & Wang, Y.M., eds, *Molecular interrelations of nutrition and cancer*. New York, Raven Press. p. 219–236

Kinlen, L.J. (1982) Meat and fat consumption and cancer mortality: a study of strict religious orders in Britain. *Lancet*, 1, 946–949

Koo, L.C. (1988) Dietary habits and lung cancer risk among Chinese females in Hong Kong who never smoked. *Nutr. Cancer*, 11, 155–172

Kosa, J. & Zola, I. (1975) *Poverty and health. A sociologic analysis* (revised edn). Cambridge, MA, Harvard University Press

Kushi, L.H., Folsom, A.R., Jacobs, D.R., Jr, Luepker, R., Elmer, P.J. & Blackburn, H. (1988) Educational attainment and nutrient consumption patterns: The Minnesota Heart Survey. *J. Am. Diet. Assoc.*, 88, 1230–1236

Kushi, L.H., Sellers, T.A., Potter J.D., Nelson, C., Munger, R., Kaye, S. & Folsom, A.(1992) Dietary fat and postmenopausal breast cancer. *J. Natl Cancer Inst.*, 84, 1092–1099

Larsson, B., Svarsudd, K., Welin, L., Wilhelmsen, L., Bjorntorp, P. & Tibblin, G. (1984) Abdominal adipose tissue distribution, obesity, and risk of cardiovascular disease and death: 13 year follow-up of participants in the study of men born in 1913. *Br. Med. J.*, 288, 1401–1404

LaVecchia, C., Decarli, A., Fasoli, M. Parazzini, F., Franceschi, S., Gentile, A. & Negri, E. (1988b) Dietary vitamin A and the risk of intraepithelial and invasive cervical neoplasia. *Gynecol. Oncol.*, 30, 187–195

LaVecchia, C., Franceschi, S., Decarli, A., Gallus, G. & Tognoni, G. (1984) Risk factors for endometrial cancer at different ages. *J. Natl Cancer Inst.*, 73, 667–671

LaVecchia, C., Negri, E., Decarli, A., D'Avanzo, B., Gallotti, A. & Franceschi, S. (1988a) A case-control study of diet and colorectal cancer in northern Italy. *Int. J. Cancer*, 41, 492–498

LaVecchia, C., Negri, E., Decarli, A., D'Avanzo, B. & Franceschi, S. (1987) A case-control study of diet and gastric cancer in northern Italy. *Int. J. Cancer*, 40, 484–489

Le, M.G., Moulton, L.H., Hill, C. & Kramer, A. (1986) Consumption of dairy produce and alcohol in a case-control study of breast cancer. *J. Natl Cancer Inst.*, 77, 633–636

Lew, E.A. & Garfinkel, L. (1979) Variations in mortality by weight among 750,000 men and women. *J. Chronic Dis.*, 32, 563–576

Li, J.Y., Taylor, P.R., Li, B., Dawsey, S., Wang, G., Ershow, A., Gao, W., Liu, S., Yang, C. & Shen, Q. (1993) Nutrition intervention trials in Linxian, China: multiple vitamin/mineral supplementation, cancer incidence, and disease-specific mortality among adults with esophageal dysplasia. *J. Natl Cancer Inst.*, 85, 1492–1498

Lubin, J.H., Burns, P.E., Blot, W.J. Ziebler, R., Lees, A. & Fraumeni, J. (1981) Dietary factors and breast cancer risk. *Int. J. Cancer*, 28, 685–689

Lubin, F., Wax, Y. & Modan, B. (1986) Role of fat, animal protein and dietary fiber in breast cancer etiology: a case-control study. *J. Natl Cancer Inst.*, 77, 605–612

Lyon, J.L., Mahoney, A.W., West, D.W., Gardner, J., Smith, K., Sorenson, A. & Stanish, W. (1987) Energy intake: its relationship to colon cancer risk. *J. Natl Cancer Inst.*, 78, 853–861

McLaughlin, J.K., Gridley, G., Block, G. Winn, D., Preston-Martin, S., Schoenberg, J., Greenberg, R., Stemhagen, A., Austin, D. & Ershow, A. (1988) Dietary factors in oral and pharyngeal cancer. *J. Natl Cancer Inst.*, 15, 1237–1243

McMichael, A.J. (1979) Alimentary tract cancer in Australia in relation to diet and alcohol. *Nutr. Cancer*, 1, 82–89

McMichael, A.J. (1981) Coffee, soya and pancreatic cancer. *Lancet*, 2, 689–690

McMichael, A.J. & Potter, J.D. (1985) Diet and colon cancer: integration of the descriptive, analytic, and metabolic epidemiology. *Natl Cancer Inst. Monogr.*, 69, 223–228

McMichael, A.J. & Potter, J.D. (1986) Dietary influences upon colon carcinogenesis: In: Hayashi, Y. *et al.*, eds, *Diet, nutrition and cancer*. Tokyo, Japan Sci. Soc. Press. pp. 275–290

Macquart-Moulin, G., Riboli, E., Cornee, J., Charnay, B., Berthezene, P. & Day, N. (1986) Case-control study on colorectal cancer and diet in Marseilles. *Int. J. Cancer*, 38, 183–191

Mele, L., West, K.P., Kusdiono, Pandji, A., Nendrawati, H., Tilden, R. & Tarwotjo, I. (1991) Nutritional and household risk factors for xerophthalmia in Aceh, Indonesia: a case-control study. *Am. J. Clin. Nutr.*, 53, 1460–1465

Micozzi, M.S. (1985) Nutrition, body size, and breast cancer. *Yearbook Phys. Anthropol.*, 28, 175–206

Miller, A.B., Kelly, A., Choi, N.W., Matthews, V., Morgan, R., Munan, L., Burch, J., Feather, J., Howe, G. & Jain, M. (1978) A study of diet and breast cancer. *Am. J. Epidemiol.*, 107, 499–509

Mirvish, S.S. (1981) Ascorbic acid inhibition of N-nitroso compound formation in chemical, food and biological systems. In: Zedeck, M.S. & Lipkin, M., eds, *Inhibition of tumor induction and development*. New York, Plenum Publishing. pp. 101–126

Muñoz, N., Wahrendorf, J., Lu, J.B., Crespi, M., Thurnham, D.I., Day, N.E., Zheng, H.J., Grassi, A., Li, W.Y., Liu, G.L., Lang, Y.Q., Zhang, C.Y., Zheng, S.F., Li, J.Y., Correa, P., O'Conor, G.T. & Bosch, X. (1985) No effect of riboflavine, retinol and zinc on prevalence of precancerous lesions of oesophagus. Randomized double-blind intervention study in high-risk population of China. *Lancet*, 2, 111–114

Murphy, S., Rose, D., Hudes, M. & Viteri, F.E. (1992) Demographic and economic factors associated with dietary quality for adults in the 1987–88 Nationwide Food Consumption Survey. *J. Am. Diet. Assoc.*, 92, 1352–1357

Myers, A.W. & Kroetsch, D. (1978) The influence of family income on food consumption patterns and nutrient intake in Canada. *Can. J. Public Health*, 69, 208–221

Neel, J.V. (1969) Current concepts of the genetic basis of diabetes mellitus and the biological significance of the diabetic predisposition. In: Ostman, J. & Milner, R.D.G., eds, *Diabetes*. Excerpta Medica Amsterdam. pp. 68–78

Newman, V., Norcross, W. & McDonald, R. (1991) Nutrient intake of low-income Southeast Asian pregnant women. *J. Am. Diet. Assoc.*, 91, 793–799

Oishi, K., Okada, K., Yoshida, O., Yamabe, H., Ohno, Y., Hayes, R. & Schroeder, F. (1988) A case-control study of prostatic cancer with reference to dietary habits. *Prostate*, 12, 179–190

Paffenbarger, R.S., Kampert, J.B. & Chang, H-G. (1980) Characteristics that predict risk of breast cancer before and after the menopause. *Am. J. Epidemiol.*, 112, 258–268

Paffenbarger, R.S., Hyde, R.T. & Wing, A.L. (1987) Physical activity and incidence of cancer in diverse populations: a preliminary report. *Am. J. Clin. Nutr.*, 45, 311–317

Pariza, M.W. & Simopoulos, A.P. (1986) Calories and energy expenditure in carcinogenesis. *Am. J. Clin. Nutr.*, 45 (Suppl.), 149–272

Peers, F., Bosch, X., Kaldon, J., Linsell, A. & Pluijmen, M. (1987) Aflatoxin exposure, hepatitis B virus infection, and liver cancer in Swaziland. *Int. J. Cancer*, 39, 545–553

Peters, R.K., Pike, M.C., Garabrant, D. & Mack, T.M. (1992) Diet and colon cancer in Los Angeles County, California. *Cancer Causes Control*, 3, 457–473

Peto, R., Doll, R., Buckley, J.D. & Sporn, M.B. (1981) Can dietary beta-carotene materially reduce human cancer rates? *Nature*, 290, 201–208

Petrakis, N.L., Maack, C.A., Lee, R.E. & Lyon, M. (1980) Mutagenic activity in nipple aspirates of human breast fluid. *Cancer Res.*, 40, 188–189

Phillips, R.L. (1975) Role of life-style and dietary habits in risk of cancer among Seventh-day Adventists. *Cancer Res.*, 35, 3513–3522

Phillips, R.L. & Snowdon, D.A. (1983) Association of meat and coffee use with cancers of the large bowel, breast and prostate among Seventh-day Adventists: preliminary results. *Cancer Res.*, 43 (Suppl.), 2403–2408

Phillips, R.L., Garfinkel, L., Kuzma, J.W., Beeson, W., Lotz, T. & Brin, B. (1980) Mortality among California Seventh-day Adventists for selected cancer sites. *J. Natl Cancer Inst.*, 65, 1097–1107

Physician Task Force on Hunger in America (1985) *Hunger in America, the growing epidemic*. Boston, Harvard University, School of Public Health

Physician Task Force on Hunger in America (1986) *Hunger counties 1986*. Boston, Harvard University, School of Public Health

Physician Task Force on Hunger in America (1987) *Hunger reaches blue collar America: an unbalanced recovery in a service economy*. Boston, Harvard University, School of Public Health

Pitot, H.C. (1986) *Fundamentals of oncology* (3rd edn). New York, Marcel Dekker

Potter, J.D. (1989) The epidemiology of fiber and colorectal cancer: why don't the analytic epidemiology data make better sense? In: Kritchevsky, D., Bonfield, C. & Anderson, J., eds, *Dietary fiber: chemistry, physiology and health effects*. New York, Plenum Press. pp. 431–445

Potter, J.D. (1990) The epidemiology and prevention of pancreas cancer. In: Zatonski, W., Boyle, P. & Tyczynski, J. eds, *The prevention of cancer: vital statistics to intervention*. Warsaw, P.A. Interpress. pp. 146–151

Potter, J.D. (1992a) Colon cancer – reconciling the epidemiology, physiology, and molecular biology. *J. Am. Med. Assoc.*, 268, 1573–1577

Potter, J.D. (1992b) The epidemiology of diet and cancer: evidence of human maladaption. In: Moon, T.E. & Micozzi, M.S., eds, *Macronutrients: investigating their role in cancer*. New York, Marcel Dekker. pp. 55–84

Potter, J.D. & McMichael, A.J. (1986) Diet and cancer of the colon and rectum. A case-control study. *J. Natl Cancer Inst.*, 76, 557–569

Potter, J.D., McMichael, A.J. & Hartshorne, J.M. (1982) Alcohol and beer consumption in relation to cancer of bowel and lung: an extended correlation analysis. *J. Chronic Dis.*, 35, 823–824

Potter J.D., Slattery M.L., Bostick R.M. & Gapstur S.M. (1993) Colon cancer: a review of the epidemiology. *Epidemiol. Rev.*, 15, 499–545

Potter, J.D., Cerhan, J.R., Sellers, T.A., McGovern, P., Drinkard, C., Kushi, L. & Folsom, A. (1994) Progesterone and estrogen receptors and mammary neoplasia: how many kinds of breast cancer are there? *Cancer Epidemiol. Biomarkers Prev.*, 4, 319–326

Selby, J.V., Friedman, G.D. & Quesenberry, C.P. (1989) Precursors of essential hypertension. The role of body fat distribution pattern. *Am. J. Epidemiol.*, 129, 43–53

Sellers, T.A., Kushi, L.H., Potter, J.D., Kaye, S., Nelson, C., McGovern, P. & Folsom, A. (1992) Effect of family history, body fat distribution and reproductive factors on risk of postmenopausal breast cancer. *New Engl. J. Med.*, 326, 1323–1329

Shea, S., Melnik, T.A., Stein, A., Zansky, S., Maylahn, C. & Basch, C. (1993) Age, sex, educational attainment, and race/ethnicity in relation to consumption of specific foods contributing to the atherogenic potential of diet. *Prev. Med.*, 22, 203–218

Shun-Zhang, Y., Rui-Fang, L., Da-Dao, X. & Howe, G.R. (1990) A case-control study of dietary and non-dietary factors for breast cancer in Shanghai. *Cancer Res.*, 50, 5017–5121

Slattery, M.L., Schumacher, M.C., Smith, K.R., West, D.W. & Abd-Elghany, N. (1988a) Physical activity, diet and risk of colon cancer in Utah. *Am. J. Epidemiol.*, 128, 989–999

Slattery, M.L., Sorenson, A.W. & Ford, M.H. (1988b) Dietary calcium intake as a mitigating factor in colon cancer. *Am. J. Epidemiol.*, 128, 504–514

Smith, A.M. & Baghurst, K.I. (1992) Public health implications of dietary differences between social status and occupational category groups. *J. Epidemiol. Community Health*, 46, 409–416

Snow, J. (1855) *On the mode of communication of cholera* (2nd edn). London, Churchill

Steele, P., Dobson, A., Alexander, H. & Russell, A. (1991) Who eats what? A comparison of dietary patterns among men and women in different occupational groups. *Aust. J. Public Health*, 15, 286–295

Steinmetz, K. & Potter, J.D. (1991a) A review of vegetables, fruit, and cancer I: epidemiology. *Cancer Causes Control*, 2, 325–357

Steinmetz, K. & Potter, J.D. (1991b) A review of vegetables, fruit, and cancer II: mechanisms. *Cancer Causes Control*, 2, 427–442

Stemmermann, G.N., Nomura, A.M.Y. & Heilbrun, L.K. (1984) Dietary fat and the risk of colorectal cancer. *Cancer Res.*, 44, 4633–4637

Stragand, J.J. & Hagemann, R.F. (1977) Effect of lumenal contents on colonic cell replacement. *Am. J. Physiol.*, 233, E208–E211

Subar, A.S., Heimendinger, J., Krebs-Smith, S.M., Patterson, B.H., Kessler, R. & Pivonka, E. (1992) *5-a-day for better health: a baseline study of Americans' fruit and vegetable consumption*. Bethesda, MD, National Cancer Institute

Sugimura, T. (1985) Carcinogenicity of mutagenic heterocyclic amines formed during the cooking process. *Mutat. Res.*, 150, 33–41

Sugimura, T. & Sato, S. (1983) Mutagens-carcinogens in foods. *Cancer Res.*, 43, 2415s–2421s

Talamini, R., LaVecchia, C., Decarli, A., Franceschi, S., Grattoni, E., Grigoletto, E., Liberati, A. & Tognoni, G. (1984) Social factors, diet and breast cancer in a northern Italian population. *Br. J. Cancer*, 49, 723–729

Taren, D.L., Clark, W., Chernesky, M. & Quirk, E. (1990) Weekly food servings and participation in social programs among low income families. *Am. J. Public Health*, 80, 1376–1378

Thurnham, D.I., Zheng, S-F., Muñoz, N., Crespi, M., Grassi, A., Hambridge, K. & Chai, T. (1985) Comparison of riboflavin, vitamin A, and zinc status of Chinese populations at high and low risk for esophageal cancer. *Nutr. Cancer*, 7, 131–143

Turesky, R.J., Lang, N., Butler, M.A., Teitel, C. & Kadlubar, F. (1991) Metabolic activation of carcinogenic heterocyclic aromatic amines by human liver and colon. *Carcinogenesis*, 12, 1839–1845

Tuyns, A.J., Kaaks, R. & Haelterman, M. (1988) Colorectal cancer and the consumption of foods: a case-control study in Belgium. *Nutr. Cancer*, 11, 189–204

Vague, J. (1956) The degree of masculine differentiation of obesities: a factor determining predisposition to diabetes, atherosclerosis, gout and uric calculus disease. *Am. J. Clin. Nutr.*, 4, 20–34

van den Brandt, P.A., van't Veer, P., Goldbohm, R.A., Dorant, E., Volovics, A., Hermus, R. & Sturmans, F. (1993) A prospective cohort study on dietary fat and the risk of postmenopausal breast cancer. *Cancer Res.*, 53, 75–82

van Rensburg, S.J. (1981) Epidemiologic and dietary evidence for a specific nutritional predisposition to esophageal cancer. *J. Natl Cancer Inst.*, 67, 243–251

Velema, J.P., Walker, A.M. & Gold, E.B. (1986) Alcohol and pancreatic cancer: insufficient epidemiological evidence for a causal relationship. *Epidemiol. Rev.*, 8, 28–41

Vena, J.E., Graham, S., Zielezny, M., Swanson, M.K., Barnes, R.E. & Nolan, J. (1985) Lifetime occupational exercise and colon cancer. *Am. J. Epidemiol.*, 122, 357–365

Vitale, J.J. & Broitman, S.A. (1981) Lipids and immune function. *Cancer Res.*, 41, 3706–3710

Wahrendorf, J., Muñoz, N., Lu, J-B., Thurnham, D., Crespi, M. & Bosch, F. (1988) Blood retinol, zinc, riboflavin status in relation to pre-cancerous lesions of the esophagus: findings for a vitamin intervention trial in the People's Republic of China. *Cancer Res.*, 48, 2280–2283

Wald, N., Idle, M., Boreham, J. & Bailey, A. (1980) Low serum-vitamin A and subsequent risk of cancer. Preliminary results of a prospective study. *Lancet*, 2, 813–815

Wattenberg, L.W. (1977) Inhibition of carcinogenic effects of polycyclic hydrocarbons by benzyl isothiocyanate and related compounds. *J. Natl Cancer Inst.*, 58, 395–398

Wattenberg, L.W. (1983) Inhibition of neoplasia by minor dietary constituents. *Cancer Res.*, 43 (Suppl.), 2448–2453

Wattenberg, L.W. (1985) Chemoprevention of cancer. *Cancer Res.*, 45, 1–8

Wattenberg, L.W. (1992) Inhibition of carcinogenesis by minor dietary constituents. *Cancer Res.*, 52 (Suppl.), 2085–2091

Weber, W.W. (1987) *The acetylator genes and drug response.* New York, Oxford University Press

Welsch, C.W. & Aylsworth, C.F. (1983) Enhancement of murine mammary tumorigenesis by feeding high levels of dietary fat: a hormonal mechanism. *J. Natl Cancer Inst.*, 70, 215–221

Willett, W.C. & Stampfer, M.J. (1986) Total energy intake: implications for epidemiologic analyses. *Am. J. Epidemiol.*, 124, 17–27

Willett, W.C., Polk, B.F., Morris, J.S., Pressel, S., Polk, B., Stampfer, M., Rosner, B., Schneider, K. & Hames, C. (1983) Prediagnostic serum selenium and risk of cancer. *Lancet*, 2, 130–134

Willett, W.C., Browne, M.L., Bain, C., Lipnick, R., Stampfer, M., Rosner, B., Colditz, G., Hennekens, C. & Speizer, F. (1985) Relative weight and risk of breast cancer among premenopausal women. *Am. J. Epidemiol.*, 122, 731–740

Willett, W.C., Stampfer, M.J., Colditz, G.A., Rosner, B., Hennekens, C. & Speizer, F. (1987) Dietary fat and risk of breast cancer. *New Engl. J. Med.*, 316, 22–28

Willett, W.C., Hunter, D.J., Stampfer, M.J., Colditz, G., Manson, J., Spiegelman, D., Rosner, B., Hennekens, C. & Speizer, F. (1992) Dietary fat and fiber in relation to risk of breast cancer. *J. Am. Med. Assoc.*, 268, 2037–2044

Williams, R.R. (1976) Breast and thyroid cancer and malignant melanoma promoted by alcohol-induced pituitary secretion of prolactin, TSH, and MSH. *Lancet*, 1, 996–999

Wood, A.W., Huang, M.T., Chang, R.L., Newmark, H., Lehr, R., Yogi, H., Sayer, J., Jerina, D. & Conney, A. (1982) Inhibition of the mutagenicity of bay-region diol epoxides of polycyclic hydrocarbons by naturally occurring plant phenols: exceptional activity of ellagic acid. *Proc. Natl Acad. Sci. USA*, 79, 5513–5517

Wu, A.H., Paganini-Hill, A., Ross, R.K. & Henderson, B.E. (1987) Alcohol, physical activity and other risk factors for colorectal cancer: a prospective study. *Br. J. Cancer*, 55, 687–694

Wynder, E.L., Kajitani, T., Ishikana, S., Dodo, H. & Tako, A. (1969) Environmental factors of cancer of the colon and rectum II. Japanese epidemiological data. *Cancer*, 23, 1210–1220

Yang, C.S. (1980) Research on esophageal cancer in China: a review. *Cancer Res.*, 40, 2633–2644

You, W-C., Blot, W.J., Chang, Y-S., Ershow, A., Yang, Z., An, Q., Henderson, B., Xu, O., Fraumeni, J., Jr. & Wang, J. (1988) Diet and high-risk of stomach cancer in Shandong, China. *Cancer Res.*, 48, 3518–3523

Young, T.B. & Wolf, D.A. (1988) Case-control study of proximal and distal colon cancer and diet in Wisconsin. *Int. J. Cancer*, 42, 167–175

Zeitlin, M.F., Megawangi, R., Kramer, E.M. & Armstrong, H.C. (1992) Mothers' and children's intakes of vitamin A in rural Bangladesh. *Am. J. Clin. Nutr.*, 56, 136–147

Ziegler, R.G. (1989) A review of epidemiologic evidence that carotenoids reduce the risk of cancer. *J. Nutr.*, 119, 116–122

J.D. Potter
Cancer Prevention Research Program,
Fred Hutchinson Cancer Research Center,
1124 Columbia St, Seattle, WA 98104, USA

Socioeconomic differences in reproductive behaviour

I. dos Santos Silva and V. Beral

There are marked socioeconomic variations in the risk of female reproductive cancers. We examine here data from the World Fertility Surveys, the Demographic and Health Surveys, and other national surveys, to assess whether these variations in cancer risk might be explained, at least in part, by socioeconomic variations in reproductive behaviour. There were marked socioeconomic differentials in achieved parity, age at first birth, final childlessness, duration of breastfeeding, and possibly also age at menopause. These differentials were present in almost all settings: countries with low and high levels of modernization, and countries with low and high levels of fertility. In general, women of higher socioeconomic status and with more education had lower fertility and later age at first birth, but a greater prevalence of childlessness, shorter duration of breastfeeding and later age at menopause. However, the size and even the direction of these differentials varied markedly from country to country according to its level of economic development and, within each country, from generation to generation of women. It is possible that some of these socioeconomic differences may be narrowing in recent generations in Western countries. There was little evidence of socioeconomic variations in age at menarche.

The observed socioeconomic differentials in most aspects of reproductive behaviour could potentially account for some of the socioeconomic variation in the risk of female reproductive cancers. However, this relationship could not be assessed directly because such analysis would require birth-cohort-specific data on socioeconomic variations in reproductive behaviour and in cancer risks. Unfortunately, these data are not available.

A woman's reproductive history plays an important role in the risk of breast, ovarian and endometrial cancers. Studies have consistently shown a higher risk of these cancers in nulliparous than parous women (Elwood et al., 1977; MacMahon et al., 1970; The Centers for Disease Control, 1983), and an inverse relationship with parity for ovarian and endometrial cancers. Age at first birth has been considered as a major risk factor for breast cancer since the large international study of MacMahon et al. (1970).

The risks of these cancers are also reduced by an early age at menopause (Hildreth et al., 1981; Elwood et al., 1977; Pike et al., 1983), and some studies, but not all, have also found a late age at menarche to be protective (Booth, 1991; Elwood et al., 1977; Pike et al., 1983). There is conflicting evidence of a duration-related protective effect of lactation against the risk of breast and ovarian cancers (Risch et al., 1983; Yuan et al., 1988).

There are large socioeconomic differences in the risk of female reproductive cancers. The general pattern is that breast, ovarian and endometrial cancers are more common in women of higher socioeconomic status. In this chapter, we examine whether these socioeconomic variations in cancer risks could be explained, at least in part, by socioeconomic variations in reproductive behaviour.

Sources of data

The World Fertility Surveys (WFS) provide a unique opportunity to examine the relationship between fertility and socioeconomic factors using comparable data from a wide variety of settings: countries with high and low socioeconomic developments and varying levels of fertility. These surveys were intended to be nationally representative (and internationally comparable) surveys of women in the reproductive ages. The WFS were conducted in about 40 developing countries and 20 developed countries mainly in the late 1970s (United Nations, 1987).

For most countries, the WFS consisted of both a household survey and an individual survey. The household survey provided a listing of persons living in the household along with some basic demographic data (sex, age, marital status, and so on), on the basis of which women eligible for the detailed individual interview on reproductive behaviour were identified. There was some variability in the criteria used to select women for the individual interview in developed countries (that is, European countries and the United States of America). For the analyses shown here it was possible, however, to select in each country a subsample of currently married women in their first marriage aged under 45 years (Jones, 1982). The only exceptions were Denmark and Poland, where all currently married women were included, and France, where women aged under 20 years were excluded. The survey in Belgium included only the Flemish part of the country.

The individual surveys conducted in developing countries usually included all women aged 15 to 49 or 50 years (United Nations, 1987). However, in Costa Rica and Panama, the youngest age for interview was 20 and in Venezuela the oldest age for interview was 44. In Asian countries, where extramarital childbearing is thought to be relatively rare, the individual interviews were restricted to ever-married women. At the other extreme, all women, regardless of marital status, were interviewed in all but one of the Latin American surveys (Peru). The pattern in Africa was mixed but in most of the sub-Saharan countries all women aged 15 to 49 years were interviewed irrespective of their marital status.

The WFS were succeeded by the Demographic and Health Surveys (DHS), which started in the late 1980s and early 1990s. In contrast to the WFS, the DHS were set up only in developing countries. Generally, they consisted of national representative samples of all women aged 15 to 49 regardless of their marital status (Westoff et al., 1994). There were, however, some exceptions to this design. Individual interviews were restricted to women aged 15 to 44 years in Brazil and Guatemala and to ever-married women in northern African and in Asian countries.

In both the WFS and the DHS, fertility patterns for all women (regardless of their marital status) were estimated for those countries in which single women were excluded from the individual interview, by using data obtained in the household survey (Hogdson & Gibbs, 1980; Westoff et al., 1994).

Both the WFS and the DHS collected data on various measures of socioeconomic status such as type of place of residence (urban versus rural), educational level of the respondents and of their husbands, family income, and husband's socio-occupational status. The definition and measurement of these variables were bounded by country-specific criteria, with content and levels varying substantially across countries. They were, however, designed with the intention of capturing the full range of socioeconomic variability in each particular country.

In this chapter, we extracted data from the WFS and the DHS to examine whether there were socioeconomic variations in reproductive behaviour. Data on socioeconomic differentials in final childlessness and age at menopause were not available from either of these two surveys and were therefore extracted from other published sources. Data on childlessness were available from national surveys carried out in Norway (Kravdal, 1992), England and Wales (Office of Population Censuses and Surveys, 1983) and the USA (Poston, 1974) and from special studies carried out in some developing countries (Poston, 1988; Poston & Rogers, 1988; Poston et al., 1985; Romaniuk, 1980). Data on socioeconomic differences in age at menopause were available from nationally representative surveys carried out in Finland (Luoto et al., 1994) and in the USA (Stanford et al., 1987); to our knowledge, no similar surveys were conducted in any other developed countries or in any of the developing countries.

Results
Parity
Tables 1–5 present WFS data on socioeconomic differentials in achieved parity for women in various European countries and for the United States. In these tables, achieved parity was calculated as the mean number of live births the respondents had had up to the date of the survey, standardized by duration of marriage. Despite the overall low fertility of these countries, there were considerable socioeconomic differentials in achieved parity. In almost all countries (except Belgium and Spain) the higher a woman's educational level the lower was her achieved parity (Table 1). This educational gradient was particularly marked in eastern European countries where the number of children born to women in the lowest educational level was 45–76% greater than the number born to those in the highest educational group. A similar relationship

was observed with husband's educational level but the differentials were not so marked (Table 2), and in some countries (Belgium, Denmark, Finland, France, Great Britain and Spain) there was some evidence that women whose husbands had post-secondary education had a slightly higher parity than women whose husbands fell in the higher-secondary category.

Similar gradients were observed with other measures of socioeconomic status. Achieved parity declined as family income rose in Bulgaria, Poland and the USA (Table 3); however, in Belgium and, to a lesser extent, in former Czechoslovakia, Denmark, France and Norway there was some suggestion of a U-shaped relationship, with families with very high and very low incomes having more births than adjacent groups. The average number of live births decreased consistently as husband's socioeconomic status rose in all the countries for which data were available (Table 4). Women living in rural areas had a higher number of live births than women living in urban areas (Table 5); this urban–rural gradient was particularly marked in eastern European countries.

There was also considerable evidence of socioeconomic differentials in parity in developing countries. Tables 6–8 show data on socioeconomic differentials in the number of children ever born to all women (regardless of their marital status) aged 40 to 49 years at the time of the survey (except Table 7 where the analysis by husband's educational level had to be restricted to ever-married women). Education differentials in parity showed an overall pattern of decreasing fertility with increasing woman's education but the size of the differentials varied substantially across countries (Table 6). In most African countries for which data were available, women with no or a few years of education usually had the highest fertility. In most Latin American countries the educational differentials were large, with parity decreasing consistently with increasing educational level. Similar patterns were observed when data were examined by husband's level of education but the magnitude of the differentials tended to be smaller (Table 7).

Parity tended to decline with increasing urbanization in most of the developing countries where the WFS were carried out (Table 8); these differentials were most marked in Latin America and considerably smaller in Africa.

Age at first birth

The WFS conducted in Europe and in the USA did not publish data on socioeconomic differentials in age at first birth, but data on socioeconomic differences in the time from first marriage to first birth were published (Ford, 1984). Table 9 shows results by woman's level of education. The median number of months from first marriage to first birth increased with educational level in most countries except Belgium, former Czechoslovakia and former Yugoslavia (Table 9). These data seem to suggest that differentials in age at first birth might have been in the same direction unless women in higher educational groups married at an earlier age than those in lower educational groups. Data from other sources indicate, however, that this might not have been the case. For instance, data from the 1971 Census in England and Wales showed that the proportion of women who married at ages under 25 years was greater among those married to men in manual occupations than among those married to men in non-manual occupations (Office of Population Censuses and Surveys, 1983). A similar pattern was observed in Norway (Central Bureau of Statistics of Norway, 1981).

Tables 10 and 11 present data from the DHS on the percentage of all women (regardless of their marital status) aged 20 to 49 years at the time of the survey who had their first birth before reaching the exact age of 20 years. The percentage of women with an early first birth was substantially higher among women with no education than among those with a high educational level, although in several countries (for example, Bolivia, Jordan, Liberia, Uganda and Zambia) the proportion among those with primary education was the same as or higher than among those with no education (Table 10). The percentages of women having had an earlier first birth were also higher in rural than in urban areas in almost every country included in this analysis (Table 11).

Childlessness

Table 12 shows results from the few studies that have examined the relationship between socioeconomic status and childlessness at an individual level. In Norway, there was a clear positive association between educational level and percentage of women who were still childless by age 35 years for cohorts born between 1935 and 1950. The magnitude of these educational differentials is so large that they are unlikely to reflect just a higher tendency of well-educated women to postpone childbearing to ages above 35.

The pattern was different for women born in England and Wales and in the USA. Final childlessness for cohorts born in England and Wales around 1920–1930 was 11% lower among women married to manual workers (social classes III-M, IV and V) than among those married to non-manual workers (social classes III-NM, II and I). Within each broad category, however, childlessness was lowest among those married to men in the highest social class (Table 12). In the USA, the relationship between childlessness and income for women born around 1924–1935 was U-shaped, with women in the highest income category having the same level of childlessness as those in the lowest income category (Table 12). It should be noted, however, that these two surveys were restricted to married women.

The relationship between childlessness and socioeconomic status in developing countries has been assessed only in ecological studies. These studies have shown that during the early stages of modernization the level of childlessness, which is mainly involuntary, tends to decline as a result of better access to health services. A decline in childlessness with modernization was observed in various developing countries such as Zaire (Romaniuk, 1980) and Mexico (Poston et al., 1985). As modernization progresses, however, the level of childlessness increases due to a rise in voluntary fertility control. This evolution has been shown in Taiwan (Poston, 1988) and in Brazil (Poston & Rogers, 1988) – countries that have been actively engaged in modernization programmes for several decades. It is conceivable that similar socioeconomic gradients in childlessness might have been present at an individual level.

Ages at menarche and menopause

Data on socioeconomic differences in age at menarche were collected in the WFS carried out in a small number of developing countries (Table 13). These data showed no clear relationship between age at menarche and woman's type of place of residence or educational level. Unfortunately, data on age at menarche were not collected in any of the WFS carried out in developed countries.

Very few studies have investigated the determinants of age at menopause. A nationally representative survey of women aged 45 to 64 years carried out in Finland in 1989 provided some evidence of socioeconomic differences in age at natural menopause (Table 14). Women in the higher socioeconomic levels had a slightly later age at natural menopause than those in the lower socioeconomic groups, but most of the estimates were based on small numbers. The age-adjusted proportion of women having had surgical menopause was also higher in women of low socioeconomic level (Table 14). Similar socioeconomic differentials in age at natural menopause were observed in a study of women who participated in a nationwide breast cancer screening programme carried out in the USA in 1973–1980 (Table 14).

Breastfeeding and postpartum amenorrhoea

Data from the WFS showed marked socioeconomic differences in duration of breastfeeding and postpartum amenorrhoea in a large number of developing countries (Tables 15–17). The WFS questions on breastfeeding and postpartum amenorrhoea applied to the live births resulting from the two last pregnancies, and therefore they refer to behaviour around the time of the surveys (Singh & Ferry, 1984; United Nations, 1987). In all countries for which the comparison could be made, breastfeeding duration decreased consistently with increasing educational level (Table 15) but the gradient was particularly marked at the higher end of the educational scale: the largest difference occurred between women with from four to six years of schooling and those with seven years or more. The educational differentials were greatest in Latin America and the Caribbean where the duration of breastfeeding of women in the lowest educational group was at least twice that for women in the highest educational group. A similar pattern was observed with woman's place of residence (Table 16). Data on postpartum amenorrhoea were available for a few WFS countries (Table 17). Although the duration of postpartum amenorrhoea was considerably smaller than the duration of breastfeeding, the socioeconomic differentials were similar.

No similar data from the WFS were available for developed countries. Recent trends in the incidence and duration of breastfeeding in the USA seem to suggest, however, that they are in the opposite direction to that observed in developing countries (United States Department of Health and Human

Services, 1984). Among White women, a significantly higher percentage of breastfeeding was observed with increasing maternal education in 1969, as well as in 1980. In sharp contrast, among Black women with newborns in 1969, there was a significant decrease in breastfeeding as the educational level of the mother increased. However, the relationship in 1980 was similar to that observed among White women.

General comments

The data presented here shows large socioeconomic differentials in reproductive behaviour. In general, women of higher socioeconomic status had fewer children and had them later than women of lower status. The direction and magnitude of the differences varied according to the level of economic development of each particular country. In general, fertility differentials were larger in the relatively more developed of the developing countries. This widening of differentials with modernization should be viewed in the context of the recent declines in overall fertility; while there might be societies where, from the beginning, fertility decline proceeded more or less at the same pace for all socioeconomic groups, decline more typically began amongst the highest socioeconomic groups and spread later to the lowest socioeconomic groups. In a few countries, the widening of the socioeconomic differentials might have been accomplished in part through small fertility increases among those women in the lower socioeconomic groups.

Socioeconomic variations in reproductive behaviour were also present in developed countries, where fertility had already reached very low levels. There is, however, evidence that the magnitude and even direction of the differentials is not constant. Table 18 clearly illustrates this point. There was relatively little socioeconomic variation in achieved fertility for women who married in England and Wales around the years 1851–1861. For women married since then, however, achieved fertility tended to be lowest among women married to men in professional and managerial occupations, and highest among the wives of unskilled manual workers. This pattern is similar to that observed now in many developing countries. But for women in England and Wales married since the 1940s, fertility rates were no longer lowest in the highest social group. The distribution of family size has been since then U-shaped, with the lowest average completed family size being recorded in group III-NM, the non-professional white-collar workers. In the manual group, fertility rises from social class III-M to social class V, and women married to men in social class V are more fertile than those in other classes of the manual group. In the non-manual group, by contrast, wives of men in social class I (professional occupations) have the largest families, and family size declines from social class I to social class III-NM.

The recent tendency for a slightly higher fertility at the upper end of the socioeconomic scale has also been observed in other developed countries. In Norway, there has been a recent trend among highly educated mothers of two children to proceed to a third birth more often than their less educated counterparts (Kravdal, 1992). A similar trend in third-birth rates has been observed in the USA (Kravdal, 1992). If this trend persists it may more than compensate for the tendency of the most highly educated women to remain childless, so that the educational gradient in achieved fertility may become positive at the higher end of the educational scale.

It is not clear whether age at first birth has changed recently in the different socioeconomic groups but it is worth noting that the differentials on age at first birth observed here (Tables 9–11) were larger than those on parity (Tables 1–8).

The observed socioeconomic differentials in most aspects of reproductive behaviour could potentially account for some of the socioeconomic variation in the risk of female reproductive cancers. Any assessment of this notion would need to take into account the marked changes in the pattern of the socioeconomic differentials in reproductive behaviour for successive generations of women. Unfortunately, most of the published data on socioeconomic differentials in fertility and in cancer risks do not relate to any specific cohort(s) of women.

There are long delays in socioeconomic differences in fertility manifesting themselves as differences in cancer rates – that is, fertility patterns at around ages 20–30 years generally manifest themselves as risk factors for cancer in women some decades later. Thus, cancer rates now reflect reproductive practices long ago and reproductive practices now will not affect cancer rates for many decades to come, and as socioeconomic differences in fertility may be changing in the West there is no reason to believe that the present socioeconomic differences in reproductive cancers will continue forever.

References

Booth, M. (1991) Aetiology and epidemiology of ovarian cancer. In: Blackledge, G.R.P., Jordan, J.A. & Shingleton, H.M., eds, *Textbook of gynaecologic oncology.* London, WB Saunders. pp. 103–113

The Centers for Disease Control Cancer and Steroid Hormone Study (1983) Oral contraceptive use and the risk of ovarian cancer. *J. Am. Med. Assoc.,* 249, 1596–1599

Central Bureau of Statistics of Norway (1981) *Fertility Survey 1977.* Oslo, Central Bureau of Statistics

Elwood, J.M., Cole, P., Rothman, K.J. & Kaplan, S.D. (1977) Epidemiology of endometrial cancer. *J. Natl Cancer Inst.,* 59, 1055–1060

Ford, K. (1984) *Timing and spacing of births* (WFS Comparative Studies. ECE Analyses of WFS Surveys in Europe and USA, No. 38). Voorburg, Netherlands, International Statistical Institute

Hildreth, N.G., Kelsey, J.L., Livolsi, V.A., Fischer, D.B., Holford, T.R., Mostow, E.D., Schwartz, P.E. & White, C. (1981) An epidemiologic study of epithelial carcinoma of the ovary. *Am. J. Epidemiol.,* 114, 398–405

Hodgson, M. & Gibbs, J. (1980) *Children ever born* (WFS Comparative Studies. Cross-National Summaries, No. 12.) Voorburg, Netherlands, International Statistical Institute

Jones, E.F. (1982) *Socio-economic differentials in achieved fertility* (WFS Comparative Studies. ECE Analysis of WFS Surveys in Europe and USA, No. 21). Voorburg, Netherlands, International Statistical Institute

Kravdal, Ø. (1992) The emergence of a positive relation between education and third birth rates in Norway with supportive evidence from the United States. *Popul. Stud.,* 46, 459–475

Luoto, R., Kaprio J. & Uutela, A. (1994) Age at natural menopause and sociodemographic status in Finland. *Am. J. Epidemiol.,* 139, 64–76

MacMahon, B., Cole, P., Lin, T.M., Lowe, C.R., Mirra, A.P., Ravnihar, B., Salber, E.J., Valaoras, V.G. & Yuasa, S. (1970) Age at first birth and breast cancer risk. *Bull. WHO,* 43, 209–211

Office of Population Censuses and Surveys (1983) *Fertility report from the 1971 census* (The Registrar General's decennial supplement for England and Wales 1971. Series DS No. 5). London, Her Majesty's Stationery Office

Pike, M.C., Krailo, M.D., Henderson, B.E., Casagrande, J.T. & Hoel, D.G. (1983) 'Hormonal' risk factors, 'breast tissue age' and the age-incidence of breast cancer. *Nature,* 303, 767–770

Poston, D.L. (1974) Income and childlessness in the United States: is the relationship always inverse? *Soc. Biol.,* 21, 296–307

Poston, D.L. (1988) Childlessness patterns in Taiwan. *J. Popul. Stud.,* 11, 55–78

Poston, D.L. & Rogers, R.G. (1988) Development and childlessness in the states and territories of Brazil. *Social Biol.,* 35, 267–284

Poston, D.L., Briody, E., Trent, K. & Browning, H.L. (1985) Modernisation and childlessness in the states of Mexico. *Econ. Dev. Cult.,* 33, 503–519

Risch, H.A., Weiss, N.S., Lyon, L.J., Daling, J.R. & Liff, J.M. (1983) Events of reproductive life and the incidence of epithelial ovarian cancer. *Am. J. Epidemiol.,* 117, 128–139

Romaniuk, A. (1980) Increase in natural fertility during the early stages of modernisation: evidence from an African case study, Zaire. *Popul. Stud.,* 34, 293

Singh, S. & Ferry, B. (1984) *Biological and traditional factors that influence fertility: results from the WFS surveys.* (WFS Comparative Studies, No. 40). Voorburg, Netherlands, International Statistical Institute.

Stanford J.L., Hartge, P., Brinton, L.A., Hoover, R.N. & Brookmeyer, R. (1987) Factors influencing the age at natural menopause. *J. Chron. Dis.,* 40, 995–1002

Stevenson, T.H.C. (1920) The fertility of various social classes in England and Wales from the middle of the nineteenth century to 1911. *J. R. Stat. Soc.,* 83, 401–444

United States Department of Health and Human Services (1984) Racial and educational factors associated with breastfeeding – United States, 1969 and 1980. *Morb. Mortal. Weekly Rep.,* 33, 153–154

United Nations, Department of International Economic and Social Affairs (1987) *Fertility behaviour in the context of development. Evidence from the World Fertility Survey* (Population Studies No. 100). New York, United Nations

Westoff, C.F., Blanc, A.K. & Nyblade, L. (1994) *Marriage and entry into parenthood* (Demographic and Health Surveys. Comparative Studies No. 10). Calverton, Maryland, Macro International

Yuan, J-M., Yu, M.C., Ross, R.K., Gao, Y-T. & Hendeson, B.E. (1988) Risk factors for breast cancer in Chinese women in Shangai. *Cancer Res.,* 48, 1949–1953

Corresponding author:
I. dos Santos Silva
Epidemiological Monitoring Unit, Department of Epidemiology and Population Sciences, London School of Hygiene and Tropical Medicine, Keppel Street, London, UK WC1E 7HT

Table 1. Average number of live births (standardized by duration of marriage) by woman's level of education[a]

Country (sample size)	Woman's level of education					% Excess of 1 relative to 5[c]
	1 Elementary not completed	2 Elementary completed	3 Lower secondary	4 Higher secondary	5 Post-secondary[d]	
Belgium (4010)	1.77[a]		1.83	1.72	(1.90)	−7
Bulgaria (6352)	2.41	1.74	1.55	1.50	1.37	+76
Czechoslovakia (2932)	2.35[a]		2.08	1.80	1.62	+45
Denmark (3129)	2.20[a]		1.87	1.86	(1.85)	+19
Finland (5349)	2.01[a]		1.80	1.74	1.64	+23
France (2290)	2.51	2.03	1.86	1.79	(1.66)	+51
Great Britain (3682)	2.15[a]		1.90	1.73	1.72	+25
Italy (5359)	2.45	1.96	1.74	1.65	(1.48)	+66
Norway (2824)	2.40[a]		2.11	1.95	1.86	+29
Poland (9799)	2.70	2.32	1.95	1.71	1.55	+74
Romania (8771)	2.25[a]		1.68	1.52	1.39	+62
Spain (4618)	2.63	2.28	2.42	2.27	(2.41)	+9
USA (5545)	2.76[a]		2.34	2.07	1.82	+52
Yugoslavia (6806)	2.43	1.81	1.57		1.40	+74

[a]World Fertility Surveys conducted in Europe and in the USA during the years 1975–1979; modified from Jones, 1982.
[b]Pooled estimate, groups 1 + 2.
[c]Calculated as $[(1 - 5) / 5] \times 100$, or as $\{[(1 + 2) - 5] / 5\} \times 100$ if the two lowest categories were combined.
[d]Estimates were placed in parentheses if either the total number of women involved was less than 50 or the number of women in any one category of the underlying distribution by duration of marriage was less than 5.

Table 2. Average number of live births (standardized by duration of marriage) by husband's level of education[a]

Country (sample size)	Husband's level of education					% Excess of 1 relative to 5[c]
	1 Elementary not completed	2 Elementary completed	3 Lower secondary	4 Higher secondary	5 Post-secondary[b]	
Belgium (4010)	1.85[d]		1.72	1.73	1.91	−3
Bulgaria (6352)	2.39	1.73	1.58	1.57	1.41	+70
Czechoslovakia (2932)	NA		NA	NA	NA	–
Denmark (3129)	2.14[d]		1.88	1.83	1.85	+16
Finland (5349)	1.97[d]		1.84	1.72	1.78	+11
France (2290)	2.31	2.06	1.96	1.75	1.80	+28
Great Britain (3682)	2.00[d]		1.91	1.71	1.79	+12
Italy (5359)	2.53	2.02	1.83	1.71	(1.66)	+52
Norway (2824)	2.22[d]		2.19	2.00	1.93	+15
Poland (9799)	2.67	2.37	2.00	1.71	1.54	+73
Romania (8771)	1.75[d]		1.95	1.66	1.46	+20
Spain (4618)	2.62	2.30	2.33	2.33	2.58	+2
USA (5545)	2.66[d]		2.27	2.05	1.93	+38
Yugoslavia (6806)	2.52	2.02	1.77[e]		1.70	+48

NA; data not available.
[a]World Fertility Surveys conducted in Europe and in the USA during the years 1975–1979; modified from Jones, 1982.
[b]Estimates in parentheses are when total number of women involved was less than 50 or the number of women in any one category of the underlying distribution of marriage was less than 5.
[c]Calculated as [(1 − 5) / 5] × 100, or as {[(1 + 2) − 5] / 5} × 100 if the two lowest categories were combined.
[d]Pooled estimate, groups 1 + 2.
[e]Pooled estimate, groups 3 + 4.

Table 3. Average number of live births (standardized by duration of marriage) by family income at the time of the survey[a]

Country	Family income (quintiles of the distribution in each national sample)					% Excess of 1 relative to 5[c]
	1 Very low[b]	2 Low	3 Medium	4 High	5 Very high	
Belgium	2.01	1.95	1.80	1.55	1.79	+12
Bulgaria	1.86	1.72	1.62	1.51	1.49	+25
Czechoslovakia	2.10	1.82	1.84	1.95	1.87	+12
Denmark	2.05	2.12	1.96	1.68	1.76	+16
Finland	1.86	1.71	1.63	1.66	1.66	+12
France	2.59	2.10	1.85	1.51	1.57	+65
Great Britain	NA	NA	NA	NA	NA	–
Italy	NA	NA	NA	NA	NA	–
Norway	(2.29)	2.07	1.98	1.65	1.68	+36
Poland	2.19	1.94	1.81	1.80	1.60	+37
Romania	1.92	2.06[d]			1.57	+22
Spain	NA	NA	NA	NA	NA	–
USA	2.53	2.21	2.08	1.91	1.84	+38
Yugoslavia	NA	NA	NA	NA	NA	–

NA, data not available.
[a]World Fertility Surveys during the years 1975–1979; modified from Jones, 1982. The analysis was restricted to urban residents except for the USA, where all respondents were included. No information on the number of urban respondents in each country was given in Jones, 1982.
[b]Estimates in brackets when either the total number of women involved was less than 50 or the number of women in any one category of the underlying distribution of marriage was less than 5.
[c]Calculated as $[(1-5)/5] \times 100$.
[d]Pooled estimate, groups 2 + 3 + 4.

Table 4. Average number of live births (standardized by duration of marriage) by husband's socio-occupational status at the time of the survey[a]

Country (sample size)	Husband's socio-occupational status			% Excess of 1 relative to 5[b]
	1) Agricultural workers	2 Manual workers	3 Non-manual workers	
Belgium (4010)	2.23	1.80	1.75	+27
Bulgaria (6352)	1.68	1.74	1.48	+14
Czechoslovakia (2932)	NA	NA	NA	–
Denmark (3129)	NA	NA	NA	–
Finland (5349)	2.07	1.86	1.76	+18
France (2290)	2.04	2.20	1.83	+11
Great Britain (3682)	2.02	1.92	1.74	+16
Italy (5359)	2.33	1.91	1.94	+20
Norway (2824)	2.26	2.14	1.93	+17
Poland (9799)	2.51	2.10	1.64	+53
Romania (8771)	2.44	2.12	1.46	+67
Spain (4618)	2.58	2.41	2.35	+10
USA (5545)	2.62	2.19	1.96	+34
Yugoslavia (6806)	2.44	2.11	1.90	+28

NA, data not available.
[a]World Fertility Surveys conducted in Europe and in the USA during the years 1975–1979; modified from Jones, 1982. Women whose husbands were unemployed were excluded from the analysis. In some countries (Finland, Poland, Romania and Spain) there was a group of 'other workers' but it was too small and too heterogenous for meaningful comparisons.
[b]Calculated as $[(1 - 3) / 3] \times 100$.

Table 5. Average number of live births (standardized by duration of marriage) by woman's place of residence at the time of the survey[a]

Country (sample size)	Place of residence		% Excess of 1 relative to 2[b]
	1 Rural	2 Urban	
Belgium (4010)	1.85	1.78	+4
Bulgaria (6352)	1.97	1.64	+20
Czechoslovakia (2932)	2.16	1.89	+14
Denmark (3129)	2.16	1.94	+11
Finland (5349)	2.09	1.71	+22
France (2290)	2.19	1.93	+13
Great Britain (3682)	NA	NA	–
Italy (5359)	1.98	1.89	+5
Norway (2824)	2.18	1.91	+14
Poland (9799)	2.47	1.82	+36
Romania (8771)	2.25	1.74	+29
Spain (4618)	2.45	2.40	+2
USA (5545)	NA	NA	–
Yugoslavia (6806)	2.45	1.96	+25

NA, data not available.
[a]World Fertility Surveys conducted in Europe and in the USA during the years 1975–1979; modified from Jones, 1982.
[b]Calculated as $[(1 - 2) / 2] \times 100$.

Table 6. Mean number of children ever born to all women aged 40–49 at the time of the surveys by woman's educational level[a]

Country (sample size)[b]	Year of survey	Approximate year of birth	1) Zero	2) 1–3	3) 4–6	4) 7+	% Excess of 1 relative to 4[c]
Africa							
Benin (4018)	1982	1932–43	6.2	NA	(5.0)	(4.8)	+29
Cameroon (8219)	1978	1928–39	5.2	5.1	4.9	(3.6)	+44
Côte d'Ivoire (5764)	1980	1930–41	6.8	(6.8)			–
Egypt (8788)	1980	1930–41	6.8	7.2	6.5	3.7	+84
Ghana (6125)	1979–80	1929–41	6.4	6.4	7.0	5.5	+16
Kenya (8100)	1977–78	1927–39	7.6	8.4	7.8	7.8	–3
Mauritania (3504)	1981	1931–42	5.9	6.0	NA	NA	–
Morocco (5801)	1979–80	1929–41	7.1	(7.3)	(6.3)	NA	–
Senegal (3985)	1978	1928–39	6.9	(6.9)			–
Sudan (3115)	1978–79	1938–40	6.1	(6.9)	(5.8)	(3.9)	+56
Asia and Oceania							
Bangladesh (6513)	1975–76	1925–37	6.9	7.0	7.6	(6.9)	0
Fiji (4298)	1974	1924–35	6.9	7.1	6.1	5.6	+23
Indonesia (9155)	1976	1926–37	5.2	6.1	5.6	4.5	+16
Jordan (3612)	1976	1926–37	8.9	9.0	7.2	6.2	+44
Korea, Rep. of (5430)	1974	1924–35	6.0	5.7	5.2	4.0	+50
Malaysia (6316)	1974	1924–35	6.3	6.2	5.9	3.7	+70
Nepal (5940)	1976	1926–37	5.7	3.9			–
Pakistan (4996)	1975	1925–36	6.9	(5.8)	6.5	(5.1)	+35
Philippines (9268)	1978	1928–39	7.0	7.4	6.9	5.2	+35
Sri Lanka (6812)	1975	1925–36	6.4	6.0	5.8	4.4	+45
Syrian Arab Rep. (4487)	1978	1928–39	7.8	6.3	6.6	4.0	+95
Thailand (3778)	1975	1925–36	6.4	6.6	6.5	4.0	+60
Latin America and the Caribbean							
Colombia (5378)	1976	1926–37	7.0	6.8	5.9	4.5	+56
Costa Rica (3935)	1976	1926–37	8.7	7.2	6.0	3.6	+42
Dominican Rep. (3115)	1975	1925–36	7.2	6.9	5.8	4.3	+67
Ecuador (6797)	1979	1929–40	7.9	7.4	6.2	3.8	+108
Guyana (4642)	1977	1927–38	7.5	6.5	6.9	5.8	+29
Haiti (3365)	1977	1927–38	5.8	6.0	5.6	(3.5)	+66
Jamaica (3096)	1975–76	1925–37	(4.4)	5.5	6.1	5.3	–17
Mexico (7310)	1976	1926–37	7.4	7.1	6.4	3.8	+95
Panama (3701)	1975–76	1925–37	7.1	7.0	5.9	4.0	+78
Paraguay (4682)	1979	1929–40	7.4	7.2	5.4	3.1	+139
Peru (5640)	1977–78	1927–38	7.4	6.6	5.8	3.9	+90
Trinidad & Tobago (4359)	1977	1927–38	7.1	6.9	6.1	5.0	+42
Venezuela (4361)	1977	1927–38	7.9	6.9	5.3	4.0	+98

[a]World Fertility Surveys conducted in selected developing countries during the years 1974–1982; modified from United Nations, 1987.
[b]These numbers represent the total sample size of each national survey; equivalent figures for women aged 40–49 years only were not given in the above publication. Values shown in parentheses are based on 10–24 cases. Categories containing fewer than 10 cases are not shown (NA) or (if this would mean suppressing more than two categories) are combined with adjacent categories.
[c]Calculated as $[(1 - 4) / 4] \times 100$.

Table 7. Mean number of children ever born to ever married women aged 40–49 at the time of the surveys by husbands's educational level[a]

Country[b]	Year of survey	Approximate year of birth of women	Years of schooling completed by the husband				% Excess of 1 relative to 4[c]
			1 Zero	2 1–3	3 4–6	4 7+	
Africa							
Benin (4018)	1982	1932–43	6.2	(7.4)	5.6	(6.2)	0
Cameroon (8219)	1978	1928–39	5.1	6.1	5.5	5.1	0
Côte d'Ivoire (5764)	1980	1930–41	6.8	7.0	6.2	6.8	0
Egypt (8788)	1980	1930–41	6.9	7.2	6.9	4.6	+50
Ghana (6125)	1979–80	1929–41	6.4	(6.4)	6.1	6.4	0
Kenya (8100)	1977–78	1927–39	7.4	8.2	8.1	7.8	–5
Mauritania (3504)	1981	1931–42	5.6	6.3	8.9	NA	–
Morocco (5801)	1979–80	1929–41	7.2	NA	6.3	(6.8)	+6
Senegal (3985)	1978	1928–39	7.0	NA	(7.7)	6.3	+11
Sudan (3115)	1978–79	1938–40	5.9	6.5	7.2	6.7	–12
Asia and Oceania							
Bangladesh (6513)	1975–76	1925–37	6.8	7.2	7.1	7.6	–11
Fiji (4298)	1974	1924–35	6.9	6.7	6.3	6.1	+13
Indonesia (9155)	1976	1926–37	5.0	5.6	5.7	5.5	–9
Jordan (3612)	1976	1926–37	8.8	9.3	9.0	7.5	+17
Korea, Rep. of (5430)	1974	1924–35	6.3	6.5	5.5	4.7	+34
Malaysia (6316)	1974	1924–35	6.1	6.8	6.3	5.0	+22
Nepal (5490)	1976	1926–37	5.7	(5.9)	5.2	5.2	+10
Pakistan (4996)	1975	1925–36	7.0	6.9	6.8	6.6	+6
Philippines (9268)	1978	1928–39	7.4	7.7	7.1	6.1	+21
Sri Lanka (6812)	1975	1925–36	6.5	6.0	6.0	5.2	+25
Syrian Arab Rep. (4487)	1978	1928–39	7.9	7.5	7.7	6.1	+30
Thailand (3778)	1975	1925–36	6.6	6.6	6.6	5.2	+27
Latin America and the Caribbean							
Colombia (5378)	1976	1926–37	7.6	7.4	6.5	5.3	+43
Costa Rica (3935)	1976	1926–37	8.9	8.1	6.5	4.3	+107
Dominican Rep. (3115)	1975	1925–36	7.8	7.4	6.5	4.5	+73
Ecuador (6797)	1979	1929–40	8.4	7.7	7.2	4.5	+87
Guyana (4642)	1977	1927–38	6.2	7.8	7.3	6.1	+2
Haiti (3365)	1977	1927–38	6.1	6.4	5.5	4.6	+33
Jamaica (3096)	1975–76	1925–37	7.2	7.4	7.0	5.2	+38
Mexico (7310)	1976	1926–37	7.8	7.6	6.6	5.0	+56
Panama (3701)	1975–76	1925–37	7.7	7.2	5.9	4.4	+75
Paraguay (4682)	1979	1929–40	8.2	7.4	6.0	3.9	+110
Peru (5640)	1977–78	1927–38	7.6	7.6	6.8	4.9	+55
Trinidad & Tobago (4359)	1977	1927–38	7.2	8.4	6.6	5.2	+38
Venezuela (4361)	1977	1927–38	8.2	6.6	5.9	4.8	+71

[a]World Fertility Surveys conducted in selected developing countries during the years 1974–1982; modified from United Nations, 1987.
[b]These numbers represent the total sample size of each national survey; equivalent figures for women aged 40–49 years only were not given in the above publication. Values shown in parentheses are based on 10–24 cases. Categories containing fewer than 10 cases are not shown (NA) or (if this would mean suppressing more than two categories) are combined with adjacent categories.
[c]Calculated as $[(1 - 4) / 4] \times 100$.

Table 8. Mean number of children ever born to all women aged 40–49 at the time of the surveys by woman's place of residence[a]

Country (sample size)[b]	Year of survey	Approximate year of birth	Place of residence			% Excess of 1 relative to 4[c]
			1 Rural	2 Other urban	3 Major urban	
Africa						
Benin (4018)	1982	1932–43	6.2	6.0	5.9	+5
Cameroon (8219)	1978	1928–39	5.3	4.7	4.9	+8
Côte d'Ivoire (5764)	1980	1930–41	6.9	6.4	6.7	+3
Egypt (8788)	1980	1930–41	6.9	6.7	5.7	+21
Ghana (6125)	1979–80	1929–41	6.6	6.0	5.8	+14
Kenya (8100)	1977–78	1927–39	7.8	5.8	7.1	+10
Mauritania (3504)	1981	1931–42	5.9	6.0	5.8	+2
Morocco (5801)	1979–80	1929–41	7.7	6.3	6.2	+24
Senegal (3985)	1978	1928–39	7.0	6.6	6.9	+1
Sudan (3115)	1978–79	1938–40	6.0	6.3	6.1	–2
Asia and Oceania						
Bangladesh (6513)	1975–76	1925–37	7.0	6.8	6.8	+3
Fiji (4298)	1974	1924–35	6.5	6.0	5.5	+18
Indonesia (9155)	1976	1926–37	5.2	5.1	5.6	–7
Jordan (3612)	1976	1926–37	8.8	8.8	8.0	+10
Korea, Rep. of (5430)	1974	1924–35	6.1	5.1	4.5	+36
Malaysia (6316)	1974	1924–35	6.3	6.0	5.4	+17
Nepal (5940)	1976	1926–37	5.7	4.4	NA	–
Pakistan (4996)	1975	1925–36	6.9	7.0	6.5	+6
Philippines (9268)	1978	1928–39	7.0	5.7	5.1	+37
Sri Lanka (6812)	1975	1925–36	5.7	4.9	5.0	+14
Syrian Arab Rep. (4487)	1978	1928–39	7.7	7.7	6.5	+18
Thailand (3778)	1975	1925–36	6.4	5.0	4.8	+33
Latin America and the Caribbean						
Colombia (5378)	1976	1926–37	7.2	6.4	4.9	+47
Costa Rica (3935)	1976	1926–37	8.0	5.5	4.7	+70
Dominican Rep. (3115)	1975	1925–36	7.9	5.4	5.0	+58
Ecuador (6797)	1979	1929–40	7.4	6.4	4.9	+51
Guyana (4642)	1977	1927–38	7.0	6.3	4.9	+43
Haiti (3365)	1977	1927–38	6.1	(4.5)	5.0	+22
Jamaica (3096)	1975–76	1925–37	6.1	5.4	4.1	+49
Mexico (7310)	1976	1926–37	7.6	6.2	5.8	+31
Panama (3701)	1975–76	1925–37	6.9	6.1	4.6	+50
Paraguay (4682)	1979	1929–40	7.2	5.4	4.1	+76
Peru (5640)	1977–78	1927–38	7.1	6.6	5.0	+42
Trinidad & Tobago (4359)	1977	1927–38	6.1	5.4	4.9	+24
Venezuela (4361)	1977	1927–38	7.8	6.2	5.0	+56

[a]World Fertility Surveys conducted in selected developing countries during the years 1974–1982; modified from United Nations, 1987.
[b]These numbers represent the total sample size of each national survey; no corresponding figures for women aged 40–49 years were given in the above publication. Values shown in parentheses are based on 10–24 cases. Categories containing fewer than 10 cases are not shown (NA) or (if this would mean suppressing more than two categories) are combined with adjacent categories.
[c]Calculated as $[(1 - 3) / 3] \times 100$.

Table 9. Median number of months from first marriage to first birth by woman's level of education[a]

Country (sample size)	Level of education[b]					% Excess of 5 relative to 1[c]
	1 Elementary not completed	2 Elementary completed	3 Lower secondary	4 Higher secondary	5 Post-secondary	
Belgium (2375)	–	15.9	16.5	16.9	–	+6
Czechoslovakia (1570)	–	16.3	14.0	16.0	–	–2
Finland (3009)	–	10.3	12.1	17.6	25.2	+145
France (2121)	14.8	14.0	16.6	17.0	24.0	+62
Great Britain (1933)	–	–	21.7	31.2	38.9	+79
Hungary (2413)	–	(10.2)[d]	13.6	17.4	–	+71
Italy (3101)	12.6	13.7	15.2	15.1	–	+20
Netherlands (4184)	–	15.6	20.7	26.0	–	+67
Norway (1702)	–	–	7.4	11.3	23.6	+219
Poland (5857)	–	10.5	10.6	11.8	15.4	+47
Spain (2849)	11.8	12.1	12.4	15.1	–	+28
USA (1730)	–	–	14.2	20.9	37.8	+166
Yugoslavia (1732)	15.1	12.3	–	11.0	(15.7)	–4

[a]World Fertility Surveys during the years 1975–1979; modified from Ford, 1984.
[b]Not all countries have data for all the five educational levels because the data supplied by some countries had fewer categories.
[c]Calculated as $[(5 - 1) / 1] \times 100$.
[d]Values shown in brackets are based on 50–100 cases. No estimates are based on less than 50 cases.

Table 10. Percentage of all women aged 20–49 at the time of the surveys who had their first birth before reaching exact age 20 by woman's level of education[a]

Country (sample size)	Year of survey	Woman's level of education			% Excess of 1 relative to 3[b]
		1 No education	2 Primary	3 Higher	
Africa					
Botswana (3430)	1988	60.3	61.3	39.3	+53
Burundi (3239)	1987	35.7	31.8	15.0	+138
Cameroon (2952)	1991	69.5	70.2	46.8	+49
Egypt (9906)[c]	1988–89	56.1	41.9	5.2	+979
Ghana (3639)	1988	59.9	54.6	19.3	+210
Kenya (5645)	1988–89	69.0	70.9	40.7	+70
Liberia (4102)	1986	54.1	72.8	63.1	−14
Madagascar (4839)	1992	64.5	64.6	32.8	+97
Mali (2677)	1987	61.7	62.5	17.6	+251
Morocco (7111)	1992	36.3	22.5	5.8	+526
Namibia (4162)	1992	49.2	46.6	27.2	+81
Niger (5124)	1992	71.7	68.2	33.8	+112
Nigeria (7170)	1990	59.2	53.4	26.6	+123
Rwanda (5087)	1992	35.8	26.8	12.9	+178
Senegal (3440)	1986	65.2	46.1	30.1	+117
Sudan (7308)[c]	1989–90	54.3	42.6	11.6	+368
Tanzania (7053)	1991–92	67.0	59.5	20.9	+221
Togo (2636)	1988	61.5	58.0	34.5	+78
Tunisia (5644)[c]	1988	26.6	16.2	6.8	+291
Uganda (3573)	1988–89	70.2	72.0	45.4	+55
Zambia (5076)	1992	67.8	75.1	47.1	+44
Zimbabwe (3180)	1988–89	62.8	61.7	32.1	+96
Asia and Oceania					
Indonesia (24 620)[c]	1991	55.6	55.4	19.4	+187
Jordan (8089)[c]	1990	43.9	51.6	24.4	+80
Pakistan (7216)[c]	1990–91	40.8	36.9	20.9	+95
Sri Lanka (7650)[c]	1987	NA	NA	NA	–
Thailand (8165)[c]	1987	41.4	30.6	6.3	+557
Latin America and the Caribbean					
Bolivia (6242)	1989	44.1	45.9	28.8	+53
Brazil (4579)	1986	43.7	34.9	7.6	+475
Colombia (6706)	1990	55.7	43.9	20.5	+172
Dominican Rep. (5610)	1991	66.2	57.6	17.1	+287
Ecuador (3672)	1987	52.7	48.2	25.5	+107
Guatemala (3978)	1987	60.4	52.2	17.7	+241
Mexico (7096)	1987	63.0	46.7	19.9	+217
Paraguay (4563)	1990	54.4	43.3	17.7	+207
Peru (12 405)	1991–92	54.8	53.6	20.4	+167
Trinidad & Tobago (3122)	1987	53.3	45.2	22.0	+142

NA, date not available.
[a]Demographic and Health Surveys conducted in selected developing countries during the years 1986–1992; data from Westoff et al., 1994.
[b]Calculated as [(1 − 3) / 3] × 100.
[c]Only ever-married women were included in the survey.

Table 11. Percentage of all women aged 20–49 at the time of the surveys who had their first birth before reaching exact age 20 by woman's place of residence[a]

Country (sample size)	Year of survey	Place of residence		% Excess of 1 relative to 2[b]
		1 Rural	2 Urban	
Africa				
Botswana (3430)	1988	56.8	54.2	+5
Burundi (3239)	1987	34.0	47.4	−28
Cameroon (2952)	1991	67.9	60.1	+13
Egypt (9916)[c]	1988–89	52.5	29.7	+77
Ghana (3639)	1988	58.1	46.5	+25
Kenya (5653)	1988–89	66.9	51.5	+30
Liberia (4101)	1986	55.9	60.9	−8
Madagascar (4840)	1992	60.6	39.4	+54
Mali (2677)	1987	61.8	59.6	+4
Morocco (7111)	1992	34.8	23.8	+46
Namibia (4162)	1992	38.2	41.7	−8
Niger (5124)	1992	71.8	64.5	+11
Nigeria (7170)	1990	55.7	43.7	+27
Rwanda (5087)	1992	30.0	23.6	+27
Senegal (3440)	1986	67.3	49.3	+37
Sudan (7333)[c]	1989–90	45.1	34.9	+29
Tanzania (7055)	1991–92	61.2	59.3	+3
Togo (2636)	1988	63.5	45.5	+40
Tunisia (5668)[c]	1988	24.6	17.0	+45
Uganda (3573)	1988–89	69.9	59.2	+18
Zambia (5076)	1992	70.1	64.1	+9
Zimbabwe (3180)	1988–89	56.9	50.3	+13
Asia and Oceania				
Indonesia (24 677)[c]	1991	51.7	35.5	+46
Jordan (8121)[c]	1990	34.6	33.1	+5
Pakistan (7231)[c]	1990–91	38.6	34.6	+12
Sri Lanka (7650)[c]	1987	22.5	15.0	+50
Thailand (8169)[c]	1987	30.3	16.3	+86
Latin America and the Caribbean				
Bolivia (6241)	1989	44.0	36.0	+22
Brazil (4587)	1986	34.2	27.0	+27
Colombia (6709)	1990	38.9	27.9	+39
Dominican Republic (5609)	1991	53.4	35.3	+51
Ecuador (3672)	1987	46.2	34.4	+34
Guatemala (3978)	1987	57.9	40.4	+43
Mexico (7098)	1987	53.4	34.9	+53
Paraguay (4564)	1990	44.8	27.5	+63
Peru (12 406)	1991–92	51.2	27.7	+85
Trinidad & Tobago (3902)	1987	50.0	32.3	+55

[a]Demographic and Health Surveys conducted in selected developing countries during the years 1986–1992; data from Westoff et al. (1994).
[b]Calculated as $[(1 - 3) / 3] \times 100$.
[c]Ony ever–married women were included in the survey.

Table 12. Percentage of childless women in certain developed countries by socioeconomic status

Norway: percentage of childless women (born 1935–1950) at age 35 years[a]

Birth cohort	Woman's level of education (years of schooling)					% Excess of 1 relative to 5[b]
	1 7–9	2 10	3 11–12	4 13–14	5 15+	
1935	8.8	9.9	13.0	17.6	18.0	−51
1940	8.2	9.1	11.8	14.6	16.5	−50
1945	7.5	8.7	11.5	13.2	16.8	−55
1950	7.4	8.8	12.3	15.3	19.0	−61

England and Wales, 1971: percentage of childless women (born around 1920–1930) at the end of their reproductive life[c]

	Husband's social class						% Excess of 1 relative to 6[d]
	1 V	2 IV	3 III-M	4 III-NM	5 II	6 I	
Number of women	6811	20 475	47 164	12 992	26 796	5780	
% Childless	11.3	11.6	10.3	14.0	11.5	10.3	+10

United States of America, 1970: percentage of childless women (born around 1924–1935) at ages 35–44 years[e]

	Family income (US$) in 1969					% Excess of 1 relative to 5[b]
	1 <6000	2 6000–9999	3 10 000–14 999	4 15 000–19 999	5 ≥20 000	
Number of women	35 977	156 338	371 461	235 701	174 025	
% Childless	18.4	14.7	15.3	17.9	18.4	0

[a]Data from Kravdal, 1992. Study population included all women born in the country regardless of their marital status. The numbers of women on which these analyses were based are not given in the original publication.
[b]Calculated as [(1 − 5 /5] × 100.
[c]Data from Office of Population Censuses and Surveys, 1983. Study population included all women married in the years 1946–1950 who had been married once only at ages under 45.
[d]Calculated as [(1 − 6) / 6] × 100.
[e]Data from Poston, 1974. Study population included all white women aged 35–44 who were married after age 22, in the labour force, married once only, and were wives of civilian household heads employed in non-farming occupations.

Table 13. Mean age at menarche (in years) for ever-married women[a] by woman's place of residence and educational level[b]

Country (sample size)	Place of residence			% Excess of 1 relative to 3[c]	Years of schooling				% Excess of 4 relative to 7[d]
	1 Rural	2 Other urban	3 Major urban		4 0	5 1–3	6 4–6	7 7+	
Africa									
Benin (3330)	14.7	14.6	15.0	−2	(14.7)[e]	14.8	14.9	(14.8)	−1
	14.8	*14.7*	*15.3*	*−3*	*14.8*	*15.2*	*15.2*	*14.9*	*−1*
Cameroon (7256)	14.2	14.2	14.1	+1	14.2	14.0	14.3	14.1	+1
	14.5	*14.5*	*14.2*	*+2*	*14.3*	*14.5*	*14.9*	*14.4*	*−1*
Côte d'Ivoire (4984)	14.2	14.1	14.2	0	14.2	13.9	13.8	13.9	+2
	14.3	*14.2*	*14.3*	*0*	*14.3*	*14.1*	*14.1*	*14.0*	*+2*
Egypt (8782)	13.6	13.3	13.0	+5	13.5	13.3	13.2	13.1	+3
Ghana (4462)	14.9	15.0	14.9	0	14.9	14.8	14.9	15.1	−1
	15.2	*15.2*	*15.1*	*+1*	*15.0*	*14.9*	*15.2*	*15.4*	*−2*
Kenya (4641)	14.4	14.3	14.3	+1	14.2	14.5	14.4	14.7	−3
	14.9	*14.4*	*14.5*	*+3*	*14.4*	*14.9*	*15.3*	*14.8*	*−3*
Mauritania (3385)	13.9	13.7	13.4	+4	13.7	13.7	13.7	13.8	−1
Sudan (2884)	13.3	13.1	13.0	+2	13.2	13.1	13.2	13.3	−1
Tunisia (2727)	13.8	13.6	13.2	+5	13.6	13.5	13.6	13.4	+1
Asia and Pacific									
Nepal (5047)	15.3	15.1	—[f]	—	15.4	(15.1)	14.8	(14.5)	+6
Philippines (9266)	14.0	13.8	13.6	+3	14.0	14.0	14.1	13.7	+2
Syrian Arab Republic (4362)	13.5	13.4	13.3	+2	13.4	13.4	13.3	13.5	−1
Americas									
Colombia (3280)	13.7	13.5	13.5	+1	13.5	13.6	13.6	13.3	+2
	13.7	*13.4*	*13.4*	*+2*	*13.6*	*13.7*	*13.6*	*13.3*	*+2*
Haiti (2263)	14.9	14.2	14.3	+4	14.8	14.8	14.2	13.8	+7
	14.9	*15.6*	*15.2*	*−2*	*15.4*	*15.5*	*14.5*	*14.1*	*+9*

[a] Values for all women regardless of their marital status are presented in italics for the countries for which these data were available.
[b] World Fertility Surveys conducted in selected developing countries during the years 1974–1982; data from Singh and Ferry, 1984.
[c] Calculated as $[(1 - 3) / 3] \times 100$.
[d] Calculated as $[(4 - 7) / 7] \times 100$.
[e] Estimates were placed in brackets if based on 50–100 cases.
[f] There were no 'Major urban' areas in Nepal.

Table 14. Age at natural menopause and proportion of women with a surgical menopause in two developed countries by woman's occupation and educational level

Finland, 1989[a]

	Sample size	Mean age at natural menopause (median age)	Age-adjusted % of women with surgical menopause
Occupation			
Upper white-collar workers	153	53.5 (52)	16.2
Lower white-collar workers	612	51.9 (52)	18.9
Farmers	182	51.8 (51)	17.0
Blue-collar service workers	234	50.5 (51)	23.4
Blue-collar factory workers	277	51.0 (51)	21.0
Housewives	47	51.0 (51)[b]	11.8
Education (years of schooling)			
≥11	158	52.7 (52)	11.2
9–10	281	51.8 (52)	22.4
≤8	1066	51.4 (51)[c]	19.6

USA, 1973–1980[d]

		Median age at natural menopause	
Income (US$)	Sample size	Years	Months
≥30 000	285	51	10
20 000–29 999	250	51	2
10 000–19 999	350	50	11
≤9999	344	50	7
Education (years of schooling)			
≥13	616	51	5
12	534	51	4
≤11	272	50	3

[a]Data from a nationally representative survey of women aged 45–64 years (Luoto et al., 1994).
[b]Mantel-Cox test of trend: $P = 0.02$.
[c]Mantel-Cox test of trend: $P = 0.03$.
[d]Data from a study of women who participated in a nationwide breast screening programme during the years 1973–1980 (Stanford et al., 1987).

Table 15. Mean duration of breastfeeding (months) by woman's educational level[a]

Country (sample size)	Year of survey	Years of schooling				% Excess of 1 relative to 4[b]
		1 Zero	2 1–3	3 4–6	4 7+	
Africa						
Benin (4018)	1982	21.6	NA	17.8	NA	–
Cameroon (8219)	1978	21.0	19.1	17.5	14.6	+44
Côte d'Ivoire (5764)	1980	20.4	18.0	14.6	11.5	+77
Egypt (8788)	1980	21.2	19.5	16.3	10.2	+108
Ghana (6125)	1979–80	21.3	NA	19.2	15.7	+36
Kenya (8100)	1977–78	19.6	17.4	15.2	12.5	+57
Mauritania (3504)	1981	17.7	17.2	15.6	15.7	+13
Morocco (5801)	1979–80	16.7	NA	9.3	NA	–
Senegal (3985)	1978	21.1	NA	16.8	NA	–
Sudan (northern only)	1978–79	17.3	16.1	NA	NA	–
Asia and Oceania						
Bangladesh (6513)	1975–76	34.4	30.4	NA	NA	–
Fiji (4298)	1974	13.0	11.1	11.8	8.7	+49
Indonesia (9155)	1976	28.4	27.0	24.7	13.7	+107
Jordan (3612)	1976	13.9	13.0	10.5	7.7	+81
Korea, Republic of (5430)	1974	21.0	17.7	18.0	13.7	+53
Malaysia (6316)	1974	7.6	5.7	5.7	3.8	+100
Nepal (5940)	1976	29.3	NA	NA	NA	–
Pakistan (4996)	1975	22.0	NA	19.8	NA	–
Philippines (9268)	1978	18.9	17.6	14.8	9.5	+99
Sri Lanka (6812)	1975	26.1	24.7	23.4	18.5	+41
Syrian Arab Republic (4487)	1978	12.9	NA	10.7	9.5	+36
Thailand (3778)	1975	20.9	NA	20.8	NA	NA
Latin America and the Caribbean						
Colombia (5378)	1976	11.9	11.4	8.3	5.3	+125
Costa Rica (3935)	1976	NA	8.1	4.6	3.2	–
Dominican Republic (3115)	1975	12.2	10.5	8.6	5.2	+135
Ecuador (6797)	1979	17.0	14.5	13.0	8.9	+91
Guyana (4642)	1977	NA	9.2	7.7	6.6	–
Haiti (3365)	1977	19.0	14.1	NA	NA	–
Jamaica (3096)	1975–76	NA	NA	8.9	6.2	–
Mexico (7310)	1976	12.9	10.9	8.3	3.8	+239
Panama (3701)	1975–76	NA	13.0	9.2	2.4	–
Paraguay (4682)	1979	15.7	14.6	11.4	6.1	+157
Peru (5640)	1977–78	19.3	16.6	12.0	7.0	+176
Trinidad and Tobago (4359)	1977	NA	NA	10.0	7.1	–
Venezuela (4361)	1977	11.6	10.0	6.7	3.5	+231

NA, data not available because of small sample sizes.
[a]World Fertility Surveys conducted in selected developing countries in 1974–1982; modified from United Nations, 1987.
[b]Calculated as $[(1 - 4) / 4] \times 100$.

Table 16. Mean duration of breastfeeding (months) by woman's place of residence at the time of the survey[a]

Country (sample size)	Year of survey	Place of residence			% Excess of 1 relative to 3[b]
		1 Rural	2 Other urban	3 Major urban	
Africa					
Benin (4018)	1982	22.1	20.3	16.5	+34
Cameroon (8219)	1978	20.4	16.3	15.3	+33
Côte d'Ivoire (5764)	1980	20.8	16.8	16.7	+25
Egypt (8788)	1980	21.4	16.7	13.9	+54
Ghana (6125)	1979–80	20.7	17.2	14.7	+41
Kenya (8100)	1977–78	17.3	12.4	12.9	+34
Mauritania (3504)	1981	17.5	16.2	17.8	–2
Morocco (5801)	1979–80	17.6	11.6	12.7	+39
Senegal (3985)	1978	22.0	18.5	17.5	+26
Sudan (3115)	1978–79	17.4	15.5	16.7	+4
Asia and Oceania					
Bangladesh (6513)	1975–76	33.7	29.6	NA	–
Fiji (4298)	1974	12.1	7.1	6.6	+83
Indonesia (9155)	1976	28.2	18.2	15.8	+78
Jordan (3612)	1976	13.1	11.5	10.3	+27
Korea, Republic of (5430)	1974	20.0	15.8	14.0	+43
Malaysia (6316)	1974	6.9	4.6	2.1	+229
Nepal (5940)	1976	29.2	NA	NA	–
Pakistan (4996)	1975	22.8	18.8	16.8	+36
Philippines (9268)	1978	15.2	10.9	7.4	+105
Sri Lanka (6812)	1975	23.4	18.5	NA	–
Syrian Arab Republic (4487)	1978	12.8	11.6	9.5	+35
Thailand (3778)	1975	21.7	NA	7.5	+189
Latin America and the Caribbean					
Colombia (5378)	1976	11.7	8.7	5.8	+102
Costa Rica (3935)	1976	6.4	4.3	3.2	+100
Dominican Republic (3115)	1975	11.8	7.9	5.0	+136
Ecuador (6797)	1979	15.3	10.8	8.9	+72
Guyana (4642)	1977	8.3	8.0	5.2	+60
Haiti (3365)	1977	19.2	NA	11.0	+75
Jamaica (3096)	1975–76	8.9	10.2	6.2	+44
Mexico (7310)	1976	12.3	8.0	6.0	+105
Panama (3701)	1975–76	10.8	5.0	4.0	+170
Paraguay (4682)	1979	13.6	12.6	5.7	+139
Peru (5640)	1977–78	18.9	13.0	8.3	+128
Trinidad and Tobago (4359)	1977	9.4	8.6	6.5	+45
Venezuela (4361)	1977	11.5	7.0	4.1	+180

NA, data not available because of small sample size.
[a]World Fertility Surveys conducted in selected developing countries in 1974–1982; data from United Nations, 1987.
[b]Calculated as $[(1 - 3) / 3] \times 100$.

Table 17. Mean duration of postpartum amenorrhoea (in months) by woman's place of residence and educational level[a]

Country (sample size)	Place of residence				Years of schooling				
	1 Rural urban	2 Other urban	3 Major	% Excess of 1 relative to 3[b]	4 0	5 1–3	6 4–6	7 7+	% Excess of 4 relative to 7[c]
Africa									
Benin (2803)	12.9	10.5	7.9	+63	12.6	(8.7)[d]	8.2	(6.0)	+110
Cameroon (4650)	12.5	9.7	7.8	+60	13.1	11.6	10.4	7.2	+82
Côte d'Ivoire (3804)	11.5	8.4	8.9	+29	11.2	11.4	7.4	5.3	+111
Egypt (5667)	10.6	7.2	6.1	+74	10.4	9.4	7.4	3.9	+167
Ghana (3335)	13.3	11.9	9.9	+34	13.5	12.8	11.4	11.0	+23
Kenya (5679)	10.3	7.3	6.7	+54	11.8	11.0	8.8	6.9	+71
Mauritania (2447)	9.8	7.8	7.4	+32	10.1	8.8	6.4	4.1	+146
Sudan (2242)	11.7	8.4	7.6	+54	11.6	9.0	8.0	(5.1)	+127
Tunisia (3021)	8.4	4.7	5.1	+65	8.1	5.3	3.5	2.4	+238
Asia and Pacific									
Bangladesh (3836)	15.0	11.2	10.5	+43	15.7	15.1	10.7	8.0	+96
Fiji (2660)	5.1	3.7	3.5	+46	5.0	4.6	5.3	4.1	+22
Philippines (6627)	8.9	6.4	4.6	+93	10.4	9.4	8.8	5.9	+76
Syrian Arab Republic (4025)	7.5	5.9	4.2	+79	7.6	4.8	4.6	4.0	+90

[a]World Fertility Surveys conducted in selected developing countries during the years 1974–1982; data from Singh & Ferry, 1984.
[b]Calculated as $[(1 - 3) / 3] \times 100$.
[c]Calculated as $[(4 - 7) / 7] \times 100$.
[d]Estimates in parentheses are based on 50–100 cases.

Table 18. Achieved fertility by husband's social class and year of marriage

Date of marriage	Duration of marriage (years)	Social class					% Excess of V relative to I[a]
		I	II	III	IV	V	
		Number of children born per 100 families at the time of the 1911 census[b]					
1851–1861	50–60	662	733	746	735	763	+15
1861–1871	40–50	607	665	696	690	715	+18
1871–1881	30–40	497	567	615	616	652	+31
1881–1886	25–30	413	481	544	550	596	+44
1886–1891	20–25	357	422	482	491	541	+52
1891–1896	15–20	303	359	405	412	463	+53
1896–1901	10–15	242	284	314	323	362	+49

Date of marriage	Duration of marriage (years)	I	II	III-NM	III-M	IV	V	% Excess of V relative to III-NM[d]
		Mean family size at the time of the 1971 census[c]						
1941–1945	26–30	2.04	1.99	1.86	2.20	2.24	2.47	+33
1946–1950	21–25	2.11	2.02	1.90	2.24	2.29	2.57	+35
1951–1955	16–20	2.25	2.17	2.00	2.34	2.36	2.66	+33
1956–1960	11–15	2.23	2.12	2.00	2.29	2.31	2.58	+29

[a]Calculated as [(V − I) /I] × 100.
[b]Rates standardized for age of wife's marriage; data from Stevenson, 1920.
[c]Data from Office of Population Censuses and Surveys, 1983.
[d]Calculated as [(V − III-NM) /III-NM] × 100.

Social differences in sexual behaviour and cervical cancer

S. de Sanjosé, F.X. Bosch, N. Muñoz and K. Shah

In this chapter we first describe the variation of cervical cancer in relation to social class. Thereafter we examine the causes for the occurrence of socioeconomic differences in invasive cervical cancer, using data from two case–control studies carried out in Colombia and Spain. Cervical cancer is the most common cancer in developing countries and the sixth most common in developed countries. In all areas, it is more frequent among women of low socioeconomic status, it is associated with multiple sexual partners and early age at first sexual intercourse, and both incidence and mortality are reduced by screening. According to population-based surveys in industrialized countries, men of low socioeconomic status report fewer sexual partners than men of high socioeconomic status but there is no clear indication that the same is true of women of low socioeconomic status. In the case–control studies in Spain and Colombia, the human papillomavirus and all other sexually transmitted diseases were more prevalent among women in low socioeconomic strata. Number of sexual partners and particularly contacts with prostitutes were higher among husbands of women of low socioeconomic status. Other potential risk factors for the disease, such as smoking and oral contraceptive use, and also cervical cancer screening (Pap smears), were more common in women of high social strata. Women with no schooling had a threefold higher risk in Spain and a fivefold higher risk in Colombia of having cervical cancer compared with women who had achieved a higher educational level. After adjustment for sexual behaviour, HPV DNA status, history of Pap smears and husband's contact with prostitutes, this association was considerably reduced. These results are indicative that socioeconomic differences in the incidence of cervical cancer can be partly explained by differences in the prevalence of HPV DNA. Men's sexual behaviour and particularly contacts with prostitutes might be a major contributor to the higher prevalence of HPV DNA among the poor.

Cervical cancer has been consistently associated with low socioeconomic status (Hakama et al., 1982; Baquet et al., 1991; Kogevinas, 1990; Tomatis, 1992). Differences in sexual and reproductive behaviour across socioeconomic groups have been advanced as explanations for the variation in the incidence rates of cervical cancer with socioeconomic status (Brown, 1984; Brinton, 1992). For many cancer sites, the involvement of multiple factors and/or a limited knowledge of the mechanisms of carcinogenesis are hampering the evaluation of the role of socioeconomic status in cancer etiology. Cervical cancer is one of the few exceptions. The identification of human papillomavirus (HPV) as a cause specific to the majority of cervical cancer cases permits a re-evaluation of previously established risk factors and provides new insights into the causes of the higher incidence of cervical cancer among the poor. The high worldwide incidence of cervical cancer (it is the second most common cancer in women and the most common in countries of the developing world) and the significant geographical differences in incidence rates add further importance to understanding the link between cervical cancer and poverty.

In this chapter, we first review the most recent surveys on sexual behaviour that provide information on its relationship with social class. We then assess the association between characteristics commonly associated with cervical cancer, including HPV infection as a basis for explaining social class differences in cervical cancer. For this purpose we use data generated by two case–control studies carried out simultaneously in Colombia and Spain in which HPV DNA was detected by a method based on amplification by the polymerase chain reaction (PCR). These countries have contrasting wealth (gross national product per capita in 1986 was US$ 1204

Table 1. Sexual partnership as reported in population-based studies

Author (country)	Sex and age	Indicator of partnership	Socioeconomic level Low No. (%)	Socioeconomic level High No. (%)	P value
Leigh et al., 1993 (USA)	Men & women: 42.8	≥5 sexual partners in last 5 years	234 (5.5)	731 (12.2)	0.016
Seidman et al., 1992 (USA)	Women: 15–44	≥ 2 sexual partners 3 months before interview	1954 (6.5)	5057 (2.6)	0.000
Laumann et al., 1994 (USA)	Men & women: 18–59	>10 sexual partners in adult lifetime	431 (14.6)	1776 (22.2)	0.0004
Johnson & Wadworth, 1994 (UK)	Men: 35–44	≥2 sexual partners 1 year before interview	254 (6.8)	932 (10.1)	0.0003
	45–59		291 (2.8)	845 (7.3)	0.0049
	Women: 35–44	>1 sexual partner 1 year before interview	403 (4.4)	1190 (4.1)	0.89
	45–59		518 (2.4)	1033 (2.2)	0.91
Spira et al., 1993 (France)	Men: 18–69	>1 sexual partner 1 year before interview	375 (7.4)	699 (15.9)	0.000
	Women: 18–69	>1 sexual partner 1 year before interview	359 (3.0)	565 (7.6)	0.004
Melbye & Beggar, 1992 (Denmark)	Men: 18–59	>1 partner 1 year before interview	483 (5.4)	796 (25.7)	0.0000

P value refers to the comparison between high and low socioeconomic status.

in Colombia and US$ 7640 in Spain) and contrasting incidence rates of cervical cancer (the age-standardized incidence rate was 42.2 per 100 000 in Colombia and 7.7 per 100 000 in Spain) (Parkin et al., 1992).

Sexual behaviour and social class

There are now extensive surveys that illustrate variations in sexual behaviour across socioeconomic strata. The results of several studies referring to sexual partnership are summarized in Table 1 (Leigh et al., 1993; Johnson & Wadsworth, 1994; Spira et al., 1993; Seidman et al., 1992; Melbye & Biggar, 1992). These are population-based studies, from industrialized countries, that characterized sexual behaviour of populations to allow better predictions for the control of the AIDS epidemic. We could not find similar data for developing countries.

Within 48 states of the United States of America, Leigh et al. (1993) interviewed 2058 randomly selected men and women. Data were obtained using a self-administered questionnaire. In this study individuals with a high level of education (some college education) were almost three times more likely to have had more than five sexual partners the year preceding the interview than individuals who had not completed high school.

Data from the National Survey Growth conducted in 1988 in the United States were used to investigate the sexual behaviour of 7011 women (Seidman et al., 1992). Having two or more recent sexual partners was associated with marital status, race, family income/poverty status, age at first sexual intercourse and religious affiliation. Women with a low family income were 2.5 times more likely to report two or more recent sexual partners than women with a higher income level. Among single Black women, formal education was inversely associated with having multiple recent partners.

Among divorced/separated White women, educational level was positively associated with having multiple partners. After controlling for the effects of age, education, residence in a metropolitan area, age at first sexual intercourse, religious affiliation and religiosity on the number of sexual partners, educational level (college education) remained an independent risk factor among divorced/separated White women of having multiple recent partners. This was not observed among Black women.

In another study in the United States, 3432 men and women were randomly selected from the national sampling frame and personally interviewed about sexual experiences (Laumann et al., 1994). Men and women who had graduated from college (or reached a higher educational level) were more likely to report 11 or more sexual partners during their adult life (since age 18) than men and women who had only high school or some college education [the adjusted predicted probabilities were 24.4 versus 19.0 ($P < 0.05$), respectively]. Men and women with less than a high-school educational level reported fewer sexual partners in their lifetimes as compared with those with high-school grades [adjusted predicted probabilities were 12.9 versus 19.0 ($P < 0.05$), respectively]. These differences were statistically significant after controlling for the effect of age, gender, marital status, ethnic group, sexual activity in early life, religious guidance, a period in jail, and having been touched before puberty. In Table 1 crude percentages are provided for comparison with other studies.

Concurrently, two large studies on sexual behaviour were carried out in the United Kingdom (Johnson & Wadworth, 1994) and in France (Spira et al., 1993). The studies aimed to provide information on sexual attitudes that could be related to transmission of HIV and other sexually transmitted infections. In the United Kingdom, marital status and age were strong predictors of sexual behaviour. Among married men and women, those in social classes I and II were more likely to report a greater number of sexual partners than those in social classes IV and V. In men, this observation was more evident among those over the age of 45. The trend with social class remained after controlling for age, marital status and age at first sexual intercourse. In France, men and women of all ages with higher education were more likely to have had more than one sexual partner in the last year than those with a lower educational level. Multiple partnership increased with decreasing age, and was higher among single people and those living in Paris compared with those living in other geographical areas of the country.

Melbye & Beggar (1992) investigated 'at risk' sexual behaviour with relation to acquiring AIDS among 3178 subjects in Denmark using a self-administered questionnaire. Sexual promiscuity increased with educational level. Also, higher educational level was a risk factor for 'sexual risk exposures' (sexual contacts with homosexual/bisexual men, intravenous drug users, sub-Saharan residents or prostitutes).

Overall, in the United Kingdom, France, Denmark and probably the United States, men in high socioeconomic strata reported higher promiscuity than men in the low socioeconomic strata. Women appeared to have a much less consistent pattern across socioeconomic strata; for example, French women in high socioeconomic strata reported a higher number of sexual partners than those in low socioeconomic strata; in the United Kingdom no difference across strata was observed; and in the United States women in low socioeconomic strata reported a higher number of recent sexual partners.

Case–control studies on invasive cervical cancer in Spain and Colombia

Two case–control studies on invasive cervical cancer have been carried out in two countries with contrasting incidence rates of cervical cancer: a low rate in Spain and a high rate in Colombia. The study methods have been described in detail (Muñoz et al., 1992). The field work was conducted in Spain from June 1985 to December 1987 and in Colombia from June 1985 to December 1988. The studies included all incident, histologically confirmed, invasive squamous cell carcinomas of the cervix identified among residents in the city of Cali, Colombia, in eight provinces of Spain (Alava, Gerona, Guipuzcoa, Murcia, Navarra, Salamanca, Seville and Vizcaya) and in the city of Zaragoza, Spain. Controls, one per case, were a five-year age-stratified random sample selected from the general population that generated the cases. Study subjects and their husbands were interviewed using a structured questionnaire that covered sociodemographic characteristics, reproduction, sexual behaviour, smoking habit, oral contraceptive use, practice of

Table 2. Odds ratios (ORs) of invasive cervical cancer and schooling attainment

	Spain				Colombia			
	Cases/controls	OR[a]	OR[b]	OR[c] (95% CI)	Cases/controls	OR[a]	OR[b]	OR[c] (95% CI)
Education								
Secondary and higher	11/23	1	1	1	13/26	1	1	1
Primary schooling	85/81	2.0	1.7	1.5 (0.5–4.5)	59/65	2.1	1.9	0.8 (0.3–2.3)
No schooling	46/26	3.3[e]	2.8[e]	2.9 (0.9–9.4)	15/7	5.3[e]	5.7[e]	2.8 (0.6–12.6)
No. of women[d]	142/130				87/98			

[a]OR adjusted for age.
[b]OR adjusted for age, number of sexual partners, age at first sexual intercourse, history of Pap smears, area of recruitment (only in Spain) and husband's contacts with prostitute.
[c]OR adjusted in addition for HPV DNA.
[d]Only those with known HPV DNA status included.
[e]95% CI does not include 1.

Papanicolaou (Pap) smears and history of common sexually transmitted diseases. Among different indicators of socioeconomic status, women's schooling and educational attainment are used in this report as the socioeconomic indicator with which to evaluate behavioural differences. Household income level, parental and husband's educational attainment and household amenities were also explored. None of these showed a statistically significant association with cervical cancer risk after controlling for HPV status (de Sanjosé et al., 1996).

HPV DNA sequences were sought in cytological specimens obtained by cervical scraping, and in males by scraping the distal urethra and the coronal sulcus of the glans. PCR amplification using HPV L1 consensus primers was carried out for 60% of the study subjects (Guerrero et al., 1992). Antibodies to *Chlamydia trachomatis, Neisseria gonorrhoea, Treponema pallidum* or Herpes virus type 2 (HSV-2) were tested in 87% of the subjects (de Sanjosé et al., 1994).

The prevalence of exposure to each cervical cancer risk factor was calculated for control women of different educational categories as percentages of women exposed. When appropriate, tests for linear trends across socioeconomic strata or the χ^2 test for heterogeneity were applied. Logistic regression techniques were used to estimate odds ratios (ORs) and 95% confidence intervals (95% CIs), adjusting for potential confounding factors (Breslow & Day, 1980).

A total of 760 women (373 cases and 387 controls) and 425 husbands were included, which represented 80% of the incident cases, 74% of the eligible controls and 55% of the eligible husbands.

Results

Of control women, 22.7% in Spain and 6% in Colombia had not attended school. Cases in both countries were less educated than their age-stratified controls. Cervical cancer was strongly associated in Spain and in Colombia with no schooling (Table 2). After adjustment for sexual and reproductive behaviour and history of Pap smears, the association between women's education and cervical cancer was reduced. Adjustment for history of Pap smears reduced the OR for low education from 5.2 to 4.6 in Colombia and from 3.3 to 3.0 in Spain. The adjustment for HPV DNA presence reduced the OR for those with no schooling from 5.7 to 2.8 in Colombia and increased the OR from 2.8 to 2.9 in Spain.

The prevalence of exposure to established risk factors for cervical cancer among control women by educational level is summarized in Table 3 (Spain) and Table 4 (Colombia). In Spain, the prevalence of ever having had a Pap test, having more than one sexual partner, ever use of oral contraceptives and smoking habit increased with educational level. Husbands of women who did not attend school or reached only primary-school level reported use of prostitutes twice as often as husbands of women

Table 3. Characteristics of Spanish control women and their husbands by women's education

	No schooling	Primary schooling	Further schooling	P value for trend
No. of women (mean age)	54 (59)	142 (53)	40 (42)	
	%	%	%	
History of Pap smear	20.4	26.1	45.2	0.009
Sexual partners >1	3.7	6.3	22.5	0.002
First intercourse <18 years old	5.6	6.3	5.0	0.942
Ever use oral contraceptives	7.4	25.3	52.5	0.000
No. of children >4	25.9	21.8	10.0	0.150
Ever smoking	3.7	2.7	35.0	0.000
Antibodies to STDs[a]	25.9	19.7	16.7	0.250
HPV-DNA-positive[b]	7.6	3.8	4.0	0.450
No. of husbands	26	89	24	
	%	%	%	
Sexual partners >5	46.2	46.1	39.1	0.636
Use of prostitutes	61.5	64.0	29.2	0.025
HPV-DNA-positive[c]	0	5.3	0	0.941

STD, sexually transmitted disease.
[a]Antibodies to any of four STDs: *N. gonorrhoea, C. trachomatis, T. pallidum* and Herpes virus type II.
[b]Due to untested samples, denominators for the three columns are 26, 79 and 25, respectively.
[c]Due to untested samples, denominators for the three columns are 17, 27 and 18, respectively.

with higher education. Although women with a lower educational level had fewer sexual partners, they were 1.9 times more likely to harbour HPV DNA than women with a higher educational level. However, differences between strata were not statistically significant.

In Colombia, the percentages of ever having had a Pap test and ever use of oral contraceptives were higher among the more educated women. Women who did not attend school or reached only primary-school level were significantly more likely to have higher parity than better-educated women. Husbands of less-educated women reported a higher use of prostitute services than husbands of more highly educated women. Women with a low educational level had a prevalence of HPV DNA 4.4 times as high as women with a higher educational level, although differences did not reach statistical significance ($P = 0.09$); their husbands were also more likely to harbour HPV DNA.

The characteristics associated with low educational attainment (no schooling or only primary schooling) in Colombia and Spain are shown in Figure 1. Results are presented as ORs where the risk among women with a low educational level (no schooling or primary school) for each potential risk factor for cervical cancer is compared with that of women with a higher educational level (secondary schooling or higher). ORs are adjusted for age and history of Pap smears. However, the control for age and history of Pap smears did not modify the association between sexual behaviour, oral contraceptive use and smoking habit shown in Tables 3 and 4.

Discussion

The association between HPV and cervical cancer offers us an etiological model similar to that of smoking habit and lung cancer. For both cancer sites, a major cause of the disease with an uneven social distribution has been identified. Prior to the advent of highly sensitive techniques to identify HPV, high number of sexual partners, early age at first sexual intercourse and low socioeconomic status were found to have independent effects in cervical cancer causation (Brinton, 1992). The results presented here add evidence that socioeconomic status in Spain and Colombia is associated

Table 4. Characteristics of Colombian control women and their husbands by women's education

	No schooling and primary schooling	Further schooling	P value for hetereogeneity test
No. of women (mean age)	110 (50)	39 (41)	
	%	%	
History of Pap smear	63.6	92.3	0.001
First intercourse <18 years old	46.4	35.9	0.257
Sexual partners >1	43.6	35.9	0.400
Ever use oral contraceptives	28.2	58.9	0.001
No. of children >4	61.8	30.8	0.001
Ever smoking	32.7	25.6	0.410
Antibodies to STDs[a]	69.1	64.1	0.566
HPV-DNA-positive[b]	16.7	3.8	0.098
No. of husbands	53	19	
	%	%	
Sexual partners >5	95.8	88.9	0.29
Use of prostitutes	92.4	73.7	0.03
HPV-DNA-positive[c]	31.0	10.0	0.188

STD, sexually transmitted disease.
[a]Antibodies to any of four STDs: *N. gonorrhoea, C. trachomatis, T. pallidum* and Herpes virus type II.
[b]Due to untested samples, denominators for the two columns are 72 and 26, respectively.
[c]Due to untested samples, denominators for the two columns are 29 and 10, respectively.

with aspects of sexual behaviour favouring the acquisition of HPV infections.

It is well accepted that the major etiological cause of cervical cancer is persistence of certain types of HPV infections (Muñoz et al., 1992). In trying to explain socioeconomic differences in the disease rates, a higher HPV DNA prevalence, a higher number of sexual partners or an earlier age at first sexual intercourse were expected among the socially less privileged. In our data for both countries, HPV prevalence decreased with increasing education. In Spain, women with a lower educational level reported fewer sexual partners, but their partners were more likely to have used prostitute services. In Colombia, the prevalence of sexually related risk factors for cervical cancer was higher than in Spain, and higher for less-educated women; also husbands of less-educated women reported use of prostitute services more often. Taken together these results indicate that social differences in cervical cancer are at least partially due to differences in HPV prevalence and that sexual contacts by the husband with prostitutes may be a more important source of HPV infection than the woman's lifetime number of sexual partners. This is in agreement with our previous observation of a higher rate of preneoplastic lesions among prostitutes relative to nonprostitutes (de Sanjosé et al., 1993). Prostitutes may be an important reservoir of HPV infection.

In addition to HPV infection, no regular practice of Pap smears probably also contributes to the higher rates of cervical cancer among women in low socioeconomic strata. In both countries, women in the low socioeconomic strata had a lower record of ever having a Pap smear than better-educated women. In Colombia, history of Pap smears was high in all social strata and almost all women of a high educational level reported having had a Pap smear at least once. The higher rate of practice of Pap smears reported in Colombia, even among less-educated women (63.6%), compared with a much lower rate of screening practice in Spain (20.4% among less-educated women; 45.2%

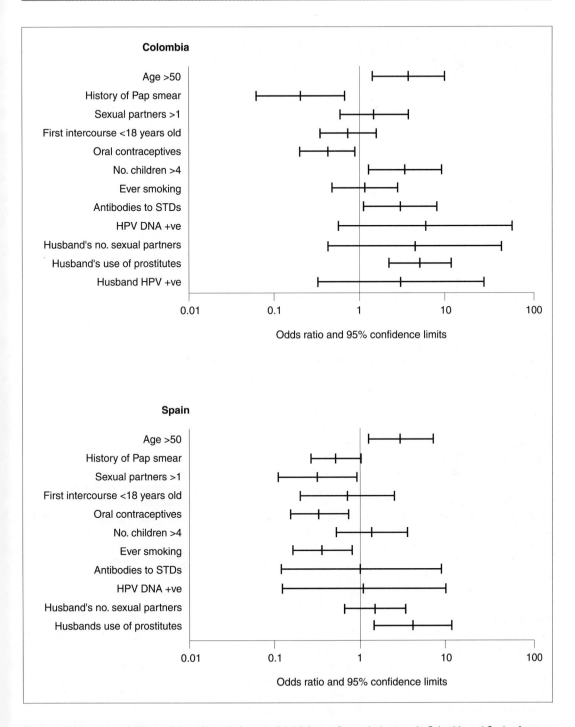

Figure 1. Odds ratios and 95% confidence intervals for potential risk factors for cervical cancer in Colombia and Spain of women with low educational attainment, relative to risks among women of high educational attainment (secondary-school level and higher). ORs are adjusted for age and history of Pap smear.

among the best-educated women) is not surprising. In Spain, cervical cancer is perceived as a minor public health problem and Pap smears are usually taken as an opportunistic adjuvant to gynaecological care. Colombia has one of the highest incidence rates of cervical cancer in the world; in Cali, there is an organized cervical cancer screening programme and in 1984 about 30% of women reported having an annual screening (Aristizabal et al., 1984).

Smoking and oral contraceptive use have been advanced as possible risk factors for cervical cancer. We had previously shown no association with smoking habit and cervical cancer but there was an association between ever use of oral contraceptives and cervical cancer among HPV-DNA-positive women (Bosch et al., 1992). Better-educated women in both countries were more likely to have used oral contraceptives and in Spain also to be smokers. This inverse relationship of smoking and oral contraceptives with socioeconomic status may well be limited to the age cohort under study, particularly for smoking. In some societies, smoking has been accepted first by those in the higher social groups, but later these are the first to give up the habit in response to public health actions (Marmot et al., 1987). Data available for one Spanish region show that the smoking habit among women is increasing and is higher among the better educated. Social differences in smoking are diminishing, however (Enquesta de Salut de Barcelona, 1994).

Probably the major potential source of classification error in these studies is under-detection of HPV infection. Even using a highly sensitive PCR technique to detect HPV DNA, we have reported that HPV DNA among cases was under-detected when using cervical scrapes rather than tissue samples (Bosch et al., 1994). The exclusion of those subjects with unknown HPV status reduced considerably the association between cervical cancer and socioeconomic indicators, even though it implied an important reduction of the sample size.

Hildesheim et al. (1993) reported a negative correlation between years of education and HPV DNA prevalence among young Hispanic women. Our study assessed older women, so it is likely that their sexual and reproductive history and that of their partners reflect almost all their lifetime experiences. Although the women investigated were cytologically normal, those who are HPV-positive are likely to be HPV carriers rather than recently infected, as is probably the case for other studies (Wheeler et al., 1993; Bauer et al., 1993; Hildesheim et al., 1993).

In conclusion, socioeconomic differences in the incidence of invasive cervical cancer could partly be explained by differences in the prevalence of HPV DNA and in the use of preventive medical care. Male behaviour may contribute to the link between socioeconomic stratus and cervical cancer through higher use of prostitute services by men in lower socioeconomic strata.

Acknowledgements

We are indebted to all study participants; to the gynaecologists, pathologists and oncologists who facilitated the identification and contribution of the participants and to the supervisors of the field work (L. Tafur, I. Izarzugaza, M. Gili, P. Viladiu, C. Navarro, C. Martos, N. Ascunce and L.C. Gonzalez); to the panel of pathologists that reviewed all cases (M. Santamaria, P. Alonso de Ruiz and N. Aristizabal); to J. Orfila, K. Reimann, N.S. Pedersen and B. Wahren for serological testing; and to M. Kogevinas for his comments. Financial support was received from the IARC, the European Community CI 1-0371-F (CD) and the Fondo de Investigaciones Sanitarias (FIS) of the Spanish Government (86/753, 87/1513, 88/ 2049, 90/0901).

References

Aristizabal, N., Cuello, C., Correa, P., Collazos, T. & Haenzel, W. (1984) The impact of vaginal cytology on cervical cancer risks in Cali, Colombia. *Int. J. Cancer*, 34, 5–9

Baquet, C.R., Horm, J.W., Gibbs, T. & Greenwald, P. (1991) Socioeconomic factors and cancer incidence among Blacks and Whites. *J. Natl Cancer. Inst.*, 83, 551–557

Bauer, H.M., Hildesheim, A., Schiffman, M.H., Glass, A.G., Rush, B.B., Scott, D.R., Cadell, D.M., Kurman, R.J. & Manos, M.M. (1993) Determinants of genital papillomavirus infection in low-risk women in Portland, Oregon. *Sex. Transm. Dis.*, 20, 279–285

Bosch, F.X., Muñoz, N., de Sanjosé, S., Izarzugaza, I., Gili, M., Viladiu, P., Tormo, M.J., Moreo, P., Ascunce, N., González, L.C., Tafur, L., Kaldor, J.M., Guerrero, E., Aristizabal, N., Santamaria, M., Alonso de Ruiz, P. & Shah, K. (1992) Risk factors for cervical cancer in Colombia and Spain. *Int. J. Cancer*, 52, 750–758

Bosch, F.X., de Sanjosé, S. & Muñoz, N. (1994) Test reliability is critically important to molecular epidemiology. *Cancer Res.*, 54, 6288–6289

Breslow, N. E. & Day, N. E. (1980) *Statistical methods in cancer research (Vol. 1): the analysis of case-control studies* (IARC Scientific Publication No. 32). Lyon, International Agency for Research on Cancer

Brinton, L. (1992) Epidemiology of cervical cancer – overview. In: Muñoz, N., Bosch, F.X., Shah, K.V. & Meheus, A., eds, *The epidemiology of human papillomavirus and cervical cancer* (IARC Scientific Publication No. 119). Lyon, International Agency for Research on Cancer. pp. 3–23

Brown, S., Vessey, M. & Harris, R. (1984) Social class, sexual habits and cancer of the cervix. *Community Med.*, 6, 281–286

de Sanjosé, S., Palacio, V., Tafur, L., Vázquez, S., Espitia, V., Vázquez, F., Muñoz, N. & Bosch, F.X. (1993) Prostitution, HIV, and cervical neoplasia: a survey in Spain and Colombia. *Cancer Epidemiol. Biomarkers Prev.*, 2, 531–535

de Sanjosé, S., Muñoz, N., Bosch, F.X., Reimann, K., Pedersen, N.S., Orfila, J., Ascunce, N., González, L.C., Tafur, L., Gili, M., Lette, I., Viladiu, P., Tormo, M.J., Moreo, P., Shah, K.V. & Wahren, B. (1994) Sexually transmitted agents and cervical neoplasia in Colombia and Spain. *Int. J. Cancer*, 56, 358–363

de Sanjosé, S., Bosch, F.X., Muñoz, N., Tafur, L., Gili, M., Izarzugaza, I., Izquierdo, A., Navarro, C., Moreo, P., Muñoz, M.T., Ascunce, N. & Shah, K.V. (1996) Socioeconomic differences in cervical cancer: two case-control studies in Colombia and Spain. *Am. J. Public Health*, 86, 1532–1538

Enquesta de Salut de Barcelona (1994) *1992–93. Resultats principals. Ajuntament de Barcelona.* Area de Salut Pública Barcelona

Guerrero, E., Daniel, R.W., Bosch, F.X., Castellsague, X., Muñoz, N., Gili, M., Viladiu, P., Navarro, C., Martos, C., Ascunce, N., Gonzalez, L.C., Tafur, L., Izarzugaza, I. & Shah, K.V. (1992) A comparison of Virapap, Southern hybridization and polymerase chain reaction methods for human papillomavirus identification in an epidemiological investigation of cervical cancer. *J. Clin. Microbiol.*, 30, 2951–2959

Hakama, M., Hakulinen, T., Pukkala, E., Saxen, E. & Teppo, L. (1982) Risk indicators of breast and cervical cancer on ecologic and individual levels. *Am. J. Epidemiol.*, 116, 990–1000

Hildesheim, A., Gravitt, P., Schiffman, M.H., Kurman, R.J., Barnes, W., Jones, S., Tchabo, J.G., Brinton, L.A., Copeland, C., Janet, E.P.P. & Manos, M.M. (1993) Determinants of genital human papillomavirus infection in low-income women in Washington, D.C. *Sex. Transm. Dis.*, 20, 279–285

Johnson, A. & Wadsworth, J. (1994) Heterosexual partnership. In: Johnson, A.M., Wadsworth, J., Wellings, K. & Field, J., eds, *Sexual attitudes and lifestyles.* Oxford, Blackwell Scientific Publications. pp. 110–182

Kogevinas, E. (1990) *Longitudinal study: socio-demographic differences in cancer survival 1971–1983* (Office of Population Censuses and Surveys: Series LS No. 5). London, Her Majesty's Stationery Office

Laumann, E.O., Gagnon, J.H., Michael, R.T. & Michaels, S. (1994) *The social organization of sexuality.* Chicago, The University of Chicago Press

Leigh, B.C., Temple, M.T. & Trocki, K.F. (1993) The sexual behavior of US adults: results from a national survey. *Am. J. Public Health*, 83, 400–1408

Marmot, M.G., Kogevinas, M. & Elston, M.A. (1987) Social/economic status and disease. *Annu. Rev. Public Health*, 8, 111–135

Melbye, M. & Biggar, R.J. (1992) Interactions between persons at risk for AIDS and the general population in Denmark. *Am. J. Epidemiol.*, 135, 593–602

Muñoz, N., Bosch, F.X., de Sanjosé, S., Tafur, L., Izarzugaza, I., Gili, M., Viladiu, P., Navarro, C., Martos, C., Ascunce, N., González, L.C., Kaldor, J.M., Guerrero, E., Lörincz, A., Santamaria, M., Alonso de Ruiz, P., Aristizabal, N. & Shah, K. (1992) The causal link between human papillomavirus and invasive cervical cancer: a population-based case-control study in Colombia and Spain. *Int. J. Cancer*, 52, 743–749

Parkin, D.M., Muir, C.S., Whelan, S.L., Gao, Y.T., Ferlay, J. & Powell, J., eds (1992) *Cancer incidence in five continents Vol. VI* (IARC Scientific Publication No. 120). Lyon, International Agency for Research on Cancer

Seidman, S.N., Mosher, W.D. & Aral, S.O. (1992) Women with multiple sexual partners: United States, 1988. *Am. J. Public Health*, 82, 1388–1394

Spira, A., Bajos, N. & ACSF group (1993) *Les comportements sexuels en France.* Paris, La Documentation Française

Tomatis, L. (1992) Poverty and cancer. *Cancer Epidemiol. Biomarkers Prev.*, 1, 167–175

Wheeler, C.M., Parmenter, A., Hunt, W.C., Becker, T.M., Greer, C.E., Hildesheim, A. & Manos, M. (1993) Determinants of genital human papillomavirus infection among cytologically normal women attending the University of New Mexico Student Health Center. *Sex. Transm. Dis.*, 20, 286–289

Corresponding author:
S. de Sanjosé
Servei d'Epidemiolog'ia i Registre del Cancer, Institut Català d'Oncologia, Ciutat Sanitaria i Universitaria de Bellvitge, Autovia de Castelldefels Km 2.7, E-08907 L'Hospitalet de Llobregat, Barcelona, Spain

Infection with hepatitis B and C viruses, social class and cancer

S.O. Stuver, C. Boschi-Pinto and D. Trichopoulos

The hepatitis B and C viruses (HBV and HCV) are major etiological factors in the occurrence of hepatocellular carcinoma (HCC) worldwide, but most especially in developing countries where the majority of liver cancer cases can be found. In parallel with the geographic distribution of HCC, high levels of HBV endemicity are concentrated in the developing world. The association between chronic infection with HBV and low social class is quite strong; socioeconomic factors such as low educational attainment, lower social stratum, and crowded urban residence have been reported to predict higher HBV chronic carrier prevalence in both developed and developing countries. More importantly, the effect of poverty on HBV endemicity is clearly evident among younger age groups, and earlier chronic HBV infection seems to increase the risk of development of HCC. As assays for detecting HCV antibodies have only recently become available, the data on the relationship between HCV infection and socioeconomic status are much fewer. However, the limited number of studies that have investigated the seroepidemiology of HCV report an association between higher prevalence of antibodies to HCV and indicators of low social class. It would appear that the striking correlation between HCC and low socioeconomic status is largely related to the impact of poverty on the spread of HBV and probably HCV.

The scope of the epidemiological and microbiological evidence that chronic infection with hepatitis B virus (HBV) causes hepatocellular carcinoma (HCC) is extensive. Based on the strength of the immense volume of collected data, the International Agency for Research on Cancer (IARC, 1994) recently concluded that HBV should be classified as a Group 1 human carcinogen. Hepatitis C virus (HCV), the newly identified virus responsible for most of non-A, non-B hepatitis, has also been etiologically linked with the development of HCC. Although much less is known about the natural history of HCV infection, the compelling nature of the existing data led the same IARC Working Group to categorize HCV similarly as carcinogenic to humans (IARC, 1994). Together, chronic infections with HBV and HCV are probably responsible for the majority of HCC cases occurring worldwide (Tomatis, 1990; Tanaka et al., 1991; Tabor & Kobayashi, 1992). Of additional importance is the possibility that HBV and HCV may act synergistically in the causation of this malignancy (Yu et al., 1990; Kaklamani et al., 1991).

Primary liver cancer (PLC), of which HCC is the dominant histological subtype, is the eighth most frequently occurring cancer in the world, and the sixth among males (Parkin et al., 1993). With respect to cancer mortality, PLC ranks even higher – the fourth most common cause of death worldwide and the third among males (Pisani et al., 1993) – due to its nearly uniform fatality. Moreover, 77% of all PLC cases and deaths occur in countries of the developing world (Parkin et al., 1993; Pisani et al., 1993). In such countries, liver cancer is the seventh most frequent malignancy, with the third highest number of deaths.

PLC's variation worldwide is greater than that for any other major tumour site. The incidence of this cancer is relatively rare in some countries, such as the United States of America, Norway, and the United Kingdom (Stuver & Trichopoulos, 1994), but is very common in parts of Africa and Asia (Parkin et al., 1993). Intermediate risk areas include southern and eastern Europe, and Micronesia (Parkin et al., 1993). Socioeconomic differences have been implicated as an important causal component in the explanation of the wide geographic differences observed in HCC's distribution, except in the case of Japan where PLC occurrence is relatively high (Pisani et al., 1993) and may be increasing (Stuver

& Trichopoulos, 1994). Within-country variation has been reported in provinces of China (Chen et al., 1990), in ethnic populations of South Africa (Muñoz & Linsell, 1982), and in Japan (Tanaka et al., 1994). In addition, incidence rates are higher among Blacks than Whites in the USA, the former group having a rate consistent with an intermediate risk pattern (Ries et al., 1994).

Studies in migrants, as discussed by Muñoz and Bosch (1987), show that population groups who migrate to other countries retain the PLC risks of their country of origin for at least one generation. A study of Chinese permanently residing in New York City (Szmuness et al., 1978) found that the age-standardized mortality rate from HCC in Chinese males was four times higher than that among Black males and 10 times higher than that among White males. This observation suggests that exposure during early life to the suspected risk factors is one of the main determinants of HCC risk in later life. Moreover, in populations with low and intermediate risk, HCC is rarely seen in people under 40 years of age, whereas in high-incidence areas a shift in the age curve towards the younger age groups can be observed (Muñoz & Bosch, 1987).

Hepatitis B virus

The etiological association of HBV, a double-stranded DNA virus, with HCC appears indisputable and is supported by the integration of HBV DNA sequences into the chromosomal DNA of hepatoma cells from HBV-positive HCC patients (Popper, 1988). In its causation of HCC, HBV probably acts as a 'complete' carcinogen (Pitot, 1982), both by initiation through HBV DNA integration and by promotion through cirrhosis-related liver regeneration (Trichopoulos et al., 1987). The proportion of HCC due to HBV infection is estimated to be between 60% and 90% in high-risk countries, but below 50% in low- or intermediate-risk countries (Bosch & Muñoz, 1989).

The strong and specific association between HBV and HCC is restricted to the chronic carrier form of infection, which can be distinguished by the presence of hepatitis B surface antigen (HBsAg) (Muñoz & Bosch, 1987; IARC, 1994). The likelihood of becoming a chronic carrier of HBV is very high for neonatal exposure (~90% for babies born to HBsAg-positive mothers), but appears to decrease with increasing age at first infection (Tabor, 1988; Thomas, 1990). Interestingly, in the study of Chinese immigrants described above (Szmuness et al., 1978), the few subjects tested who were born in the USA had a lower prevalence of markers of HBV infection than those born in Asia, which implied that most of the migrants positive for HBsAg were infected before they came to the USA. A number of epidemiological studies also strongly support the importance of chronic HBV infection early in life for the subsequent development of HCC (Chang et al., 1989; Muñoz et al., 1989; Hsieh et al., 1992).

The World Health Organization classifies HBV endemicity according to the proportion of the adult population who are chronic HBV carriers (that is, positive for HBsAg). Populations with 0–2% carriers are considered as having low endemicity for HBV; those with 2–7%, intermediate endemicity; and those with 8% or more, high endemicity (IARC, 1994). In general, low prevalences of HBV carriers are found in North America, western Europe, Australia, and South America with the exception of the Amazon region. Intermediate levels occur in eastern and southern Europe, the Middle East, Japan, and South Asia. Finally, high prevalences are found in China, South-East Asia, and sub-Saharan Africa (IARC, 1994). The striking graphical correlation in the geographic distribution of the prevalence of chronic carriers of HBV and the occurrence of liver cancer has been pointed out by Maupas and Melnick (1981). Their mapping also effectively characterized the concentration in the developing world of both elevated HCC incidence and HBsAg endemicity. Thus, South-East Asia and sub-Saharan Africa, which have very high HBsAg prevalence, also have the highest rates of HCC, and most populations in the USA and Europe that have low rates of HCC have a low prevalence of HBsAg carriers (Muñoz & Bosch, 1987).

HBV infection is strongly associated with low socioeconomic status in both developed and developing countries (IARC, 1994). As reported by Szmuness (1975), a nationwide survey carried out in the USA in 1971 showed an important variation in the prevalence of HBsAg between geographic regions, with 0.5/1000 in north–central USA versus 3.4/1000 in Puerto Rico. Moreover, the differences in HBsAg detection between ethnic groups found among volunteer blood donors in the USA probably reflected socioeconomic factors. Szmuness also noted that both Whites and non-

Whites with a lower socioeconomic status showed significantly higher prevalences of HBsAg than individuals of a higher educational level (Szmuness, 1975). In the Middle East, a study of seroprevalence of HBV infection carried out in 1985 revealed that HBV carrier status was inversely associated with socioeconomic status, being sixfold greater in the lower socioeconomic stratum compared with the upper stratum (Toukan et al., 1990). Moreover, Phoon et al. (1987) reported a slightly higher risk of HBsAg positivity among male agricultural and factory workers and unskilled labourers than among men in professional, technical and administrative occupations; the investigators felt that the difference was due to socioeconomic factors.

In addition, volunteer blood donors in Jordan who lived in high-class, uncrowded urban areas were shown to have an HBsAg prevalence of 0.7%; those living in intermediate socioeconomic level urban areas had a prevalence of 1.7%; and those from poor, crowded urban areas, refugee camps, or rural areas had a prevalence of 6.9% (Awidi et al., 1984). A relationship between HBsAg prevalence and low socioeconomic status was also reported among blood donors in Thailand (Nuchprayoon & Chumnijarakij, 1992).

The association between poverty and HBsAg positivity seems to hold among younger age groups for whom the risk of developing HBV-related HCC is likely to be greatest. Several studies have reported associations between the prevalence of HBsAg among pregnant women, who represent an important reservoir for HBV transmission to their offspring, and indicators of socioeconomic status. In France, 1.5% of women attending the Centre d'Hémobiologie Périnatale in Paris were chronic carriers, a level sixfold higher than that for the general female population of France; 78% of these women with HBsAg were from outside the metropolitan area (many from Africa), and 80% were of a low socioeconomic level (Soulié, 1984). In Venezuela, as well, none of the pregnant women tested in a private clinic used by a medium to high social class population was found to be positive for HBsAg, whereas 3.8% of women tested in the maternity unit serving a low-income population were HBV carriers (Pujol et al., 1994). In addition, a study of the coverage rate for a neonatal vaccination programme in New York City involving 830 HBsAg-positive mothers reported that infants of mothers covered by Medicaid (government-sponsored) insurance or of uninsured Black and Hispanic women were significantly less likely to have completed the HBV vaccine series (Henning et al., 1992).

Finally, seroepidemiological studies in Greece (Papaevangelou & Roumeliotou-Karayannis, 1988) and Italy (D'Amelio et al., 1992) both observed recent reductions in the prevalence of HBsAg among military recruits, which were attributed by the authors to improvements in societal and economic conditions over the past 10 to 20 years. Moreover, the incidence of HBV seroconversion was reported to be shifting from childhood to older age groups where the probability of becoming a chronic carrier is lower (Papaevangelou & Roumeliotou-Karayannis, 1988; D'Amelio et al., 1992). Decreases in HBsAg positivity have also been reported among children in Japan, again ascribed to better environmental, hygienic and nutritional conditions (Matsuto et al., 1990). It seems likely that decreasing trends in the occurrence of PLC in some countries may be a direct result of such socioeconomic changes (Stuver & Trichopoulos, 1994).

Hepatitis C virus

HCV is a single-stranded RNA virus that has no reverse transcriptase and is non-integrating (Tabor, 1992). Its role in hepatocarcinogenesis is believed to lie in the realm of progression and growth enhancement through the pathogenesis of cirrhosis (Kaklamani et al., 1991; Tabor, 1992; Tabor & Kobayashi, 1992). Because specific tests for HCV infection became available only in 1989, studies investigating the seroepidemiology of this infection are scarce. However, in most populations of the world, 0.5–2% of individuals have serological evidence of past or current HCV infection (IARC, 1994). Ngatchu et al. (1992) carried out a study in Cameroon on children 4–14 years old to evaluate the seroprevalence of HCV and its relationship with sociodemographic factors. The overall prevalence was found to be 14.5%, and a significant elevation of anti-HCV seroprevalence was observed in the lowest social class.

A number of studies of volunteer blood donors also have noted an association between positivity for anti-HCV and indicators of low socioeconomic status. As for HBsAg prevalence in blood donors in the USA (Szmuness, 1975), anti-HCV was detected more frequently in Black and Hispanic donors than in White donors (Stevens et al., 1990). In a study of Finnish blood donors, the investigators reported an inverse relationship between increasing level of education and positivity for HCV antibodies as confirmed by recombinant immunoblot assay (RIBA) (Kolho & Krusius, 1992). A similar trend was found for RIBA-confirmed, anti-HCV-positive blood donors seen at a Cairo hospital in 1992 (Darwish et al., 1993). Furthermore, Patiño-Sarcinelli et al. (1994) reported a decreased anti-HCV prevalence in Brazil among those with an education beyond the college level compared with those with a lower educational attainment. In that study, non-White race was a significant risk factor for a positive anti-HCV test, which the authors felt reflected a relationship between lower socioeconomic conditions and exposure to HCV infection (Patiño-Sarcinelli et al., 1994).

HCC and socioeconomic status

Evidence linking HCC with lower socioeconomic status is primarily based on the studies of the geographic variation of liver cancer (Maupas & Melnick, 1981; Muñoz & Linsell, 1982), with higher rates observed in developing countries, for the most part (Parkin et al., 1993). In addition, the United Kingdom Registrar General's Decennial Supplement (Office of Population Census and Surveys, 1978) on occupational mortality for England and Wales reported that the standardized mortality ratio for HCC was highest for the lowest social classes. Further support of the association between HCC and socioeconomic status is provided by a small number of analytical studies. A case–control study in Taiwan (Pan et al., 1993) showed that poor education and occupation as a farmer or labourer, in addition to heavy alcohol consumption and smoking, were risk factors for HBV-related HCC. In northern Italy, an inverse association was reported between lower social class and fewer years of education on the one hand and the risk of HCC on the other (La Vecchia et al., 1988); in that study, information on HBV and HCV seromarkers was not available. Furthermore, in a Greek case–control study, cases of HCC, particularly those linked to HBV infection, were more likely to be of lower socioeconomic status than were controls (Trichopoulos, 1981). Finally, a study by Ross et al. (1992) found an inverse association between higher level of education and HCC incidence in China.

Although the relation of low socioeconomic class and cancer cuts across a wide spectrum of malignancies, the association of poverty with HCC is both more striking than most and better understood. Low socioeconomic level facilitates transmission of many infectious agents, including HBV, thereby shifting the age of first exposure towards younger groups and increasing the likelihood of the development of the chronic carrier state. Moreover, HBV is frequently transmitted perinatally. Thus, to the extent that early establishment of the carrier state increases the risk of subsequent liver cancer occurrence, the existing socioeconomic class gradient of chronic HBV infection among young people indicates that HCC will continue to be a malignancy disproportionately concentrated among poor populations and individuals. Other important and common factors in the etiology of HCC, such as HCV infection, tobacco smoking, and heavy alcohol drinking, tend also to be more prevalent among the poor. Lastly, efforts to vaccinate against HBV have been less successful in lower socioeconomic status groups and have not even been systematically attempted in several poor countries at high risk of HCC.

It would appear that HCC is at present, and is likely to remain, a paradigm of a poverty-related cancer, both among countries on an international scale and among individuals within most countries.

References

Awidi, A.S., Tarawneh, M.S., El-Khateeb, M., Hijazi, S. & Shahrouri, M. (1984) Incidence of hepatitis B antigen among Jordanian volunteer blood donors. *Public Health*, 98, 92–96

Bosch, F.X. & Muñoz, N. (1989) Epidemiology of hepatocellular carcinoma. In: Bannash, P., Keppler, D. & Weber, G., eds, *Liver cell carcinoma* (Falk Symposium No. 51). Dordrecht, Kluwer Academic. pp. 3–14

Chang, M-H., Chen, D-S., Hsu, H-C., Hsu, H-Y. & Lee, C-Y. (1989) Maternal transmission of hepatitis B virus in childhood hepatocellular carcinoma. *Cancer*, 64, 2377–2380

Chen, J., Campbell, T.C., Li, J. & Peto, R. (1990) *Diet, lifestyle and mortality in China. A study of the characteristics of 65 Chinese counties.* Oxford, Oxford University Press

D'Amelio, R., Matricardi, P.M., Biselli, R., Stroffolini, T., Mele, A., Spada, E., Chionne, P., Rapicetta, M., Ferrigno,

L. & Pasquini, P. (1992) Changing epidemiology of hepatitis B in Italy: public health implications. *Am. J. Epidemiol.*, 135, 1012–1018

Darwish, M.A., Raouf, T.A., Rushdy, P., Constantine, N.T., Rao, M.R. & Edelman, R. (1993) Risk factors associated with a high seroprevalence of hepatitis C virus infection in Egyptian blood donors. *Am. J. Trop. Med. Hyg.*, 49, 440–447

Henning, K.J., Pollack, D.M. & Friedman, S.M. (1992) A neonatal hepatitis B surveillance and vaccination program: New York City, 1987 to 1988. *Am. J. Public Health*, 82, 885–888

Hsieh, C-C., Tzonou, A., Žavitsanos, X., Kaklamani, E., Lan, S-J. & Trichopoulos, D. (1992) Age at first establishment of chronic hepatitis B virus infection and hepatocellular carcinoma risk. A birth order study. *Am. J. Epidemiol.*, 136, 1115–1121

IARC (1994) *IARC monographs on the evaluation of carcinogeic risk to humans* (Vol. 59): *hepatitis viruses*. Lyon, International Agency for Research on Cancer

Kaklamani, E., Trichopoulos, D., Tzonou, A., Zavitsanos, X., Koumantaki, Y., Hatzakis, A., Hsieh, C.C. & Hatziyannis, S. (1991) Hepatitis B and C viruses and their interaction in the origin of hepatocellular carcinoma. *J. Am. Med. Assoc.*, 265, 1974–1976

Kolho, E.K. & Krusius, T. (1992) Risk factors for hepatitis C virus antibody positivity in blood donors in a low-risk country. *Vox Sang.*, 63, 192–197

La Vecchia, C., Negri, E., Decarli, A., D'Avanzo, B. & Franceschi, S. (1988) Risk factors for hepatocellular carcinoma in northern Italy. *Int. J. Cancer*, 42, 872–876

Matsuto, A., Kusumoto, Y., Ohtsuka, E., Ohtsuru, A., Nakamura, Y., Tajima, H., Shima, M., Nakata, K., Muro, T., Satoh, A., Ishii, N., Kohji, T. & Nagataki, S. (1990) Changes in HBsAg carrier rate in Goto Islands, Nagasaki Prefecture, Japan. *Lancet*, 335, 955–957

Maupas, P. & Melnick, J.L. (1981) Hepatitis B infection and primary liver cancer. *Prog. Med. Virol.*, 27, 1–5

Muñoz, N. & Bosch, F.X. (1987) Epidemiology of hepatocellular carcinoma. In: Okuda, K. & Ishaki, K.G., eds, *Neoplasms of the liver*. Tokyo, Springer. pp. 3–19

Muñoz, N. & Linsell, A. (1982) Epidemiology of primary liver cancer. In: Correa, P. & Haenszel, W., eds, *Epidemiology of cancer of the digestive tract*. The Hague, Martinus Nijhoff. pp. 161–195

Muñoz, N., Lingao, A., Lao, J., Estève, J., Viterbo, G., Domingo, E.O. & Lansang, M.A. (1989) Patterns of familial transmission of HBV and the risk of developing liver cancer: a case-control study in the Philippines. *Int. J. Cancer*, 44, 981–984

Ngatchu, T., Stroffolini, T., Rapicetta, M., Chionne, P., Lantum, D. & Chiaramonte, M. (1992) Seroprevalence of anti-HCV in an urban child population: a pilot survey in a developing area, Cameroon. *J. Trop. Med. Hyg.*, 95, 57–61

Nuchprayoon, T. & Chumnijarakij, T. (1992) Risk factors for hepatitis B carrier status among blood donors of the National Blood Center, Thai Red Cross Society. *Southeast Asian J. Trop. Med. Public Health*, 23, 246–253

Office of Population Census and Surveys (1978) *Occupational mortality. The Registrar General's decennial supplement for England and Wales, 1970–72* (Series DS No. 1). London, Her Majesty's Stationery Office

Pan, W-H., Wang, C-Y., Huang, S-M., Yeh, S-Y., Lin, W-G., Lin, D-I. & Liaw, Y-F. (1993) Vitamin A, vitamin E or beta-carotene status and hepatitis B-related hepatocellular carcinoma. *Ann. Epidemiol.*, 3, 217–224

Papaevangelou, G.J. & Roumeliotou-Karayannis, A. (1988) Reduction of HBV infections and mass immunisation. *Lancet*, 1, 53–54

Parkin, D.M., Pisani, P. & Ferlay, J. (1993) Estimates of the worldwide incidence of eighteen major cancers in 1985. *Int. J. Cancer*, 54, 594–606

Patiño-Sarcinelli, F., Hyman, J., Camacho, L.A.B., Linhares, D.B. & Azevedo, J.G. (1994) Prevalence and risk factors for hepatitis C antibodies in volunteer blood donors in Brazil. *Transfusion*, 34, 138–141

Phoon, W.O., Fong, N.P., Lee, J. & Leong, H.K. (1987) A study on the prevalence of hepatitis B surface antigen among Chinese adult males in Singapore. *Int. J. Epidemiol.*, 16, 74–78

Pisani, P., Parkin, D.M. & Ferlay, J. (1993) Estimates of the worldwide mortality from eighteen major cancers in 1985. Implications for prevention and projections of future burden. *Int. J. Cancer*, 55, 891–903

Pitot, H.C. (1982) The natural history of neoplastic development: the relation of experimental models to human cancer. *Cancer*, 49, 1206–1211

Popper, H. (1988) Relation between hepatocellular carcinoma and persistent hepatitis B infection. *Appl. Pathol.*, 6, 64–72

Pujol, F.H., Rodriguez, I., Martinez, N., Borberg, C., Favorov, M.O., Fields, H.A. & Liprandi, F. (1994) Viral hepatitis serological markers among pregnant women in Caracas, Venezuela: implications for perinatal transmission of hepatitis B and C. *GEN*, 48, 25–28

Ries, L.A.G., Miller, B.A., Hankey, B.F., Kosary, C.L., Hanas, A. & Edwards, B.K. (1994) *SEER cancer statistics review, 1973–1991. Tables and graphs*, National Cancer Institute (NIH Publication No. 94-2789). Bethesda, MD, NIH

Ross, R.K., Yuan, J-M., Yu, M.C., Wogan, G.N., Qian, G-S., Tu, J-T., Groopman, J.D., Gao, Y-T. & Henderson, B.E. (1992) Urinary aflatoxin biomarkers and risk of hepatocellular carcinoma. *Lancet*, 339, 943–946

Soulié, J.C. (1984) Prévention de la transmission "verticale" du virus de l'hépatite B. Expérience du Centre d'Hémobiologie Périnatale. *Revue Française de Transfusion et Immuno-hématologie*, 27, 169–180

Stevens, C.E., Taylor, P.E., Pindyck, J., Choo, Q-L., Bradley, D.W., Kuo, G. & Houghton, M. (1990) Epidemiology of hepatitis C virus. A preliminary study in volunteer blood donors. *J. Am. Med. Assoc.*, 263, 49–53

Stuver, S.O. & Trichopoulos, D. (1994) Liver cancer. In: Doll, R., Fraumeni, J. & Muir, C., eds, *Trends in cancer incidence and mortality* (Cancer Surv. 19/20). London, Imperial Cancer Research Fund. pp. 99–124

Szmuness, W. (1975) Recent advances in the study of the epidemiology of hepatitis B. *Am. J. Pathol.*, 81, 629–650

Szmuness, W., Stevens, C.E., Ikram, H., Much, M.I., Harley, E.J. & Hollinger, B. (1978) Prevalence of hepatitis B virus infection and hepatocellular carcinoma in Chinese-Americans. *J. Infect. Dis.*, 137, 822–829

Tabor, E. (1988) Etiology, diagnosis, and treatment of viral hepatitis in children. *Adv. Pediatr. Infect. Dis.*, 3, 19–46

Tabor, E. (1992) Hepatitis C virus and hepatocellular carcinoma. *AIDS Res. Hum. Retroviruses*, 8, 793–796

Tabor, E. & Kobayashi, K. (1992) Hepatitis C virus, a causative infectious agent of non-A, non-B hepatitis: prevalence and structure – summary of a conference on hepatitis C virus as a cause of hepatocellular carcinoma. *J. Natl Cancer Inst.*, 84, 86–90

Tanaka, K., Hirohata, T., Koga, S., Sugimachi, K., Kanematsu, T., Ohryohji, F., Nawata, H., Ishibashi, H., Maeda, Y., Kiyokawa, H., Tokunaga, K., Irita, Y., Takeshita, S., Arase, Y. & Nishino, N. (1991) Hepatitis C and hepatitis B in the etiology of hepatocellular carcinoma in the Japanese population. *Cancer Res.*, 51, 2842–2847

Tanaka, H., Hiyama, T., Okubo, Y., Kitada, A. & Fujimoto, I. (1994) Primary liver cancer incidence-rates related to hepatitis-C virus infection: a correlational study in Osaka, Japan. *Cancer Causes Control*, 5, 61–65

Thomas, H.C. (1990) The hepatitis B virus and the host response. *J. Hepatol.*, 11, S83–S89

Tomatis, L., ed. (1990) *Cancer: causes, occurrence and control* (IARC Publications No. 100). Lyon, International Agency for Research on Cancer

Toukan, A.U., Sharaiha, Z.K., Abu-El-Rub, O.A., Hmoud, M.K., Dahbour, S.S., Abu-Hassan, H., Yacoub, S.M., Hadler, S.C., Margolis, H.S., Coleman, P.J. & Maynard, J.E. (1990) The epidemiology of hepatitis B virus among family members in the Middle East. *Am. J. Epidemiol.*, 132, 220–232

Trichopoulos, D. (1981) The causes of primary hepatocellular carcinoma in Greece. *Prog. Med. Virol.*, 27, 14–25

Trichopoulos, D., Day, N.E., Kaklamani, E., Tzonou, A., Muñoz, N., Zavitsanos, X., Koumantaki, Y. & Trichopoulou, A. (1987) Hepatitis B, tobacco smoking and ethanol consumption in the etiology of hepatocellular carcinoma. *Int. J. Cancer*, 39, 45–49

Yu, M.C., Tong, M.J., Coursaget, P., Ross, R.K., Govindarajan, S. & Henderson, B.E. (1990) Prevalence of hepatitis B and C viral markers in black and white patients with hepatocellular carcinoma in the United States. *J. Natl Cancer Inst.*, 82, 1038–1041

Corresponding author:
S.O. Stuver
Center for Cancer Prevention and Department of Epidemiology, Harvard School of Public Health, 677 Huntington Avenue, Boston, MA 02115, USA

Infection with *Helicobacter pylori* and parasites, social class and cancer

P. Boffetta

Three genera of parasites are known or suspected risk factors for cancer in humans: Schistosoma, Opisthorchis and Clonorchis. No adequate information is available on the determinants of infections related to social class. Infection with the bacterium *Helicobacter pylori* is an important cause of stomach cancer. Studies, in particular from the United Kingdom and the United States of America, strongly suggest that social class factors, especially those acting during childhood, are determinants of the infection, with odds ratios of seroprevalence of the order of 1.5–5 for lower social class as compared with higher social class. A conservative estimate of the contribution of social class, acting through an increased prevalence of *H. pylori* infection, to the burden of stomach cancer gives a figure of over 50 000 stomach cancers per year worldwide, or 8% of all stomach cancers. In countries with both high and low prevalence of infection with *H. pylori*, it is likely that a sizeable proportion of this difference is due to social-class-related risk factors of infection.

Infection with biological agents is an important risk factor for cancer, in particular in developing countries. As the probability and severity of infection may be determined by factors related to social class, such as housing conditions, it is conceivable that at least part of the social gradient observed for cancers associated with infection is due to social class differences in infection patterns. In this chapter, I discuss the evidence linking human cancer, social class and infection with *Helicobacter pylori* and parasites. Infection by viruses is discussed elsewhere in this book (see chapter by Stuver *et al.*).

Parasites

Chronic infection with *Schistosoma haematobium* is a risk factor for bladder cancer, and infection with other schistosomes may be responsible for other cancers in humans (IARC, 1994a). It has been estimated that over 600 million people in 74 countries are exposed to the risk of schistosomal infection, and 200 million are currently infected (WHO, 1993). Contact with untreated water is the major risk factor of infection (Jordan & Webbe, 1993). Prevalence of infection seems to be highest during the second decade of life; the decline in later age is believed to be due mainly to the gradual acquisition of immunity (IARC, 1994a). Rural populations, and those with limited or no access to treated water for domestic and recreational purposes, are at increased risk of infection, in particular during childhood; in many countries, these populations are at the bottom end of the social spectrum. Direct evidence of a difference in prevalence of infection in relation to social class, however, is very limited. In a study from the state of Minas Gerais, Brazil, infection from *Schistosoma mansoni* was higher in Black children aged 6–15 than in Mulatto and in White children (91%, 84% and 73%, respectively), and this difference was interpreted by the authors as evidence of an effect of socioeconomic level (Tavares-Neto *et al.*, 1991).

Infection with liver flukes from the *Opisthorchis* and the *Clonorchis* genera has been linked with increased risk of liver cancer (IARC, 1994b). Infection takes place from ingestion of raw fish. No data on the social class pattern of infection from these agents are available in the scientific literature.

Helicobacter pylori

Infection with *Helicobacter pylori* is a risk factor for gastric cancer (IARC, 1994c). The relative risk is of the order of 2–4, and the prevalence of infection in many countries is of the order of 10–50% (IARC, 1994c) but it can be as high as 90% in areas of China (Forman et al., 1990) and Japan (Tsugane et al., 1993). It should be noted that a relative risk of 3 in a population with 30% of exposed individuals gives an attributable risk of the order of 46%, and that stomach cancer is the second most frequent cancer in the world (Parkin et al., 1993), with approximately 750 000 new cases worldwide each year.

H. pylori has been discovered and characterized only recently, and its role in the etiology of chronic gastritis, peptic ulcer disease and stomach cancer has been clarified during the last few years. It is therefore not surprising that, despite the very large public health importance infection with H. pylori is likely to have, relatively few studies are available on its determinants.

Epidemiological studies of H. pylori infection and social class

The association between H. pylori infection and social-class-related factors has been addressed by several epidemiological studies. Results on prevalence of infection according to social class are presented in Table 1.

Table 1. Prevalence of infection with *Helicobacter pylori* by social class indicators – results of selected studies

Study	Population	Social class indicator	Category	P(%)[a]
Sitas et al., 1991[b]	749 adults, UK	Social class from occupation	I, II	49
			III	57
			IV, V	62
Fiedorek et al., 1991	245 children, USA	Income of family	<US$ 5000/year	39
			US$ 5000–25 000/year	27
			>US$ 25 000/year	16
Mendall et al., 1992	215 adults, UK	Social class from occupation	I, II	25
			III-NM	32
			III-M	46
			IV, V	34
		Persons per room in childhood	<0.70	15
			0.70–0.99	17
			1.00–1.29	43
			>1.30	54
Webb et al., 1994[b]	471 adults, UK	Persons per room in childhood	<0.76	34
			0.76–1.00	29
			1.01–1.50	49
			>1.51	49
Patel et al., 1994	554 children, UK	Persons per room	<0.5	10
			0.5–1.0	9
			>1.0	23
		Social class from occupation	I, II	8
			III	14
			IV, V	10
		Housing tenure	Owned	9
			Rented	16

[a]Prevalence of seropositivity to H. pylori.
[b]Age-adjusted prevalence.

A correlation study from 46 counties in China estimated the correlation between several causes of death, serological markers (including IgG antibody measured in a sample of the population of each county), and dietary, lifestyle and demographic variables (Junshi et al., 1990). Three indicators of education were included: proportion of university graduates, proportion of junior-/middle-school graduates, and literacy. A non-significant negative correlation was found between *H. pylori* seroprevalence and each of the three indicators of education ($r = -0.06$, -0.05 and -0.12, respectively).

In a survey of 246 individuals from Saudi Arabia aged 21–50, those with college education had a significantly lower prevalence of infection than less-educated individuals; however, no difference was observed between illiterate individuals and people with primary or secondary education (Al-Moagel et al., 1990).

A study of seroprevalence of IgG antibody in the United Kingdom included 749 adults randomly selected from the population of Caerphilly, South Wales (overall prevalence: 56.8%) (Sitas et al., 1991). A trend was shown between age-adjusted prevalence of infection and social class.

In a study of factors influencing *H. pylori* infection in 245 healthy children and young adults aged 3–21 from Little Rock, Arkansas, United States of America (overall infection prevalence: 31%), the only factors significantly associated with infection besides age in a multivariate analysis were Black race and low family income (Fiedorek et al., 1991).

Prevalence of *H. pylori* IgG antibody was determined in 215 healthy individuals from London, United Kingdom (overall prevalence: 33%) (Mendall et al., 1992). In a mutually adjusted multivariate analysis no effect of current social class was found [odds ratios (ORs) of infection for social classes III, IV and V, as compared with I and II, were 0.74, 2.15 and 0.61, respectively], while strong associations were found for use of hot water in childhood [OR for shared or no use versus sole use was 4.34; 95% confidence interval (CI) = 1.87–10.0], persons per room in childhood [ORs were 1.00 (reference) for <0.70 persons, 1.36 for 0.70–0.99 persons, 4.04 for 1.00–1.29 persons, and 6.15 for ≥1.30 persons] and number of children currently living in the household (ORs were 1.85 for one child and 5.53 for two or more children versus none). Age was the only other seropositivity risk factor.

In a survey of 1727 Chinese individuals from one city and three rural areas (overall infection prevalence: 44%), density of living was one of the significant factors retained in a stepwise multivariate model (Mitchell et al., 1992). Other factors retained in the model were age, place of residence, antibiotic usage, smoking, and liver and ulcer disease.

In a study based on blood samples from 985 blood donors in North Wales, United Kingdom, 41% of the subjects were positive for *H. pylori* IgG antibody (Whitaker et al., 1993). The OR of positivity was 1.53 (95% CI = 1.14–2.05) for manual versus non-manual social class in adulthood, and 1.50 (95% CI = 1.12–2.02) for sharing versus not sharing a bed during childhood. Other factors significantly associated with antibody positivity were age, childhood house density, and area of birth and residence. Other childhood factors such as social class, number of people in the bedroom and presence of an indoor toilet were not significantly associated with antibody positivity in a multivariate analysis.

A seroprevalence study of 471 male volunteers from Stoke on Trent, United Kingdom (overall seropositivity: 37.4%) revealed positive associations in a mutually adjusted analysis between IgG seropositivity and manual occupation of the subject (OR = 2.21; 95% CI = 1.07–4.57), manual occupation of the father (2.05; 0.85–4.93), sharing a bed (1.41; 0.83–2.40), and having more than one person per room (1.54; 0.87–2.75) (Webb et al., 1994). The only additional variable associated with seropositivity in this study was age.

In a study on 554 schoolchildren from Edinburgh, United Kingdom, sampled at age 7 and followed up to age 11 (overall infection prevalence: 11%), prevalence of infection at age 11 was significantly associated with house crowding, housing tenure, and proportion of housing rented in the school catchment area. Children of social classes I and II had a lower prevalence of infection, but no trend was found among lower social classes (Patel et al., 1994).

Additional evidence comes from two studies of institutionalized individuals. A study from New South Wales, Australia, found a higher prevalence of IgG antibody positivity among inmates of a residential institution for the mentally retarded than among blood donors or hospital controls, irrespective of the age group (Berkowicz & Lee, 1987). In a study from Thailand, children from an orphanage in Bangkok had a higher seroprevalence than non-institutionalized children from rural areas (Perez-Perez *et al.*, 1990).

Discussion
The evidence from the relatively few studies available strongly points towards an association between low social class and *H. pylori* infection (Veldhuyzen van Zanten, 1995). Factors linked to social class acting during childhood seem to play a particularly important role. However, it is too early to identify specific factors, like sharing a bed during childhood, as responsible for the increased risk of infection. The relatively few studies available may have tended to replicate the findings of early investigations, and the analysis of different factors linked to social class, which may act either during childhood or during early adulthood, is far from being complete.

Despite the lack of knowledge about the exact mechanism of infection and its relationship with social class, it is likely that the causal chain of social-class-related factors leading to *H. pylori* infection leading to stomach cancer represents one of the most important known explanations for social difference in cancer incidence and mortality.

In many studies, the social class difference for stomach cancer incidence and mortality is in the range of 50–400% excess risk for low social class as compared with high social class (see the chapter by Faggiano *et al.* in this book). In a conservative estimate, the proportion of *H. pylori* positivity attributable to low social class (assuming an OR of infection of 1.5 for that half of the population of lower social class) is of the order of 17%. The estimated number of new cases of stomach cancer in the world is approximately 750 000 (Parkin *et al.*, 1993) and the percentage attributable to infection with *H. pylori* is 46%. The total number of stomach cancers occurring yearly worldwide attributable to social-class-related *H. pylori* infection would therefore be 58 650 (750 000 x 0.46 x 0.17), or about 8% of all stomach cancers.

In countries with both high and low prevalence of infection to *H. pylori*, it is likely that a sizeable proportion of this difference is due to social-class-related risk factors of infection. The identification of such factors, together with a reduction in their prevalence and a general improvement of the social condition of life, would be important steps towards the prevention of stomach cancer in future generations.

Acknowledgement
I thank Dr N. Muñoz for helpful comments and suggestions.

References
Al-Moagel, M.A., Evans, D.G., Abdulghani, M.E., Adam, E., Evans, D.J., Jr, Malaty, H.M. & Graham, D.Y. (1990) Prevalence of *Helicobacter pylori* (formerly *Campylobacter pylori*) infection in Saudi Arabia, and comparison of those with and without upper gastrointestinal symptoms. *Am. J. Gastroenterol.*, 85, 944–948

Berkowicz, J. & Lee, A. (1987) Person-to-person transmission of *Campylobacter pylori*. *Lancet*, ii, 680–681

Fiedorek, S.C., Malaty, H.M., Evans, D.L., Pumphrey, C.L., Casteel, H.B., Evans, D.J., Jr & Graham, D.Y. (1991) Factors influencing the epidemiology of *Helicobacter pylori* infection in children. *Pediatrics*, 88, 578–582

Forman, D., Sitas, F., Newell, D.G., Stacey, A.R., Boreham, J., Peto, R., Campbell, T.C., Li, J. & Chen, J. (1990) Geographic association of *Helicobacter pylori* antibody prevalence and gastric cancer mortality in rural China. *Int. J. Cancer*, 46, 608–611

IARC (1994a) Infection with schistosomes (*Schistosoma haematobium*, *S. mansoni* and *S. japonicum*). In: *IARC monographs on the evaluation of carcinogenic risks to humans* (Vol. 61): *Schistosomes, liver flukes and Helicobacter pylori*. Lyon, International Agency for Research on Cancer. pp. 45–119

IARC (1994b) Infection with liver flukes (*Opisthorchis viverrini*, *O. felineus* and *Clonorchis sinensis*). In: *IARC monographs on the evaluation of carcinogenic risks to humans* (Vol. 61): *Schistosomes, liver flukes and Helicobacter pylori*. Lyon, International Agency for Research on Cancer. pp. 121–175

IARC (1994c) Infection with *Helicobacter pylori*. In: *IARC monographs on the evaluation of carcinogenic risks to humans* (Vol. 61): *Schistosomes, liver flukes and Helicobacter pylori*. Lyon, International Agency for Research on Cancer. pp. 177–240

Jordan, P. & Webbe, G. (1993) Epidemiology. In: Jordan, P., Webbe, G. & Sturrock, R.F., eds, *Human schistosomiasis*. Wallingford, CAB International. pp. 87–158

Junshi, C., Campbell, T.C., Junyao, L. & Peto, R. (1990) *Diet, life-style, and mortality in China: a study of the characteristics of 65 Chinese counties*. Oxford, Oxford University Press

Mendall, M.A., Goggin, P.M., Molineaux, N., Levy, J., Toosy, T., Strachan, D. & Northfield, T.C. (1992) Childhood living conditions and *Helicobacter pylori* seropositivity in adult life. *Lancet*, 339, 896–897

Mitchell, H.M., Li, Y.Y., Hu, P.J., Liu, Q., Chen, M., Du, G.G., Wang, Z.J., Lee, A. & Hazell, S.L. (1992) Epidemiology of *Helicobacter pylori* in southern China: identification of early childhood as the critical period for acquisition. *J. Infect. Dis.*, 166, 149–153

Parkin, D.M., Pisani, P. & Ferlay, J. (1993) Estimates of the worldwide incidence of eighteen major cancers in 1985. *Int. J. Cancer*, 54, 594–606

Patel, P., Mendall, M.A., Khulusi, S., Northfield, T.C. & Strachan, D.P. (1994) *Helicobacter pylori* infection in childhood: risk factors and effect on growth. *Br. Med. J.*, 309, 1119–1123

Perez-Perez, G.I., Taylor, D.N., Bodhidatta, L., Wongsrichanalai, J., Baze, W.B., Dunn, B.E., Echeverria, P.D. & Blaser, M.J. (1990) Seroprevalence of *Helicobacter pylori* infections in Thailand. *J. Infect. Dis.*, 161, 1237–1241

Sitas, F., Forman, D., Yarnell, J.W.G., Burr, M.L., Elwood, P.C., Pedley, S. & Marks, K.J. (1991) *Helicobacter pylori* infection rates in relation to age and social class in a population of Welsh men. *Gut*, 32, 25–28

Tavares-Neto, J., dos Santos, S.B. & Prata, A. (1991) Schistosomiasis infection and race of carriers. *Rev. Latinoam. Microbiol.*, 33, 49–54

Tsugane, S., Kabuto, M., Imai, H., Gey, F., Tei, Y., Hanaoka, T., Sugano, K. & Watanabe, S. (1993) *Helicobacter pylori*, dietary factors, and atrophic gastritis in five Japanese populations with different gastric cancer mortality. *Cancer Causes Control*, 4, 297–305

Veldhuyzen van Zanten, S.J.O. (1995) Do socio-economic status, marital status and occupation influence the prevalence of *Helicobacter pylori* infection? *Aliment. Pharmacol. Ther.*, 9 (Suppl. 2), 41–44

Webb, P.M., Knight, T., Greaves, S., Wilson, A., Newell, D.G., Elder, J. & Forman, D. (1994) Relation between infection with *Helicobacter pylori* and living conditions in childhood: evidence for person to person transmission in early life. *Br. Med. J.*, 308, 750–753

Whitaker, C.J., Dubiel, A.J. & Galpin, O.P. (1993) Social and geographical risk factors in *Helicobacter pylori* infection. *Epidemiol. Infect.*, 111, 63–70

WHO (1993) *The control of schistosomiasis – second report of the WHO expert committee* (WHO Technical Report No. 830). Geneva, World Health Organization

P. Boffetta
Unit of Environmental Cancer Epidemiology,
International Agency for Research on Cancer,
150 cours Albert Thomas, 69372 Lyon cedex 08,
France

Exposure to occupational carcinogens and social class differences in cancer occurrence

P. Boffetta, M. Kogevinas, P. Westerholm and R. Saracci

It has been estimated that occupational exposures are responsible for about 4% of all human cancers in industrialized countries. These cancers are concentrated among manual workers and in the lower social classes, thus contributing to the social class gradient in cancer incidence and mortality. On the basis of the 1971 cancer mortality data from England and Wales, it was estimated that occupational cancer is responsible for about a third of the total cancer difference between high (I, II and III-NM) and low (III-M, IV and V) social classes, and for about half of the difference for lung and bladder cancer. However, direct evidence on the extent of the contribution of occupational exposure to carcinogens to social class differences is lacking, and several problems, such as the possible interaction between carcinogens and the effect of extraoccupational confounding factors, add further elements of uncertainty.

The analysis of social class differences in cancer occurrence by occupation involves aspects of circular reasoning, since in many cases people are classified by social class according to the job they hold or have held at sometime during their life (see the chapter by Berkman and Macintyre in this book), and main occupation is highly correlated with other indicators of social class, such as education and income.

Occupational exposure to carcinogens has some peculiar characteristics compared with other causes of cancer related to social class, such as tobacco smoking and alcohol drinking. First, the exposure is to a large extent involuntary. Although some aspects of personal choice exist, such as the decision not to use protective equipment, the determinants of the exposure to carcinogens in the workplace are mainly inherent in the job tasks. Second, the cancer hazard may not be known to the worker, such as in the case of complex mixtures with variable composition (for example, mineral oil mist). Third, occupational cancer can be prevented by means other than (or in addition to) changes in personal behaviour. Finally, more data are available on the interaction between specific occupational exposures and other risk factors, both occupational and extraoccupational, than in other areas of cancer epidemiology (Saracci & Boffetta, 1994).

Occupational causes of cancer

Over the last two decades, the *Monographs* programme of the International Agency for Research on Cancer (IARC) has systematically evaluated the carcinogenic risk to humans from exposure to chemical, physical and biological agents and mixtures (IARC 1972–1996), and the majority of known occupational or suspected occupational carcinogens have now been assessed[*]. At present, 21 chemicals, groups of chemicals or mixtures for which exposures are mostly occupational (excluding pesticides and drugs) are classified as human carcinogens (IARC Group 1; Table 1). While some agents, such as asbestos, benzene and heavy metals, are currently widely used in many countries, others are mainly of historical interest (for example, mustard gas and 2-naphthylamine). An additional 20 agents are classified as probably carcinogenic to humans (IARC Group 2A): these are mainly agents carcinogenic in experimental animals with limited evidence of carcinogenicity in humans from epidemiological studies (Table 2). Exposures to some of these agents, such as crystalline silica, formaldehyde and

[*]Although the IARC Monographs programme has covered most of the known or suspected causes of cancer, there are some important groups of occupational agents that have not yet been evaluated by IARC – namely ionizing radiation, and electrical and magnetic fields.

Table 1. Chemicals, groups of chemicals or mixtures for which exposures are mostly occupational, and industrial processes and occupations, evaluated in the IARC *Monographs* (Vols 1–63) as carcinogenic to humans (IARC Group 1)[a]

Agents[b]	Human target organ(s)/cancer	Main industry/use
Chemicals and groups of chemicals		
4-Aminobiphenyl [92-67-1]	Bladder	Rubber manufacture
Arsenic [7440-38-2] and arsenic compounds[c]	Lung, skin	Glass, metals, pesticides
Asbestos [1332-21-4]	Lung, pleura, peritoneum	Insulation, filter material, textiles
Benzene [71-43-2]	Leukaemia	Solvent, fuel
Benzidine [92-87-5]	Bladder	Dye/pigment manufacture, laboratory agent
Beryllium [7440-41-7] and beryllium compounds (1993)	Lung	Aerospace industry/metals
Bis(chloromethyl)ether [542-88-1] and chloromethyl methyl ether [107-30-2] (technical-grade)	Lung	Chemical intermediate, by-product
Cadmium [7440-43-9] and cadmium compounds (1993)	Lung	Dye/pigment manufacture
Chromium[VI] compounds (1990)	Nasal cavity, lung	Metal plating, dye/pigment manufacture
Coal-tar pitches [65996-93-2]	Skin, lung, bladder	Building material, electrodes
Coal-tars [8007-45-2]	Skin, lung	Fuel
Ethylene oxide [75-21-8]	Leukaemia	Chemical intermediate, sterilant
Mineral oils (untreated and mildly treated)	Skin	Lubricants
Mustard gas (sulphur mustard) [505-60-2]	Pharynx, lung	War gas
2-Naphthylamine [91-59-8]	Bladder	Dye/pigment manufacture
Nickel compounds (1990)	Nasal cavity, lung	Metallurgy, alloys, catalyst
Shale-oils [68308-34-9]	Skin	Lubricants, fuels
Soots	Skin, lung	Pigments
Talc containing asbestiform fibres	Lung	Paper, paints
Vinyl chloride [75-01-4]	Liver, lung, blood vessels	Plastics, monomer
Wood dust (1994)	Nasal cavity	Wood industry
Industrial processes and occupations		
Aluminium production	Lung, bladder	
Auramine manufacture	Bladder	
Boot and shoe manufacture and repair	Nasal cavity, leukaemia	
Coal gasification	Skin, lung, bladder	
Coke production	Skin, lung, kidney	
Furniture and cabinet making	Nasal cavity	
Haematite mining (underground) with exposure to radon	Lung	
Iron and steel founding	Lung	
Isopropanol manufacture (strong-acid process)	Nasal cavity	
Magenta manufacture (1993)	Bladder	

Table 1. (Contd) Chemicals, groups of chemicals or mixtures for which exposures are mostly occupational, and industrial processes and occupations, evaluated in the IARC Monographs (Vols 1–63) as carcinogenic to humans (IARC Group 1)[a]

Agents[b]	Human target organ(s)/cancer	Main industry/use
Industrial processes and occupations		
Painter (occupational exposure as a) (1989)	Lung	
Rubber industry (certain occupations)	Bladder, leukaemia	
Strong-inorganic-acid mists containing sulphuric acid (occupational exposure to) (1992)	Lung, larynx	

[a]Several drugs used in cancer chemotherapy are classified as human carcinogens; occupational exposure can occur in manufacturing, pharmacies and hospitals.
[b]Year in parenthesis, year in which the evaluation was made subsequent to the 1987 Supplement 7 Working Group for agents, mixtures or exposure circumstances considered in Vols 43–63 of the Monographs. Number in square brackets, CAS Registry No.
[c]This evaluation applies to the group of chemicals as a whole and not necessarily to all individual chemicals within the group.

1,3-butadiene, are currently prevalent in many countries. A large number of agents are classified as possible human carcinogens (IARC Group 2B), such as acetaldehyde, DDT, inorganic lead compounds and man-made mineral fibres. For the majority of these chemicals the evidence of carcinogenicity comes from studies in experimental animals.

In addition, the IARC Monographs programme has evaluated the evidence of carcinogenic risk from employment in specific industries and occupations for which data existed from epidemiological studies, although the exposures responsible for the risk could not be identified with certainty. At present, 13 industries or occupations are classified as entailing a carcinogenic risk (Group 1; Table 1), and four additional industries or occupations are classified as probably entailing a risk (Group 2A; Table 2). Three points should be considered when looking at these tables. First, there is a certain degree of duplication between the list of agents and that of occupations and industries, which is partially due to historical reasons. For example, employment in certain wood industries such as furniture and cabinet making was classified in Group 1 in 1981, and at that time the data did not allow a conclusion to be made about the role of specific exposures, such as to wood dust. However, in 1994 exposure to wood dust was evaluated on the basis of additional evidence that had become available in recent years and was in turn classified in Group 1, so the early classification based on industry has now lost most of its relevance. Second, in contrast to the case for individual chemical and physical agents, there was no attempt in the Monographs programme to evaluate occupations and industries systematically, and the lists of these in Tables 1 and 2 are therefore by no means exhaustive (for a more complete discussion of industries and occupations entailing a carcinogenic risk, besides the Monographs evaluations, see Boffetta et al., 1995). Finally, as the classifications are based on incomplete knowledge of exposures, such evaluations do not necessarily apply to all workers employed in a given industry, and differences (although not detectable by the evaluation) are likely to exist between time periods of employment, countries, factories, and even departments and jobs within a factory. For example, employment in the rubber industry has been classified in Group 1 on the basis of an excess risk of bladder cancer that was mainly reported in studies conducted during the 1960s and 1970s of exposures between the 1930s and the 1950s; subsequent studies from the same and other plants, however, have shown much smaller, if any, excess risk, suggesting that changes in the technological process may have greatly reduced, if not abolished, the risk (Swerdlow, 1990).

Several environmental agents are known or suspected to cause cancer in humans (Table 3), and although exposure to such agents is not primarily

Table 2. Chemicals, groups of chemicals or mixtures for which exposures are mostly occupational, and industrial processes and occupations, evaluated in the IARC *Monographs* (Vols 1–63) as probably carcinogenic to humans (IARC Group 2A)

Agents[a]	Suspected human target organ(s)/cancer	Main industry/use
Chemicals and groups of chemicals		
Acrylonitrile [107-13-1]	Lung, prostate, lymphoma	Plastics, rubber, textiles, monomer
Benzidine-based dyes	Bladder	Paper, leather, textile dyes
1,3-Butadiene [106-99-0] (1992)	Leukaemia, lymphoma	Plastics, rubber, monomer
para-Chloro-ortho-toluidine [95-69-2]	Bladder	Dye/pigment manufacture, textiles and its strong acid salts (1990)
Creosotes [8001-58-9]	Skin	Wood preservation
Diethyl sulphate [64-67-5] (1992)	–	Chemical intermediate
Dimethylcarbamoyl chloride [79-44-7]	–	Chemical intermediate
Dimethyl sulphate [77-78-1]	–	Chemical intermediate
Epichlorohydrin [106-89-8]	–	Plastics/resins monomer
Ethylene dibromide [106-93-4]	–	Chemical intermediate, fumigant, fuels
Formaldehyde [50-00-0] (1995)	Nasopharynx	Plastics, textiles, laboratory agent
4,4´-Methylene bis(2-chloroaniline) (MOCA) [101-14-4] (1993)	Bladder	Rubber manufacture
Polychlorinated biphenyls [1336-36-3]	Liver, bile ducts, leukaemia, lymphoma	Electrical components
Silica [14808-60-7], crystalline	Lung	Stone cutting, mining, glass, paper
Styrene oxide [96-09-3]	–	Plastics, chemical intermediate
Tetrachloroethylene [127-18-4] (1995)	Oesophagus, lymphoma	Solvent, dry cleaning
Trichloroethylene [79-01-6] (1995)	Liver, lymphoma	Solvent, dry cleaning, metal
Tris(2,3-dibromopropyl)phosphate [126-72-7]	–	Plastics, textiles, flame retardant
Vinyl bromide [593-60-2]	–	Plastics, textiles, monomer
Vinyl fluoride [75-02-5] (1995)	–	Chemical intermediate
Industrial processes and occupations		
Art glass, glass containers and pressed ware, manufacture of (1993)	Lung, stomach	
Hairdresser or barber (occupational exposure as a) (1993)	Bladder, lung	
Non-arsenical insecticides (occupational exposures in spraying and application of) (1991)	Lung, myeloma	
Petroleum refining (occupational exposures in) (1989)	Leukaemia, skin	

[a]Year in parenthesis, year in which the evaluation was made subsequent to the 1987 Supplement 7 Working Group for agents, mixtures or exposure circumstances considered in Vols 43–63 of the *Monographs*. Number in square brackets, CAS Registry No.

occupational, there are groups of individuals exposed to them because of their work. For example, hospital workers are exposed to hepatitis B virus, food processors are exposed to aflatoxins from contaminated foodstuff, outdoor workers are exposed to ultraviolet radiation or diesel engine exhaust, and bar staff are exposed to environmental tobacco smoke.

Occupation may exert an indirect effect on cancer risk. For example, employed women, in particular those in high social classes, may have fewer pregnancies and be older at their first pregnancy than unemployed women – two factors that are linked to risk of breast cancer.

It is important to note that the known or highly suspected occupational carcinogens exert their effects on a limited number of cancer sites – namely the organs of the respiratory tract, the urinary bladder, the liver, the skin and the lymphatic and haematopoietic system. Cancers of these organs, with the exception of the lymphatic and haematopoietic system, are among those showing the strongest social class gradients (see the chapter by Faggiano et al. in this book), suggesting an important role of occupation in social class differences.

Estimates of cancers due to occupational risk factors

Different estimates of cancer risk attributable to occupation vary greatly. A summary of existing estimates is shown in Table 4. The large variability in the estimates arises from the differences in the data sets used and on the assumptions applied.

Most of the published estimates of the fraction of cancers attributable to occupational risk factors are not based on accurate measures of the proportions of exposed subjects and the degree of exposure. An exception is the paper by Vineis and Simonato (1991), which provided estimates of the number of cases of lung and bladder cancer attributable to occupation derived from a detailed review of case–control studies, and demonstrated that in specific populations located in industrial areas the proportion of lung cancer due to occupational exposures may be as high as 40%; these estimates were dependent not only on the local prevailing exposures, but also to some extent on the method of defining and assessing exposure.

The most generally accepted estimates of cancers attributable to occupations, however, are those

Table 3. Environmental agents and exposures that may be encountered in occupational settings evaluated in IARC Monographs (Vols 1–63) as carcinogenic (IARC Group 1) or probably carcinogenic (IARC Group 2A) to humans

Agent/exposure[a]	Target organ[b]
IARC Group 1	
Aflatoxins [1402-68-2] (1993)	Liver
Chronic infection with hepatitis B virus (1993)	Liver
Chronic infection with hepatitis C virus (1993)	Liver
Erionite [66733-21-9]	Lung, pleura
Radon [10043-92-2] and its decay products (1988)	Lung
Infection with Schistosoma haematobium (1994)	Bladder
Solar radiation (1992)	Skin
Tobacco smoke	Lung, bladder, oral cavity, pharynx, larynx, oesophagus, pancreas
IARC Group 2A	
Benz[a]anthracene [56-55-3]	–
Benzo[a]pyrene [50-32-8]	–
Dibenz[a,h]anthracene [52-70-3]	–
Diesel engine exhaust (1989)	(Lung, bladder)
IQc (2-Amino-3-methylimidazo[4,5-f]quinoline) [76180-96-6] (1993)	–
N-Nitrosodimethylamine [62-75-9]	–
Ultraviolet radiation A (1992)	(Skin)
Ultraviolet radiation B (1992)	(Skin)
Ultraviolet radiation C (1992)	(Skin)

[a]Year in parenthesis, year in which the evaluation was made subsequent to the 1987 Supplement 7 Working Group for agents, mixtures or exposure circumstances considered in Vols 43–63 of the Monographs. Number in square brackets, CAS Registry No.
[b]Suspected target organs are given in parentheses.

Table 4. Estimated proportions of cancer attributable (PAR) to occupations in selected studies

Study	Population	PAR and cancer site	Comments
Higginson & Muir, 1976	Not stated	1–3% total cancer	No detailed presentation of assumptions
Doll & Peto, 1981	United States, early 1980	4% (range 2–8%) total cancer. In men: 6.8% all cancers, 4% liver, 2% larynx, 15% lung, 25% nose, 25% pleura, 4% bone, 10% skin (non-melanoma), 1% prostate, 10% bladder, 10% leukaemia. In women: 1.2% all cancers	Based on all studied cancer sites; reported as 'tentative' estimate
Hogan & Hoel, 1981	United States	3% (range 1.4–4.4%) total cancer	Risk associated with occupational asbestos exposure
Vineis & Simonato, 1991	Various	1–5% lung cancer, 16–24% bladder cancer	Calculations on the basis of data from case–control studies.

presented in a detailed review by Doll and Peto (1981) on the causes of cancer in the population of the United States of America in 1980. These authors concluded that about 4% of all deaths due to cancer may be caused by occupational carcinogens, with 'acceptable limits' (that is, still plausible in view of all the evidence at hand) of 2% and 8%. These authors also provided an estimate of this proportion for specific cancer sites (Table 4), with pleural, sinonasal and lung cancer having the highest proportions.

These proportions are dependent on how causes other than occupational exposures contribute to the development of cancers. For example, the proportion of lung cancer attributable to occupational exposures would be higher in a population of lifetime non-smokers than in a population containing the same proportion of exposed workers and a higher proportion of smokers. Furthermore, if one considered not the whole population (to which most of the estimates refer) but the segments of the adult population in which exposure to occupational carcinogens almost exclusively occur (manual workers in mining, agriculture and industry, broadly taken – who in the USA numbered 31 million out of a population aged 20 and over of 158 million), the proportion of cancer deaths attributable to occupational exposure would be substantially higher than the 4% in the overall population.

Estimates of the role of occupational cancer in social class differences in cancer occurrence

One possible approach to estimate the contribution of occupational exposure to carcinogens to differences in cancer occurrence by social class is to calculate a measure of association between social class and cancer risk after excluding those cancers that may be attributable to occupation, and to compare it with the same measure based on all cancers. In the case of the comparison between two social classes, this approach can be formalized as:

$$c = \frac{R_c - R_a}{R_c - 1} \times 100,$$

where R_c is the 'crude' measure of association, equal to the ratio of the rate of cancer in the lower class over the rate in the upper class (r_1 / r_0), R_a is the 'adjusted' measure once the contribution of occupation is accounted for ($R_a = r_1'/r_0'$), and c is the percentage of the difference explained by the adjustment. One can calculate r_1' and r_0' as follows:

$$r_1' = (r_1 p_1 - (r_1 p_1 + r_0 p_0)db)/p_1, \text{ and}$$
$$r_0' = [r_0 p_0 - (r_1 p_1 + r_0 p_0)d(1 - b)]/p_0,$$

where p_1 and p_0 are the proportion of subjects in the two classes, d is the estimate of the proportion of cancers attributable to occupation and b is the proportion of such cancers occurring in the lower social class. When $b = 1$ (that is, it is assumed that

Table 5. Ratios of cancer mortality between manual and non-manual social classes with and without excluding cancers attributable to occupational exposures (England and Wales, 1971)[a]

Cancer site	Crude rate ratio (R_c)[b]	Rate ratio for the proportion of cancers not attributable to occupation (R_a)[c]	Excess risk (%) attributable to occupation[d]
Liver	1.16	1.09	42
Larynx	1.76	1.71	5
Lung	1.71	1.37	48
Nose	1.38	0.90	100
Skin (non-melanoma)	1.77	1.55	29
Prostate	1.19	1.17	9
Bladder	1.36	1.17	52
All cancers	1.40	1.27	32

[a]Based on 25–64 years cumulative rates reported by Logan (1982). Only cancer sites that have been strongly related with occupational exposures are reported. Proportions of cancers attributable to occupation were derived from Doll & Peto (1981); all cancers related to occupation were assumed to occur among manual workers.
[b]Ratio of the rate among manual workers to the rate among non-manual workers.
[c]As crude rate ratio, after excluding cancers attributable to occupation (see text for details).
[d]Percentage of the crude rate ratio accounted for by cancers attributable to occupation, or $[(R_c - R_a)/(R_c - 1)] \times 100$.

all cancers attributable to occupation occur in the lower social class), $r_0' = r_0$. This method can be easily expanded to a comparison of more than two classes.

This approach has been applied to the 1971 cancer mortality data of England and Wales reported by Logan (1982; see the chapter by Faggiano et al. in this book for detailed results). Cumulative mortality rates between ages 25 and 64 were calculated for the combined social classes I, II and III-NM (r_0) and III-M, IV and V (r_1). The values for 1971 of $p_1 = 0.65$ and $p_0 = 0.35$ were also derived from Logan (1982). The proportions of cancer attributable to occupational exposures (d) proposed by Doll and Peto (1981) were used (Table 4). Table 5 shows the results among males for those cancer sites, included in the review by Doll and Peto (1981), that showed a social class gradient ($r_1 > r_0$) in the data reported by Logan (1982), assuming $b = 1$. Occupational exposures were estimated to account for about a third of the difference in total cancer mortality, for the whole difference for sinonasal cancer and for about half of the difference for lung and bladder cancer.

Key elements in these estimates are the parameters b and d. In particular, the values for d were derived from the review of Doll and Peto (1981). These authors stress that their values may represent an overestimation of the true proportion of cancers due to occupational exposures; in addition, they attempted an estimate for the United States, and this proportion in other countries may be smaller. The effect of the assumptions on the value of b and d was addressed by repeating the analysis on lung cancer with different values for these two parameters (Figure 1). Although the figure suggests the role of occupation in social class differences in cancer is small if the proportion of lung cancers attributable to occupation is below 10% or if the proportion of such cancers occurring in manual workers is below 90%, the most reliable estimates for the percentage of the social class difference due to occupation are in the range 20–50%.

Direct evidence from epidemiological studies

While in many epidemiological studies socioeconomic status has been treated as a potential confounder in the analysis of occupational cancer risk

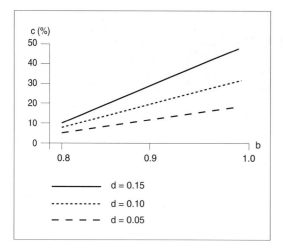

Figure 1. Percentage of difference in lung cancer between manual and non-manual workers attributable to occupational exposure (c) according to the proportion of cancers attributable to occupation occurring among manual workers (b), and the proportion of cancers attributable to occupation (d).

factors, no study has addressed this relationship from the other point of view – that is, to what extent occupational exposure confounds the association between social class and cancers. In the traditional approach, the increased cancer risk among certain occupational groups is seen as resulting from the combined effect of an 'occupational' factor, related to workplace exposures, and a 'social' factor, related to lifestyle or other determinants of cancer risk. It has therefore been proposed that the latter be adjusted to obtain an unbiased estimate of the former (Fox & Adelstein, 1978). Different methods have been proposed for such adjustment, such as comparison of cancer mortality of men and their wives (Office of Population Censuses and Surveys, 1978), standardization for social class (Fox & Adelstein, 1978; Milham 1985), and multivariate regression including a social class term in the model (Miettinen & Rossiter, 1990).

A number of studies have addressed this issue analytically. For example, Siemiatycki et al. (1988) have calculated the degree of confounding introduced by tobacco smoking, ethnicity (French versus other) and family income (used as an indicator of social class) in the association between lung cancer and 25 occupations in a large case–control study conducted in Montreal, Canada. The results did not suggest a large confounding effect of family income: for five of the 25 occupations, family income exerted a stronger confounding effect than either smoking or ethnicity, and only in the same number of occupations the confounding bias was greater than 1.10.

The approach of adjusting for social class when studying occupational carcinogens has been criticized, as it will lead to an underestimation of the risk if the group chosen for comparison (for example, other occupations in the same social class stratum) also has job-related carcinogenic exposures (Brisson et al., 1987).

Although no direct evidence can therefore be drawn from analytical studies, it is clear that a confounding effect exists, which is likely to act in both directions, and that the difference between social groups defined by occupation (for example, manual and non-manual workers) in cancer risk cannot be viewed as indicating solely an effect of occupational exposures.

Interaction between occupational exposures and other cancer risk factors

An important aspect to take into account when considering occupational exposures as a cause of social class differences in cancer is the possibility of an interaction between these exposures and other risk factors in determining cancer risk. Although interaction between contributory factors may be a general characteristic of carcinogenesis – and the interactions between alcohol drinking and tobacco smoking in the etiology of cancer of the upper aerodigestive tract (Boyle et al., 1992) and between aflatoxin intake and chronic infection with hepatitis B virus in the etiology of liver cancer (Ross et al., 1992) have been extensively studied – it may be particularly important when considering occupational exposures, as it offers a particularly strong argument in favour of prevention.

Strictly speaking, interaction occurs when the combined effect of two exposures differs from the sum (additive model) or the product (multiplicative model) of the effect of each exposure, or:

$R_{AB} \neq R_A + R_B - 1$ (additive model), and
$R_{AB} \neq R_A \times R_B$ (multiplicative model),

where R_A and R_B are the relative risk among those exposed to binary exposure variables A and B, and R_{AB} is the relative risk of those exposed to both. In practice, however, one also speaks of interaction when $R_{AB} = R_A \times R_B$, and the example below refers to this situation.

Let us consider an example of two populations of manual and non-manual workers differing only either for exposure to tobacco smoking (with relative risk of lung cancer of 10 among smokers as compared with non-smokers) or for an occupational exposure, say asbestos (with relative risk of 5), assuming no interaction according to a multiplicative model (relative risk among those exposed to both factors of 50), and a rate of lung cancer of 1/1000 among those exposed to neither factor. If 40% of nonmanual workers smoke and no non-manual workers are exposed to asbestos, their overall lung cancer rate would be 4.6/1000 (1/1000 × 0.6 + 10/1000 × 0.4). The rate among manual workers with a 20% higher proportion of smokers would be 6.4/1000 (1/1000 × 0.4 + 10/1000 × 0.6) while the rate with 40% smokers but 10% of the workers exposed to asbestos would be 6.44/1000 (1/1000 × 0.54 + 10/1000 × 0.36 + 5/1000 × 0.06 + 50/1000 × 0.04). Therefore, a smaller increase in the proportion of those exposed to a weaker risk factor has a similar or greater effect than a larger increase in the proportion of those exposed to a stronger risk factor, because of the very strong risk in the small group of workers exposed to both factors. Note that no association was assumed between the two exposures – that is, the proportion of smokers was considered to be the same among workers exposed and unexposed to asbestos. Had such an association been present, the results would have been even more extreme.

This example shows that when groups, such as social class groups, differ in their exposure to more than one factor, small differences in risk in one factor that interacts with other factor(s) may have unexpectedly large effects on the overall difference in risk. In most cases, however, information on the distribution of exposures and their pattern of interaction is lacking, and one can only speculate about the relative contributions of each factor and of their interaction.

A problem related to the interaction between occupational exposures and other risk factors is exposure to mixed occupational agents; this may be through work in an environment where several carcinogenic agents are present or through exposure to complex mixtures. In these situations, one should consider that, even if the exposure to each component of the mixture may have a relatively small effect on cancer risk, the exposure to the whole mixture, resulting from the sum of the individual effects and from the interactions, may be large. Again, detailed data are rarely available to evaluate the relevance of this problem both in specific situations and in global estimates such as those presented above (Vainio et al., 1990).

Short-term workers

Short-term workers are particularly interesting with respect to the association between low social class, occupational exposure to carcinogens and cancer risk. In many occupational epidemiological studies, the characterization of job tasks, exposures and social class indicators for each member of the cohort is problematic, but information on duration of employment is available. A higher cancer risk has been frequently observed in short-term workers (Stewart et al., 1990). Short-term workers are usually defined as those with less than six months or one year of employment and they constitute a group with high job mobility. For example, Figure 2 shows as an example the standardized mortality ratios for several types of cancer among workers producing man-made vitreous fibres, followed-up between 1950 and 1990 in seven European countries, by duration of employment. Two explanations are possible for this difference. First, short-term workers may be at increased risk of cancer because of employment in particularly hazardous or dirty jobs. Second, the fact that mortality from most cancers was higher among workers with less than one year of employment as compared with longer-term workers suggests that extraoccupational factors may play an important role. Short-term workers may differ from other workers in personal habits entailing a higher risk of cancer. In this sense, specific groups of manual workers such as short-term workers may contribute disproportionately to social class differences, even in those cancers that are not directly caused by occupational exposures.

Conclusions

The complete assessment of the confounding effect of occupational exposures on the association between social class and cancer risk is complicated by a number of factors: incomplete knowledge about occupational carcinogens; possible interaction

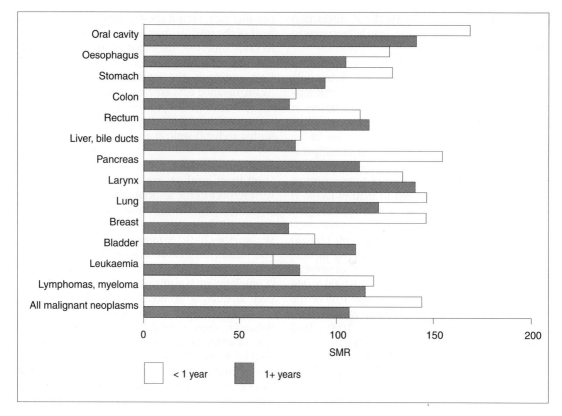

Figure 2. Standardized cancer mortality ratios (SMRs) of workers in man-made vitreous fibre production, by duration of employment.

between occupational carcinogens and other risk factors linked with social class (for example, tobacco smoking); and lack of information on the distribution of other socially determined risk factors.

A full assessment and control of confounding could be achieved only if all carcinogenic exposures (both occupational and extraoccupational) were known and measured; in practice, however, this is not possible because of both ignorance about the carcinogenicity of many agents and incomplete information on exposure to known carcinogens.

On the basis of the available evidence, occupational exposures have been estimated to be responsible, in developed countries, for approximately a third of the excess of all cancers occurring in lower social classes as compared with higher social classes, and for approximately half of this difference for important occupationally related cancers such as lung and bladder cancer. Their contribution to social differences in cancer risk in women and in people from developing countries is likely to be smaller. The figure of a third of excess cancers among men from industrialized countries in the lower social classes caused by occupational exposures may represent an overestimation as it is based on the results of epidemiological studies that investigated the effect of relatively high exposure levels occurring in the past (this argument, however, would not apply to developing countries). Other considerations suggest the possibility of an underestimation – in particular, the lack of knowledge on the interaction among different occupational exposures and between those and other cancer risk factors, and the possibility that yet undetected occupational carcinogens have operated and are still operating.

References

Boffetta, P., Kogevinas, M., Simonato, L., Wilbourn, J. & Saracci, R. (1995) Current perspectives on occupational cancer risk. *Int. J. Occup. Environ. Health*, 1, 315–325

Boyle, P., Macfarlane, G.J., Zheng, T., Maisonneuve, P., Evstifeeva, T. & Scully, C. (1992) Recent advances in epidemiology of head and neck cancer. *Curr. Opin. Oncol.*, 4, 471–477

Brisson, C., Loomis, D. & Pearce, N. (1987) Is social class standardisation appropriate in occupational studies? *J. Epidemiol. Community Health*, 41, 290–294

Doll, R. & Peto, R. (1981) *The causes of cancer. Quantitative estimates of avoidable risk of cancer in the United States today*. New York, Oxford University Press

Fox, A.J. & Adelstein, A.M. (1978) Occupational mortality: work or way of life? *J. Epidemiol. Community Health*, 32, 73–78

Higginson, J. & Muir, C.S. (1976) The role of epidemiology in elucidating the importance of environmental factors in human cancer. *Cancer Detect. Prev.*, 1, 79–105

Hogan, M.D. & Hoel, D.G. (1981) Estimated cancer risk associated with occupational asbestos exposure. *Risk Analysis*, 1, 67–76

IARC (1972–1995) *IARC monographs on the evaluation of carcinogenic risks to humans* (Vols 1–63). Lyon, International Agency for Research on Cancer

Logan, W.P.D. (1982) *Mortality from cancer in relation to occupation and social class* (IARC Scientific Publication No. 36). Lyon, International Agency for Research on Cancer

Miettinen O.S. & Rossiter, C.E. (1990) Man-made mineral fibers and lung cancer. Epidemiologic evidence regarding the causal hypothesis. *Scand. J. Work Environ. Health*, 16, 221–231

Milham, S. (1985) Improving occupational standardized proportionate mortality ratio analysis by social class stratification. *Am. J. Epidemiol.*, 121, 472–475

Office of Population Censuses and Surveys (1978) *Occupational mortality decennial supplement, 1970–1972, England and Wales*. London, Her Majesty's Stationery Office

Ross, R.K., Yuan, J.M., Yu, M.C., Wogan, G.N., Qian, G.S., Tu, J.T., Groopman, J.D., Gao, Y.T. & Henderson, B.E. (1992) Urinary aflatoxin biomarkers and risk of hepatocellular carcinoma. *Lancet*, 339, 943–946

Saracci, R. & Boffetta, P. (1994) Interactions of tobacco smoking with other causes of lung cancer. In: Samet, J.M., ed., *Epidemiology of lung cancer*. New York, Dekker. pp. 465–493

Siemiatycki, J., Wacholder, S., Dewar, R., Cardis, E., Greenwood, C. & Richardson, L. (1988) Degree of confounding bias related to smoking, ethnic group, and socioeconomic status in estimates of the associations between occupation and cancer. *J. Occup. Med.*, 30, 617–625

Stewart, P.A., Schairer, C. & Blair, A. (1990) Comparison of jobs, exposures, and mortality risks for short-term and long-term workers. *J. Occup. Med.*, 32, 703–708

Swerdlow, A.J. (1990) Effectiveness of primary prevention of occupational exposures on cancer risk. In: Hakama, M., Beral, V., Cullen, J.W. & Parkin, D.M., eds, *Evaluating effectiveness of primary prevention of cancer* (IARC Scientific Publication No. 103). Lyon, International Agency for Research on Cancer. pp. 23–56

Vainio, H., Sorsa, M. & McMichael, A.J. (1990) *Complex mixtures and cancer risk* (IARC Scientific Publication No. 104). Lyon, International Agency for Research on Cancer

Vineis, P. & Simonato, L. (1991) Proportion of lung and bladder cancers in males resulting from occupation: a systematic approach. *Arch. Environ. Health*, 46, 6–15

Corresponding author:
P. Boffetta
Unit of Environmental Cancer Epidemiology,
International Agency for Research on Cancer,
150 Cours Albert Thomas, 69372 Lyon Cedex 08,

I: Unemployment and cancer: a literature review

E. Lynge

With a tenth of the labour force involuntarily out of work, unemployment has become an important element among the socioeconomic determinants of health in the rich countries. Unemployed men have an excess cancer mortality of close to 25% compared with that of all men in the labour force. The available data from various countries indicate that this excess risk is found both in periods when the unemployment rate is about 1% and in periods when it is about 10%. Furthermore, it persists long after the start of unemployment and it does not disappear when social class, smoking, alcohol intake, and previous sick days are controlled for. The excess cancer mortality comes mainly from lung cancer, and the increased risk of lung cancer does not disappear when social class and number of previous sick days are controlled for. Unemployment does not increase smoking, but unemployed men have a slightly higher smoking prevalence before unemployment. However, as the excess lung cancer risk among unemployed men remains after controlling for social class, it seems unlikely that it can be explained only by differences in smoking prior to unemployment.

The populations of the countries of the Organization for Economic Cooperation and Development on average experienced an unemployment rate of 6.5% in 1993. This percentage increased to 9.2% when discouraged workers and those involuntarily working part time were classified as unemployed (OECD, 1994). The unemployment rate in the United States of America has fluctuated between 5% and 10% during the past 20 years, and was 7% in 1993. The unemployment rate in Japan increased from close to 1% in the early 1970s to more than 2% in 1993. A very dramatic increase was seen in the unemployment rate in countries of the European Union. The rate was below 3% in the early 1970s but went up to 11% in 1985, and although it decreased to close to 8% in 1990, it increased again to close to 11% in 1993 (European Commission, 1994) (Figure 1).

With a tenth of the labour force involuntarily out of work, unemployment has thus become an important element among the socioeconomic determinants of health in the rich countries. This chapter focuses on unemployment as a potential risk factor for developing cancer. The problems faced by cancer patients in continuing working after diagnosis and treatment are not discussed.

The chapter reviews available studies on cancer mortality and morbidity among unemployed men. Due to the considerable changes in the participation of women in the labour market during the last 30 years, data on health in relation to unemployment among women are often difficult to interpret and are therefore not discussed here.

Data sources

The available studies on cancer mortality and cancer morbidity among unemployed men are all cohort studies, where populations classified as unemployed at the start of the study period are followed-up for cancer death and/or incident cancer cases over a certain period of time. The observed numbers of cancer deaths and/or incident cancer cases are compared with the expected numbers based on the accumulated person years at risk and rates for either all men or all economically active men in the respective populations. The results are reported as standardized mortality ratios (SMRs), standardized incidence ratios (SIRs) or relative risks (RRs). Cancer mortality among unemployed men has been studied in the United Kingdom, Finland, Italy (Turin), the United States, and Denmark. Cancer morbidity among unemployed men has been studied in the United Kingdom and Denmark.

United Kingdom

Longitudinal study 1971 In the Office of Population Censuses and Surveys longitudinal study, routinely

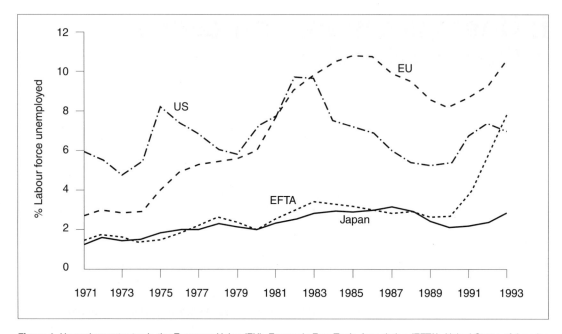

Figure 1. Unemployment rates in the European Union (EU), Economic Free Trade Association (EFTA), United States of America (US) and Japan 1971 to 1993. Redrawn from European Commission, 1994.

collected census, vital status, and other data are brought together continuously in time for a 1% sample of individuals in the United Kingdom. The study started with a 1% sample of the 1971 census population, which included 530 000 persons. They were followed-up for mortality in the National Health Service Central Register for the 10-year period 1971–1981, and 513 000 were successfully traced. In the analysis, the unemployed group comprised those men who indicated in response to the 1971 census question on economic position that they were seeking work or waiting to take up a job in the week before the census. Among 161 699 men aged 15–64, 5961 (3.6%) were seeking work, and during the 10-year follow-up period 1971–1981, 102 cancer deaths occurred in this group (Moser et al., 1990). The population was also followed-up for incident cancer cases during the same period by linkage to data from the National Cancer Registration Scheme, and 267 incident cancer cases occurred (Kogevinas, 1990).

Longitudinal study 1981 When the longitudinal study was updated with information from the 1981 census, 14 675 men in the age group 15–64 were recorded as seeking work. They were followed-up for mortality in the National Health Service Central Register for the three-year period 1981–1983. Cancer deaths were reported only for 1983, when 25 cancer deaths occurred in this group (Moser et al., 1987).

British Regional Heart Study In 1978–1980, men aged 40–59 were randomly selected from one general practice in each of 24 towns in England, Wales and Scotland to form the study population of the British Regional Heart Study. At the time of recruitment, research nurses administered a standard questionnaire includimg questions on occupational history, employment status, and medical history. After five years (1983–1985), a postal questionnaire was sent to all surviving men still resident in the United Kingdom and information was obtained from 7275 (98%) on their employment status during the five years before recruitment and the five years between the two questionnaires. Included in the analysis were 6191 men with continuous employment during the five years preceding recruitment; 923 (15%) of these men experienced unemployment not due to illness during the next five years. Mortality was followed through the National Health Service Central Register up to January 1990, and 27 cancer deaths occurred (Morris et al., 1994).

Table 1. Populations covered by studies of cancer mortality and/or morbidity among unemployed men

Country	Population	Age group	Number of All men	Number of Unemployed men	Percentage unemployed	Reference
United Kingdom	1971 census	15–64	161 699	5661	4%	Moser et al., 1990
	1981 census	16–64	NA	14 675	6–14%	Moser et al., 1987
	1978–1980 British Regional Heart Study	40–59	6191	923	15%	Morris et al., 1994
Finland	1980 census	30–54	NA	NA	7%	Martikainen, 1990
Italy	1981 census	15–59	NA	13 462	4%	Costa & Segnan, 1987
USA	1979–1983 Current Population Surveys	White, 25–64	137 274	7501	5%	Sorlie & Rogot, 1990
Denmark	1970 census	20–64	1 292 337	15 340	1%	Lynge & Andersen, 1996
	1986 census	20–64	1 352 932	183 184	14%	Lynge & Andersen, 1996

NA, not applicable.

Finland

The Finnish study was based on the 1980 census records in Finland linked with all deaths during 1981–1985 and certain variables from the 1970 and 1975 censuses. In addition, variables that measure health and income were linked from the data files of the Social Insurance Institution and the National Board of Taxation. The analysis concerned men in the labour force aged 30–54 at the time of the census. Only wage earners were included. The duration of employment during the year before the census was known for all subjects. During the study period, 2.7 million person years were accumulated, of which 200 000 (7%) were among the unemployed. The number of observed cancer deaths was not reported (Martikainen, 1990).

Italy

A longitudinal study in Italy followed the mortality in a cohort of more than a million people who were residents of Turin at the time of the 1981 census. This study included 13 462 men aged 15–59 who, having lost their job, were seeking a new one in the week before the 1981 census (4%). Their mortality was followed through 1985, and 78 cancer deaths occurred (Costa & Segnan, 1987).

United States

The National Longitudinal Mortality Study in the United States comprises records assembled from the United States Bureau of Census Current Population Surveys. The surveys used for the analysis were obtained in March 1979, April 1980, August 1980, December 1980, March 1981, March 1982, and March 1983. In all, 137 274 White men aged 25–64 were identified. Of these, 7501 (5.4%) were recorded as unemployed. They were matched to the National Death Index to ascertain death for a five-year follow-up period, and 20 cancer deaths occurred (Sorlie & Rogot, 1990).

Denmark

1970 census In total, 2.8 million persons aged 20–64 participated in the 1970 census in Denmark.

Table 2. Total cancer mortality and morbidity, and lung cancer mortality and morbidity, in unemployed men

Country and population	Follow-up period	Age group	All cancer			Lung cancer			Reference
			Obs.	SMR/RR	Adj. SMR/RR	Obs.	SMR/RR	Adj. SMR/RR	
United Kingdom 1971 census, mortality	1971–1981 1971–1975 1976–1981	15–64	102 52 50	1.44 1.42 1.48	1.28[a] NA NA	56 33 23	1.89 2.06 1.70	1.62[a] 1.74[a] 1.45[a]	Moser et al., 1990
1971 census, incidence	1971–1981 1971–1975 1976–1981	15–64	267 103 164	1.29 1.18 1.37	NA NA NA	96 43 53	1.50 1.56 1.46	NA NA NA	Kogevinas, 1990
1981 census, mortality	1983	16–64	25	1.38	NA	14	2.09	NA	Moser et al., 1987
1978–1980 British Regional Heart Study, mortality	1978/80–1989	40–59	27	1.74	1.59[b]	NA	NA	NA	Morris et al., 1994
Finland 1980 census, mortality	1981–1985	30–54	NA	1.39	1.17[c]	NA	2.05	1.43[c]	Martikainen, 1990
Denmark 1970 census, mortality	1970–1980 1970–1975 1976–1980	20–64	NA 163 NA	1.33 1.24 1.40	NA NA NA	NA 70[d] NA	NA 1.54[d] NA	NA NA NA	Iversen et al., 1987 Lynge & Andersen, 1996 Iversen et al., 1987
1970 census, incidence	1970–1975		291	1.25	NA	97[d]	1.64[d]	NA	Lynge & Andersen, 1996
1986 census, mortality	1986–1990	20–64	1204	1.23	NA	464[d]	1.44[d]	NA	Lynge & Andersen, 1996
Italy 1981 census, mortality	1981–1985	15–59	78	1.75	NA	NA	NA	NA	Costa & Segnan, 1987

Table 2. (contd.) Total cancer mortality and morbidity, and lung cancer mortality and morbidity, in unemployed men

Country and population	Follow-up period	Age group	All cancer			Lung cancer			Reference
			Obs.	SMR/RR	Adj. SMR/RR	Obs.	SMR/RR	Adj. SMR/RR	
USA 1979–1983 Current Population Survey, mortality	1979–1983	25–64	20	0.86	NA	NA	NA	NA	Sorlie & Rogot, 1990

Obs., observed number; Adj., adjusted.
[a] Age and social class controlled for.
[b] Age, town, social class, smoking, alcohol intake, and pre-existing disease at initial screening controlled for.
[c] Age, socioeconomic status, education, marital status, use of reimbursable medicines, and number of sick allowance days controlled for.
[d] Cancer of respiratory organs (International Classification of Diseases and Causes of Death, revision 8, 160–163).

They were classified by status of work based on the information from self-administered census questionnaires on their status of work on 9 November 1979. Among 1 292 337 economically active men, 15 340 (1.2%) were recorded as unemployed. Their mortality was followed for a 10-year period by linkage to the National Population and Death Registers (Iversen et al., 1987; Lynge & Andersen, 1996); there were 163 cancer deaths during the first five years. Their cancer incidence was followed for a five-year period by linkage to the National Cancer Register; there were 291 incident cancer cases during these five years (Lynge & Andersen, 1996).

1986 census In total, 3 million persons aged 20–64 were included in the 1986 census in Denmark. They were classified by status of work based on information in public administrative registers for the period 1981–1985. Persons classified as unemployed had been unemployed for at least 30% of the year for each of the years during the period 1981–1985. Among 1 352 932 economically active men, 183 184 (13.5%) were unemployed. Their mortality was followed for a five-year period by linkage to the National Population and Death Registers, and 1204 cancer deaths occurred (Lynge & Andersen, 1996).

Results

Table 1 shows that the proportion of unemployed men varied from 1% (1970 cohort from Denmark) to 14% (1981 cohort from the United Kingdom and 1986 cohort from Denmark) in the studied cohorts. Table 2 shows that all of these cohorts of unemployed men had an excess cancer mortality, except for the 1979–1983 cohort from the United States.

In the United Kingdom, men who described themselves as 'seeking work' at the time of the 1971 census had an SMR of 1.44 for dying of cancer during the period 1971–1981. The SMR was reduced to 1.28 when social class was controlled for. The excess was seen both in the period 1971–1975 (SMR = 1.42) and in the period 1976–1981 (SMR = 1.48) (Moser et al., 1990). The unemployed men also had an excess cancer morbidity with an SIR of 1.28 for the period 1971–1981 (Kogevinas, 1990). Men who described themselves as 'seeking work'

Table 3. Cancer mortality and morbidity for sites other than lung in unemployed men

Country and population	Follow-up period	Age group	Cancer site	Obs.	SMR/RR	Adj. SMR/RR	Reference
United Kingdom 1971 census, incidence	1971–1981	15–64	Stomach	14	0.82	NA	Kogevinas, 1990
			Colon	16	1.40	NA	
			Rectum	12	1.14	NA	
			Pancreas	7	1.17	NA	
			Prostate	12	1.04	NA	
			Bladder	17	1.33	NA	
			Leukaemia	6	1.54	NA	
Finland 1980 census, mortality	1981–1985	30–54	Stomach	NA	1.18	1.06[a]	Martikainen, 1990
			Colon and rectum	NA	1.42	1.60[a]	
Denmark 1970 census, mortality	1970–1975	20–64	Digestive organs	47	1.15	NA	Lynge & Andersen, 1996
			Male genital and urinary organs	22	1.27	NA	
			Haematopoietic system	8	0.70	NA	
Denmark 1970 census, incidence	1970–1975	20–64	Digestive organs	77	1.34	NA	Lynge & Andersen, 1996
			Male genital and urinary organs	48	0.98	NA	
			Haematopoietic system	14	0.79	NA	
Denmark 1986 census, mortality	1986–1990	20–64	Digestive organs	306	1.14	NA	Lynge & Andersen, 1996
			Male genital and urinary organs	138	1.10	NA	
			Haematopoietic system	89	0.91	NA	

Obs., observed number; Adj., adjusted.
[a]Age, socioeconomic status, education, marital status, use of reimbursable medicines, and number of sick allowance days controlled for.

at the time of the 1981 census in the United Kingdom had an SMR of 1.38 for dying of cancer in 1983 (Moser et al., 1987).

Men who were employed both when they were recruited to the British Regional Heart Study in 1978–1980 and during the previous five years but who lost employment during the five subsequent years had a RR of 1.74 for dying of cancer between recruitment and 1989. The RR was reduced to 1.59 when town, social class, smoking, alcohol intake, and pre-existing disease at the initial screening were controlled for (Morris et al., 1994).

The duration of unemployment in the year before the census was known for all subjects in the 1981 census in Finland. During the period 1981–1985, unemployed men had a RR of 1.39 for dying of cancer. This was reduced to 1.17 when socioeconomic status, education, marital status, and use of reimbursable medicines and number of sick allowance days in the previous year were controlled for (Martikainen, 1990).

An SMR of 1.75 for dying of cancer during the years 1981–1985 was found for men who reported seeking work at the time of the 1981 census in Turin, Italy (Costa & Segnan, 1987).

In the USA an SMR of 0.86 was found for cancer mortality among men registered as unemployed at any of the Current Population Surveys in

1979–1983. Mortality was followed through 1983 only (Sorlie & Rogot, 1990).

In Denmark, unemployed men had an excess cancer mortality of close to 25% during the first five years of follow-up both in the 1970 cohort (SMR = 1.24) and in the 1986 cohort (SMR = 1.23). A similarly increased cancer morbidity was found for the 1970 cohort (SIR = 1.25) (Lynge & Andersen, 1996). The observation from the 1971 census in the United Kingdom of an excess cancer mortality among unemployed men during both the first five years and the next five years of follow-up was also seen in the Danish 1970 cohort, where the RR for dying of cancer was 1.25 during the years 1970–1975 and 1.40 during the years 1976–1980 (Iversen et al., 1987).

Table 2 shows that in all countries with available data, the excess cancer mortality came mainly from an excess risk of lung cancer. The SMR for lung cancer in the United Kingdom 1971 census data was 1.89, and 1.62 when social class was controlled for (Moser et al., 1990). The SIR for lung cancer in this cohort was 1.50 (Kogevinas, 1990). The SMR for lung cancer in the United Kingdom 1981 census data was 2.09 (Moser et al., 1978). The RR for lung cancer for unemployed men in the Finnish 1981 census data was 2.05, and 1.43 when background variables such as socioeconomic status and number of sick days were controlled for (Martikainen, 1990). In Denmark, the SMRs for cancer of the respiratory organs were 1.54 for the 1970 cohort and 1.44 for the 1986 cohort. The SIR for cancer of the respiratory organs for the 1970 cohort was 1.64 (Lynge & Andersen, 1996).

The data are sparse on cancer mortality and morbidity for sites other than lung among unemployed men. The available data are listed in Table 3. There might be an indication of an excess risk for cancer of the colon, rectum and bladder among unemployed men. However, more detailed tabulations than those presently published are needed for the data to be informative.

Discussion

In summary, the data reviewed in this chapter point to an excess cancer risk among unemployed men. This excess risk exists when unemployed men constitute 1% or 10% of all men; it persists long after the unemployment started; and it does not disappear even when social class, smoking, alcohol intake, and previous sick days are controlled for.

The key question concerning the excess cancer mortality and morbidity seen in unemployed men is whether this excess is purely an effect of selection of unhealthy persons into unemployment or whether the unemployment in itself increases the risk of cancer.

Unemployment has in many studies been associated with minor psychological disorders such as negative effects on happiness, life satisfaction, self-esteem and an increase in general distress, anxiety and depressed mood (Hammerström, 1994). Social activity and participation – and therefore social support – also fall dramatically for many unemployed people (Bartley, 1994).

It is difficult to study the association be-tween psychosocial factors and the risk of later developing cancer. However, some follow-up studies have been undertaken of cancer mortality and cancer morbidity in populations in which personality, extent of social network, and so on have been assessed at the start of the study period. Although a positive association between a depressed mood and later risk of cancer has been reported (Persky et al., 1987), most studies have failed to find such an association (Kaplan & Reynolds, 1988; Zonderman et al., 1989; Linkins & Comstock, 1990; Vogt et al., 1992). Site-specific cancer data are available from only some of the studies. In the 12-year follow-up study of cancer incidence in persons from Washington County, Maryland, USA (Linkins & Comstock, 1990), an 18-fold risk for cancer sites associated with smoking was found among heavy smokers with a depressed mood compared with never smokers without a depressed mood. The sparse data in the other studies do not indicate an excess risk of such cancers in depressed persons (Persky et al., 1987; Kaplan & Reynolds, 1988).

The results presented in this chapter showed an excess cancer incidence among unemployed men, but this excess risk was not equally shared among all cancers. The excess risk came mainly from lung cancer. The SMR for lung cancer in the United Kingdom 1971 census data was 1.89, and 1.62 when social class was controlled for. The SMR for lung cancer in the United Kingdom 1981 census data was 2.09. The RR for lung cancer for unemployed men in the Finnish 1981 census data was 2.05, and 1.43 when background variables such as socio-

Table 4. Changes in smoking habits when continuously employed men become unemployed

Country and population	Year of recruitment	Year of follow-up	Age at recruitment	Smoking at recruitment			Smoking at follow-up		Reference
				Number of men	Current smoker	Heavy smoker	Current smoker	Heavy smoker	
United Kingdom British Regional Heart Study	1978–1980	1983–1985	40–59						Morris et al., 1992
– continuously employed[a]				4401	37%	13%	29%	8%	
– unemployed for other reasons[b]				376	45%	16%	35%	10%	
Denmark MONICA I	1982	1987	30, 40 50, 60						Osler, 1995
– continuously employed[c]				1083	57%	36%	51%	34%	
– not continuously employed[d]				96	63%	37%	61%	34%	

[a] Men employed at initial screening, without unemployment in previous five years and with continuous employment throughout the five years after initial screening.
[b] Men employed at initial screening, without unemployment in previous five years and with unemployment for reasons other than illness during the five years after initial screening.
[c] Men employed at initial screening and continuously employed throughout the five years of follow-up.
[d] Men employed at initial screening and not continuously employed throughout the five years of follow-up.

economic status and number of sick days were controlled for. The SMR of 1.54 for cancer of respiratory organs among unemployed men in the Danish 1971 census data was above the SMRs of 1.35 and 1.13 found for skilled and unskilled workers, respectively (Lynge, 1979).

Unemployed men thus seem to have an excess lung cancer risk. This exists when unemployed men constitute 1% or more than 10% of all men; it persists long after the unemployment started; and it does not disappear when social class and number of sick days are controlled for.

As smoking is the main risk factor for lung cancer it is interesting to look at the available data on unemployment and smoking habits. Data from the longitudinal studies on risk factors for heart disease from the 1980s from both the United Kingdom (Morris et al., 1992) and Denmark (Osler, 1995) show that the slightly higher prevalence of smoking among unemployed men than among employed men in the 1980s was due to differences in smoking habits established before unemployment started (Table 4). The available data therefore do not support the hypothesis that smoking increases as a result of unemployment. A survey of smoking habits among men in Denmark in 1986–1987 showed that less than 50% of salaried employees with a higher education smoked, whereas 60% of unskilled workers and more than 70% of unemployed men smoked. A similar survey in 1990–1991 showed, however, a prevalence of smoking between 50% and 60% in all of the three groups (Osler, 1992).

Moser et al. (1990) argued that 'to explain an excess of lung cancer mortality in 1971–1981, exposure to a risk factor before 1971 would seem likely to have been necessary'. None of the available studies provides data on lung cancer among the unemployed in which smoking has been controlled for. However, in the British Regional Heart Study the excess risk (RR) of all cancer mortality changed only from 1.74 to 1.59 when social class, smoking and so on at the time of recruitment were controlled for, and the excess risk of lung cancer prevailed after controlling for social class both in the United Kingdom 1971 and 1981 census and in the Finnish 1981 census studies. It therefore seems unlikely that differences in smoking habits prior to unemployment alone can explain the excess lung cancer risk among unemployed men.

Maybe men who became unemployed had been exposed to occupational carcinogens more frequently, or maybe development of lung cancer was actually accelerated during unemployment. It is interesting in this context that the combination of heavy smoking and depressed mood was associated with a high risk of smoking-associated cancers in one study (Linkins & Comstock, 1990). From the available data it is not possible to distinguish between such possibilities. But it remains an observation that unemployed men compared with employed men have a 40% to 70% excess risk of lung cancer, which can probably not be explained by excessive smoking alone.

References

Bartley, M. (1994) Unemployment and ill health: understanding the relationship. *J. Epidemiol. Community Health*, 48, 333–337

Costa, G. & Segnan, N. (1987) Unemployment and mortality. *Br. Med. J.*, 294, 1550–1551

European Commission (1994) *Employment in Europe 1994*. Luxembourg, Office for Official Publications of the European Communities

Hammerström, A. (1994) Health consequences of youth unemployment – review from a gender perspective. *Soc. Sci. Med.*, 38, 699–709

Iversen, L., Andersen, O., Andersen, P.K., Christoffersen, K. & Keiding, N. (1987) Unemployment and mortality in Denmark 1970–80. *Br. Med. J.*, 295, 879–884

Kaplan, G.A. & Reynolds, P. (1988) Depression and cancer mortality and morbidity: prospective evidence from the Alameda County Study. *J. Behav. Med.*, 11, 1–13

Kogevinas, M. (1990) *Longitudinal study: socio-demographic differences in cancer survival 1971–1983*. London, Her Majesty's Stationery Office

Linkins, R.W. & Comstock, G.W. (1990) Depressed mood and development of cancer. *Am. J. Epidemiol.*, 132, 962–972

Lynge, E. (1979) Occupational mortality in Denmark 1970–75 (in Danish) (Statistiske undersøgelser nr. 37). København, Danmarks Statistik

Lynge, E. & Anderson, O. (1997) Unemployment and cancer in Denmark 1970–1975 and 1986–1990. In: *Social Inequalities and Cancer*, Kogevinas, M., Pearce, N., Susser, M. & Boffetta, P., eds (IARC Scientific Publication No. 138) Lyon, International Agency for Research on Cancer pp. 353–359

Martikainen, P. (1990) Unemployment and mortality among Finnish men. *Br. Med. J.*, 301, 407–411

Morris, J.K., Cook, D.G. & Shaper, A.G. (1992) Non-employment and changes in smoking, drinking, and bodyweight. *Br. Med. J.*, 304, 536–541

Morris, J.K., Cook, D.G. & Shaper, A.G. (1994) Loss of employment and mortality. *Br. Med. J.*, 308, 1135–1139

Moser, K.A., Goldblatt, P.O., Fox, A.J. & Jones, D.R. (1987) Unemployment and mortality: comparison of the 1971 and 1981 longitudinal study samples. *Br. Med. J.*, 294, 86–90

Moser, K., Goldblatt, P., Fox, J. & Jones, D. (1990) Unemployment and mortality. In: Goldblatt, P.O., ed., *Longitudinal study: mortality and social organisation*. London, Her Majesty's Stationery Office. pp. 81–97

OECD (1994) Employment outlook. Paris, Organization for Economic Cooperation and Development

Osler, M. (1992) Smoking habits in Denmark from 1953 to 1991: a comparative analysis of results from three nationwide health surveys among adult Danes in 1953–1954, 1986–1987 and 1990–1991. *Int. J. Epidemiol.*, 21, 862–871

Osler, M. (1995) Unemployment and change in smoking behaviour among Danish adults. *Tobacco Control*, 4, 53–56

Persky V.W., Kempthorne-Rawson J. & Shekelle, R.B. (1987) Personality and risk of cancer: 20-year follow-up of the Western Electric Study. *Psychosomatic Med.*, 49, 435–449

Sorlie, P.D. & Rogot, E. (1990) Mortality by employment status in the national longitudinal mortality study. *Am. J. Epidemiol.*, 132, 983–992

Vogt, T.M., Mullooly, J.P., Ernst, D., Pope, C.R. & Hollas, J.F. (1992) Social networks as predictors of ischemic heart disease, cancer, stroke and hypertension: incidence, survival and mortality. *J. Clin. Epidemiol.*, 45, 659–666

Zonderman, A.B., Costa, P.T. & McCrae, R.R. (1989) Depression as a risk for cancer morbidity and mortality in a national representative sample. *J. Am. Med. Assoc.*, 262, 1191–1195

E. Lynge
Danish Cancer Society, Strandboulevarden 49, DK-2100 Copenhagen Ø, Denmark

II: Unemployment and cancer in Denmark, 1970–1975 and 1986–1990

E. Lynge and O. Andersen

We have analysed cancer mortality and cancer incidence among unemployed persons identified from the Danish linkage studies based on the 1970 census and the 1986 register-based census. In 1970, 1% of Danish men were unemployed; in 1986, 14% were unemployed. In both periods, unemployed men had an excess cancer mortality of close to 25% when they were followed-up for a five-year period and their mortality was compared with that of all men in the labour force. Unemployed women in the 1970 cohort also had an excess cancer mortality of 25%. Cancer incidence data were not available for the 1986 cohort. For both cohorts, the excess risk came mainly from lung cancer. Survey data from Denmark in the 1980s indicated that unemployed men had a slightly higher smoking prevalence before unemployment than men who continued working, and that unemployment did not increase smoking. It is therefore unlikely that the excess lung cancer risk among unemployed men is explained by differences in smoking habits alone.

In Denmark, several national linkage studies are available on mortality and diseases, and it is therefore possible to study the disease pattern even in small subgroups of the population. Using this methodology, a study has previously been undertaken of the mortality of unemployed persons in 1970 (Iversen *et al.*, 1987). At this time, 1% of men in the economically active age groups were unemployed. Since then unemployment has become a more common phenomenon in Danish society, and 14% of men were unemployed in 1986. We have analysed the cancer mortality and cancer incidence among unemployed persons identified from the Danish linkage studies based on the 1970 census and the 1986 register-based census.

A key question for the interpretation of a possible excess cancer incidence and mortality among unemployed persons is whether this excess is purely an effect of selection of unhealthy persons into unemployment or whether unemployment in itself increases the risk of cancer. To shed light on this problem we have also included data for cancer incidence and mortality among inactive persons. In the economically active age groups, by far the majority of inactive men have been granted an early pension for health reasons.

Material and methods
Our study is based on two cohorts of national census populations, each followed-up for deaths and emigrations for a five-year period. The first of the two cohorts was also followed-up for incident cancer cases for a five-year period.

The first cohort included 2 804 192 persons aged 20–64 years at the Danish census on 9 November 1970. They were followed-up until 8 November 1975 (Lynge, 1979; Andersen, 1985). They were classified by status of work based on the information on their status of work on 9 November 1970 from the self-administered census questionnaires. The total population was divided into those who were economically active on the census date and those who were not (housewives, pensioners, students, and so on). The economically active were divided into those who were employed on the census date and those who were unemployed.

The second cohort included 3 018 942 persons aged 20–64 years on 1 January 1986. They were followed-up until 31 December 1990 (Ingerslev *et al.*, 1994). They were classified by status of work based on information in public administrative registers for the period 1981–1985. Based on tax information from 1985, the population was divided into those who were economically active and whose who were not. The economically active persons were divided into those with unemployment insurance and those without. A distinction between employed and unemployed persons was possible only for the insured persons. Persons classified as unemployed had been unemployed for at least

Table 1. Men and women in Denmark aged 20–64 years in 1970 and 1986, by status of work

	Men		Women	
	1970	1986	1970	1986
Total	1 401 967	1 522 560	1 402 225	1 496 382
Total active	1 292 337	1 352 932	730 545	1 157 464
Insured	NA	1 021 992	NA	910 117
Not insured	NA	330 940	NA	247 347
Employed	1 276 997	838 808	723 849	709 510
– at work	1 177 489	NA	648 638	NA
– not at work	99 508	NA	75 211	NA
Unemployed	15 340	183 184	6700	200 607
Inactive	109 630	169 628	671 676	338 918
Unemployed as % of active	1%	14%	1%	17%
Inactive as % of total	8%	11%	48%	23%

NA, not applicable.

30% of the year for each of the years with insurance during the period 1981–1985.

Data on deaths and emigrations during the follow-up period were retrieved from the Central Population Register, and data on cause of death from the Death Certificate Register. Linkage based on unique personal identification numbers used in all of the registers ensured complete follow-up. Each person contributed with person years at risk from the entry date until emigration, death or end of follow-up – whichever came first.

Standardized mortality ratios (SMRs) were calculated by dividing the observed numbers of deaths in a given work status group with the expected number based on multiplication of person years at risk in each five-year age group (with age defined

Table 2. Total mortality in Denmark in 1970–1975 for persons aged 20–64 in 1970, and in 1986–1990 for persons aged 20–64 in 1986, by status of work

	Men				Women			
	1970–1975		1986–1990		1970–1975		1986–1990	
	Obs.	SMR	Obs.	SMR	Obs.	SMR	Obs.	SMR
Total	46 869	1.17	46 147	1.30	28 661	1.35	29 114	1.42
Total active	36 037	1.00	29 284	1.00	9036	1.00	12 743	1.00
Insured	NA	NA	21 677	1.04	NA	NA	9543	1.02
Not insured	NA	NA	7607	0.90	NA	NA	3200	0.95
Employed	35 324	0.99	16 828	0.96	8920	0.99	7546	0.97
– at work	31 013	0.93	NA	NA	7561	0.93	NA	NA
– not at work	4311	1.88	NA	NA	1359	1.62	NA	NA
Unemployed	713	1.62	4849	1.49	116	1.71	1997	1.25
Inactive	10 832	2.81	16 863	2.69	19 625	1.61	16 371	2.10

NA, not applicable; Obs., observed; SMR standardized mortality ratio.

Table 3. Cancer mortality in Denmark in 1970–1975 for persons aged 20–64 in 1970, and in 1986–1990 for persons aged 20–64 in 1986, by status of work

	Men				Women			
	1970–1975		1986–1990		1970–1975		1986–1990	
	Obs.	SMR	Obs.	SMR	Obs.	SMR	Obs.	SMR
Total	12688	1.07	13093	1.14	11999	1.11	13073	1.15
Total active	10676	1.00	9267	1.00	4577	1.00	6961	1.00
Insured	NA	NA	6760	1.04	NA	NA	5174	1.02
Not insured	NA	NA	2507	0.90	NA	NA	1787	0.95
Employed	10513	1.00	5556	1.01	4535	1.00	4269	1.01
– at work	9090	0.92	NA	NA	3800	0.92	NA	NA
– not at work	1423	2.14	NA	NA	735	1.75	NA	NA
Unemployed	163	1.24	1204	1.23	42	1.25	905	1.08
Inactive	2012	1.65	3826	1.70	7422	1.19	6112	1.38

NA, not applicable; Obs., observed; SMR, standardized mortality ratio

as age at the time of entry) by the mortality rate for all economically active persons in that age group. 95% confidence intervals (95% CIs) were calculated under the assumption that the observed number of cases follow a Poisson distribution if under 30 and a normal distribution if above 30.

Data on incident cancer cases during the follow-up period for the 1970 cohort were retrieved from the Danish Cancer Register (Lynge & Thygesen, 1990). Each person contributed with person years at risk from the entry date until emigration, first cancer diagnosis, death or end of follow-up – whichever came first. As person years were counted only up until first cancer diagnosis, the number of person years at risk was slightly lower for all cancers than for each cancer site. Standardized incidence ratios (SIRs) were calculated following the same procedure as used for calculation of SMRs.

Results

Table 1 shows the number of persons aged 20–64 in the two study populations. There was a dramatic increase in the proportion of unemployed men from 1% in 1970 to 14% in 1986. A marginal increase was seen in the proportion of inactive men (mainly early pensioners) from 8% in 1970 to 11% in 1986. The proportion of unemployed women increased from 1% in 1970 to 17% in 1986. At the same time, the proportion of inactive women (mainly housewives) decreased from 48% to 23%.

Table 2 shows the total mortality recorded for the five years of follow-up in each of the two study populations. In the 1970 cohort of unemployed men there were 713 deaths, giving an SMR of 1.62 (95% CI = 1.51–1.74). In the 1986 cohort of unemployed men there were 4849 deaths, giving an SMR of 1.49 (95% CI = 1.45–1.53). The SMR values for inactive men in the 1970 and 1986 cohorts were 2.81 and 2.69, respectively. There were only 116 deaths in the 1970 cohort of unemployed women, giving an SMR of 1.71 (95% CI = 1.43–2.05), while the number of deaths in unemployed women in the 1986 cohort was 1997, with an SMR of 1.25 (95% CI = 1.20–1.31). The SMR values for inactive women in the 1970 and 1986 cohorts were 1.61 and 2.10, respectively.

Table 3 shows the cancer mortality for the study populations. In the 1970 cohort of unemployed men there were 163 cancer deaths, giving an SMR of 1.24 (95% CI = 1.06–1.45), and in the 1986 cohort of unemployed men there were 1204 cancer deaths, giving an SMR of 1.23 (95% CI = 1.16–1.30). The SMR values for inactive men in the 1970 and 1986 cohorts were 1.65 and 1.70, respectively. Among unemployed women there were 42 cancer deaths in the 1970 cohort (SMR = 1.25; 95% CI = 0.92–1.69) and 905 cancer deaths in the 1986 cohort

Table 4. Cancer incidence and mortality in Denmark in 1970–1975 for unemployed persons aged 20–64 in 1970, and cancer mortality in 1986–1990 for unemployed persons aged 20–64 in 1986

	Men						Women					
	1970–1975				1986–1990		1970–1975				1986–1990	
	Incidence		Mortality		Mortality		Incidence		Mortality		Mortality	
	Obs.	SIR	Obs.	SMR	Obs.	SMR	Obs.	SIR	Obs.	SMR	Obs.	SMR
Total mortality	–	–	713	1.62	4849	1.49	–	–	116	1.71	1997	1.25
Cancer[a]	291	1.25	163	1.24	1204	1.23	90	1.14	42	1.25	905	1.08
Digestive organs[b]	77	1.34	47	1.15	306	1.14	13	1.09	10	1.36	168	1.09
Respiratory organs[c]	97	1.64	70	1.54	464	1.44	6	1.46	7	2.30	212	1.48
Breast[d]	1	2.94	0	–	1	1.04	21	0.97	7	0.89	206	0.97
Female genital organs[e]	0	–	0	–	0	–	33	1.49	11	1.36	139	1.01
Male genital and urinary organs[f]	48	0.98	22	1.27	138	1.10	1	0.35	0	–	22	0.86
Haematopoietic system[g]	14	0.79	8	0.70	89	0.91	3	0.75	4	1.69	49	0.95
Other sites[h]	57	1.09	16	0.96	206	1.23	13	0.98	3	0.83	109	0.98

Obs., observed; SMR, standardized mortality ratio; SIR, standardized incidence ratio
Codes (International Classification of Diseases and Causes of Death) for cancer mortality (first) and cancer incidence (second):
[a]ICD-8 140-209; ICD-7 140-205.
[b]ICD-8 150-159; ICD-7 150-159.
[c]ICD-8 160-163; ICD-7 160-164.
[d]ICD-8 174; ICD-7 170.
[e]ICD-8 180-184; ICD-7 171-176.
[f]ICD-8 185-189; ICD-7 177-181.
[g]ICD-8 200-209; ICD-7 200-205.
[h]ICD-8 140-149, 170-173, 190-199; ICD-7 140-148, 190-199.

(SMR = 1.08; 95% CI = 1.01–1.15). The SMR values for inactive women in the 1970 and 1986 cohorts were 1.19 and 1.38, respectively.

Table 4 shows that the excess cancer mortality among unemployed men came in both cohorts mainly from cancer of the respiratory system, with SMRs of 1.54 (95% CI = 1.22–1.95) and 1.44 (95% CI = 1.31–1.58), respectively. Cancers of the digestive organs, with SMRs of 1.15 (95% CI = 0.86–1.53) and 1.14 (95% CI = 1.02–1.28), and cancers of the male genital and urinary organs, with SMRs of 1.27 (95% CI = 0.77–1.92) and 1.10 (95% CI = 0.93–1.18), also contributed to the excess cancer mortality. An excess risk of other cancers was seen in the 1986 cohort (SMR = 1.23; 95% CI = 1.07–1.41), but not in the 1970 cohort (SMR = 0.96; 95% CI = 0.55–1.56). The excess cancer mortality among unemployed women in the 1970 cohort came from several cancer sites, but the numbers were small. Among unemployed women in the 1986 cohort, excesses were seen only for cancer of the digestive organs (SMR = 1.09; 95% CI = 0.94–1.27) and for cancer of the respiratory organs (SMR = 1.48; 95% CI = 1.29–1.69).

Table 4 also shows the cancer incidence in the 1970 cohort of unemployed persons. There were 291 incident cancer cases among men, giving an SIR of 1.25 (95% CI = 1.11–1.40). As with cancer mortality, the excess risk came from cancer of the respiratory organs (SIR = 1.64; 95% CI = 1.34–2.00) and from cancer of the digestive organs (SIR = 1.34; 95% CI = 1.07–1.68). A marginal excess risk was seen for the incidence of cancers of other sites (SIR = 1.09; 95% CI = 0.84–1.41). There were 90 incident cancer cases in women, giving an SIR of

Table 5. Cancer incidence and mortality in Denmark in 1970–1975 for inactive persons aged 20–64 in 1970, and cancer mortality in 1986–1990 for inactive persons aged 20–64 in 1986

	Men						Women					
	1970–1975				1986–1990		1970–1975				1986–1990	
	Incidence		Mortality		Mortality		Incidence		Mortality		Mortality	
	Obs.	SIR	Obs.	SMR	Obs.	SMR	Obs.	SIR	Obs.	SMR	Obs.	SMR
Total mortality	–	–	10832	2.81	16863	2.69	–	–	19625	1.61	16371	2.10
Cancer[a]	2581	1.25	2012	1.65	3826	1.70	13581	1.00	7422	1.19	6112	1.38
Digestive organs[b]	653	1.22	552	1.42	902	1.41	2533	1.05	1803	1.18	1249	1.27
Respiratory organs[c]	839	1.52	792	1.82	1546	1.91	828	1.06	677	1.16	1296	1.56
Breast[d]	4	1.24	3	2.91	7	2.42	3448	0.94	1727	1.22	1291	1.35
Female genital organs[e]	0	–	0	–	0	–	3559	1.01	1706	1.19	942	1.29
Male genital and urinary organs[f]	508	1.15	275	1.69	518	1.54	637	1.14	324	1.25	268	1.63
Haematopoietic system[g]	189	1.27	182	1.85	275	1.55	657	0.97	478	1.16	361	1.38
Other sites[h]	434	1.04	208	1.57	578	2.00	2103	1.01	707	1.17	705	1.38

Obs., observed; SMR, standardized mortality ratio; SIR, standardized incidence ratio
Codes (International Classification of Diseases and Causes of Death) for cancer mortality (first) and cancer incidence (second):
[a] ICD-8 140-209; ICD-7 140-205.
[b] ICD-8 150-159; ICD-7 150-159.
[c] ICD-8 160-163; ICD-7 160-164.
[d] ICD-8 174; ICD-7 170.
[e] ICD-8 180-184; ICD-7 171-176.
[f] ICD-8 185-189; ICD-7 177-181.
[g] ICD-8 200-209; ICD-7 200-205.
[h] ICD-8 140-149, 170-173, 190-199; ICD-7 140-148, 190-199.

1.14 (95% CI = 0.93–1.40). The numbers of specific cancer sites were small, but there was a significant excess risk for cancer of the female genital organs (SIR = 1.49; 95% CI = 1.06–2.10).

Table 5 shows that the excess cancer mortality among inactive men and women came from all cancer sites in both the 1970 cohort and the 1986 cohort. The table also shows that for inactive persons the SIR values were systematically below the SMR values.

Discussion

Cancer in unemployed women
In 1970, only half of the Danish women aged 20–64 were working outside their homes; the other half were housewives. In 1986, only a quarter of women were housewives.

The unemployment rate among women increased from 1% in 1970 to 17% in 1986, and it should be kept in mind that as the workforce increased during the same period the 1% in 1970 represented 6700 women whereas the 17% in 1986 represented more than 200 000 women. The analysis of mortality and cancer incidence in the unemployed women showed:

- The small group of unemployed women in the 1970 cohort had an excess total mortality of 70%, whereas the much larger group of unemployed women in the 1986 cohort had an excess total mortality of only 25%.
- Similar differences were seen in the cancer mortality, where there was a 25% excess in the 1970 cohort (based on small numbers) and only an 8% excess in the 1986 cohort.

- The moderate excess cancer mortality in the 1986 cohort came from a 50% excess risk of cancer of the respiratory organs and a 10% excess risk of cancer of the digestive organs. The numbers for specific cancer sites for the 1970 cohort of unemployed women were too small for analysis.
- The unemployed women in the 1970 cohort had an excess cancer incidence of 14%, which was less than their excess cancer mortality of 25%; but the numbers were small, and the 95% CIs overlapped.
- Inactive women in the 1970 cohort (which constituted half of the women) had no excess cancer incidence compared with economically active women, but a 19% excess cancer mortality.

Owing to the dramatic change in the proportion of women working outside homes from 1970 to 1986, it is very difficult to interpret the data on mortality and cancer incidence for unemployed women in Denmark, and further discussion is therefore restricted to men.

Cancer in unemployed men
By far the majority of Danish men aged 20–64 were economically active both in 1970 and in 1986. However, the proportion of unemployed men increased from 1% in 1970 to 14% in 1986. The analysis of the mortality and cancer incidence in these men showed:

- The small group of unemployed men in the 1970 cohort had an excess total mortality of 60% during the following five years. The much larger group of unemployed men in the 1986 cohort had a similar excess total mortality of 50% during the following five years.
- The unemployed men in both the 1970 cohort and the 1986 cohort had an excess cancer mortality of 25%.
- The excess cancer mortality for the unemployed men came in both the 1970 cohort and the 1986 cohort mainly from an excess risk of death from cancer of the respiratory organs of close to 50%; in addition, the excess risk of death from cancer of the digestive organs of 15% and the excess risk from cancer of the male genital and urinary organs of 10–25% made minor contributions to the excess cancer mortality.
- The unemployed men in the 1970 cohort had an excess cancer incidence of 25%, which was equivalent to their excess cancer mortality. As for mortality, the excess cancer incidence came primarily from cancer of the respiratory organs and to a lesser extent from cancer of the digestive organs.
- Among the unemployed men in the 1970 cohort the excess cancer incidence was thus of the same magnitude as the excess cancer mortality. However, this was not the case for the health-selected, inactive men in the 1970 cohort; their excess cancer incidence was systematically below their excess cancer mortality.

It thus seems reasonable to conclude that men who became unemployed or inactive developed cancer with a frequency that is 25% above that of working men. As a major contributor to this excess risk is lung cancer, which has a relatively short average survival time, this increased frequency of developing cancer is reflected in an increased frequency of dying from cancer. In addition, inactive men have an additional excess cancer mortality, which probably results from cancer cases developed before the time of retirement.

It is difficult to explain the 25% excess cancer mortality among unemployed men as a result of selection of unhealthy men into the group, because the excess is the same in the 1970 cohort, when the unemployed men constituted 1%, and in the 1986 cohort, when the unemployed men constituted 14%. Furthermore, if the excess risk was explained by a selection effect, we would expect this excess to disappear over time. In the 1970 cohort of unemployed men, however, the relative risk for dying of cancer was 1.25 during the years 1970–1975 but 1.40 during the years 1976–1980 (Iversen *et al.*, 1987).

As smoking is the main risk factor for lung cancer it is interesting to look at the available data on unemployment and smoking habits. A survey of smoking habits among men in Denmark in 1986–1987 showed that fewer than 50% of salaried employees with a higher education smoked, whereas 60% of unskilled workers and more than 70% of unemployed men smoked. The proportion of smokers among unemployed men in 1986–1987 was thus only slightly higher than the proportion of smokers among unskilled workers. Furthermore, a similar survey in 1990–1991 of salaried employees, unskilled workers and unemployed men showed a prevalence of smoking between 50% and 60% for all three groups (Osler, 1992).

Data from the longitudinal studies on risk factors for heart disease from the 1980s from Denmark (Osler, 1995) show that the slightly higher prevalence of smoking among unemployed men than among employed men in the 1980s was due to differences in smoking habits established before the unemployment started. The available data thus do not support the hypothesis that smoking increases as a result of unemployment.

We can therefore conclude that unemployed men have a 25% excess risk of developing cancer during their unemployment, and it is unlikely that this excess risk is explained by selection of unhealthy men into the ranks of the unemployed. Furthermore, a major contribution to this excess risk comes from lung cancer, and it seems unlikely that excessive smoking habits either before or during unemployment can explain this.

References

Andersen, O. (1985) Occupational mortality in Denmark 1970–80 (in Danish) (Statistiske undersøgelser nr. 41). København, Danmarks Statistik

Ingerslev, O., Madsen, M. & Andersen, O. (1994) Mortality by socio-economic groups (in Danish). København, Middellevetidsudvalget, Sundhedsministeriet

Iversen, L., Andersen, O., Andersen, P.K., Christoffersen, K. & Keiding, N. (1987) Unemployment and mortality in Denmark 1970–80. *Br. Med. J.*, 295, 879–884

Lynge, E. (1979) Occupational mortality in Denmark 1970–75 (in Danish) (Statistiske undersøgelser nr. 37). København, Danmarks Statistik

Lynge, E. & Thygesen, L. (1990) Occupational cancer in Denmark. Cancer incidence in the 1970 census population. *Scand. J. Work Environ. Health*, 16 (Suppl. 2), 1–35

Osler, M. (1992) Smoking habits in Denmark from 1953 to 1991: a comparative analysis of results from three nationwide health surveys among adult Danes in 1953–1954, 1986–1987 and 1990–1991. *Int. J. Epidemiol.*, 21, 862–871

Osler, M. (1995) Unemployment and change in smoking behaviour among Danish adults. *Tobacco Control*, 4, 53–56

Corresponding author:
E. Lynge
Danish Cancer Society, Strandboulevarden 49, DK-2100 Copenhagen Ø, Denmark

Environmental exposure, social class, and cancer risk

A. Woodward and P. Boffetta

Exposure to a variety of environmental factors associated with cancer occurrence varies by social class. These factors include air pollutants (SO_2, NO_2, total suspended particulates, etc.), toxic waste hazards, and ionizing and other radiation. Heavy environmental pollution has been associated with an increased risk of some cancers and in particular lung cancer. There is limited evidence suggesting that individuals from lower social classes are exposed to higher levels of environmental pollutants than are individuals from higher social classes. This may be due to the placement of new sources of pollution or of toxic processes in disadvantaged areas, or to the selective migration of the poorer sectors of society to these areas. The available data do not allow any conclusion on the possible contribution of exposure to environmental pollution to social class differences in cancer occurrence. Exposure to ultraviolet (UV) radiation, principally from sunlight, is modified strongly by personal behaviours such as choice of recreation and use of protective clothing. Those in outdoor occupations are likely to receive the highest cumulative exposure to UV radiation. There is no clear evidence from recent survey research in Australia and North America that socioeconomic factors are strongly related to non-occupational exposure to UV radiation. Information is lacking on the influence of socioeconomic status on sun exposure in other parts of the world. There is little information on the social distribution of exposure to ionizing radiation.

The concept of environment is often used in a broad sense, to comprise all factors unrelated to the genetic make-up of an individual, such as occupation, nutrition, lifestyle and reproductive habits, infections and so on. In this broad sense, the environment is likely to be responsible for the majority of the cases of cancer in humans (Tomatis et al., 1990). Most commonly, however, the concept of environment is used in a narrower sense, to cover only factors related to the place where people live, and over which each individual has little control. This paper addresses the relationship between cancer risk, social class and environmental factors considered in this narrow sense.

At present, our knowledge of the role of the environment in human cancer covers four major groups of factors: air pollutants, water pollutants, non-ionizing [mainly ultraviolet (UV)] radiation and ionizing radiation (Table 1). In this chapter, the evidence linking cancer risk, social class and these groups of factors is discussed in detail.

Environmental pollution
Cancer risk from environmental pollution

Most of the evidence on the association between environmental (mainly air and water) pollution and cancer comes from descriptive or ecological studies comparing cancer rates in populations exposed to different levels of pollution, such as urban and rural populations (Simonato & Pershagen, 1993).

The interpretation of these epidemiological studies is complicated by the low quality of information on past exposure – a problem common to most investigations on the effects of environmental exposures (Hatch & Thomas, 1993). Most studies on air pollution and cancer suffer in addition from lack of specificity of outcome, since the organ most heavily affected is the lung, and the overwhelming cause of lung cancer in most populations is tobacco smoking; the latter may therefore be an important confounder of the association between air pollution and cancer.

A clear association between environmental pollution and human cancer has been shown in analytical epidemiological studies in several cases of heavy pollution, such as drinking-water contamination from arsenic in Taiwan (Chiou et al., 1995), air pollution from residential and industrial sources (mainly metal smelting) in Upper Silesia,

Poland (Jedrychowski et al., 1990), and contamination with dioxins in Seveso, Italy (Bertazzi et al., 1993). At present, however, the evidence of an increased risk of cancer (mainly for the lung) following exposure to light or moderate levels of pollution, such as those to which many industrial populations are exposed, is still inconclusive (Simonato & Pershagen, 1993). Indoor air pollution may also represent a cancer hazard; in particular, non-smokers exposed to environmental tobacco smoke (Tredaniel et al., 1993) and people (mainly women) exposed to large amounts of combustion fumes from cooking and heating (Xu et al., 1989) have been shown to be at increased risk of lung cancer.

Table 1. Environmental agents and exposures known or suspected to cause cancer in humans[a]

Agent/exposure	Target organ[b]	Strength of evidence[c]
Air pollutants		
Erionite	Lung, pleura	1
Asbestos	Lung, pleura	1
Polycyclic aromatic hydrocarbons[d]	–	E
Coal-tar pitches	Skin, lung, bladder	1
Coal tars	Skin, lung	1
Mineral oils (untreated and mildly treated)	Skin	1
Shale oils	Skin	1
Soots	Skin, lung	1
Creosotes	(Skin)	2A
Diesel engine exhaust	(Lung, bladder)	2A
Bitumens, extracts of steam-refined and air-refined	(Lung, skin)	2B
Carbon-black extracts	(Bladder)	2B
Engine exhaust, gasoline	(Lung)	2B
Fuel oils, residual (heavy)	(Skin, lung)	2B
Water pollutants		
Arsenic	Skin, lung	1
Chlorination by-products	(Bladder)	S
Nitrate and nitrite	(Oesophagus, stomach)	S
Ionizing radiation		
Radon and its decay products	Lung	1
Radium, thorium	Bone	E
X-rays	Leukaemia, breast, thyroid, others	E
Non-ionizing radiation		
Solar radiation	Skin	1
Ultraviolet radiation A	(Skin)	2A
Ultraviolet radiation B	(Skin)	2A
Ultraviolet radiation C	(Skin)	2A
Use of sunlamps and sunbeds	(Skin)	2A
Electric and magnetic fields	(Leukaemia)	S

[a]Agents and exposures occurring mainly in occupational settings, as well as medicines, are excluded.
[b]Suspected target organs are given in parentheses.
[c]IARC Monographs evaluations (IARC, 1972–1995) are reported wherever available (1, human carcinogen; 2A, probable human carcinogen; 2B, possible human carcinogen); otherwise 'E' and 'S' are used (E, established carcinogen; S, suspected carcinogen).
[d]Only mixtures of polycyclic aromatic hydrocarbons are listed separately; several individual hydrocarbons have been classified in IARC Groups 2A and 2B.

Table 1 shows the carcinogenicity of environmental pollutants, some of which have been evaluated within the IARC *Monographs* programme as established or probable carcinogens (IARC, 1972–1995). Many estimates of the proportion of cancers due to air and water pollution are in the range of 0.5–2% of all cancers; for example, Doll and Peto (1981) estimated that pollution was responsible for 2% of cancers in the United States of America, with air pollution causing half of these. In specific situations, however, the share of cancers due to pollution may be higher; for example, in the study conducted in Upper Silesia, the risk of lung cancer attributed to air pollution among women was 10% (Jedrychowski et al., 1990).

Role of social class

It is likely that exposure to environmental pollution is higher among lower social classes than among higher social classes; some of the differences in cancer risk among social classes (see the chapter by Faggiano et al. in this book) may therefore be attributed to environmental pollution. Increased environmental exposure to carcinogens in lower social classes may result from residence in neighbourhoods with higher air pollution and lack of unpolluted (for example, bottled) drinking-water, more time spent in outdoor polluted workplaces, difficulty in moving from contaminated areas of the cities, use of less efficient cooking and heating systems (Goldstein et al., 1986, 1988), and higher probability of living with a smoker. Although differences in living style such as these are obviously related to social class, there is little direct evidence on their existence and magnitude.

Data from the USA on social-class-related differences in air pollution exposure have been reviewed by Sexton et al. (1993). At the 1990 USA census, Blacks (86%) and Hispanics (91%) were more likely to live in urban areas than Whites (70%); the same ethnic difference was seen in the proportion of the population living in areas out of compliance with the USA Environmental Protection Agency air quality criteria – for particulate matter this proportion was 15% for Whites, 17% for Blacks and 34% for Hispanics (Sexton et al., 1993). In an ecological study of 34 areas of the United Kingdom, a strong positive correlation was found between social class index and domestic air pollution (Nixon & Carpenter, 1974). The opposite result, however, was

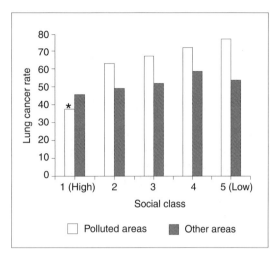

Figure 1. Age-adjusted annual lung cancer rates (per 100 000) in white men in Los Angeles county, USA, for 1968–1970 (mortality) and 1972 (incidence). Modified from Henderson et al., 1975. The data indicated by an asterisk are based on five cases or deaths.

found in an analysis of total suspended particulate (TSP) and socioeconomic status in census tracts in Harris County, Texas, USA, in which a strong negative correlation was found (Buffler et al., 1989).

Henderson et al. (1975) analysed lung cancer incidence (1972) and mortality (1968–1970) in areas of Los Angeles county, California, USA with high mortality from lung cancer and high air pollution, and in the rest of the county. This study showed consistently higher rates in lower social classes and in the most polluted parts of the county (Fig. 1), the difference between the polluted and the less polluted parts being greatest in the lowest social class. The overall rate ratios were 1.33 for air pollution and 1.26 for social class (classes 4 and 5 versus classes 1 and 2); when men from classes 1 and 2 living in the less polluted areas were taken as reference, men from classes 4 and 5 living in more polluted areas had a rate ratio of 1.53. Overall, this study suggests that air pollution had in this population an effect on lung cancer rate as large as that of other factors related to social class.

More data are available on social class or ethnic group differences in exposure to non-carcinogenic air pollutants, such as indoor lead (Agency for Toxic Substance and Disease Registry, 1988) and indoor carbon monoxide (Schwab, 1990). These studies consistently showed a higher exposure in lower social classes, although the magnitude of the difference varied greatly.

Limited data are available on differences in exposure to water pollution; for example, in the USA, hazardous waste sites and in particular sites classified in the National Priority List, many of which entail contamination of ground water, occur more frequently in counties classified as rural poor counties at the 1980 census than in other counties (Calderon et al., 1993). Similar data were not available on populations exposed to carcinogens from contamination of soil or other media.

UV radiation

Cancer risk from UV radiation

UV radiation is the major environmental cause of skin cancers, both melanocytic and non-melanocytic (squamous-cell and basal-cell carcinomas). Radiation in the middle of the UV range (UVB 280–320 nm) is the wavelength principally responsible for sunburning and skin cancer (IARC, 1992).

Melanoma is one of the few neoplasms that occurs more commonly in groups of higher socioeconomic status than in less advantaged groups. This pattern has been observed in many countries (see the chapter by Faggiano et al. in this book), and is thought to be principally due to differences in the quality and timing of exposures to intense sunlight (Elwood et al., 1985). In contrast, squamous and basal cell cancers of the skin are associated with cumulative exposures to UV radiation. These cancers are generally not included in disease registers, and there are few good incidence studies of socioeconomic-related risk factors. However, research in Australia, where rates of non-melanocytic skin cancers are particularly high (the lifetime risk among white-skinned Australians is more than two in three), has shown that increasing age, male sex, tendency to sunburn and outdoor occupation are positively related to disease risk (Marks et al., 1989).

Photographs and paintings show that in European cultures those who could afford to avoid the sun generally did so, by choice of clothing and recreation, until the second and third decades of this century. Then the prevailing models of attractiveness changed, tanning was favoured, and the health-giving properties of sunlight were emphasized. Increasingly, a suntan was regarded as a symbol of success and well-being (Arthey & Clarke, 1995). There is some evidence that attitudes are changing again – a survey of Australian women's magazines between 1982 and 1991 found a reduction in the proportion of models with a deep tan (Chapman et al., 1992).

Role of social class

In many parts of the world agriculture remains the livelihood and means of subsistence for the bulk of the population. As a consequence, exposure to UV radiation in these settings may be said to be still predominantly occupational. Elsewhere there has been a marked reduction in the number of people employed in outdoor work, such as agriculture, construction and transport, with increasing mechanization and growth of the manufacturing, service and information industries. However, there remains a substantial proportion of the workforce that is exposed to UV radiation as part of their job. For example, the New Zealand census findings for 1986 showed that approximately 24% of males and 9% of females in the labour force were classified as full-time outdoor workers (New Zealand Department of Statistics, 1988). There are few detailed data on sun exposures due to work outdoors, but in mid-latitude countries such as New Zealand and Australia unprotected workers may experience high levels of UV radiation – a survey of telephone-line staff in Australia in 1990 found that 48% did not use hats and 20% commonly wore sleeveless shirts, singlets or no shirt at all when working (Borland et al., 1991). In general, outdoor workers are more often from low social class than are other workers.

In industrialized and developing countries, exposure to sunlight outside work – so-called recreational exposure – has become increasingly important as a source of UV radiation for most socioeconomic groups. The relation of sun exposure to social factors is likely to vary between countries, and over time, depending on fashions, income differentials and opportunities for outdoor recreation. Most of the studies available have come from Australia, where skin cancer rates are the highest in the world, and good research on suntanning and sun exposure has been carried out. However, the findings on social class differentials (or their absence) should be extrapolated to other countries with caution.

In Victoria, Australia, household surveys are carried out each year to obtain information on knowledge and attitudes to suntanning and sun exposure, and sun-protection behaviour (Hill *et al.*, 1990, 1993). The surveys include a representative sample of the population aged 14 years and over, and socioeconomic status is measured by the occupational status of the chief income earner in the household. The studies have found little difference in sun-protective behaviours (use of shade, hats and covering clothing) by socioeconomic status, although those from lower blue-collar and lower white-collar households were less likely to report using topical sunscreens (Segan & Borland, 1994). Awareness of public health programmes concerned with sun exposure and skin cancer was generally greater among upper white-collar and upper blue-collar households, and the strongest pro-tan beliefs were expressed by those from households of lower socioeconomic status. However, these differences were not reflected in sun-protective behaviours or inferred doses of UV radiation. For example, there were no differences apparent by socioeconomic status or educational level in the proportion of respondents who chose to stay out of the sun during the summer. Nor were there differences in the frequency of self-reported sunburn; there was a strong age effect (younger people were most likely to report sunburn in the preceding summer) and men reported sunburn more often that women, but there were no associations with socioeconomic status or educational attainment.

In South Australia, where similar research has been carried out, groups of higher socioeconomic status were found to better informed in some regards, but the association of socioeconomic status and knowledge of sun exposure and appropriate protection behaviours was not strong (South Australian Omnibus Health Survey, unpublished data). Some differences in the use of topical sunscreens and protective clothing were apparent (for example, 57% of individuals in the highest socioeconomic status group reported that they usually or almost always wore a hat outdoors in the summer, compared with 43% of persons in the lowest socioeconomic status group). Except for the lowest rank socioeconomic status group, no consistent associations between social factors and sun exposure were observed.

Sun exposure in childhood and adolescence may be particularly important in the etiology of melanoma. Studies in New South Wales found that only a minority of high-school students took any deliberate measures to restrict sun exposure during summer, such as wearing a hat, applying a sunscreen, or seeking shade (Cockburn *et al.*, 1989). Smoking status and area of residence were predictors of sun-protection behaviour, suggesting an association of socioeconomic status and UV radiation exposure. The number of melanocytic nevi on the skin is thought to be a guide to the frequency of exposure to intense sunlight in early life, and is related to age, ethnic background, hair colour, ability to tan and propensity to sunburn. After adjustment for these factors, a Western Australian study found no association between nevi count in children aged 5–12 years and mother's level of schooling (English & Armstrong, 1994).

The survey in Victoria has found that socioeconomic status is weakly related to knowledge and attitudes concerning what is regarded as 'healthy' sun exposure, but there is no clear indication of differences between social groups in terms of how peoples behave, or the frequency of intense sun exposures (Segan & Borland, 1994). This is not due to a lack of sunshine or rarity of health outcomes – approximately 50% of Victoria respondents reported sunburn, and about a third of these reported burns that led to persisting tenderness and blistering. The results are unlikely to be a chance outcome, as consistent findings have been observed in multiple surveys involving thousands of respondents. It may be that the apparent differences in beliefs and knowledge are a reporting artifact, or that the responses elicited by surveys of this kind bear no strong relation to actual sun-protective behaviour. Alternatively, there may be a relation, but one that is not immediately apparent, and differences in behaviours may emerge in the future as individuals of upper socioeconomic status move more rapidly to adopt new norms.

Little has been published on other exposures to UV radiation, such as from use of suntanning beds and UV lamps, although these may be significant sources of skin-damaging radiation (IARC, 1992). A survey of adults in the USA found that tanning-bed users were more likely to be female and to be more knowledgeable about the long-term effects of UV radiation, but knowledge of hazards was unrelated to educational attainment (Mawn & Fleischer, 1993). A survey in Victoria, Australia in 1993 enquired about the use of 'tan accelerators', skin products that are reputed to decrease the sun exposure required to obtain a suntan: 8% of the sample reported using such an agent, women more so than men, but there were no clear relations with age, socioeconomic status or education (Segan & Borland, 1994).

Ionizing radiation

Ionizing radiation is a cause of leukaemia, myeloma and several solid tumours, notably breast, lung, bone and thyroid cancer (Tomatis, 1990). Apart from lung cancer, whose pattern is mainly due to characteristics of tobacco smoking, these tumours show a weaker social gradient than other neoplasms, and in some cases the rates are greater in high than in low socioeconomic groups (see the chapter by Faggiano et al. in this book).

The information on radiation and cancer comes chiefly from studies on four groups of exposed individuals: atomic bomb survivors, occupationally exposed workers (mainly miners and nuclear and medical workers), patients treated with radiotherapy for malignant or benign conditions, and subjects living in houses with high exposure to radon and its decay products.

These studies have a better definition of dose than studies done in other areas of epidemiology; this is mainly due to the availability of historical exposure records for many groups known to be exposed to ionizing radiation. A consequence of this situation is that less attention has traditionally been paid in these studies as compared with other areas of research to other factors that may influence the exposure, such as social class. Direct information is lacking on differences in radiation exposure by social class, except for a study on lung cancer risk following domestic exposure to radon, which suggested that education may act as a confounder (Latourneau et al., 1994).

The pattern of occupations entailing exposure to ionizing radiation, however, provides some indirect evidence against the hypothesis of a strong role of exposure to ionizing radiation in determining social-class-related differences in cancer risk. The proportion of miners exposed to high levels of radon is relatively low (IARC, 1988), and many other occupations entailing exposure to ionizing radiation, such as nuclear industry worker, radiologist and radiological nurse, include a high proportion of white-collar occupations. Indoor exposure to radon depends on the geological characteristics of the soil underlying houses, on their distance from the soil and on the ventilation of the rooms. Although direct evidence is lacking, it is plausible that people from lower socioeconomic groups in many countries live in dwellings with higher natural ventilation than those of people from higher socioeconomic groups, suggesting an inverse relationship between social class and radon exposure.

References

Agency for Toxic Substance and Disease Registry (1988) *Proceedings of the national minority health conference.* Atlanta, GA

Arthey, S. & Clarke, V.A. (1995) Suntanning and sun protection: a review of the psychological literature. *Soc. Sci. Med.,* 40, 265–274

Bertazzi, P.A., Pesatori, A.C., Consonni, D., Tironi, A., Landi, M.T. & Zocchetti, C. (1993) Cancer incidence in a population accidentally exposed to 2,3,7,8,-tetra-chlorodibenzo-para-dioxin. *Epidemiology,* 4, 398–406

Borland, R., Hocking, B., Godkin, G.A., Gibbs, A.F. & Hill, D.J. (1991) The impact of a skin cancer control education package for outdoor workers. *Med. J. Aust.,* 154, 686–688

Buffler, P.A., Cooper, S.P., Stinnett, S., Contant, C., Shirt, S., Hardy, R.J., Agu, V., Gehan, B. & Buray, K. (1988) Air pollution and lung cancer mortality in Harris County, Texas, 1979–1981. *Am. J. Epidemiol.,* 128, 683–699

Calderon, R.L., Johnson, C.C., Dufour, A.P., Karlin, R.J., Sinks, T. & Valentine, J.L. (1993) Health risks from contaminated water: do class and race matter? *Toxicol. Ind. Health,* 9, 879–900

Chapman, S., Marks, R. & King, M. (1992) Trends in tans and skin protection in Australian fashion magazines. *Am. J. Public Health,* 82, 1677–1680

Chiou, H.Y., Hsueh, Y.M., Liaw, K.F., Hong, S.F., Chiang, M.H., Pu, Y.S., Lin, J.S., Huang, C.H. & Chen, C.J. (1995) Incidence of internal cancers and ingested inorganic arsenic: a seven-year follow-up study in Taiwan. *Cancer Res.,* 55, 1296–1300

Cockburn, J.M., Hennrikus, D., Scott, R. & Sanson-Fisher, R. (1989) Adolescent use of sun-protection measures. *Med. J. Aust.,* 151, 136–140

Doll, R. & Peto, R. (1981) *The causes of cancer.* New York, Oxford University Press

Elwood, J.M., Gallagher, R.P. & Pearson, J.C.G. (1985) Cutaneous melanoma in relation to intermittent and constant sun exposure – the Western Canada Melanoma Study. *Int. J. Cancer,* 35, 427–433

English, D.R. & Armstrong, B.K. (1994) Melanocytic nevi in children. I. Anatomic sites and demographic and host factors. *Am. J. Epidemiol.,* 139, 390–401

Goldstein, I.F., Andrews, L.R. & Weinstein, A.L. (1986) Indoor air pollution exposures of low-income inner-city residents. *Environ. Int.*, 12, 211–219

Goldstein, I.F., Andrews, L.R. & Hartel, D. (1988) Assessment of human exposure to nitrogen dioxide, carbon monoxide and respirable particulates in New York inner-city residences. *Atmos. Environ.*, 22, 2127–2139

Hatch, M. & Thomas, D. (1993) Measurement issues in environmental epidemiology. *Environ. Health Perspect.*, 101 (Suppl. 4), 49–57

Henderson, B.E., Gordon, R.J., Menck, H., Soohoo, J., Martin, S.P. & Pike, M.C. (1975) Lung cancer and air pollution in south-central Los Angeles County. *Am. J. Epidemiol.*, 101, 477–488

Hill, D., Theobald, T., Borland, R., White, V. & Marks, R. (1990) *Summer activities, sunburn, sun-related attitudes and precautions against skin cancer. A survey of Melbourne residents in the summer of 1987/1988*. Melbourne, Anti-Cancer Council of Victoria

Hill, D., White, V., Marks, R. & Borland, R. (1993) Changes in sun-related attitudes and behaviours, and reduced sunburn prevalence in a population at high risk of melanoma. *Eur. J. Cancer Prev.*, 2, 447–456

IARC (1972–1995) *IARC monographs on the evaluation of carcinogenic risks to humans* (Vols 1–64). Lyon, International Agency for Research on Cancer

IARC (1988) *IARC monographs on the evaluation of carcinogenic risks to humans* (Vol. 43): *man-made mineral fibres and radon*. Lyon, International Agency for Research on Cancer

IARC (1992) *IARC monographs on the evaluation of carcinogenic risks to humans* (Vol. 55): *solar and ultraviolet radiation*. Lyon, International Agency for Research on Cancer

Jedrychowski, W., Becher, H., Wahrendorf, J. & Basa-Cierpialek, Z. (1990) A case-control study of lung cancer with special reference to the effect of air pollution in Poland. *J. Epidemiol. Community Health*, 44, 114–120

Latourneau, E.G., Krewski, D., Choi, N.W., Goddard, M.J., McGregor, R.G. & Zielinski, J.M. (1994) Case-control study of residential radon and lung cancer in Winnipeg, Manitoba, Canada. *Am. J. Epidemiol.*, 140, 310–322

Marks, R., Jolley, D., Dorevitch, A.P. & Selwood, T.S. (1989) The incidence of non-melanocytic skin cancers in an Australian population: results of a five-year prospective study. *Med. J. Aust.*, 150, 475–478

Mawn, V.B. & Fleischer, A.B. (1993) A survey of attitudes, beliefs and behavior regarding tanning bed use, sun bathing and suscreen use. *J. Am. Acad. Dermatol.*, 29, 959–962

New Zealand Department of Statistics (1988) New Zealand Census of Populations and Dwellings: Labour Force. Wellington, Goverment Printer.

Nixon, J.M. & Carpenter, R.G. (1974) Mortality in areas containing natural fluoride in their water supplies, taking account of socioenvironmental factors and water hardness. *Lancet*, ii, 1068–1071

Schwab, M. (1990) The influence of daily activity patterns and differential exposure to carbon monoxide among social groups. In: Strak, T.H., ed., *Proceedings of the research planning conference on human activity patterns* (EPA Publication 600/4–89/004). Washington, DC, US Environmental Protection Agency

Segan, C. & Borland, R. (1994) Public reaction to the Sunsmart Campaign and reported sun protection behaviour over the summers of 1990/1991 and 1991/1992. In: *The Anti-Cancer Council's skin cancer control program 1990/1991 1991/1992 1992/1993 and related research and evaluation*. Melbourne, Anti-Cancer Council of Victoria

Sexton, K., Gong, H., Bailar, J.C., Ford, J.G., Gold, D.R., Lambert, W.E. & Utell, M.J. (1993) Air pollution heath risks: do class and race matter? *Toxicol. Ind. Health*, 9, 843–878

Simonato, L. & Pershagen, G. (1993) Epidemiological evidence on indoor air pollution and cancer. In: Tomatis, L., ed., *Indoor and outdoor air pollution and human cancer*. Berlin, Springer-Verlag. pp. 119–148

Tomatis, L., ed. (1990) *Cancer: causes, occurrence and control* (IARC Scientific Publication No. 100). Lyon, International Agency for Research on Cancer

Tredaniel, J., Boffetta, P., Saracci, R. & Hirsch, A. (1993) Environmental tobacco smoke and the risk of cancer in adults. *Eur. J. Cancer*, 29, 2058–2068

Xu, Z.Y., Blot, W.J., Xiao, H.P., Wu, A., Feng, Y.P., Stone, B.J., Sun, J., Ershow, A.G., Henderson, B.E. & Fraumeni, J.F. (1989) Smoking, air pollution, and the high rates of lung cancer in Shenyang, China. *J. Natl Cancer Inst.*, 81, 1801–1806

Corresponding author:
A. Woodward
Wellington School of Medicine, Wellington, New Zealand

Socioeconomic status and cancer screening

N. Segnan

> The only widely applied cancer screening programmes are those for cancers of the cervix and female breast. Participation in breast cancer screening has been shown to depend on income and education, health insurance and type of health service. Women in low social classes tend to have lower screening participation rates than those in higher classes. Socioeconomic differences in screening practices tend to decrease when participation is promoted, cultural and economic barriers are removed, and social support is offered. In both developed and developing countries, women of low socioeconomic status have a higher than average risk of cervical cancer, and a lower than average participation in Pap smear screening.

There is substantial agreement within the scientific community on the efficacy of mammographic screening for breast cancer and of cytological screening for cancer of the cervix uteri. It is usually recommended that mammography is carried out every two years in women of age 50–70 years, and that cytology is performed every three years in women of age 25–64 years, although different policies have been proposed and adopted by scientific societies or national institutions.

The experimental evidence of the efficacy of faecal blood screening (Mandel et al., 1993), as well as the observational evidence of the efficacy of rectosigmoidoscopy and total colonoscopy (Atkin et al., 1992; Newcomb et al., 1992; Selby et al., 1992; Winawer et al., 1993), in reducing colorectal cancer mortality are convincing but there is yet no consensus over efficient screening policies. There is still little evidence of efficacy of screenings for other cancers such as those of the lung, stomach and urinary bladder, and recommendations and policies vary across countries.

In this article I examine the association between socioeconomic status and screening for cancers of the breast and the cervix uteri. Differences in screening by social class have also been reported for other cancers, such as colon cancer and melanomas (Myers et al., 1990; Vernon et al., 1989; Meade et al., 1994; Koh et al., 1991; Anderson et al., 1995).

Systematic comparisons of mortality and use of screening across countries is of limited value, particularly if the national income, health care system, local screening policies and quality of screening are not taken into account. Moreover, participation rates often are not measured uniformly.

The association of socioeconomic status and cancer screening attendance could, theoretically, be estimated by measuring the differences in cancer mortality (or incidence) by socioeconomic status, since variation in disease occurrence could be attributed to differential participation in screening programmes. In fact, it is difficult to provide such an estimate because attendees of screening programmes have different cancer risks compared with non-attendees, and data on both socioeconomic status and cancer risk are rarely available simultaneously for screened and unscreened population groups.

Cancer of the breast
Cancer of the breast is the most frequent tumour in the female population worldwide (Coleman et al., 1993a). The populations at highest risk for breast cancer are those of western Europe and North America, where there is a more than fivefold higher incidence of the disease than in low-risk Asian countries.

Mortality and compliance in randomized trials
In the Health Insurance Plan (HIP) trial for breast cancer screening (Fink & Shapiro, 1990), the incidence in non-compliers of the intervention group after 10 years from entry was 84% of that in compliers (1.89 per 1000 person years versus 2.24 per

1000, respectively) and the mortality in non-compliers was 96% of that in compliers (0.49 per 1000 person years versus 0.51 per 1000, respectively). Women who refused to participate were less educated and older than participants. In the Edinburgh trial (Roberts et al., 1990), breast cancer mortality increased with increasing socioeconomic status in the control population. In the study population, the findings were less clear: mortality was decreased by 44% in women of the highest socio-economic status, increased by 29% in women of medium status, and decreased by 10% in women in the lowest socioeconomic category. These data are compatible with a differential effect of screening in groups of different socioeconomic status, with most benefit observed for those of the highest socioeconomic status.

Compliance to mammographic screening
In a review of the participation in breast cancer screening programmes in North America (Vernon et al., 1990), socioeconomic status measured through education and through occupation of the head of household were positively associated with participation in mammographic screening. The social class and age distributions of invited attendees and non-attendees were more similar to each other and to census data than to the distributions for women who came as a result of self-referral.

In the United States of America, a survey by the NCI Breast Cancer Screening Consortium (1990) found that women with less than high-school education or an income less than US$ 15 000 had fewer mammograms than their better-educated or wealthier counterparts. In the seven studies examined, the differences ranged between 11% and 18% for education, and 5% and 12% for income.

In two other studies in the USA, adherence to screening guidelines was found to vary with income: it was increased significantly among women with an annual household income between US$ 25 000 and US$ 49 000 and equal to or over US$ 50 000 (Romans, 1993), or over US$ 15 000 (Zapka et al., 1989), compared with women of lower income.

In Quebec, Canada in 1987, income was not associated with the recency of mammography, but education above high-school level was a significant predisposing factor [odds ratio (OR) = 1.6; 98% confidence interval (CI) = 1.1–2.4] (Potvin et al., 1995).

The absence of an association by income is, in part, explained by the universal third-party payer health care system adopted in Quebec. However, despite a similar health care system in Ontario, Canada, the OR for mammographic screening within the previous two years in the higher-income (at least US$ 45 600) group was 1.8 (95% CI = 1.3–2.6) in 1990 compared with the lower-income group In the USA in the same year and for the same income the OR was 2.8 (95% CI = 2.3–3.2).

In the USA, cost is an important barrier for the use of mammography. Uninsured women, smokers, and women who do not visit a physician regularly may have relatively low rates of use of mammography (Urban et al., 1994), even in the presence of mandatory coverage of screening.

In a study carried out in Northern Ireland (Kee et al., 1992), a higher proportion of attendees (53%) than non-attendees (39%) of mammographic screening used a private car.

Data from a population-based case–control study in Costa Rica (Irwin et al., 1991) indicate that 59% of the most educated or 25% of wealthy women underwent physical examination of the breast compared with 21% of illiterate and 35% of poor women.

In Canada, women living in rural areas had the same basic knowledge of breast cancer screening and similar access to physician care as women in urban areas. Despite this, the prevalence rate ratio, adjusted for age, education, income and marital status, was 0.47 (95% CI = 0.37–0.62) among rural women for having had a mammographic examination in the past two years, compared with urban women (Bryant & Mah, 1992).

The opposite pattern is seen in Europe. In rural areas of Italy, the participation in organized screening programmes is over 70% compared with around 50% in cities (Giordano et al., 1996). In the United Kingdom, participation in the National Breast Screening Programme is lower in urban areas. In a population sample of inner London, the compliance was 42% versus more than 70% in the country (Sutton et al., 1994).

In the USA, findings of the National Health Interview Surveys indicate that race does not seem to be associated with the use of mammography (Breen & Kessler, 1994): 33% of White women reported being screened in 1990 versus 32% of Black and 31% of Hispanic women. In contrast, highest educational and income levels are significantly associated with mammographic screening as well as urban area of residence compared with rural area of residence: the differences are 18% between more than 12 years of schooling and less than 12 years of schooling, 15% between an income of more than US$ 20 000 and an income of less than this amount, and 9% between urban and rural residence.

Reported breast cancer screening in the USA increased between 1987 and 1992 from 23% to 49% (Anderson et al., 1995). The difference in reported mammography between Black and White women disappeared (50.4% versus 48.8% in 1992, and 19% versus 23.2% in 1987, respectively), but participation in mammography by educational levels still ranged in 1992 from 61% for more than 12 years of schooling, to 52% for 12 years of schooling, and to 35.5% for less than 12 years of schooling (Anderson et al., 1995).

In Arizona, where a Medicaid-type programme was proposed to poor women, 54% of such women had had a mammogram within the last two years, compared with 50% of women with other types of health insurance and 9% of women with no insurance (Kirkman-Liff & Kronenfeld, 1992). Differences by level of education and ethnicity were still important, with the less educated and Hispanic women being less screened. Similar results were obtained in New York, USA among women users of county-funded health centres: the screening rates of such women were not lower than those in the community sample, despite the significantly lower socioeconomic level of the former group (Lane et al., 1992).

No differences in mammographic screening have been detected for education and race among a random sample of working women who were members of a health maintenance organization (HMO) (Glanz et al., 1992). Overall, 85% had had a mammogram.

In Italy, the compliance to mammographic screening by socioeconomic status and educational level tended to be similar in women who were actively invited by organized screening programmes (Ciatto et al., 1992). Compliance may sometimes be higher in less-educated women: Donato et al. (1991) reported that the proportion of women with only an elementary-school level of education was 60% among attendees and 49% among nonattendees. In contrast, the higher the education, the higher is self-referral to mammographic screening: 19% of women with high-school education or a university degree came to screening without invitation versus 11% of women with less than a high-school education (Segnan et al., 1990).

Increasing the participation
Many projects have been promoted with the aim of increasing participation in breast screening programmes. The targets have been African-American women (Kang et al., 1994), rural Black women (Eng, 1993), Hispanic American women (Zapka et al., 1993), poor elderly Black women (Mandelblatt et al., 1993a, 1993b), and entire communities (Fletcher et al., 1993). Generally, all these intervention studies were successful in increasing participation in breast cancer screening, although the coverage was lower than the 80% objective set by the United States National Cancer Institute for breast cancer screening. However, in the Florida project, sponsored by the American Cancer Society for breast cancer detection, promotion through the media and provision of low-cost mammograms missed members of minority groups and the socioeconomically disadvantaged, who have a higher incidence of late-stage disease at diagnosis (Roetzheim et al., 1992). These groups were underrepresented, and a significantly higher proportion of White, high-school- or college-educated, higher-income women had a mammography for the first time.

Preventable breast cancer deaths
An estimate of the effect of breast cancer screening participation in reducing socioeconomic status differences in breast cancer mortality can be only empirical. Data on compliance to breast cancer screening by socioeconomic status could be used for estimating the risk of breast cancer that can be attributed to lack of participation in screening across different socioeconomic status categories. We would have to consider, however, that the risk for breast cancer and the stage at diagnosis are different by socioeconomic status, and that increases of breast cancer incidence due to screening have been observed. Moreover, it is likely that the values of these factors are not constant within risk and socioeconomic status strata. Assuming a potential 30% mortality reduction in a population that is fully screened, we can estimate that out of every

Table 1. Preventable fraction of invasive cervical cancer cases by different screening policies and by compliance

Screening interval	Age cycles	Preventable fraction[a]	Compliance						
		1.00	0.30	0.40	0.50	0.60	0.70	0.80	0.90
1 year	20–64	0.93	0.28	0.37	0.47	0.56	0.65	0.75	0.90
3 years	20–64	0.91	0.27	0.36	0.46	0.55	0.64	0.73	0.82
3 years	25–64	0.90	0.27	0.36	0.45	0.54	0.63	0.72	0.81
5 years	25–64	0.82	0.25	0.33	0.41	0.49	0.57	0.65	0.74

[a]From Day, 1996.

1000 deaths from breast cancer among women aged 50 or over at diagnosis, 300 would be preventable. Therefore, every 1% of compliance would prevent three deaths. This empirical estimate indicates that the differences in breast cancer screening compliance observed between women of different socioeconomic status can have an appreciable effect on socioeconomic status differences in breast cancer mortality.

Cancer of the cervix uteri

Cervical cancer is the second most common cancer in females, representing 15% of all cancers. However, 80% of cervical cancers are diagnosed in developing countries (Coleman et al., 1993b). The risk of cervical cancer varies by a factor of 20, the cumulative 30–74 years risk ranging from 7% in Cali, Colombia to 0.35% in the non-Jewish population of Israel. In 1986, it was estimated that in developing countries less than 5% of women, mainly women under 35 years of age, had been screened within the previous five years (WHO, 1986). In industrialized and developing countries, an inverse correlation exists between socioeconomic status and the incidence of cervical cancer (see the chapter by Faggiano et al. in this book).

Preventable invasive cervical cancers

Data from eight screening programmes in developed countries indicate that the relative protection provided in women with two or more negative tests at three to five years of follow-up is a three to five times lower risk than in unscreened women (Day, 1986). An overall estimate of the preventable fraction of invasive cervical cancer in developed countries (Day, 1986) ranges from 82% for screening eight times every five years between the ages of 25 and 64 years to 93% for screening every year between the ages of 20 and 64 years (Table 1). Yearly testing increases the percentage of preventable invasive cancers only marginally, and much less so than an increase in compliance does. The percentage of preventable invasive cancer cases is a third lower (47% versus 73%) with yearly testing and 50% compliance compared with testing every three years with 80% compliance. In the latter scenario, the number of tests are almost half (80 versus 150 per 100 women per three years) that in the former (Table 1).

The effect of different screening policies on the cumulative incidence of cervical cancer at ages 20–64 years was estimated using data from 30 cancer registries with the world's highest rates (Parkin et al., 1992) and the estimates of rate reductions of the IARC Working Group (Day, 1986) (Table 2). Even very low-intensity programmes with two to four tests per lifetime every 10 years can reduce the incidence of cervical cancer by 40–60%, if compliance is 100%. Some European countries or regions and North American subpopulations show a high cumulative incidence, suggesting low coverage and/or ineffective screening policies.

Cervical screening availability and attendance

A survey undertaken in 1991 indicated that the number of Papanicolaou (Pap) tests in European Community countries was sufficient to screen all women of age 25–64 years (Coleman et al., 1993a). Despite this, 22 000 new cancer cases are diagnosed

Table 2. Effects of different screening policies on the 20–64 years cumulative incidence (CI) of cervical cancer per 10^5 women in the areas of 30 cancer registries with the highest world rates

Interval Age (years) No. of tests per lifetime Reduction in rate (%)	– – None	10 years 45-64 2 43[a]		10 years 25-64 4 61[a]		5 years 20-64 9 84[a]	
Registry	CI[b] 20–64 per 10^5	CI 20–64 per 10^5	No. of cases prevented by 10^5 Pap tests	CI 20–64 per 10^5	No. of cases prevented by 10^5 Pap tests	CI 20–64 per 10^5	No. of cases prevented by 10^5 Pap tests
India, Madras	4050	2309	871	1580	618	648	378
Peru, Trujillo	3900	2223	839	1521	595	624	364
Paraguay, Asuncion	3860	2200	830	1505	589	618	360
Brazil, Goiania	3350	1910	720	1307	511	536	313
Colombia, Cali	3110	1773	669	1213	474	498	290
Ecuador, Quito	2630	1499	565	1026	401	421	245
India, Bangalore	2540	1448	546	991	387	406	237
Thailand, Chiang Mai	2400	1368	516	936	366	384	224
Brazil, Porto Alegre	2310	1317	497	901	352	370	216
New Zealand: Maori	2190	1248	471	854	334	350	204
Canada, Manitoba	2140	1220	460	835	326	342	200
Philippines, Manila	2030	1157	436	792	310	325	189
Mali, Bamako	2000	1140	430	780	305	320	187
Thailand, Khon Kaen	1969	1122	423	768	300	315	184
Costa Rica	1900	1083	409	741	290	304	177
France, Martinique	1880	1072	404	733	287	301	175
Poland, Lower Silesia	1860	1060	400	725	284	298	174
India, Ahmedabad	1790	1020	385	698	273	286	167
Bermuda: Black	1770	1009	381	690	270	283	165
East Germany	1740	992	374	679	265	278	162
India, Bombay	1580	901	340	616	241	253	147
Cuba	1570	895	338	612	239	251	147
Philippines, Rizal	1550	884	333	605	236	248	145
Poland, Cracow City	1550	884	333	605	236	248	145
Hong Kong	1550	884	333	605	236	248	145
Portugal, V N de Gaia	1470	838	316	573	224	235	137
USA, Los Angeles	1460	832	314	569	223	234	136
Singapore: Chinese	1410	804	303	550	215	226	132
Japan, Hiroshima	1390	792	299	542	212	222	130
Canada, Newfoundland	1380	787	297	538	210	221	129
Czech., Boh. & Morav.	1370	781	295	534	209	219	128

[a]From Day, 1986.
[b]From Parkin et al., 1992.

every year in these countries. It has been estimated that in a Danish county, 42% of the screening resources are not used in accordance with the national recommendations, indicating a waste of resources (Coleman et al., 1993a). It is estimated that over 80% of patients with invasive cervical cancer in northern England have never been screened (Gillam, 1991).

In the USA, the proportion of women older than 17 years who self-reported having had a Pap smear within the last 12 months was two-thirds, with an additional 10% reporting a Pap smear within the last 13–24 months. A higher proportion of Blacks than Whites had had the test in the previous year, as well as younger women and women with higher household income (Ackermann et al., 1992). In the National Health Interview Surveys in 1987 and 1992 (Anderson et al., 1995), women with more than high-school education, women living above the poverty level, and women residing in urban areas reported more cervical cancer screening in the last three years than did their counterparts. In 1987, screening was reported by 47% of women with less than 12 years of education, compared with 74% of women with more than 12 years education. In 1992 these percentages were 49% and 76%, respectively.

Among a random sample of women from an area in North America with 20% higher mortality rates of cervical cancer, 44.1% reported not having received adequate Pap testing during the previous four years. These women were more likely to be old, to be without medical insurance, to have never been to an obstetrician/gynaecologist and to have less knowledge of risk factors for cervical cancer (Mamon et al., 1990).

In Arizona, where a Medicaid-type programme was proposed to poor women, 73% of such women had had a Pap smear within the last two years, compared with 70% of women with other types of health insurance and 44% of women with no insurance (Kirkman-Liff & Kronenfeld, 1992). However, Latino women, of age 60–70 years, had made significantly lower use of Pap smear tests (OR = 0.56).

In a random sample of women of age 18–69 years in Turin, Italy, the ORs for having ever had a Pap test, adjusted for marital status, age and place of birth, were 7.6, 6.0 and 2.6 ($P < 0.05$) for different levels of education, ranked from high to low (Ronco et al., 1991). In the area of Milan, Italy, the proportion of hospital controls who had had more than three Pap tests during their life was 12% higher in the highest social class ($P = 0.05$) compared with the lowest (Parazzini et al., 1990).

In Costa Rica (Irwin et al., 1991), 77% of educated women had had at least one Pap smear taken in their life compared with 54% of illiterate women. In four different Latin American countries in a case–control study on screening effectiveness for cervical cancer (Herrero et al., 1992), the proportion of hospital controls never screened ranged from 45% in illiterate women to 15% in women with 10 years of school education or more.

A proportion of women as high as 32% are lost to follow-up before assessment of a detected cytological abnormality (Robertson, 1988). In an observational study in Victoria, Australia two groups of women were advised to have a repeat smear after three or six months (Mitchell & Medley, 1989). The non-compliers were 10% and 18%, respectively. In the three-month group, 17% of women of lowest social class did not comply, versus 7–8% of women in other social classes. No difference with social class was detected in the six-month group.

Conclusions

Differences by socioeconomic status in breast cancer screening practices tend to disappear or to decrease when participation is promoted, cultural and economic barriers are removed, and social support is offered. Organized screening programmes, adopting personal invitation instead of self-referral are more successful in this regard and are more cost-effective. Increased participation in breast cancer screening by women of low socioeconomic status in developed countries can be partly achieved by removing economic barriers (that is, the need for women to pay totally or partially for mammography), and in all countries by tailored programmes that promote knowledge and awareness of potential benefits and disadvantages of breast cancer mammographic screening. A high priority in developing countries seems to be to reduce the proportion of late-stage breast cancers at diagnosis and to offer adequate care to symptomatic cases: without an effective therapy, mammographic screening not only would add extra costs but also would be ineffective.

For cervical cancer, women of low socioeconomic status have a higher than average risk and a lower than average participation in screening. Increasing universal coverage is not sufficient to overcome the large disparities in the use of cancer screening procedures. Nevertheless, increasing attendance, and introducing rational screening policies, will help to reduce health inequalities. In developing countries, low-intensity, highly effective screening policies should be introduced before spontaneous – and usually irrational and frequent – cervical screenings become rooted; such screening practices already exist in richer areas and reproduce discrimination by socioeconomic status as well as waste resources.

References

Ackermann, S.P., Brackbill, R.M., Bewerse, B.A., Cheal, N.E. & Sanderson, L.M. (1992) Cancer screening behaviors among U.S. women: breast cancer, 1987–1989, and cervical cancer, 1988–1989. *MMWR CDC Surveill. Summ.*, 41, 17–25

Anderson, L.M. & May, D.S. (1995) Has the use of cervical, breast, and colorectal cancer screening increased in the United States? *Am. J. Public Health*, 85, 840–842

Atkin, W.S., Morson, B.C. & Cuzick, J. (1992) Long term risk of colorectal cancer after excision of rectosigmoid adenomas. *New Engl. J. Med.*, 326, 658–662

Breen, N. & Kessler, L. (1984) Changes in the use of screening mammography: evidence from the 1987 and 1990 National Health Interview Surveys. *Am. J. Public Health*, 84, 62–67

Bryant, H. & Mah, Z. (1992) Breast cancer screening attitudes and behaviors of rural and urban women. *Prev. Med.*, 21, 405–418

Ciatto, S., Cecchini, S., Isu, A., Maggi, A. & Cammelli, S. (1992) Determinants of non-attendance to mammographic screening. Analysis of a population sample of the screening program in the District of Florence. *Tumori*, 78, 22–25

Coleman, M.P., Esteve, J., Damiecki, P., Arslan, A. & Renard, H. (1993a) *Trends in cancer incidence and mortality* (IARC Scientific Publication No. 121). Lyon, International Agency for Research on Cancer

Coleman, D., Day, N., Douglas, G., Lynge, E., Philip, J. & Segnan, N. (1993b) European guidelines for quality assurance in cervical cancer screening. *Eur. J. Cancer*, 29A, S1–S37

Day, N.E., Hakama, M. & Miller, A.B., eds (1986) *The epidemiological basis for evaluating different screening policies in screening for cancer of the uterine cervix*. (IARC Scientific Publication No. 76). Lyon, International Agency for Research on Cancer

Donato, F., Bollani, A., Spiazzi, R., Soldo, M., Pasquale, L., Monarca, S., Lucini, L. & Nardi, G. (1991) Factors associated with non-participation of women in a breast cancer screening programme in a town in northern Italy. *J. Epidemiol. Comm. Health*, 45, 59–64

Eng, E. (1993) The Save our Sisters Project. A social network strategy for reaching rural black women. *Cancer*, 72, 1071–1077

Fink, R. & Shapiro, S. (1990) Significance of increased efforts to gain participation in screening for breast cancer. *Am. J. Prev. Med.*, 6, 34–41

Fletcher, S.W., Harris, R.P., Gonzalez, J.J., Degnan, D., Lannin, D.R., Strecher, V.J., Pilgrim, C., Quade, D., Earp, J.A. & Clark, R.L. (1993) Increasing mammography utilization: a controlled study. *J. Natl. Cancer Inst.*, 85, 112–120

Gillam, S.J. (1991) Understanding the uptake of cervical cancer screening: the contribution of the health belief model. *Br. J. Gen. Pract.*, 41, 510–513

Giordano, L., Giorgi, D., Fasolo, G., Segnan, N. & Rosselli del Turco, M. (1996) Breast cancer screening : characteristic and results of the Italian programmes in the Italian group for planning and evaluating breast cancer screening programmes (GISMA). *Tumori*, 82, 1–8

Glanz, K., Resch, N., Lerman, C., Blake, A., Gorchov, P.M. & Rimer, B.K. (1992) Factors associated with adherence to breast cancer screening among working women. *J. Occup. Med.*, 34, 1071–1078

Herrero, R., Brinton, L.A. & Reeves, W.C. (1992) Screening for cervical cancer in Latin America: a case-control study. *Int. J. Epidemiol.*, 21, 1050–1056

Irwin, K.L., Oberle, M.W. & Rosero-Bixby, L. (1991) Screening practices for cervical and breast cancer in Costa Rica. *Bull. Pan Am. Health Org.*, 25, 16–26

Kang, H.S., Bloom, J.R. & Romano, P.S. (1994) Cancer screening among African-American women: their use of tests and social support. *Am. J. Public Health*, 84, 101–103

Katz, S.J. & Hofer, T.P. (1994) Socioeconomic disparities in preventive care persist despite universal coverage. *J. Am. Med. Assoc.*, 272, 530–534

Kee, F., Telford, A.M., Donaghy, P. & O'Doherty, A. (1992) Attitude or access: reasons for not attending mammography in Northern Ireland. *Eur. J. Cancer Prev.*, 1, 311–315

Kirkman-Liff, B. & Kronenfeld, J.J. (1992) Access to cancer screening services for women. *Am. J. Public Health*, 82, 733–735

Koh, H.K., Geller, A.C., Miller, D.R., Caruso, A., Gage, I. & Lew, R.A. (1991) Who is being screened for melanoma/skin cancer? Characteristics of persons screened in Massachusetts. *J. Am. Acad. Dermatol.*, 24, 271–277

Lane, D.S., Polednak, A.P. & Burg, M.A. (1992) Breast cancer screening practices among users of county-funded health centers vs women in the entire community. *Am. J. Public Health*, 82, 199–203

Mamon, J.A., Shediac, M.C., Crosby, C.B., Sanders, B., Matanoski, G.M. & Celentano, D.D. (1990) Inner-city women at risk for cervical cancer: behavioral and utilization factors related to inadequate screening. *Prev. Med.*, 19, 363–367

Mandel, J.S., Bond, J.H., Church, T.R., Snover, D.C., Bradley, G.M., Schuman, L.M. & Ederer, F. (1993) Reducing mortality from colorectal cancer by screening for fecal occult blood. *New Engl. J. Med.*, 328, 1365–1371

Mandelblatt, J., Traxler, M., Lakin, P., Thomas, L., Chauhan, P. & Matseoane, S. (1993a) A nurse practitioner intervention to increase breast and cervical cancer screening for poor, elderly black women. *J. Gen. Intern. Med.*, 8, 173–178

Mandelblatt, J., Traxler, M., Lakin, P., Kanetsky, P. & Kao, R. (1993b) Targeting breast and cervical cancer screening to elderly poor black women: who will participate? The Harlem study team. *Prev. Med.*, 22, 20–33

Meade, C.D., McKinney, W.P. & Barnas, G.P. (1994) Educating patients with limited literacy skills: the effectiveness of printed and videotaped materals about colon cancer. *Am. J. Public Health*, 84, 119–121

Mitchell, H. & Medley, G. (1989) Adherence to recommendations for early repeat cervical smear tests. *Br. Med. J.*, 298, 1605–1607

NCI Breast Cancer Screening Consortium and National Health Interview Survey Studies (1990) Screening mammography: a missed clinical opportunity? *J. Am. Med. Assoc.*, 264, 54–58

Newcomb, P.A., Norfleet, R.G., Storer, B.E., Surawicz, T.S. & Marcus, P.M. (1992) Screening sigmoidoscopy and colorectal cancer mortality. *J. Natl Cancer Inst.*, 84, 1572–1575

Parazzini, F., Negri, E. & La Vecchia, C. (1990) Characteristics of women reporting cervical screening. *Tumori*, 76, 585–589

Parkin, D.M., Muir, C.S., Whelan, S.L., Gao, Y.T., Ferlay, J. & Powell, J., eds (1992) *Cancer incidence in five continents* (IARC Scientific Publication No. 120). Lyon, International Agency for Research on Cancer

Potvin, L., Camirand, J. & Beland, F. (1995) Patterns of health service utilization and mammography use among women aged 50 to 59 years in the Quebec medicare system. *Med. Care*, 33, 515–530

Roberts, M.M., Alexander, F.E., Anderson, T.J., Chetty, U., Donnan, P.T., Forrest, P., Hepburn, W., Huggins, A., Kirkpatrick, A.E., Lamb, J., Muir, B.B. & Prescott, R.J. (1990) Edinburgh trial of screening for breast cancer: mortality at seven years. *Lancet*, 335, 241–246

Robertson, J.H., Woodend, B.E., Crozier, E.H., & Hutchinson, J. (1988) Risk of cervical cancer associated with mild dyskaryosis. *Br. Med. J.*, 297, 18–21

Roetzheim, R.G., Vandurme, D.J., Brownlee, H.J., Herold, A.H., Pamies, R.J., Woodard, L. & Blair, C. (1992) Reverse targeting in a media-promoted breast cancer screening project. *Cancer*, 70, 1152–1158

Romans, M.C. (1993) Utilization of mammography. *Cancer*, 72, 1475–1477

Ronco, G., Segnan, N. & Ponti, A. (1991) Who has Pap tests? Variables associated with the use of Pap tests in absence of screening programmes. *Int. J. Epidemiol.*, 20, 349–353

Segnan, N., Ronco, G. & Ponti, A. (1990) Practice of early diagnosis of breast and uterine cervix cancer in a northern Italian town. *Tumori*, 76, 227–233

Selby, J.V. & Friedman, G.D. (1992) A case-control study of screening sigmoidoscopy and mortality from colorectal cancer. *New Engl. J. Med.*, 326, 653–657

Sutton, S., Bickler, G. & Sancho-Aldridge (1994) Prospective study of predictors of attendance for breast screening in inner London. *J. Epidemiol. Community Health*, 48, 65–73

Urban, N., Anderson, G.L. & Peacock, S. (1994) Mammography screening: how important is cost as a barrier to use? *Am. J. Public Health*, 84, 50–55

Vernon, S.W., Acquavella, J.F., Douglass, T.S. & Hughes, J.I. (1989) Factors associated with participation in an occupational program for colorectal cancer screening. *J. Occup. Med.*, 31, 458–463

Vernon, S.W., Laville, E.A. & Jackson, G.L. (1990) Participation in breast screening programs: a review. *Soc. Sci. Med.*, 30, 1107–1118

WHO (1986) Control of cancer of the cervix uteri. *Bull. WHO*, 6, 607–618

Winawer, S.J. & Zauber, A.G. (1993) Randomized comparison of surveillance intervals after colonscopic removal of newly diagnosed adenomatous polyps. *New Engl. J. Med.*, 328, 901–906

Zapka, J.G., Stoddard, A.M., Costanza, M.E. & Greene, H.L. (1989) Breast cancer screening by mammography: utilization and associated factors. *Health Serv. Rev.*, 28, 223–235

Zapka, J.G., Harris, D.R., Hosmer, D., Costanza, M.E., Mas, E. & Barth, R. (1993) Effect of a community health center intervention on breast cancer screening among Hispanic American women. *Am. J. Public Health*, 79, 1499–1502

N. Segnan
Unit of Cancer Epidemiology, Department of Oncology, S. Giovanni Hospital, Via S. Francesco da Paola 31, 10123 Turin, Italy

Possible explanations for social class differences in cancer patient survival

A. Auvinen and S. Karjalainen

Social class differences in cancer patient survival have been reported for most cancer types and for a number of countries. The etiology of these differences has been studied less thoroughly and less systematically than social class differences in cancer occurrence. Stage of disease at diagnosis appears to be the most important factor contributing to the social class differences in cancer patient survival. This has been observed most clearly for gastrointestinal and gynaecological cancers. Social class differences in survival are generally wider for patients diagnosed with cancer at local stages than for those diagnosed with cancer at advanced stages. The reasons why cancers are more frequently diagnosed at a local stage in high than in low social classes is not properly understood at the moment. Of other potential contributing factors, the role of treatment and psychosocial factors has scarcely been studied. Biological indicators of tumour aggressiveness have failed to explain the social class differences.

There is a great deal of evidence that cancer patients from lower social classes* do worse than those who are more privileged (see the chapter by Faggiano et al. in this book). These differences seem to exist in all societies where comparisons of survival rates by social class have been made. Differences in cancer patient survival are not due to the direct effect of social class, but rather to intervening factors, i.e. intermediate causal steps. Most studies have failed to reveal the reasons for the differences in survival. There are two main explanations for this. First, the concept of social class is complex. It cannot be observed or measured directly but only by using surrogate measures. The different operational definitions used succeed, at best, in covering only a few aspects of the concept of social class. Further discussion on this topic can be found in chapters elsewhere in this book. Second, many studies have been limited to describing the existence of survival differences, and have not examined their causes.

Stage at diagnosis is the most important factor contributing to social class differences in cancer patient survival. Nevertheless, its role seems to vary widely by type of tumour and possibly also by country. Stage at diagnosis has a clear influence on social class differences in survival from stomach and colon cancer, whereas its role is more modest in rectum, breast, cervix and bladder cancer. Social class differences in survival appear to be more prominent among patients diagnosed with cancer at a local stage than among those diagnosed at more advanced stages. The factors explaining differences in stage distribution by social class are not well known. Delay of diagnosis does not appear to contribute substantially to social class differences in cancer survival. Nor do tumour characteristics explain survival differences by social class.

Social class differences in cancer treatment have also been reported and, in some instances, they have contributed to survival differences. However, a prerequisite for studying the effect of treatment has been that there are residual differences after adjustment for stage. Thus, the handful of studies published on the role of treatment should be interpreted in a wider context.

Empirical identification of the factors contributing to social class differences in survival should be a research priority, as it is prerequisite to developing interventions that diminish such inequalities.

---

*In our article, the term social class is used to cover several aspects, including occupation, income and education, which are also sometimes referred to as socioeconomic status or sociodemographic status.

Comprehensive explanations

Berg et al. (1977), Vågerö and Persson (1987), and Leon and Wilkinson (1989) have provided lists of possible explanations for the etiology of social class differences in cancer survival. Berg et al. (1977), the first to offer such a systematic account, suggested that when there are no differences in treatment, social class differences in survival might be due to differences in when medical help is sought, in the general health and life expectancy of the patients, or in the cancer–host interaction and behaviour of the cancer.

Vågerö and Persson (1987) have provided the most complete list of explanations for survival differences. Their list – cited here – covers both causal and non-causal explanations:

(1) Early detection of cancer (without real improvement of prognosis) is more common in white collar workers.
(2) Early detection of cancer (with some real improvement of prognosis) is more common in white collar workers.
(3) Differential treatment resulting in differential prognosis favours white collar workers.
(4) Differences attributable to host factors influence body susceptibility or body response to cancer.
(5) There are differences attributable to the biological properties of the tumours compared, for instance, with the distribution of histologic types for a particular cancer localization.

Subsequent lists of possible explanations, presented by a number of authors, have been more or less modifications of the account by Vägerö and Persson (1987). To summarize, these theories suggest that the differences, if they are not artifactual, are related to the tumour, patient and/or health care.

Artifactual explanations

Lead-time bias The claim that social class differences in cancer patient survival are artifactual implies that patients having a higher social status have not really gained a true survival advantage. As Vågerö and Persson (1987) indicated, at least part of the survival advantage of earlier diagnosis is artificial: the time of diagnosis is advanced but the death is not delayed – that is, treatment does not alter the natural history of the disease. In its general form, the problem of lead-time bias was pointed out for the first time more than 30 years ago (Saxén & Hakama 1964; Hutchinson & Shapiro, 1968) and was already discussed at length (Enstrom & Austin, 1977) in the 1970s. Efforts to overcome the effects of the problem have been made (Jacques et al., 1981), but the problem has been comprehensively addressed mainly in the field of screening (Miller, 1985). Obviously, earlier diagnosis can give a real survival advantage by changing the natural history of the disease and, thus, postponing death.

The problem of lead-time bias could be tackled by calculating survival time from the first symptom instead of the verification of diagnosis. This approach cannot, however, be applied when dealing with cancers detected at a symptomless stage (for example, cancers detected through organized screening programmes, in medical check-ups or as a chance finding resulting from investigation due to an unrelated disease). Furthermore, it is possible that perception, recall and reporting of symptoms may vary by social class (Mechanic, 1972; Funch, 1988). Another approach is to use stage of disease at diagnosis as a proxy for lead-time and stratify by stage.

The role of lead-time – that is, delay in diagnosis – in the occurrence of social class differences in cancer survival has been assessed by Savage et al. (1984) in myeloma and by Auvinen (1992). Time from first symptom was longer among lower social classes in both studies, but the differences were not statistically significant and delay did not contribute materially to social class differences in survival. In a study in Italy among colon cancer patients, delay in diagnosis did not seem to account for the differences in distribution of stage at diagnosis with social class (Vineis et al., 1993). Furthermore, no prognostic impact of delay has been observed in some studies on breast cancer (Neave et al., 1990; Porta et al., 1991) and colorectal cancer (Barillari et al., 1989; Porta et al., 1991), although in breast cancer contrasting results have also been reported (Wilkinson et al., 1979; Machiavelli et al., 1989).

Diagnostic practices may affect the survival differences between social classes in another way. People from higher social classes tend to use health care services more regularly. It is possible that frequent and extensive check-ups result in rather benign tumours being diagnosed more often in higher than lower social classes. Since these tumours do not cause deaths, they contribute to high survival rates in higher social classes.

Confounding Careful consideration is needed to determine whether a prognostic factor should be considered a confounder, an effect modifier or an explanatory variable in analyses. In the early literature, even stage of disease has been taken as a confounder instead of an explanatory variable (for example: Linden, 1969; Lipworth et al., 1972; Keirn & Metter, 1985).

Clearly, factors such as age and calendar date of diagnosis should be treated as confounders. However, studying their possible effect modification might be of interest. For example, Karjalainen and Pukkala (1990) reported more marked social class differences in cancer survival among patients at older ages. The social class differences could also change over time: either by calendar period of diagnosis or with period of follow-up after diagnosis. There is some evidence that as the excess risk of death due to cancer decreases in longer follow-up, the social class differences also diminish or disappear (Karjalainen & Pukkala, 1990).

The roles of other demographic factors in addition to age – for example, marital status, race, urbanity, and place of residence – require further consideration. They are often correlated to both social class and survival, and thus easily regarded as confounders at first glance. The relationship of race and social class provides an example of this dilemma. African/American ethnic origin is correlated with low social class in the United States of America. It is difficult to establish which factor is of primary importance in analysis of cancer patient survival. Empirical results have been contradictory. In some studies, racial differences have disappeared after controlling for social class; this has been observed in breast cancer (Dayal et al., 1982; Bassett & Krieger, 1986; Stavraky et al., 1987; Gordon et al., 1992) and in prostate cancer (Dayal & Chiu, 1982; Dayal et al., 1985). The reverse pattern has been reported in colorectal cancer: controlling for race abolished social class differences (Dayal et al., 1987). In yet another study, social class explained some of the racial differences, but not all of them (Wegner et al., 1982). The confusion is not made any clearer by the fact that in some studies race has even been used as a surrogate for social class (Page & Kuntz, 1980).

However, the role of the demographic factors in relation to social class cannot be assessed solely on the basis of correlation between the variables. Whatever the final decision about the role of a given factor (confounder, modifier or explanatory factor), it needs to be justified on the basis of an underlying causal model. If the effects of social class are under study, it is noteworthy that race is liable to affect social class, whereas social class will not affect race. However, if there is evidence that there are biological differences in the disease that are not accounted for by social factors such as lifestyle or treatment, stratification by race may be the preferable option. In general, if a factor is part of the causal chain of events between the exposure and outcome variables, adjustment for it leads to over-adjustment and dilution of the real effect. If, for example, place of residence is determined to some extent by social class, one should not adjust for place of residence in a study exploring social class effects.

Causal explanations

General explanation – mortality from other causes
The crude (observed) survival rate reflects the mortality from all causes of death. Not all cancer patients die of the cancer they have and intensity of mortality from other causes affects the proportion of cancer deaths; for example, the proportion of deaths from other causes is large in studies of cancer patients in very old age (over 80 years of age). The social class differences in observed survival may, therefore, be due to variation in mortality from other causes of death (Linden, 1969). A great deal of evidence on social class differences in overall mortality has been gathered (see, for example: Antonovsky, 1967; Townsend & Davidson, 1982; Marmot et al., 1984; Fox et al., 1985). Also, mortality from specific causes of death other than cancer is well documented (Fox & Goldblatt, 1982; Davey Smith et al., 1990; Valkonen et al., 1993). In most cancer types, however, mortality from cancer is far more important than mortality from other causes of death. The exceptions may be cancers with low case fatality – for example, squamous cell cancer of the skin. Also, the excess mortality tends to wear off among cancer patients surviving a long time period.

In studies mainly concerned with social class differences in the extraneous mortality due to cancer, survival rates corrected for mortality from other causes of death should be used. These can be obtained by using information on actual causes of death (Dorn, 1950; Parkin & Hakulinen, 1991). An alternative way is to calculate relative survival rates and use social-class-specific expected survival rates (Linden, 1969). The use of general life tables in the calculation of relative survival rates leads to overestimation of the social class differences in cancer-specific mortality. The expected survival rate is too high for lower social classes and, thus, the relative survival rate becomes too low, overestimating the cancer-specific mortality. The opposite is true for the survival rates of higher social classes. This is especially important when the proportion of deaths from other causes is large – that is, in cancers with good prognosis, among older patients and with long follow-up. However, use of corrected survival rates has been an exception rather than rule in studies of social class differences in cancer patient survival; corrected rates have been employed in only a small number of studies (Berg et al., 1977; Bonett et al., 1984; Karjalainen & Pukkala, 1990; Auvinen, 1992).

The contribution of other causes of death was thoroughly evaluated by Berg et al. (1977). They estimated that depending on the type of cancer, 25–50% of the survival difference between the indigent and private patients in their study was due to deaths from causes other than cancer. Also, in the studies by Kogevinas (1990) and Auvinen et al. (1995), a social class gradient among cancer patients has been observed in deaths from cancer as well as from other causes.

The validity of corrected survival rates depends on the assumption that there are no major social class differences in accuracy of death certificates. This assumption has not, however, been empirically confirmed. Only one study on the subject has been published (Samphier et al., 1988), and this suggests differences in the accuracy of death certificates. If all death certificates were based on pathological diagnoses the social class gradient in cancer mortality would be slightly steeper. A smaller proportion of microscopically confirmed diagnosis among lower social classes has also been reported elsewhere (Auvinen et al., 1995).

Thus, while it is true that the use of observed survival rates leads to overestimation of the social class differences, it is probable that the use of corrected survival rates affects the results in the opposite direction – that is, underestimates them.

In theory, the true estimates could be obtained by calculating relative survival rates by using social-class-specific expected mortality rates. This would allow control for other causes of death without assuming similar accuracy of death certificates. Unfortunately, social-class-specific mortality data required for this method are not widely available.

It is our view that the choice of survival measure may not be of crucial importance after all. Relative or observed rates have been compared in some studies. In the study by Karjalainen and Pukkala (1990), the differences in results obtained by using relative or corrected rates were small in ages 25–54 years, but increased in older ages. Auvinen et al. (1995) observed slightly larger social class differences when using observed rates compared with corrected rates and concluded that the difference was probably due to overestimation of the differences by observed rates.

An issue related to measurement of survival is the eligibility of cases diagnosed at autopsy. There is some evidence that diagnosis at autopsy is more common in lower social classes and that in the majority of these cases cancer is the underlying cause of death (Auvinen et al., 1995). This suggests that cases first diagnosed at autopsy (at least those with cancer as a cause of death) should be included in the material with zero survival. The obvious justification for this is the fact that the probability of autopsy diagnosis may be determined by social class (through health behaviour).

Specific explanatory factors

To contribute to the prevention of poorer cancer survival in lower social classes, studies of social class differences in cancer survival should explore the etiology of the differences instead of merely describing them. In principle, all prognostic factors need to be considered as potential explanatory factors in the social class differences. This means that, ideally, information on several characteristics of both patient and tumour should be collected. This approach was pioneered by Berg et al. (1977), who was already attempting to quantify the individual contributions of different factors in the 1970s. Because stage of disease at diagnosis and mode of treatment are the principal determinants of outcome in most cancers, they are the most plausible explanatory candidates.

Stage Stage at diagnosis has been suggested as an explanation in most studies discussing the social class differences in cancer patient survival. Information on stage at diagnosis is available and has been analysed in a number of studies. Stage at diagnosis has been shown to be the most important factor contributing to social class differences in cancer patient survival. Nevertheless the importance of stage seems to vary by type of tumour and by country. An association between social class and stage of cancer at diagnosis has also been reported in studies not addressing survival differences (Mandelblatt et al., 1991; Richardson et al., 1992; Wells & Horm, 1992).

The evaluation of the role of stage at diagnosis presented here is superficial out of necessity. A detailed analysis of findings of the early studies is not feasible, primarily due to incomplete reporting of the studies. In many instances, only statistical significance has been reported and not the parameter estimates. It has also been common to present only the results of the univariate analysis and of the final model with all the significant parameters. This makes it impossible to assess the individual effects of the prognostic factors on social class differences.

There are also other considerations that complicate an evaluation of the contribution of stage at diagnosis to social class differences in cancer survival. First, cancer is not a uniform disease entity and there are differences in staging systems and the prognostic impact of stage by primary site. Second, differences in staging systems exist for some primary sites; for example, both Dukes' and TNM staging are currently used in cancers of the colon and rectum. Third, accuracy of staging information varies between studies. The varying degree of misclassification that results from this also makes direct comparison of results difficult. In addition, it is possible that some of the social class differences in survival among patients with the same stage of disease at diagnosis could be due to more accurate staging in the higher social classes. If, say, micrometastases in regional lymph nodes are detected with greater probability in higher social classes than in other groups, this would lead to apparently improved prognosis in this group for patients classified as having local-stage disease as well as for those with regional-stage disease. This artefact is due to the differential accuracy of staging known as the Will Rogers phenomenon (Feinstein et al., 1985; Greenberg et al., 1991).

In a number of studies, the social class differences observed in the univariate analysis have persisted even after controlling for stage (by stratification or modelling) (Table 1). This has been reported consistently in cancers of the uterine corpus, prostate and bladder. Most studies suggest a similar social class effect that is independent of stage in stomach, lung, kidney and both melanocytic and non-melanocytic skin cancer. Also quite consistently, social class differences have not been detectable after controlling for stage in cancers of the pancreas and ovary, which suggests that stage accounts for most or all of the social class differences in survival of cancer at these primary sites. Results regarding the role of stage are conflicting in breast, colorectal and cervical cancers.

In a study by Chirikos et al. (1984), adjustment for age, stage and primary site accounted for a quarter of the differences in survival between patients with white-collar occupations and those with blue-collar occupations, but the differences were no longer statistically significant after the adjustment. A similar pattern was observed when income was used as an indicator of socioeconomic status.

In cancer of the uterine corpus, Steinhorn et al. (1986) showed that stage of disease accounted for most of the survival differences between patients from areas with lower mean income and smaller proportion of high-school graduates. The differences remained statistically significant, however. This was not the case for uterine sarcoma, as the gradient between educational groups was steeper after adjustment for the above mentioned factors. Income was not a significant prognostic factor in sarcomas.

In a study of lung cancer, the differences in risk of poor outcome at first year of follow-up between patients with low and high levels of education disappeared after adjustment for sex, comorbidity, histology and stage at diagnosis (Stavraky et al., 1987). Unfortunately, the individual contributions of each of these attributes on differences between educational groups were not reported. Karjalainen and Pukkala (1990) reported larger social class differences in non-localized than in localized stages of breast cancer using both relative and corrected five-year survival rates. Introduction of stage in the multivariate model accounted for a minor part of the social class differences.

Table 1. Differences in stage of cancer at diagnosis by socioeconomic status of patients, and the effect of these differences in cancer patient survival differentials

Reference; country	Proportion of local cases	Method of adjustment	Main results	Comments
Oesophageal cancer				
Linden, 1969; USA	Not reported	Stratification	Differences equally small within local stage and overall	Relative five-year survival rates
Stomach cancer				
Linden, 1969; USA	Not reported	Stratification	Large differences within local stage among men and women	Relative five-year survival rates
Lipworth et al., 1970; USA	<5000 US$ 4% >5000 US$ 9%	Rates standardized by proportion of local cases	Small effect among men; adjustment not feasible among women	Observed three-year survival rates; no simultaneous adjustment for age
Lipworth et al., 1972; USA	Non-private 16% Private 15%	Rates standardized by proportion of local cases	Social class differences pronounced after adjustment	Observed ten-month survival rates; unadjusted results not presented
Kato et al., 1992; Japan	Non-employed 29% Service 34% Production 32% Clerical 44% Professional 46%	Stratification and regression modelling	Social class differences confined to local stage; differences remained significant after adjustment for stage	Observed five-year survival rates; simultaneous adjustment for age, marital status and urbanity
Colon cancer				
Lipworth et al., 1970; USA	<5000 US$ 9% >5000 US$ 10%	Standardization by proportion of local cases	Small differences among men, unaffected by adjustment; among women, high income associated with lower survival rate overall and after adjustment	Relative three-year survival rates; no simultaneous adjustment for age
Lipworth et al., 1972; USA	Non-private 34% Private 30%	Standardization by proportion of local cases	Survival advantage for private patients among men and women after adjustment for stage	Transverse colon. Observed ten-month survival rates; unadjusted results not presented; no simultaneous adjustment for age
Berg et al., 1977; USA	Indigent 35% Non-indigent 37%	Standardization by proportion of local cases	Differences between indigent, clinic pay and private patients diminished but did not disappear after adjustment for stage	Corrected five-year survival rates; no simultaneous adjustment for age

Table 1. (contd) Differences in stage of cancer at diagnosis by socioeconomic status of patients and the effect of these differences in cancer patient survival differentials

Reference; country	Proportion of local cases	Method of adjustment	Main results	Comments
Colon cancer				
Wegner et al., 1982; USA	Not reported	Regression modelling	A non-significant trend by socioeconomic status after adjustment for stage; unadjusted results not presented	Colorectal cancer. Observed seven-year survival rates; simultaneous adjustment for race, age and sex
Keirn & Metter, 1985; USA	Indigent 34% Non-indigent 29%	Stratification	No significant differences between indigent and non-indigent patients within stage; unadjusted results not presented	Colorectal cancer. Median observed survival time
Dayal et al., 1987; USA	Low 34% Medium 42% High 44%	Regression modelling	No clear differences by socioeconomic status after adjustment for stage in colon cancer; unadjusted results not presented	Colon and rectal cancer. Observed survival time; simultaneous adjustment for age and sex
Brenner et al., 1991; Germany	Low 54% Medium 58% High 51% (colorectal)	Stratification	The lowest social class had poorest survival within all stages	Colorectal cancer. Observed survival time and ten-year survival rates
Kato et al., 1992; Japan	Non-employed 33% Service 36% Production 42% Clerical 46% Professional 38%	Stratification and modelling	Some indication of social class differences within regional stage; overall differences no longer significant after adjustment for stage	Colorectal cancer. Observed five-year survival rates; simultaneous adjustment for age, marital status and urbanity
Auvinen, 1992; Finland	Class I (high) 44% Class II 41% Class III 42% Class IV (low) 37%	Regression modelling	Stage accounted for half of the social class differences	Corrected five-year survival rates; adjusted for age
Rectal cancer				
Lipworth et al., 1970; USA	<5000 US$ 9% >5000 US$ 16%	Standardization by proportion of local cases	Among men, modest differences unaffected by adjustment; among women, clear differences increased by adjustment	Relative three-year survival rates; no simultaneous adjustment for age
Lipworth et al., 1972; USA	Non-private 53% Private 36%	Standardization by proportion of local cases	Clear survival advantage for private patients over non-private after adjustment for stage	Observed ten-month survival rates; no simultaneous adjustment for age

Table 1. (contd) Differences in stage of cancer at diagnosis by socioeconomic status of patients and the effect of these differences in cancer patient survival differentials

Reference; country	Proportion of local cases	Method of adjustment	Main results	Comments
Rectal cancer				
Dayal et al., 1987; USA	Low 34% Medium 42% High 44%	Regression modelling	A non-significant trend by socioeconomic status after adjustment for stage; unadjusted results not presented	Observed survival time; simultaneous adjustment for age, sex and race
Brenner et al., 1991; Germany	Low 54% Medium 58% High 51% (colorectal)	Stratification	Differences equally large in all stages	Colorectal cancer; results not presented separately for rectal cancer. Ten-year survival rates
Pancreas cancer				
Linden, 1969; USA	Not reported	Stratification	Differences observed within local stage	Relative five-year survival rates
Larynx cancer				
Linden, 1969; USA	Not reported	Stratification	Large difference within local stage	Relative five-year survival rates
Berg et al., 1977; USA	Not reported	Standardization by proportion of local cases	Differences slightly reduced by adjustment for stage	Corrected five-year survival rates
Lung cancer				
Linden, 1969; USA	Not reported	Stratification	Marked differences within local stage among men	Relative five-year survival rates
Lipworth et al., 1970; USA	<5000 US$ 2% >5000 US$ 8%	Standardization by proportion of local cases	No clear differences overall nor within stage	Relative three-year survival rates; simultaneous stratification by sex, but not by age
Lipworth et al., 1972; USA	Non-private 13% Private 24%	Standardization by proportion of local cases	More favourable survival for private than non-private patients among men, but not among women	Observed ten-month survival rates; no simultaneous adjustment for age
Berg et al., 1977; USA	Indigent 15% Non-indigent 19%	Standardization by proportion of local cases	Modest overall differences disappeared after adjustment for stage	Median corrected survival time; no simultaneous adjustment for age
Keirn & Metter, 1985; USA	Indigent 34% Non-indigent 29%	Stratification	Non-significantly more favourable survival for non-indigent than indigent patients within local and regional stage	Median and 75th percentile of observed survival time; unadjusted survival data not presented by economic status; no adjustment for age

Table 1. (contd) Differences in stage of cancer at diagnosis by socioeconomic status of patients and the effect of these differences in cancer patient survival differentials

Reference; country	Proportion of local cases	Method of adjustment	Main results	Comments
Lung cancer				
Stavraky et al., 1987; Canada	Not reported	Regression modelling	Modest overall differences by education disappeared after adjustment for stage	Odds ratio of death at one year of follow-up; simultaneous adjustment for age, sex, histology, employment status and comorbidity
Breast cancer in women				
Linden, 1969; USA	County 82% Private 83%	Statification (local versus all)	Differences between county and private hospital patients smaller within local stage than among all patients	Relative ten-year survival rates; age group 55–64 years
Lipworth et al., 1972; USA	Non-private 31% Private 40%	Standardization by proportion of local cases	More favourable survival among private patients compared with nonprivate after adjustment for stage	Observed 10-month survival rate; no simultaneous adjustment for age; unadjusted rates not provided
Berg et al., 1977; USA	Indigent 35% Non-indigent 38%	Standardization by proportion of local cases	Differences between indigent, clinic pay and private patients diminished but did not disappear after adjustment for stage	Corrected five-year survival rates
Dayal et al., 1982; USA	Not reported	Regression modelling	Social class differences remained significant after adjustment for stage; point estimates not reported	Observed survival time; no simultaneous adjustment for age
Keirn & Metter, 1985; USA	Indigent 45% Non-indigent 35%	Stratification	Non-significantly more favourable survival for non-indigent than indigent patients within local and regional stage	Median, 75th and 80th percentile of observed survival time; unadjusted survival data not presented by economic status; no adjustment for age
Bassett & Krieger, 1986; USA	Not reported	Regression modelling	Statistically significant differences after adjustment for stage	Observed survival time; adjustment for age, race and histological type
Karjalainen & Pukkala, 1991; Finland	Class I (high) 51% Class II 51% Class III 48% Class IV (low) 47%	Stratification and modelling	Differences apparent mainly within nonlocal stage and remained significant after adjustment for stage	Corrected and relative five-year survival rates; simultaneous adjustment for age and year of diagnosis

Table 1. (contd) Differences in stage of cancer at diagnosis by socioeconomic status of patients and the effect of these differences in cancer patient survival differentials

Reference; country	Proportion of local cases	Method of adjustment	Main results	Comments
Breast cancer in women				
Gordon et al., 1992; USA	Not reported	Regression modelling	Differences remained significant after adjustment for tumour diameter and number of positive lymph nodes	Observed and disease-free survival time; adjustment for race and estrogen receptors
Cancer of the uterine cervix				
Linden, 1969; USA	Not reported	Stratification	No differences within local stage	Five-year cumulative survival rate
Lipworth et al., 1970; USA	<5000 US$ 35% >5000 US$ 45%	Standardization by proportion of local cases	Differences decreased slightly but did not disappear after adjustment for stage	Three-year relative survival rate; no simultaneous adjustment for age
Lipworth et al., 1972; USA	Non-private 38% Private 35%	Standardization by proportion of local cases	Survival advantage for private patients compared with non-private after adjustment for stage	Ten-month survival rates; unadjusted rates not presented; no simultaneous adjustment for age
Berg et al., 1977; USA	Indigent 69% Non-indigent 75%	Standardization by proportion of local cases	Differences between indigent, clinic pay and private patients diminished but did not disappear after adjustment for stage	Five-year corrected survival rates; no simultaneous adjustment for age
Cancer of the uterine corpus				
Linden, 1969; USA	Not reported	Stratification	Differences observed also within local stage	Relative five-year survival rate
Lipworth et al., 1970; USA	<5000 US$ 35% >5000 US$ 46%	Standardization by proportion of local cases	No differences overall nor after adjustment for stage	Relative three-year survival rates
Lipworth et al., 1972; USA	Non-private 48% Private 60%	Standardization by stage distribution	More favourable survival for private than non-private patients after adjustment for stage	Observed 10-month survival rates; unadjusted rates not presented; no simultaneous adjustment for age
Berg et al., 1977; USA	Not reported	Standardization by stage distribution	More favourable survival for non-indigent than indigent patients after adjustment for stage	Corrected five-year survival rates; unadjusted corrected rates not presented; adjusted for age

Table 1. (contd) Differences in stage of cancer at diagnosis by socioeconomic status of patients and the effect of these differences in cancer patient survival differentials

Reference; country	Proportion of local cases	Method of adjustment	Main results	Comments
Cancer of the uterine corpus				
Steinhorn et al., 1986; USA	Not reported	Regression modelling	Stage accounted for most of the differences, but they remained statistically significant also after adjustment for stage. For uterine sarcoma the differences between educational groups were larger after adjustment for stage	Observed survival time and five-year rates; simultaneous adjustment for age, race and area
Cancer of the ovary				
Linden, 1969; USA	Local only	Blocking	No differences within local stage	Relative five-year survival rates
Lipworth et al., 1970; USA	Non-private 27% Private 21%	Standardization by proportion of local cases	No differences after adjustment for stage	Observed ten-month survival rates
Lipworth et al., 1972; USA	Non-private 13% Private 30%	Standardization by proportion of local cases	No differences overall nor after adjustment for stage	Relative three-year survival rates
Kidney cancer				
Linden, 1969; USA	Not reported	Stratification	Differences observed within local stage	Relative five-year survival rates
Bladder cancer				
Linden, 1969; USA	Local only	Blocking	Clear differences observed within local stage among both sexes, although not as large as the overall differences	Relative five-year survival rates
Lipworth et al., 1970; USA	<5000 US$ 39% >5000 US$ 36%	Standardization by proportion of local cases	Adjusted differences larger than unadjusted among men; direction of differences reversed after adjustment among women	Relative three-year survival rates; no simultaneous adjustment for age
Lipworth et al., 1972; USA	Non-private 66% Private 63%	Standardization by proportion of local cases	Clear differences after adjustment for stage among men and women	Observed ten-month survival rates
Berg et al., 1977; USA	Indigent 70% Non-indigent 75%	Standardization by proportion of local cases	Differences reduced only slightly by adjustment for stage	Corrected five-year survival rates; no simultaneous adjustment for age

Table 1. (contd) Differences in stage of cancer at diagnosis by socioeconomic status of patients and the effect of these differences in cancer patient survival differentials

Reference; country	Proportion of local cases	Method of adjustment	Main results	Comments
Prostate cancer				
Lipworth et al., 1970; USA	<5000 US$ 27% >5000 US$ 16%	Standardization by proportion of local cases	Differences remained equally large after adjustment for stage	Relative three-year survival rates
Lipworth et al., 1972; USA	Non-private 57% Private 52%	Standardization by proportion of local cases	Differences observed after adjustment for stage	Observed ten-month survival rates
Berg et al., 1977; USA	Indigent 49% Non-indigent 44%	Standardization by proportion of local cases	Differences almost equally large after adjustment for stage	Observed seven-year survival rates; no adjustment for age
Dayal et al., 1985; USA	Not reported	Regression modelling	Differences observed in all stages	Observed survival time; adjusted for age; stage-adjusted point estimates not reported
Clark & Thompson, 1994; USA	Enlisted 75% Officers 80% (clinical stage)	Stratification	No differences overall or within stage	Observed five-year survival rates; no adjustment for age
Melanoma of the skin				
Linden, 1969; USA	Not reported	Stratification	Analysis by stage not feasible because of small numbers of cases	Corrected five-year survival rates
Berg et al., 1977; USA	Not reported	Standardization by proportion of local cases	Differences almost disappeared after adjustment for stage	Corrected five-year survival rates
Shaw et al., 1981; Australia	Stage I only	Blocking	Differences considerably smaller, yet significant within stage I	Observed five-year survival rates
Skin cancer (excluding melanoma)				
Linden, 1969; USA	Not reported	Stratification	Differences within local stage as large as overall differences	Relative five-year survival rates
Berg et al., 1977; USA	Indigent 96% Non-indigent 95%	Standardization by stage distribution	Adjustment for stage did not alter the differences	Observed, relative and corrected five-year survival rates; differences mostly due to other causes of death

Brenner *et al.* (1991) observed poorer survival for colorectal cancer patients of low social class than for patients of middle or high social class in local, regional and distant stages. The difference in five-year observed survival rate between the lowest and highest social class was between 5 and 10% in all stages.

In a study of breast cancer patients by Gordon *et al.* (1992) adjustment for stage did not materially diminish the socioeconomic differences. In the analysis of disease-free survival, the excess risk of death associated with residence in an area with low education diminished only slightly after adjustment for number of positive axillary lymph nodes, tumour diameter and estrogen receptor status. Similar results were obtained in the analysis of overall survival. In a study of colon cancer conducted in Finland, introduction of stage in the model explained half of the social class differences (Auvinen, 1992). After the adjustment, the social class differences were no longer statistically significant.

In a Japanese study with almost 4500 cancer patients, most favourable survival was observed in the highest occupational class (professional and managerial). The trend by occupational class was no longer significant after adjustment for stage in gastric cancer and in colorectal cancer (Kato *et al.*, 1992). In a study conducted in France, the effect of socioeconomic status (assessed on the basis of housing) was studied among 771 colorectal cancer patients (Monnet *et al.*, 1993). There were statistically significant differences in survival by socio-economic status among patients with localized disease, but not among patients with advanced disease.

It should be noted that stage is in turn determined by a number of other factors related to both tumour and host. In fact, stage is sometimes regarded as an expression rather than as a determinant of prognosis *per se*. These factors are discussed under 'Determinants of stage and treatment'.

Treatment Outcome of treatment depends on the treatment modality, quality of treatment, and characteristics of the tumour and the patient. To avoid confusion, these should be addressed separately. Most of the research in the field has concentrated on the choice of treatment, which is obviously based on feasibility. As concerns equity, it is important to know whether social class differences in survival are due to differences in access to treatment or in the quality of treatment.

Choice of treatment. Choice of treatment modality is a matter of critical importance in the outcome of cancer. In many cases, there are treatment protocols depending on, for example, primary site, histological type and stage of the tumour, as well as on age and health status of the patient. Differences in choice of treatment have been reported depending on insurance coverage and marital status (Greenberg *et al.*, 1988) as well as on urbanity (Howe *et al.*, 1992; Launoy *et al.*, 1992). In countries where the patient is responsible for a substantial part of the cost of treatment, economic factors may also be important in the feasibility of some costly treatments.

The potential contribution of treatment to the occurrence of social class differences in cancer patient survival was first directly assessed by Linden (1969) and Lipworth *et al.* (1970, 1972), but the effect of treatment has been directly addressed in only a small number of subsequent studies (Table 2). The findings of some studies indicate that social class differences in treatment have contributed to survival differences.

Linden (1969) stratified his material of 1662 breast cancer patients by age, race, stage and treatment. The social class differences in survival were equally large among breast cancer patients with surgically treated localized tumours as in the whole material. In the first study by Lipworth *et al.* (1970), the proportion of patients receiving neither surgery nor radiotherapy was larger among patients residing in a low-income area than among those residing in a high-income area in bladder cancer among both sexes and in rectal cancer among men, but there was a reverse association in cancers of the stomach and colon, and among women with rectal or lung cancer. Statistical significance of the differences was not assessed. The authors concluded that the social class differences in the assignment of treatment cannot explain the differences in survival. In another study, Lipworth *et al.* (1972) analysed patients with localized or regional stage combining several primary sites (standardizing the survival rates on the site). The proportion of patients dying within four months was higher among non-private than private patients in all age and treatment strata.

Berg *et al.* (1977) found differences in survival from several types of cancer between two groups defined by socioeconomic status, although the patients were treated by the same staff and should have received equal treatment. Opposite results have also been obtained. Page and Kuntz (1980) studied survival among Veterans Administration male cancer patients, and found no differences in survival by race or income except in bladder cancer. They concluded that the lack of differences was due to the fact that all the patients they studied received the same treatment with no distinctions, whereas most American hospitals placed their patients into categories on the basis of ability to pay.

In childhood leukaemia, no social class differences in treatment modality were detected by McWhirter et al. (1983). Keirn and Metter (1985) found that there were no differences in survival by economic status among lung, breast and colorectal cancer patients treated in a hospital that accepted patients regardless of race or ability to pay. Dayal and Chiu (1982) noted that the lack of a significant racial difference in the Veterans Administration study (Page & Kuntz, 1980) may have been due to the selective nature of the patient population. The income level of both Blacks and Whites who used the Veterans Administration hospital was relatively

Table 2. Treatment and social class differences in cancer patient survival

Reference	Primary site	Treatment distribution	Method of analysis	Main results	Comments
Linden, 1969; USA	Breast	Surgical treatment for 82% of county hospital and 83% of private hospital patients	Blocking	Differences observed also among surgically treated patients	Relative and corrected 10-year survival rates; local stage only
Lipworth et al., 1972; USA	Several sites combined	No surgery or radiotherapy for 7% of private and 21% of non-private patients with locoregional disease	Blocking	Differences observed within both treatment groups	Observed 20-day survival rate
Chirikos et al., 1984; USA	Several sites combined	Surgery versus radio/chemotherapy without surgery versus others – distribution not reported	Regression modelling	Differences diminished and no longer significant after adjustment for treatment	Observed survival time; simultaneous adjustment for 'severity'
Chirikos & Horner, 1985; USA	Digestive tract, colorectal	Surgery for 42% of high-income, 40% of middle-income and 41% of low-income patients in colorectal cancer	Regression modelling	Differences persist after adjustment for treatment	Observed survival time; simultaneous adjustment for age and stage
Auvinen, 1992; Finland	Colon	Curative surgery for 54% of patients in social class I, 53% in classes II and III, and 44% in class IV (lowest)	Regression modelling	Treatment accounted for the remaining differences after adjustment for stage	Corrected five-year survival rates; simultaneous adjustment for stage

low. Therefore the study population may have been too homogeneous with respect to socioeconomic status. The same criticism might be applied to the study of Keirn and Metter (1985).

McWhorter and Mayer (1987) studied the association between race and type of initial treatment in 1978–1982 in the USA and found that Blacks received less radical treatment. Black patients also had lower survival rates. In the study by Chirikos et al. (1984), social class differences were no longer significant after adjustment for age, stage and primary site. Further adjustment for treatment did not have an effect on the point estimate nor significance. In the small sample of Chirikos and Horner (1985), social class differences in survival between colorectal cancer patients with high and low income remained statistically significant even after adjustment for both stage and surgical treatment. In the study of colon cancer patients by Auvinen (1992), all social class differences in the risk of cancer death were diminished by adjustment for stage and disappeared after further adjustment for surgical treatment.

Choice of treatment may also depend on tumour characteristics, such as histology, and on host factors, such as age or comorbidity. These are discussed under 'Determinants of stage and treatment'.

Quality of treatment. It has been suggested that even when patients are given the same type of treatment, there may be 'differences in treatment efficacy' by social class (Vågerö & Persson, 1987). Quality of treatment received may vary between hospitals. The fact that social class differences have tended to be smaller in studies based on one hospital only than in studies covering several treatment centres (Weston et al., 1987 versus Savage et al., 1984 in myeloma; Berg et al., 1977 versus Linden, 1969 in stomach and oesophagus cancer and leukaemia; Keirn & Metter, 1985 versus Linden, 1969 in lung, breast and colorectal cancers; Chirikos & Horner, 1985 versus Dayal et al., 1987 in colorectal cancer) also provides some support for the role of treating hospital. It is, however, also possible that the socioeconomic background of the patients is similar within the single hospitals and the differences are not observed because of a narrow spectrum of socioeconomic status rather than uniform treatment.

Choice and implementation of treatment are also affected by compliance/patient involvement. It may be easier for patients from higher social classes to communicate with doctors (Epstein et al., 1985), which may affect choice of treatment, compliance and follow-up. Also, a direct effect of social class on quality of care has been suggested in a study by Burstin et al. (1992), who reported greater risk of medical injury due to substandard care among uninsured patients.

Access to and quality of treatment seem to be associated with place of residence (West & Lowe, 1976; Stiller, 1988), which is, in turn, associated with social class. Geographic area was not a prognostic factor in the multivariate analysis and did not account for the social class differences in cancer of the uterine corpus in the study by Steinhorn et al. (1986). Similar results were obtained for cancers of the colon and rectum by Brenner et al. (1991). Urban residence has not accounted for the social class differences in colon cancer (Bonett et al., 1984; Brenner et al., 1991; Auvinen, 1992) nor in breast cancer (Bonett et al., 1984).

It is not clear, however, how information on treatment could or should be taken into account in observational survival studies (Hakulinen, 1983). In observational studies, the treatment distribution is influenced not only by treatment practices, but also by stage distribution. As noted by Morrison et al. (1976), 'treatment tended to be selected according to apparent prognosis'. This emphasizes the importance of not regarding stage and treatment as 'independent' prognostic factors, but as hierarchical parts of a causal chain.

Determinants of stage and treatment

Tumour characteristics. Different tumour biology is one explanation proposed for the social class gradient in survival (Lipworth et al., 1970; Berg et al., 1977; McWhirter et al., 1983; Chirikos & Horner, 1985). Tumour characteristics possibly associated with response to a given mode of therapy are, in part, the same as those affecting stage of disease at diagnosis. Grade of differentiation is a principal determinant of aggressiveness of a tumour in several primary sites. The probability of achieving a remission by chemotherapy or radiotherapy may depend on growth rate of the tumour or presence or absence of a specific genetic trait, such as the *MDR2*, *p53*, or *c-erbB2* gene. However, the relatively little information available suggests that biological indicators of tumour aggressiveness are not important determinants of social class differences in cancer survival.

It is possible that the characteristics of the tumours differ across social classes as a function of etiological factors and these are reflected in differences in survival. For instance, Ramot and Magrath (1982) have suggested that the increased incidence of childhood leukaemia in higher social classes is mainly due to excess of common acute lymphoblastic leukaemia, which has a better prognosis. However, Dayal and Chiu (1982) are against the hypothesis that there are differences in tumour biology by social class, and claim that the elements of social status influencing etiology are different from those predicting survival. Ewertz et al. (1991) did not find a significant association between survival of breast cancer patients and the most important risk factors for this disease. If the differences in exposure to risk factors led to differences in tumour characteristics (important for prognosis), this would have been detected in this study. Furthermore, tumour aggressiveness is associated with the characteristics of the patients, such as the immune system, which may play a more important role than the tumour factors themselves.

Histological type is associated with aggressiveness of the tumour in a number of cancers. In the study by Bassett and Krieger (1986), adjustment for histology (ductal, lobular or other) did not account for the social class differences in breast cancer survival. In the study of Steinhorn et al. (1986), social class differences in survival from adenocarcinoma of the uterine corpus were of the same magnitude as in uterine sarcomas. No clear social class differences were observed either before or after adjustment for histological type (squamous cell versus others) among lung cancer patients in the study by Stavraky et al. (1987). Auvinen et al. (1995) reported effect modification by cell type in lung cancer and leukaemia among men in their study based on more than 100 000 cancer patients. Social class differences were not observed among all patients, but they were confined to non-small-cell lung cancer and acute leukaemia.

Other tumour characteristics that have been studied in relation to social class differences in cancer patient survival include primary-site-specific prognostic factors such as initial white cell count in childhood leukaemia (McWhirter et al., 1983), serum albumin and haematocrite in myeloma (Savage et al., 1984), hormone receptors in breast cancer (Gordon et al., 1992), tumour thickness in melanoma (Shaw et al., 1981) and grade of differentiation in soft tissue sarcoma (Ciccone et al., 1992). These factors have not accounted for the social class differences in survival. In the study by Savage et al. (1984), overcrowding remained the most important determinant of outcome even after adjustment for biological prognostic factors such as tumour burden, serum albumin and haematocrite. Carnon et al. (1994) assessed in a recent study the role of tumour size, nodal status, histological grade and estrogen receptor concentration in relation to social class differences in breast cancer survival. None of the biological indicators of prognosis was associated with social class.

Host factors. Host factors include both biological factors, such as presence of other chronic diseases, and psychosocial factors, such as health behaviour before or after diagnosis.

Several host factors are related to stage of disease at diagnosis. The importance of host factors for the occurrence of survival differences has scarcely been studied. Host resistance or host–tumour relationship has frequently been suggested as a mediator of the effect of social class on cancer patient survival. However, the exact meaning of the terms has remained unclear apart from the remarks that host resistance may be influenced by, for example, nutrition and that immunological mechanisms may be involved. Thus it is unclear how they should be operationalized.

Presence of any other chronic disease did not appear an independent prognostic factor in the study by Stavraky et al. (1987) and was not used in the multivariate analyses.

Health behaviour affects the diagnosis of a cancer and often does vary between social classes. Differences in health behaviour have been proposed as potential explanatory factors for social class differences in cancer patient survival. An example of a behavioural factor is the delay between first symptom and diagnosis of cancer. This time period depends on how the symptoms are observed and interpreted by the patient, as well as on the pattern of seeking medical attention. The first phase – observation and interpretation of symptoms – is essentially psychological and depends on knowledge and awareness. Hackett et al. (1973) have suggested that the longer delay among patients from lower social classes may be due to fear and denial. Patients who recognized their condition as a possible cancer had a shorter delay in seeking medical advice than those using a more general or vague expression for it. Taking action – that is, making an appointment with a doctor – may depend on economic resources and also on earlier experiences with health care. Longer delay among patients from lower social classes has been reported in breast cancer (Richardson et al., 1992; Vineis et al., 1993), and in colorectal cancer (MacArthur & Smith, 1984).

Host factors commonly interfere with implementation of already chosen treatment through complications caused by, for example, bone marrow, heart or neural toxicity of chemotherapy. There are individual differences in susceptibility to such complications, but their relationship to factors such as specific diseases, general health status and nutritional status, or lifestyle factors such as smoking or alcohol consumption, is not clear.

Psychosocial factors. A direct effect of psychosocial factors in cancer survival has also been suggested but results are contradictory. Furthermore, correlation of the psychosocial factors with social class, not to mention empirical assessment of their contribution to social class differences, has rarely been assessed.

In breast cancer, social network (Waxler-Morrison et al., 1991) and certain traits of personality (Hislop et al., 1987) were associated with favourable survival after controlling for clinical prognostic factors. This could not be confirmed in another study (Cassileth et al., 1988). Similarly, the initial findings on the effects of adverse life events on relapse of breast cancer (Ramirez et al., 1989) have not been supported by subsequent research (Barraclough et al., 1992). It has also been reported that quality of life predicts survival in breast (Coates et al., 1992) and lung cancer (Ganz et al., 1991) as well as melanoma (Coates et al., 1993).

It may nevertheless be worthwhile to consider these factors when searching for reasons for the social class differences in cancer survival. A number of methodological problems remain: the studies have mostly been based on materials with insufficient sample size, instruments used for measurement of psychosocial factors have been diverse and the control for conventional prognostic factors has been inadequate. Not only have the validity and reliability of the studies been different, but also they have also been developed for various purposes. Also, the fact that quality of life scores correlate with survival is not sufficient to prove an independent prognostic effect since adjustment for other prognostic factors has been inadequate in some studies. It is plausible that quality of life is affected by extent of disease and this should be taken into account carefully before accepting a reverse relationship. Studies with intervention to improve quality of life have yielded contrasting results (Spiegel et al., 1989; Gellert et al., 1993).

Final remarks

Social class differences in cancer patient survival have been extensively described in the literature. It seems that cancer is diagnosed at an advanced stage more often in lower than in higher social classes. However, the survival differences by social class have persisted even after adjustment for stage in most studies. In a small number of studies, the contribution of treatment to the survival differences has also been assessed. The results are somewhat contradictory: some of the studies showed social class differences even after controlling for treatment, while in other studies the differences disappeared after adjustment for treatment.

The conflicting results are understandable because there is probably real variation in the extent of the social class differences and in the relative importance of factors among different primary sites and countries. Furthermore, the social class indicator used influences the results, because different indicators measure different dimensions of social stratification and because the relative sizes of each class may vary. Hence, one must be careful in generalizing the results from a single study.

It is apparent, however, that the understanding of the phenomenon is still superficial. In most of the studies, a descriptive approach has been adopted. Thus most of the research has concentrated on simple hypothesis testing (the questions addressed being directed at existence versus non-existence of the phenomenon and the conclusion based on statistical significance) and/or quantification of the effect.

A descriptive approach to research on the subject can hardly yield valuable new information now, at least in the industrialized countries where a number of studies have already demonstrated the extent of the problem. For the development of strategies to diminish the differences, there is an urgent need to understand the etiology of such differences – that is, which factors are involved in the genesis of the differences and what is their relative contribution.

These factors cover several domains: behavioural (values and attitudes of the patient, health behaviour), social (social support, economic resources) and clinical (functional status and comorbidity, choice of treatment, response to treatment, complications, relapse, cause of death). Also, the temporal dimension of these factors ranges from the first symptom to the diagnosis, treatment and the eventual death of the patient. This makes it impossible to obtain information on all relevant aspects from a single source. The relevant set of variables also differs between different types of cancer. The complexity is real – that is, no improvements in, say, measurement of social class or survival can be expected to decrease it.

Some potential approaches for intervention may be outlined already. Even if there is not sufficient proof of their efficacy in decreasing social-class differences in cancer patient survival in lower social classes, their implementation can be justified by benefit for the general population. If the importance of some of them is proven in the future, they could be directed especially to the lower social classes. Health education programmes to increase awareness of early symptoms might improve stage at diagnosis. Screening, if accessible, could have the same effect, although efficacy has not been proven for types of cancer other than cancers of the breast and uterine cervix (for further discussion, see the chapter by Segnan). Furthermore, efforts to diminish economic barriers to utilization of health care services are warranted. One can also consider whether patients from lower social classes should be regarded as a high-risk group and consequently allocated an intensified treatment regime. It is equally plausible, however, that they are more vulnerable to complications of cancer treatment and a conservative line of treatment should be the preferred option.

References

Antonovsky, A. (1965) Social class, life expectancy and overall mortality. *Millbank Mem. Fund Q.*, 45, 31–73

Auvinen, A. (1992) Social class and colon cancer survival in Finland. *Cancer*, 70, 402–409

Auvinen, A., Karjalainen, S. & Pukkala, E. (1995) Social class and cancer patient survival in Finland. *Am. J. Epidemiol.*, 142, 1089–1102

Barillari, P., de Angelis, R., Valabrega, S., Indinnimeo, M., Gozzo, P., Ramacciato, G. & Fegiz, G. (1989) Relationship of symptom duration and survival in patients with colorectal carcinoma. *Eur. J. Surg. Oncol.*, 15, 441–445

Barraclough, J., Pinder, P., Cruddas, M., Osmond, C., Taylor, I. & Perry, M. (1992) Life events and breast cancer prognosis. *Br. Med. J.*, 304, 1078–1081

Bassett, M.T. & Krieger, N. (1986) Social class and black-white differences in breast cancer survival. *Am. J. Public Health*, 76, 1400–1403

Berg, J.W., Ross, R. & Latourette, H.B. (1977) Economic status and survival of cancer patients. *Cancer*, 39, 467–477

Bonett, A., Roder, D. & Esterman, A. (1984) Determinants of case survival for cancers of the lung, colon, breast and cervix in South Australia. *Med. J. Aust.*, 141, 705–709

Brenner, H., Mielck, A., Klein, R. & Ziegler, H. (1991) The role of socioeconomic factors in the survival of patients with colorectal cancer in Saarland, Germany. *J. Clin. Epidemiol.*, 44, 807–815

Burstin, H.R., Lipsitz, S.R. & Brennan, T.A. (1992) Socioeconomic status and risk of substandard medical care. *J. Am. Med. Assoc.*, 268, 2383–2387

Carnon, A.G., Ssemwogerere, A., Lamont, D.W., Hole, D.J., Mallon, E.A., George, W.D. & Gillis, C.R. (1994) Relation between socioeconomic deprivation and pathological prognostic factors in women with breast cancer. *Br. Med. J.*, 309, 1054–1057

Cassileth, B.R., Walsh, W.P. & Lusk, E.J. (1988) Psychosocial correlates of cancer survival: a subsequent report 3 to 8 years after cancer diagnosis. *J. Clin. Oncol.*, 6, 1753–1759

Chirikos, T.N. & Horner, R.D. (1985) Economic status and survivorship in digestive system cancers. *Cancer*, 56, 210–217

Chirikos, T.N., Reiches, N.A. & Moeschberger, M.L. (1984) Economic differentials in cancer survival: a multivariate analysis. *J. Chronic Dis.*, 37, 183–193

Ciccone, G., Magnani, C., Delsedime, L. & Vineis, P. (1991) Socioeconomic status and survival from soft-tissue sarcomas: a population-based study in northern Italy. *Am. J. Public Health*, 81, 747–749

Coates, A.S., Gebski, V., Signorini, D., Murray, P., McNeil, D., Byrne, M. & Forbes, J.F. (1992) Prognostic value of quality of life scores during chemotherapy for advanced breast cancer. *J. Clin. Oncol.*, 10, 1833–1838

Coates, A., Thompson, D., McLeod, G.R.M., Hersey, P., Gill, P.G., Olver, I.N., Kefford, R., Lowenthal, R.M., Beadle, G. & Walpole, E. (1993) Prognostic value of quality of life scores in a trial of chemotherapy with or without interferon in patients with metastatic malignant melanoma. *Eur. J. Cancer*, 29A, 1731–1734

Davey Smith, G., Shipley, M.J. & Rose, G. (1990) Magnitude and causes of socioeconomic differences in mortality: further evidence from the Whitehall study. *J. Epidemiol. Community Health*, 44, 265–270

Dayal, H.H. & Chiu, C. (1982) Factors associated with racial differences in survival for prostatic carcinoma. *J. Chronic Dis.*, 35, 553–560

Dayal, H.H., Power, R.N. & Chiu, C. (1982) Race and socio-economic status in survival from breast cancer. *J. Chronic Dis.*, 35, 675–683

Dayal, H.H., Polissar, L. & Dahlberg, S. (1985) Race, socioeconomic status and other prognostic factors for survival from prostate cancer. *J. Natl Cancer Inst.*, 74, 1001–1006

Dayal, H., Polissar, L., Yang, C.Y. & Dahlberg, S. (1987) Race, socioeconomic status and other prognostic factors for survival from colo-rectal cancer. *J. Chronic Dis.*, 40, 857–864

Dorn, H.F. (1950) Methods of analysis of follow-up studies. *Hum. Biol.*, 22, 238–248

Enstrom, J.E. & Austin, D.F. (1977) Interpreting cancer survival rates. *Science*, 195, 847–851

Epstein, A.M., Taylor, W.C. & Seage, G.R., III (1985) Effect of patients' socioeconomic status and physicians' training and practice on patient-doctor communication. *Am. J. Med.*, 78, 101–106

Ewertz, M., Gillanders, S., Meyer, L. & Zedeler, K. (1991) Survival of breast cancer patients in relation to factors which affect the risk of developing breast cancer. *Int. J. Cancer*, 49, 526–530

Feinstein, A.R., Sosin, D.M. & Wells, C.K. (1985) The Will Rogers phenomemon. Stage migration and new diagnostic techniques as a source of misleading statistics for cancer survival. *N. Engl. J. Med.*, 312, 1604–1608

Fox, A.J. & Goldblatt, P. (1982) Longitudinal study: sociodemographic mortality differentials 1971–1975 (Series LS No. 1). London, Office of Population Censuses and Surveys

Fox, A.J., Goldblatt, P.O. & Jones, D.R. (1985) Social class mortality differences: artefact, selection or life circumstances? *J. Epidemiol. Community Health*, 39, 1–8

Funch, D.P. (1988) Predictors and consequences of symptom reporting behaviors in colorectal cancer patients. *Med. Care*, 26, 1000–1008

Ganz, P.A., Lee, J.J. & Siau, J. (1991) Quality of life assessment. An independent prognostic variable for survival in lung cancer. *Cancer*, 67, 3131–3135

Gellert, G.A., Maxwell, R.M. & Siegel, B.S. (1993) Survival of cancer patients receiving adjunctive psychosocial support therapy: a 10-year follow-up study. *J. Clin. Oncol.*, 11, 66–69

Gordon, N.H., Crowe, J.P., Brumberg, D.J. & Berger, N.A. (1992) Socioeconomic factors and race in breast cancer recurrence and survival. *Am. J. Epidemiol.*, 135, 609–618

Greenberg, E.R., Baron, J.A., Dain, B.J., Freeman, D.J., Yates, J.W. & Korson, R. (1991) Cancer staging may have different meaning in academic and community hospitals. *J. Clin. Epidemiol.*, 44, 505–512

Greenberg, E.R., Chute, C.C., Stukel, T., Baron, J.A., Freeman, D.H., Yates, J. & Korson, R. (1988) Social and economic factors in the choice of lung cancer treatment. *New Engl. J. Med.*, 318, 612–617

Hackett, T.P., Cassem, N.H.A. & Raker, J.W. (1973) Patient delay in cancer. *New Engl. J. Med.*, 289, 14–20

Hakulinen, T. (1983) A comparison of nationwide cancer survival statistics in Finland and Norway. *World Health Stat. Q.*, 36, 35–46

Hislop, T.G., Waxler, N.E., Coldman, A.J., Elwood, M.J. & Kan, L. (1987) The prognostic significance of psychosocial factors in women with breast cancer. *J. Chronic Dis.*, 40, 729–735

Howe, H.L., Katterhagen, J.G., Yates, J. & Lehnherr, M. (1992) Urban-rural differences in the management of breast cancer. *Cancer Causes Control*, 3, 533–539

Hutchinson, G.B. & Shapiro, S. (1968) Lead time gained by diagnostic screening for breast cancer. *J. Natl. Cancer Inst.*, 43, 665–681

Jacques, P.F., Hartz, S.C., Tuthill, R.W. & Hollingsworth, C. (1981) Elimination of "lead time" bias in assessing the effect of early breast cancer diagnosis. *Am. J. Epidemiol.*, 113, 93–97

Karjalainen, S. & Pukkala, E. (1990) Social class as a prognostic factor in breast cancer survival. *Cancer*, 66, 819–826

Kato, I., Tominaga, S. & Ikari, A. (1992) The role of socioeconomic factors in the survival of patients with gastrointestinal cancers. *Jpn. J. Clin. Oncol.*, 22, 270–277

Keirn, W. & Metter, G. (1985) Survival of cancer patients by economic status in a free care setting. *Cancer*, 55, 1552–1555

Kogevinas, M. (1990) Longitudinal study 1971–1983. Socio-economic differences in cancer survival (Series LS No. 5). London, Office of Population Censuses and Surveys

Launoy, G., Le Coutour, X., Gignoux, M., Pottier, D. & Dugleux, G. (1992) Influence of rural environment on diagnosis, treatment and prognosis of colorectal cancer. *J. Epidemiol. Community Health*, 46, 365–367

Leon, D. & Wilkinson, R.G. (1989) Inequalities in prognosis: socio-economic differences in cancer and heart disease survival. In: Fox, J., ed., *Health inequalities in European countries*. Aldershot, Gower. pp. 280–300

Linden, G. (1969) The influence of social class in the survival of cancer patients. *Am. J. Public Health*, 59, 267–274

Lipworth, L., Abelin, T. & Connelly, R.R. (1970) Socio-economic factors in the prognosis of cancer patients. *J. Chronic Dis.*, 23, 105–115

Lipworth, L., Bennett, B. & Parker, P. (1972) Prognosis of nonprivate cancer patients. *J. Natl Cancer Inst.*, 48, 11–16

MacArthur, C. & Smith, A. (1984) Factors associated with speed of diagnosis, referral, and treatment in colorectal cancer. *J. Epidemiol. Community Health*, 38, 122–126

McWhirter, W.R., Smith, H. & McWhirter, K.M. (1983) Social class as a prognostic variable in acute lymphoblastic leukaemia. *Med. J. Aust.*, 2, 319–321

McWhorter, A.P. & Mayer, W.J. (1987) Black and white differences in type of initial breast cancer treatment and implications for survival. *Am. J. Public Health*, 77, 1515–1517

Machiavelli, M., Leone, B., Romero, A., Perez, J., Vallejo, C., Bianco, A., Rodriguez, R., Estevez, R., Chacon, R., Dansky, C., Alvarez, L., Xynos, F. & Rabinovich, M. (1989) Relation between delay and survival in 596 patients with breast cancer. *Oncology*, 46, 78–82

Mandelblatt, J., Andrews, H., Kerner, J., Zauber, A. & Burnett, W. (1991) Determinants of late stage diagnosis of breast and cervical cancer: the impact of age, race, social class and hospital type. *Am. J. Public Health*, 81, 646–649

Marmot, M.G., Shipley, M.J. & Rose, G. (1984) Inequalities in death – specific explanations of a general pattern? *Lancet*, I, 1003–1006

Mechanic, D. (1972) Social psychologic factors affecting the perception of bodily complaints. *N. Engl. J. Med.*, 286, 1132–1139

Miller, A.B., ed. (1985) *Screening for cancer*. Orlando, Academic Press

Monnet, E., Boutron, M.C., Faivre, J. & Milan, C. (1993) Influence of socioeconomic status on prognosis of colorectal cancer. A population-based study in Cote d'Or, France. *Cancer*, 72, 1165–1170

Morrison, A.S., Lowe, C.R., MacMahon, B., Ravnihar, B. & Yuasa, S. (1976) Some international differences in treatment and survival in breast cancer. *Int. J. Cancer*, 18, 269–273

Neave, L.M., Mason, B.H. & Kay, R.G. (1990) Does delay in diagnosis of breast cancer affect survival? *Breast Cancer Res. Treat.*, 15, 103–108

Page, W.F. & Kuntz, A.J. (1980) Racial and socio-economic factors in cancer survival. A comparison of Veterans Administration Results with selected studies. *Cancer*, 45, 1029–1040

Parkin, D.M. & Hakulinen, T. (1991) Analysis of survival. In: Jensen, O.M., Parkin, D.M., MacLennan, R., Muir, C.S., & Skeet R.G. eds, *Cancer registration: principles and methods* (IARC Scientific Publications No. 95). Lyon, International Agency for Research on Cancer. pp. 159–176

Porta, M., Gallen, M., Malats, N. & Planas, J. (1991) Influence of "diagnostic delay" upon cancer survival: an analysis of five tumour sites. *J. Epidemiol. Community Health*, 45, 225–230

Ramirez, A.J., Craig, T.K., Watson, J.P., Fentiman, I.S., North, W.R.S. & Rubens, R.D. (1989) Stress and relapse of breast cancer. *Br. Med. J.*, 298, 291–293

Ramot, B. & Magrath, I. (1982) Hypothesis: the environment is a major determinant of the immunological subtype of lymphoma and acute lymphoblastic leukaemia in children. *Br. J. Haematol.*, 52, 183–189

Richardson, J.L., Langholtz, B., Bernstein, L., Danley, K. & Ross, R.K. (1992) Stage and delay in breast cancer diagnosis by race, socioeconomic status, age and year. *Br. J. Cancer*, 65, 922–926

Samphier, M.L., Robertson, C. & Bloor, M.J. (1988) A possible artefactual component in specific cause mortality gradients. *J. Epidemiol. Community Health*, 42, 138–143

Savage, D., Lindenbaum, J., van Ryzin, J., Struening, E. & Garrett, T.J. (1984) Race, poverty and survival in multiple myeloma. *Cancer*, 54, 3085–3094

Saxén, E. & Hakama, M. (1964) Cancer illness in Finland with a note on the effects of age adjustment and early diagnosis. *Ann. Med. Exp. Biol Fenniae*, 42, 1–28

Shaw, H.M., McGovern, V.J., Milton, G.W. & Farago, G.A. (1981) Cutaneous malignant melanoma: occupation and prognosis. *Med. J. Aust.*, 1, 37–38

Spiegel, D., Bloom, J.R., Kraemer, H.C. & Gottheil, E. (1989) Effect of psychosocial treatment on survival of patients with metastatic breast cancer. *Lancet*, II, 888–891

Stavraky, K.M., Kincade, J.E., Stewart, M.A. & Donner, A.P. (1987) The effect of socioeconomic factors on the early prognosis of cancer. *J. Chronic Dis.*, 40, 237–244

Steinhorn, S.C., Myers, M.H., Hankey, B.F. & Pelham, V.F. (1986) Factors associated with survival differences between black and white women with cancer of the uterine corpus. *Am. J. Epidemiol.*, 124, 85–93

Stiller, C.A. (1988) Centralisation of treatment and survival rates for cancer. *Arch. Dis. Child.*, 63, 23–30

Townsend, P. & Davidson, N. (1982) *Inequalities in health. The Black report*. Harmondsworth, Penguin Books

Vågerö, D. & Persson, G. (1987) Cancer survival and social class in Sweden. *J. Epidemiol. Community Health*, 41, 204–209

Valkonen, T., Martelin, T., Rimpelä, A., Notkola, V. & Savela, S. (1993) *Socio-economic mortality differences in Finland 1981–90. Population 1*. Helsinki, Statistics Finland

Vineis, P., Fornera, G., Magnino, A., Giacometti, R. & Ciccone, G. (1993) Diagnostic delay, clinical stage and social class: a hospital based study. *J. Epidemiol. Community Health*, 47, 229–231

Waxler-Morrison, N., Hislop, T.G., Mears, B. & Kan, L. (1991) Effects of social relationship on survival for women with breast cancer: a prospective study. *Soc. Sci. Med.*, 33, 177–183

Wegner, E.L., Kolonel, L.N., Nomura, A.M.Y. & Lee, J. (1982) Racial and socioeconomic differences in survival of colorectal cancer patients in Hawaii. *Cancer*, 49, 2208–2216

Wells, B.L. & Horm, J.W. (1992) Stage at diagnosis in breast cancer: race and socioeconomic factors. *Am. J. Public Health*, 82, 1383–1385

West, R.R. & Lowe, C.R. (1976) Regional variations in need for and provision and use of child health services in England and Wales. *Br. Med. J.*, II, 843–846

Weston, B., Grufferman, S., MacMillan, J.P. & Cohen, H.J. (1987) Effect of socioeconomic and clinical factors on survival in multiple myeloma. *J. Clin. Oncol.*, 5, 1977–1984

Wilkinson, G.S., Edgerton, F., Wallace, H., Jr, Reese, P., Patterson, J. & Priore, R. (1979) Delay, stage of disease and survival from breast cancer. *J. Chron. Dis.*, 32, 365–373

Corresponding author:
A. Auvinen
National Cancer Institute, Radiation Epidemiology Branch, Executive Plaza North, Suite 408, Bethesda, Maryland 20892, USA

IARC Monographs on the Evaluation of Carcinogenic Risks to Humans

Volume 1
Some Inorganic Substances, Chlorinated Hydrocarbons, Aromatic Amines, N-Nitroso Compounds, and Natural Products
1972; 184 pages; ISBN 92 832 1201 0
(out of print)

Volume 2
Some Inorganic and Organometallic Compounds
1973; 181 pages; ISBN 92 832 1202 9
(out of print)

Volume 3
Certain Polycyclic Aromatic Hydrocarbons and Heterocyclic Compounds
1973; 271 pages; ISBN 92 832 1203 7
(out of print)

Volume 4
Some Aromatic Amines, Hydrazine and Related Substances, N-Nitroso Compounds and Miscellaneous Alkylating Agents
1974; 286 pages; ISBN 92 832 1204 5

Volume 5
Some Organochlorine Pesticides
1974; 241 pages; ISBN 92 832 1205 3
(out of print)

Volume 6
Sex Hormones
1974; 243 pages; ISBN 92 832 1206 1
(out of print)

Volume 7
Some Anti-Thyroid and Related Substances, Nitrofurans and Industrial Chemicals
1974; 326 pages; ISBN 92 832 1207 X
(out of print)

Volume 8
Some Aromatic Azo Compounds
1975; 357 pages; ISBN 92 832 1208 8

Volume 9
Some Aziridines, N-, S- and O-Mustards and Selenium
1975; 268 pages; ISBN 92 832 1209 6

Volume 10
Some Naturally Occurring Substances
1976; 353 pages; ISBN 92 832 1210 X
(out of print)

Volume 11
Cadmium, Nickel, Some Epoxides, Miscellaneous Industrial Chemicals and General Considerations on Volatile Anaesthetics
1976; 306 pages; ISBN 92 832 1211 8
(out of print)

Volume 12
Some Carbamates, Thiocarbamates and Carbazides
1976; 282 pages; ISBN 92 832 1212 6

Volume 13
Some Miscellaneous Pharmaceutical Substances
1977; 255 pages; ISBN 92 832 1213 4

Volume 14
Asbestos
1977; 106 pages; ISBN 92 832 1214 2
(out of print)

Volume 15
Some Fumigants, the Herbicides 2,4-D and 2,4,5-T, Chlorinated Dibenzodioxins and Miscellaneous Industrial Chemicals
1977; 354 pages; ISBN 92 832 1215 0
(out of print)

Volume 16
Some Aromatic Amines and Related Nitro Compounds – Hair Dyes, Colouring Agents and Miscellaneous Industrial Chemicals
1978; 400 pages; ISBN 92 832 1216 9

Volume 17
Some N-Nitroso Compounds
1978; 365 pages; ISBN 92 832 1217 7

Volume 18
Polychlorinated Biphenyls and Polybrominated Biphenyls
1978; 140 pages; ISBN 92 832 1218 5

Volume 19
Some Monomers, Plastics and Synthetic Elastomers, and Acrolein
1979; 513 pages; ISBN 92 832 1219 3
(out of print)

Volume 20
Some Halogenated Hydrocarbons
1979; 609 pages; ISBN 92 832 1220 7
(out of print)

Volume 21
Sex Hormones (II)
1979; 583 pages; ISBN 92 832 1521 4

Volume 22
Some Non-Nutritive Sweetening Agents
1980; 208 pages; ISBN 92 832 1522 2

Volume 23
Some Metals and Metallic Compounds
1980; 438 pages; ISBN 92 832 1523 0
(out of print)

Volume 24
Some Pharmaceutical Drugs
1980; 337 pages; ISBN 92 832 1524 9

Volume 25
Wood, Leather and Some Associated Industries
1981; 412 pages; ISBN 92 832 1525 7

Volume 26
Some Antineoplastic and Immunosuppressive Agents
1981; 411 pages; ISBN 92 832 1526 5

Volume 27
Some Aromatic Amines, Anthraquinones and Nitroso Compounds, and Inorganic Fluorides Used in Drinking Water and Dental Preparations
1982; 341 pages; ISBN 92 832 1527 3

Volume 28
The Rubber Industry
1982; 486 pages; ISBN 92 832 1528 1

Volume 29
Some Industrial Chemicals and Dyestuffs
1982; 416 pages; ISBN 92 832 1529 X

Volume 30
Miscellaneous Pesticides
1983; 424 pages; ISBN 92 832 1530 3

Volume 31
Some Food Additives, Feed Additives and Naturally Occurring Substances
1983; 314 pages; ISBN 92 832 1531 1

Volume 32
Polynuclear Aromatic Compounds, Part 1: Chemical, Environmental and Experimental Data
1983; 477 pages; ISBN 92 832 1532 X

Volume 33
Polynuclear Aromatic Compounds, Part 2: Carbon Blacks, Mineral Oils and Some Nitroarenes
1984; 245 pages; ISBN 92 832 1533 8
(out of print)

Volume 34
Polynuclear Aromatic Compounds, Part 3: Industrial Exposures in Aluminium Production, Coal Gasification, Coke Production, and Iron and Steel Founding
1984; 219 pages; ISBN 92 832 1534 6

Volume 35
Polynuclear Aromatic Compounds: Part 4: Bitumens, Coal-Tars and Derived Products, Shale-Oils and Soots
1985; 271 pages; ISBN 92 832 1535 4

Volume 36
Allyl Compounds, Aldehydes, Epoxides and Peroxides
1985; 369 pages; ISBN 92 832 1536 2

Volume 37
Tobacco Habits Other than Smoking; Betel-Quid and Areca-Nut Chewing; and Some Related Nitrosamines
1985; 291 pages; ISBN 92 832 1537 0

Volume 38
Tobacco Smoking
1986; 421 pages; ISBN 92 832 1538 9

Volume 39
Some Chemicals Used in Plastics and Elastomers
1986; 403 pages; ISBN 92 832 1239 8

Volume 40
Some Naturally Occurring and Synthetic Food Components, Furocoumarins and Ultraviolet Radiation
1986; 444 pages; ISBN 92 832 1240 1

Volume 41
Some Halogenated Hydrocarbons and Pesticide Exposures
1986; 434 pages; ISBN 92 832 1241 X

Volume 42
Silica and Some Silicates
1987; 289 pages; ISBN 92 832 1242 8

Volume 43
Man-Made Mineral Fibres and Radon
1988; 300 pages; ISBN 92 832 1243 6

Volume 44
Alcohol Drinking
1988; 416 pages; ISBN 92 832 1244 4

Volume 45
Occupational Exposures in Petroleum Refining; Crude Oil and Major Petroleum Fuels
1989; 322 pages; ISBN 92 832 1245 2

Volume 46
Diesel and Gasoline Engine Exhausts and Some Nitroarenes
1989; 458 pages; ISBN 92 832 1246 0

Volume 47
Some Organic Solvents, Resin Monomers and Related Compounds, Pigments and Occupational Exposures in Paint Manufacture and Painting
1989; 535 pages; ISBN 92 832 1247 9

Volume 48
Some Flame Retardants and Textile Chemicals, and Exposures in the Textile Manufacturing Industry
1990; 345 pages; ISBN: 92 832 1248 7

Volume 49
Chromium, Nickel and Welding
1990; 677 pages; ISBN: 92 832 1249 5

Volume 50
Some Pharmaceutical Drugs
1990; 415 pages; ISBN: 92 832 1259 9

Volume 51
Coffee, Tea, Mate, Methylxanthines and Methylglyoxal
1991; 513 pages; ISBN: 92 832 1251 7

Volume 52
Chlorinated Drinking-Water; Chlorination By-products; Some other Halogenated Compounds; Cobalt and Cobalt Compounds
1991; 544 pages; ISBN: 92 832 1252 5

Volume 53
Occupational Exposures in Insecticide Application, and Some Pesticides
1991; 612 pages; ISBN 92 832 1253 3

Volume 54
Occupational Exposures to Mists and Vapours from Strong Inorganic Acids; and other Industrial Chemicals
1992; 336 pages; ISBN 92 832 1254 1

Volume 55
Solar and Ultraviolet Radiation
1992; 316 pages; ISBN 92 832 1255 X

Volume 56
Some Naturally Occurring Substances: Food Items and Constituents, Heterocyclic Aromatic Amines and Mycotoxins
1993; 600 pages; ISBN 92 832 1256 8

Volume 57
Occupational Exposures of Hairdressers and Barbers and Personal Use of Hair Colourants; Some Hair Dyes, Cosmetic Colourants, Industrial Dyestuffs and Aromatic Amines
1993; 428 pages; ISBN 92 832 1257 6

Volume 58
Beryllium, Cadmium, Mercury and Exposures in the Glass Manufacturing Industry
1994; 444 pages; ISBN 92 832 1258 4

Volume 59
Hepatitis Viruses
1994; 286 pages; ISBN 92 832 1259 2

Volume 60
Some Industrial Chemicals
1994; 560 pages; ISBN 92 832 1260 6

Volume 61
Schistosomes, Liver Flukes and *Helicobacter pylori*
1994; 280 pages; ISBN 92 832 1261 4

Volume 62
Wood Dusts and Formaldehyde
1995; 405 pages; ISBN 92 832 1262 2

Volume 63
Dry cleaning, Some Chlorinated Solvents and Other Industrial Chemicals
1995; 558 pages; ISBN 92 832 1263 0

Volume 64
Human Papillomaviruses
1995; 409 pages; ISBN 92 832 1264 9

Volume 65
Printing Processes, Printing Inks, Carbon Blacks and Some Nitro Compounds
1996; 578 pages; ISBN 92 832 1265 7

Volume 66
Some Pharmaceutical Drug
1996; 514 pages; ISBN 92 832 1266 5

Volume 67
Human Immunodeficiency Viruses and Human T-cell Lymphotropic Viruses
1996; 424 pages; ISBN 92 832 1267 3

Volume 68
Silica, Some Silicates, Coal Dust and para-Aramid Fibrils
1997; 506 pages; ISBN 92 832 1268 1

Volume 69
Polychlorinated Dibenzo-para-dioxins and Dibenzofurans
1997; c. 600 pages; ISBN 92 832 1269 X

Supplements

Supplement No.1
Chemicals and Industrial Processes Associated with Cancer in Humans (IARC Monographs, Volumes 1 to 20)
1979; 71 pages; ISBN 92 832 1404 8
(out of print)

Supplement No. 2
Long-Term and Short-Term Screening Assays for Carcinogens: A Critical Appraisal
1980; 426 pages; ISBN 92 832 1404 8

Supplement No. 3
Cross Index of Synonyms and Trade Names in Volumes 1 to 26
1982; 199 pages; ISBN 92 832 1405 6
(out of print)

Supplement No.4
Chemicals, Industrial Processes and Industries Associated with Cancer in Humans (Volumes 1 to 29)
1982; 292 pages; ISBN 92 832 1407 2
(out of print)

Supplement No. 5
Cross Index of Synonyms and Trade Names in Volumes 1 to 36
1985; 259 pages; ISBN 92 832 1408 0
(out of print)

Supplement No. 6
Genetic and Related Effects: An Updating of Selected IARC Monographs from Volumes 1 to 42
1987; 729 pages; ISBN 92 832 1409 9

Supplement No. 7
Overall Evaluations of Carcinogenicity: An Updating of IARC Monographs Volumes 1 to 42
1987; 440 pages; ISBN 92 832 1411 0

Supplement No. 8
Cross Index of Synonyms and Trade Names in Volumes 1 to 46
1989; 346 pages; ISBN 92 832 1417 X

IARC Scientific Publications

No. 1
Liver Cancer
1971; 176 pages; ISBN 0 19 723000 8

No. 2
Oncogenesis and Herpesviruses
Edited by P.M. Biggs, G. de Thé and L.N. Payne
1972; 515 pages; ISBN 0 19 723001 6

No. 3
N-Nitroso Compounds: Analysis and Formation
Edited by P. Bogovski, R. Preussman and E.A. Walker
1972; 140 pages; ISBN 0 19 723002 4

No. 4
Transplacental Carcinogenesis
Edited by L. Tomatis and U. Mohr
1973; 181 pages; ISBN 0 19 723003 2

No. 5/6
Pathology of Tumours in Laboratory Animals. Volume 1: Tumours of the Rat
Edited by V.S. Turusov
1973/1976; 533 pages; ISBN 92 832 1410 2

No. 7
Host Environment Interactions in the Etiology of Cancer in Man
Edited by R. Doll and I. Vodopija
1973; 464 pages; ISBN 0 19 723006 7

No. 8
Biological Effects of Asbestos
Edited by P. Bogovski, J.C. Gilson, V. Timbrell and J.C. Wagner
1973; 346 pages; ISBN 0 19 723007 5

No. 9
N-Nitroso Compounds in the Environment
Edited by P. Bogovski and E.A. Walker
1974; 243 pages; ISBN 0 19 723008 3

No. 10
Chemical Carcinogenesis Essays
Edited by R. Montesano and L. Tomatis
1974; 230 pages; ISBN 0 19 723009 1

No. 11
Oncogenesis and Herpes-viruses II
Edited by G. de-Thé, M.A. Epstein and H. zur Hausen
1975; Two volumes, 511 pages and 403 pages; ISBN 0 19 723010 5

No. 12
Screening Tests in Chemical Carcinogenesis
Edited by R. Montesano, H. Bartsch and L. Tomatis
1976; 666 pages; ISBN 0 19 723051 2

No. 13
Environmental Pollution and Carcinogenic Risks
Edited by C. Rosenfeld and W. Davis
1975; 441 pages; ISBN 0 19 723012 1

No. 14
Environmental N-Nitroso Compounds. Analysis and Formation
Edited by E.A. Walker, P. Bogovski and L. Griciute
1976; 512 pages; ISBN 0 19 723013 X

No. 15
Cancer Incidence in Five Continents, Volume III
Edited by J.A.H. Waterhouse, C. Muir, P. Correa and J. Powell
1976; 584 pages; ISBN 0 19 723014 8

No. 16
Air Pollution and Cancer in Man
Edited by U. Mohr, D. Schmähl and L. Tomatis
1977; 328 pages; ISBN 0 19 723015 6

No. 17
Directory of On-Going Research in Cancer Epidemiology 1977
Edited by C.S. Muir and G. Wagner
1977; 599 pages; ISBN 92 832 1117 0
(out of print)

No. 18
Environmental Carcinogens. Selected Methods of Analysis. Volume 1: Analysis of Volatile Nitrosamines in Food
Editor-in-Chief: H. Egan
1978; 212 pages; ISBN 0 19 723017 2

No. 19
Environmental Aspects of N-Nitroso Compounds
Edited by E.A. Walker, M. Castegnaro, L. Griciute and R.E. Lyle
1978; 561 pages; ISBN 0 19 723018 0

No. 20
Nasopharyngeal Carcinoma: Etiology and Control
Edited by G. de Thé and Y. Ito
1978; 606 pages; ISBN 0 19 723019 9

No. 21
Cancer Registration and its Techniques
Edited by R. MacLennan, C. Muir, R. Steinitz and A. Winkler
1978; 235 pages; ISBN 0 19 723020 2

No. 22
Environmental Carcinogens: Selected Methods of Analysis. Volume 2: Methods for the Measurement of Vinyl Chloride in Poly(vinyl chloride), Air, Water and Foodstuffs
Editor-in-Chief: H. Egan
1978; 142 pages; ISBN 0 19 723021 0

No. 23
Pathology of Tumours in Laboratory Animals. Volume II: Tumours of the Mouse
Editor-in-Chief: V.S. Turusov
1979; 669 pages; ISBN 0 19 723022 9

No. 24
Oncogenesis and Herpesviruses III
Edited by G. de-Thé, W. Henle and F. Rapp
1978; Part I: 580 pages, Part II: 512 pages; ISBN 0 19 723023 7

xvii

No. 25
Carcinogenic Risk: Strategies for Intervention
Edited by W. Davis and C. Rosenfeld
1979; 280 pages; ISBN 0 19 723025 3

No. 26
Directory of On-going Research in Cancer Epidemiology 1978
Edited by C.S. Muir and G. Wagner
1978; 550 pages; ISBN 0 19 723026 1
(out of print)

No. 27
Molecular and Cellular Aspects of Carcinogen Screening Tests
Edited by R. Montesano, H. Bartsch and L. Tomatis
1980; 372 pages; ISBN 0 19 723027 X

No. 28
Directory of On-going Research in Cancer Epidemiology 1979
Edited by C.S. Muir and G. Wagner
1979; 672 pages; ISBN 92 832 1128 6
(out of print)

No. 29
Environmental Carcinogens. Selected Methods of Analysis. Volume 3: Analysis of Polycyclic Aromatic Hydrocarbons in Environmental Samples
Editor-in-Chief: H. Egan
1979; 240 pages; ISBN 0 19 723028 8

No. 30
Biological Effects of Mineral Fibres
Editor-in-Chief: J.C. Wagner
1980; Two volumes, 494 pages & 513 pages; ISBN 0 19 723030 X

No. 31
N-Nitroso Compounds: Analysis, Formation and Occurrence
Edited by E.A. Walker, L. Griciute, M. Castegnaro and M. Börzsönyi
1980; 835 pages; ISBN 0 19 723031 8

No. 32
Statistical Methods in Cancer Research. Volume 1: The Analysis of Case-control Studies
By N.E. Breslow and N.E. Day
1980; 338 pages; ISBN 92 832 0132 9

No. 33
Handling Chemical Carcinogens in the Laboratory
Edited by R. Montesano, H. Bartsch, E. Boyland, G. Della Porta, L. Fishbein, R.A. Griesemer, A.B. Swan and L. Tomatis
1979; 32 pages; ISBN 0 19 723033 4
(out of print)

No. 34
Pathology of Tumours in Laboratory Animals. Volume III: Tumours of the Hamster
Editor-in-Chief: V.S. Turusov
1982; 461 pages; ISBN 0 19 723034 2

No. 35
Directory of On-going Research in Cancer Epidemiology 1980
Edited by C.S. Muir and G. Wagner
1980; 660 pages; ISBN 0 19 723035 0
(out of print)

No. 36
Cancer Mortality by Occupation and Social Class 1851–1971
Edited by W.P.D. Logan
1982; 253 pages; ISBN 0 19 723036 9

No. 37
Laboratory Decontamination and Destruction of Aflatoxins B1, B2, G1, G2 in Laboratory Wastes
Edited by M. Castegnaro, D.C. Hunt, E.B. Sansone, P.L. Schuller, M.G. Siriwardana, G.M. Telling, H.P. van Egmond and E.A. Walker
1980; 56 pages; ISBN 0 19 723037 7

No. 38
Directory of On-going Research in Cancer Epidemiology 1981
Edited by C.S. Muir and G. Wagner
1981; 696 pages; ISBN 0 19 723038 5
(out of print)

No. 39
Host Factors in Human Carcinogenesis
Edited by H. Bartsch and B. Armstrong
1982; 583 pages;
ISBN 0 19 723039 3

No. 40
Environmental Carcinogens: Selected Methods of Analysis. Volume 4: Some Aromatic Amines and Azo Dyes in the General and Industrial Environment
Edited by L. Fishbein, M. Castegnaro, I.K. O'Neill and H. Bartsch
1981; 347 pages; ISBN 0 19 723040 7

No. 41
N-Nitroso Compounds: Occurrence and Biological Effects
Edited by H. Bartsch, I.K. O'Neill, M. Castegnaro and M. Okada
982; 755 pages; ISBN 0 19 723041 5

No. 42
Cancer Incidence in Five Continents Volume IV
Edited by J. Waterhouse, C. Muir, K. Shanmugaratnam and J. Powell
1982; 811 pages; ISBN 0 19 723042 3

No. 43
Laboratory Decontamination and Destruction of Carcinogens in Laboratory Wastes: Some N-Nitrosamines
Edited by M. Castegnaro, G. Eisenbrand, G. Ellen, L. Keefer, D. Klein, E.B. Sansone, D. Spincer, G. Telling and K. Webb
1982; 73 pages; ISBN 0 19 723043 1

No. 44
Environmental Carcinogens: Selected Methods of Analysis.
Volume 5: Some Mycotoxins
Edited by L. Stoloff, M. Castegnaro, P. Scott, I.K. O'Neill and H. Bartsch
1983; 455 pages; ISBN 0 19 723044 X

No. 45
Environmental Carcinogens: Selected Methods of Analysis.
Volume 6: N-Nitroso Compounds
Edited by R. Preussmann, I.K. O'Neill, G. Eisenbrand, B. Spiegelhalder and H. Bartsch
1983; 508 pages; ISBN 0 19 723045 8

No. 46
Directory of On-going Research in Cancer Epidemiology 1982
Edited by C.S. Muir and G. Wagner
1982; 722 pages; ISBN 0 19 723046 6
(out of print)

No. 47
Cancer Incidence in Singapore 1968–1977
Edited by K. Shanmugaratnam, H.P. Lee and N.E. Day
1983; 171 pages; ISBN 0 19 723047 4

No. 48
Cancer Incidence in the USSR (2nd Revised Edition)
Edited by N.P. Napalkov, G.F. Tserkovny, V.M. Merabishvili, D.M. Parkin, M. Smans and C.S. Muir
1983; 75 pages; ISBN 0 19 723048 2

No. 49
Laboratory Decontamination and Destruction of Carcinogens in Laboratory Wastes: Some Polycyclic Aromatic Hydrocarbons
Edited by M. Castegnaro, G. Grimmer, O. Hutzinger, W. Karcher, H. Kunte, M. Lafontaine, H.C. Van der Plas, E.B. Sansone and S.P. Tucker
1983; 87 pages; ISBN 0 19 723049 0

No. 50
Directory of On-going Research in Cancer Epidemiology 1983
Edited by C.S. Muir and G. Wagner
1983; 731 pages; ISBN 0 19 723050 4
(out of print)

No. 51
Modulators of Experimental Carcinogenesis
Edited by V. Turusov and R. Montesano
1983; 307 pages; ISBN 0 19 723060 1

No. 52
Second Cancers in Relation to Radiation Treatment for Cervical Cancer: Results of a Cancer Registry Collaboration
Edited by N.E. Day and J.C. Boice, Jr
1984; 207 pages; ISBN 0 19 723052 0

No. 53
Nickel in the Human Environment
Editor-in-Chief: F.W. Sunderman, Jr
1984; 529 pages; ISBN 0 19 723059 8

No. 54
Laboratory Decontamination and Destruction of Carcinogens in Laboratory Wastes: Some Hydrazines
Edited by M. Castegnaro, G. Ellen, M. Lafontaine, H.C. van der Plas, E.B. Sansone and S.P. Tucker
1983; 87 pages; ISBN 0 19 723053

No. 55
Laboratory Decontamination and Destruction of Carcinogens in Laboratory Wastes: Some N-Nitrosamides
Edited by M. Castegnaro, M. Bernard, L.W. van Broekhoven, D. Fine, R. Massey, E.B. Sansone, P.L.R. Smith, B. Spiegelhalder, A. Stacchini, G. Telling and J.J. Vallon
1984; 66 pages; ISBN 0 19 723054 7

No. 56
Models, Mechanisms and Etiology of Tumour Promotion
Edited by M. Börzsönyi, N.E. Day, K. Lapis and H. Yamasaki
1984; 532 pages; ISBN 0 19 723058 X

No. 57
N-Nitroso Compounds: Occurrence, Biological Effects and Relevance to Human Cancer
Edited by I.K. O'Neill, R.C. von Borstel, C.T. Miller, J. Long and H. Bartsch
1984; 1013 pages; ISBN 0 19 723055 5

No 58
Age-related Factors in Carcinogenesis
Edited by A. Likhachev, V. Anisimov and R. Montesano
1985; 288 pages; ISBN 92 832 1158 8

No. 59
Monitoring Human Exposure to Carcinogenic and Mutagenic Agents
Edited by A. Berlin, M. Draper, K. Hemminki and H. Vainio
1984; 457 pages; ISBN 0 19 723056 3

No. 60
Burkitt's Lymphoma: A Human Cancer Model
Edited by G. Lenoir, G. O'Conor and C.L.M. Olweny
1985; 484 pages; ISBN 0 19 723057 1

No. 61
Laboratory Decontamination and Destruction of Carcinogens in Laboratory Wastes: Some Haloethers
Edited by M. Castegnaro, M. Alvarez, M. Iovu, E.B. Sansone, G.M. Telling and D.T. Williams
1985; 55 pages; ISBN 0 19 723061 X

No. 62
Directory of On-going Research in Cancer Epidemiology 1984
Edited by C.S. Muir and G. Wagner
1984; 717 pages; ISBN 0 19 723062 8
(out of print)

No. 63
Virus-associated Cancers in Africa
Edited by A.O. Williams, G.T. O'Conor, G.B. de Thé and C.A. Johnson
1984; 773 pages; ISBN 0 19 723063 6

No. 64
Laboratory Decontamination and Destruction of Carcinogens in Laboratory Wastes: Some Aromatic Amines and 4-Nitrobiphenyl
Edited by M. Castegnaro, J. Barek, J. Dennis, G. Ellen, M. Klibanov, M. Lafontaine, R. Mitchum, P. van Roosmalen, E.B. Sansone, L.A. Sternson and M. Vahl
1985; 84 pages; ISBN: 92 832 1164 2

No. 65
Interpretation of Negative Epidemiological Evidence for Carcinogenicity
Edited by N.J. Wald and R. Doll
1985; 232 pages; ISBN 92 832 1165 0

No. 66
The Role of the Registry in Cancer Control
Edited by D.M. Parkin, G. Wagner and C.S. Muir
1985; 152 pages; ISBN 92 832 0166 3

No. 67
Transformation Assay of Established Cell Lines: Mechanisms and Application
Edited by T. Kakunaga and H. Yamasaki
1985; 225 pages; ISBN 92 832 1167 7

No. 68
Environmental Carcinogens: Selected Methods of Analysis. Volume 7: Some Volatile Halogenated Hydrocarbons
Edited by L. Fishbein and I.K. O'Neill
1985; 479 pages; ISBN 92 832 1168 5

No. 69
Directory of On-going Research in Cancer Epidemiology 1985
Edited by C.S. Muir and G. Wagner
1985; 745 pages; ISBN 92 823 1169 3
(out of print)

No. 70
The Role of Cyclic Nucleic Acid Adducts in Carcinogenesis and Mutagenesis
Edited by B. Singer and H. Bartsch
1986; 467 pages; ISBN 92 832 1170 7

No. 71
Environmental Carcinogens: Selected Methods of Analysis. Volume 8: Some Metals: As, Be, Cd, Cr, Ni, Pb, Se, Zn
Edited by I.K. O'Neill, P. Schuller and L. Fishbein
1986; 485 pages; ISBN 92 832 1171 5

No. 72
Atlas of Cancer in Scotland, 1975–1980: Incidence and Epidemiological Perspective
Edited by I. Kemp, P. Boyle, M. Smans and C.S. Muir
1985; 285 pages; ISBN 92 832 1172 3

No. 73
Laboratory Decontamination and Destruction of Carcinogens in Laboratory Wastes: Some Antineoplastic Agents
Edited by M. Castegnaro, J. Adams, M.A. Armour, J. Barek, J. Benvenuto, C. Confalonieri, U. Goff, G. Telling
1985; 163 pages; ISBN 92 832 1173 1

No. 74
Tobacco: A Major International Health Hazard
Edited by D. Zaridze and R. Peto
1986; 324 pages; ISBN 92 832 1174 X

No. 75
Cancer Occurrence in Developing Countries
Edited by D.M. Parkin
1986; 339 pages; ISBN 92 832 1175 8

No. 76
Screening for Cancer of the Uterine Cervix
Edited by M. Hakama, A.B. Miller and N.E. Day
1986; 315 pages; ISBN 92 832 1176 6

No. 77
Hexachlorobenzene: Proceedings of an International Symposium
Edited by C.R. Morris and J.R.P. Cabral
1986; 668 pages; ISBN 92 832 1177 4

No. 78
Carcinogenicity of Alkylating Cytostatic Drugs
Edited by D. Schmähl and J.M. Kaldor
1986; 337 pages; ISBN 92 832 1178 2

No. 79
Statistical Methods in Cancer Research. Volume III: The Design and Analysis of Long-term Animal Experiments
By J.J. Gart, D. Krewski, P.N. Lee, R.E. Tarone and J. Wahrendorf
1986; 213 pages; ISBN 92 832 1179 0

No. 80
Directory of On-going Research in Cancer Epidemiology 1986
Edited by C.S. Muir and G. Wagner
1986; 805 pages; ISBN 92 832 1180 4
(out of print)

No. 81
Environmental Carcinogens: Methods of Analysis and Exposure Measurement. Volume 9: Passive Smoking
Edited by I.K. O'Neill, K.D. Brunnemann, B. Dodet and D. Hoffmann
1987; 383 pages; ISBN 92 832 1181 2

No. 82
Statistical Methods in Cancer Research. Volume II: The Design and Analysis of Cohort Studies
By N.E. Breslow and N.E. Day
1987; 404 pages; ISBN 92 832 0182 5

No. 83
Long-term and Short-term Assays for Carcinogens: A Critical Appraisal
Edited by R. Montesano, H. Bartsch, H. Vainio, J. Wilbourn and H. Yamasaki
1986; 575 pages; ISBN 92 832 1183 9

No. 84
The Relevance of N-Nitroso Compounds to Human Cancer: Exposure and Mechanisms
Edited by H. Bartsch, I.K. O'Neill and R. Schulte-Hermann
1987; 671 pages; ISBN 92 832 1184 7

No. 85
Environmental Carcinogens: Methods of Analysis and Exposure Measurement. Volume 10: Benzene and Alkylated Benzenes
Edited by L. Fishbein and I.K. O'Neill
1988; 327 pages; ISBN 92 832 1185 5

No. 86
Directory of On-going Research in Cancer Epidemiology 1987
Edited by D.M. Parkin and J. Wahrendorf
1987; 685 pages; ISBN: 92 832 1186 3
(out of print)

No. 87
International Incidence of Childhood Cancer
Edited by D.M. Parkin, C.A. Stiller, C.A. Bieber, G.J. Draper. B. Terracini and J.L. Young
1988; 401 page; ISBN 92 832 1187 1
(out of print)

No. 88
Cancer Incidence in Five Continents, Volume V
Edited by C. Muir, J. Waterhouse, T. Mack, J. Powell and S. Whelan
1987; 1004 pages; ISBN 92 832 1188 X

No. 89
Methods for Detecting DNA Damaging Agents in Humans: Applications in Cancer Epidemiology and Prevention
Edited by H. Bartsch, K. Hemminki and I.K. O'Neill
1988; 518 pages; ISBN 92 832 1189 8
(out of print)

No. 90
Non-occupational Exposure to Mineral Fibres
Edited by J. Bignon, J. Peto and R. Saracci
1989; 500 pages; ISBN 92 832 1190 1

No. 91
Trends in Cancer Incidence in Singapore 1968–1982
Edited by H.P. Lee, N.E. Day and K. Shanmugaratnam
1988; 160 pages; ISBN 92 832 1191 X

No. 92
Cell Differentiation, Genes and Cancer
Edited by T. Kakunaga, T. Sugimura, L. Tomatis and H. Yamasaki
1988; 204 pages; ISBN 92 832 1192 8

No. 93
Directory of On-going Research in Cancer Epidemiology 1988
Edited by M. Coleman and J. Wahrendorf
1988; 662 pages; ISBN 92 832 1193 6
(out of print)

No. 94
Human Papillomavirus and Cervical Cancer
Edited by N. Muñoz, F.X. Bosch and O.M. Jensen
1989; 154 pages; ISBN 92 832 1194 4

No. 95
Cancer Registration: Principles and Methods
Edited by O.M. Jensen, D.M. Parkin, R. MacLennan, C.S. Muir and R. Skeet
1991; 296 pages; ISBN 92 832 1195 2

No. 96
Perinatal and Multigeneration Carcinogenesis
Edited by N.P. Napalkov, J.M. Rice, L. Tomatis and H. Yamasaki
1989; 436 pages; ISBN 92 832 1196 0

No. 97
Occupational Exposure to Silica and Cancer Risk
Edited by L. Simonato, A.C. Fletcher, R. Saracci and T. Thomas
1990; 124 pages; ISBN 92 832 1197 9

No. 98
Cancer Incidence in Jewish Migrants to Israel, 1961-1981
Edited by R. Steinitz, D.M. Parkin, J.L. Young, C.A. Bieber and L. Katz
1989; 320 pages; ISBN 92 832 1198 7

No. 99
Pathology of Tumours in Laboratory Animals, Second Edition, Volume 1, Tumours of the Rat
Edited by V.S. Turusov and U. Mohr
1990; 740 pages; ISBN 92 832 1199 5
For Volumes 2 and 3
(Tumours of the Mouse and Tumours of the Hamster), see
IARC Scientific Publications
Nos. 111 and 126.

No. 100
Cancer: Causes, Occurrence and Control
Editor-in-Chief: L. Tomatis
1990; 352 pages; ISBN 92 832 0110 8

No. 101
Directory of On-going Research in Cancer Epidemiology 1989–1990
Edited by M. Coleman and J. Wahrendorf
1989; 828 pages; ISBN 92 832 2101 X

No. 102
Patterns of Cancer in Five Continents
Edited by S.L. Whelan, D.M. Parkin and E. Masuyer
1990; 160 pages; ISBN 92 832 2102 8

No. 103
Evaluating Effectiveness of Primary Prevention of Cancer
Edited by M. Hakama, V. Beral,
J.W. Cullen and D.M. Parkin
1990; 206 pages; ISBN 92 832 2103 6

No. 104
Complex Mixtures and Cancer Risk
Edited by H. Vainio, M. Sorsa and
A.J. McMichael
1990; 441 pages; ISBN 92 832 2104 4

No. 105
Relevance to Human Cancer of N-Nitroso Compounds, Tobacco Smoke and Mycotoxins
Edited by I.K. O'Neill, J. Chen and H. Bartsch
1991; 614 pages; ISBN 92 832 2105 2

No. 106
Atlas of Cancer Incidence in the Former German Democratic Republic
Edited by W.H. Mehnert, M. Smans,
C.S. Muir, M. Möhner and D. Schön
1992; 384 pages; ISBN 92 832 2106 0

No. 107
Atlas of Cancer Mortality in the European Economic Community
Edited by M. Smans, C. Muir and P. Boyle
1992; 213 pages + 44 coloured maps;
ISBN 92 832 2107 9

No. 108
Environmental Carcinogens: Methods of Analysis and Exposure Measurement. Volume 11: Polychlorinated Dioxins and Dibenzofurans
Edited by C. Rappe, H.R. Buser,
B. Dodet and I.K. O'Neill
1991; 400 pages; ISBN 92 832 2108 7

No. 109
Environmental Carcinogens: Methods of Analysis and Exposure Measurement. Volume 12: Indoor Air
Edited by B. Seifert, H. van de Wiel,
B. Dodet and I.K. O'Neill
1993; 385 pages; ISBN 92 832 2109 5

No. 110
Directory of On-going Research in Cancer Epidemiology 1991
Edited by M.P. Coleman and J. Wahrendorf
1991; 753 pages; ISBN 92 832 2110 9

No. 111
Pathology of Tumours in Laboratory Animals, Second Edition. Volume 2: Tumours of the Mouse
Edited by V. Turusov and U. Mohr
1994; 800 pages; ISBN 92 832 2111 1

No. 112
Autopsy in Epidemiology and Medical Research
Edited by E. Riboli and M. Delendi
1991; 288 pages; ISBN 92 832 2112 5

No. 113
Laboratory Decontamination and Destruction of Carcinogens in Laboratory Wastes: Some Mycotoxins
Edited by M. Castegnaro, J. Barek,
J.M. Frémy, M. Lafontaine, M. Miraglia,
E.B. Sansone and G.M. Telling
1991; 63 pages; ISBN 92 832 2113 3

No. 114
Laboratory Decontamination and Destruction of Carcinogens in Laboratory Wastes: Some Polycyclic Heterocyclic Hydrocarbons
Edited by M. Castegnaro, J. Barek, J. Jacob,
U. Kirso, M. Lafontaine, E.B. Sansone,
G.M. Telling and T. Vu Duc
1991; 50 pages; ISBN 92 832 2114 1

No. 115
Mycotoxins, Endemic Nephropathy and Urinary Tract Tumours
Edited by M. Castegnaro, R. Plestina, G. Dirheimer, I.N. Chernozemsky and
H. Bartsch
1991; 340 pages; ISBN 92 832 2115 X

No. 116
Mechanisms of Carcinogenesis in Risk Identification
Edited by H. Vainio, P. Magee,
D. McGregor and A.J. McMichael
1992; 615 pages; ISBN 92 832 2116 8

No. 117
Directory of On-going Research in Cancer Epidemiology 1992
Edited by M. Coleman, E. Demaret and
J. Wahrendorf
1992; 773 pages; ISBN 92 832 2117 6

No. 118
Cadmium in the Human Environment: Toxicity and Carcinogenicity
Edited by G.F. Nordberg, R.F.M. Herber
and L. Alessio
1992; 470 pages; ISBN 92 832 2118 4

No. 119
The Epidemiology of Cervical Cancer and Human Papillomavirus
Edited by N. Muñoz, F.X. Bosch,
K.V. Shah and A. Meheus
1992; 288 pages; ISBN 92 832 2119 2

No. 120
Cancer Incidence in Five Continents, Vol. VI
Edited by D.M. Parkin, C.S. Muir, S.L. Whelan,
Y.T. Gao, J. Ferlay and J. Powell
1992; 1020 pages; ISBN 92 832 2120 6

No. 121
Time Trends in Cancer Incidence and Mortality
By M. Coleman, J. Estéve, P. Damiecki,
A. Arslan and H. Renard
1993; 820 pages; ISBN 92 832 2121 4

No. 122
International Classification of Rodent Tumours.
Part I. The Rat
Editor-in-Chief: U. Mohr
1992–1996; 10 fascicles of 60–100 pages; ISBN 92 832 2122 2

No. 123
Cancer in Italian Migrant Populations
Edited by M. Geddes, D.M. Parkin,
M. Khlat, D. Balzi and E. Buiatti
1993; 292 pages; ISBN 92 832 2123 0

No. 124
Postlabelling Methods for the Detection of DNA Damage
Edited by D.H. Phillips,
M. Castegnaro and H. Bartsch
1993; 392 pages; ISBN 92 832 2124 9

No. 125
DNA Adducts: Identification and Biological Significance
Edited by K. Hemminki, A. Dipple,
D.E.G. Shuker, F.F. Kadlubar,
D. Segerbäck and H. Bartsch
1994; 478 pages; ISBN 92 832 2125 7

No. 126
Pathology of Tumours in Laboratory Animals, Second Edition. Volume 3: Tumours of the Hamster
Edited by V. Turosov and U. Mohr
1996; 464 pages; ISBN 92 832 2126 5

No. 127
Butadiene and Styrene: Assessment of Health Hazards
Edited by M. Sorsa, K. Peltonen, H. Vainio and K. Hemminki
1993; 412 pages; ISBN 92 832 2127 3

No. 128
Statistical Methods in Cancer Research. Volume IV. Descriptive Epidemiology
By J. Estéve, E. Benhamou and L. Raymond
1994; 302 pages; ISBN 92 832 2128 1

No. 129
Occupational Cancer in Developing Countries
Edited by N. Pearce, E. Matos,
H. Vainio, P. Boffetta and M. Kogevinas
1994; 191 pages; ISBN 92 832 2129 X

No. 130
Directory of On-going Research in Cancer Epidemiology 1994
Edited by R. Sankaranarayanan,
J. Wahrendorf and E. Démaret
1994; 800 pages; ISBN 92 832 2130 3

No. 132
Survival of Cancer Patients in Europe: The EUROCARE Study
Edited by F. Berrino, M. Sant,
A. Verdecchia, R. Capocaccia,
T. Hakulinen and J. Estève
1995; 463 pages; ISBN 92 832 2132 X

No. 134
Atlas of Cancer Mortality in Central Europe
W. Zatonski, J. Estéve, M. Smans,
J. Tyczynski and P. Boyle
1996; 300 pages; ISBN 92 832 2134 6

No. 135
Methods for Investigating Localized Clustering of Disease
Edited by F.E. Alexander and P. Boyle
1996; 235 pages; ISBN 92 832 2135 4

No. 136
Chemoprevention in Cancer Control
Edited by M. Hakama, V. Beral,
E. Buiatti, J. Faivre and D.M. Parkin
1996; 160 pages; ISBN 92 832 2136 2

No. 137
Directory of On-going Research in Cancer Epidemiology 1996
Edited by R. Sankaranarayan, J. Warendorf and E. Démaret
1996; 810 pages; ISBN 92 832 2137 0

No. 138
Social Inequalities and Cancer
Edited by M. Kogevinas, N. Pearce, M. Susser and P. Boffetta
1997; 412 pages; ISBN 92 832 2138 9

No. 139
Principles of Chemoprevention
Edited by B.W. Stewart, D. McGregor and P. Kleihues
1996; 358 pages; ISBN 92 832 2139 7

No. 140
Mechanisms of Fibre Carcinogenesis
Edited by A.B. Kane, P. Boffetta, R. Saracci and J.D. Wilbourn
1996; 135 pages; ISBN 92 832 2140 0

IARC Technical Reports

No. 1
Cancer in Costa Rica
Edited by R. Sierra, R. Barrantes,
G. Muñoz Leiva, D.M. Parkin,
C.A. Bieber and N. Muñoz Calero
1988; 124 pages;
ISBN 92 832 1412 9

No. 2
SEARCH: A Computer Package to Assist the Statistical Analysis of Case-Control Studies
Edited by G.J. Macfarlane, P. Boyle and P. Maisonneuve
1991; 80 pages; ISBN 92 832 1413 7

No. 3
Cancer Registration in the European Economic Community
Edited by M.P. Coleman and E. Démaret
1988; 188 pages; ISBN 92 832 1414 5

No. 4
Diet, Hormones and Cancer: Methodological Issues for Prospective Studies
Edited by E. Riboli and R. Saracci
1988; 156 pages; ISBN 92 832 1415 3

No. 5
Cancer in the Philippines
Edited by A.V. Laudico, D. Esteban and D.M. Parkin
1989; 186 pages; ISBN 92 832 1416 1

No. 6
La genèse du Centre international de recherche sur le cancer
By R. Sohier and A.G.B. Sutherland
1990, 102 pages; ISBN 92 832 1418 8

No. 7
Epidémiologie du cancer dans les pays de langue latine
1990, 292 pages; ISBN 92 832 1419 6

No. 8
Comparative Study of Anti-smoking Legislation in Countries of the European Economic Community
By A. J. Sasco, P. Dalla-Vorgia and P. Van der Elst
1992; 82 pages; ISBN: 92 832 1421 8
Etude comparative des Législations de Contrôle du Tabagisme dans les Pays de la Communauté économique européenne
1995; 82 pages; ISBN 92 832 2402 7

No. 9
Epidémiologie du cancer dans les pays de langue latine
1991; 346 pages; ISBN 92 832 1423 4

No. 10
Manual for Cancer Registry Personnel
Edited by D. Esteban, S. Whelan,
A. Laudico and D.M. Parkin
1995; 400 pages; ISBN 92 832 1424 2

No. 11
Nitroso Compounds: Biological Mechanisms, Exposures and Cancer Etiology
Edited by I. O'Neill and H. Bartsch
1992; 150 pages; ISBN 92 832 1425 X

No. 12
Epidémiologie du cancer dans les pays de langue latine
1992; 375 pages; ISBN 92 832 1426 9

No. 13
Health, Solar UV Radiation and Environmental Change
By A. Kricker, B.K. Armstrong,
M.E. Jones and R.C. Burton
1993; 213 pages; ISBN 92 832 1427 7

No. 14
Epidémiologie du cancer dans les pays de langue latine
1993; 400 pages; ISBN 92 832 1428 5

No. 15
Cancer in the African Population of Bulawayo, Zimbabwe, 1963–1977
By M.E.G. Skinner, D.M. Parkin,
A.P. Vizcaino and A. Ndhlovu
1993; 120 pages; ISBN 92 832 1429 3

No. 16
Cancer in Thailand 1984–1991
By V. Vatanasapt, N. Martin,

H. Sriplung, K. Chindavijak, S. Sontipong, S. Sriamporn, D.M. Parkin and J. Ferlay
1993; 164 pages; ISBN 92 832 1430 7

No. 18
Intervention Trials for Cancer Prevention
By E. Buiatti
1994; 52 pages; ISBN 92 832 1432 3

No. 19
Comparability and Quality Control in Cancer Registration
By D.M. Parkin, V.W. Chen, J. Ferlay, J. Galceran, H.H. Storm and S.L. Whelan
1994; 110 pages plus diskette; ISBN 92 832 1433 1

No. 20
Epidémiologie du cancer dans les pays de langue latine
1994; 346 pages; ISBN 92 832 1434 X

No. 21
ICD Conversion Programs for Cancer
By J. Ferlay
1994; 24 pages plus diskette; ISBN 92 832 1435 8

No. 22
Cancer in Tianjin
By Q.S. Wang, P. Boffetta, M. Kogevinas and D.M. Parkin
1994; 96 pages; ISBN 92 832 1433 1

No. 23
An Evaluation Programme for Cancer Preventive Agents
By Bernard W. Stewart
1995; 40 pages; ISBN 92 832 1438 2

No. 24
Peroxisome Proliferation and its Role in Carcinogenesis
1995; 85 pages; ISBN 92 832 1439 0

No. 25
Combined Analysis of Cancer Mortality in Nuclear Workers in Canada, the United Kingdom and the United States of America
By E. Cardis, E.S. Gilbert, L. Carpenter, G. Howe, I. Kato, J. Fix, L. Salmon, G. Cowper, B.K. Armstrong, V. Beral, A. Douglas, S.A. Fry, J. Kaldor, C. Lavé, P.G. Smith, G. Voelz and L. Wiggs
1995; 160 pages; ISBN 92 832 1440 4

No. 26
Mortalité par Cancer des Imigrés en France, 1979-1985
By C. Bouchardy, M. Khlat, P. Wanner and D.M. Parkin
1997; 150 pages; ISBN 92 832 2404 3

No. 27
Cancer in Three Generations of Young Israelis
By J. Iscovich and D.M. Parkin
1997; 150 pages; ISBN 92 832 2441 2

No. 29
International Classification of Childhood Cancer 1996
By E. Kramarova, C.A. Stiller, J. Ferlay, D.M. Parkin, G.J. Draper, J. Michaelis, J. Neglia and S. Qurechi
1996; 48 pages + diskette; ISBN 92 832 1443 9

IARC CancerBase

No. 1
EUCAN90: Cancer in the European Union (Electronic Database with Graphic Display)
By J. Ferlay, R.J. Black, P. Pisani, M.T. Valdivieso and D.M. Parkin
1996; Computer software on 3.5" IBM diskette + user's guide (50 pages); ISBN 92 832 1450 1

All IARC Publications are available directly from
IARCPress, 150 Cours Albert Thomas, F-69372 Lyon cedex 08, France
(Fax: +33 4 72 73 83 02; E-mail: press@iarc.fr).

IARC Monographs and Technical Reports are also available from the
World Health Organization Distribution and Sales, CH-1211 Geneva 27
(Fax: +41 22 791 4857)
and from WHO Sales Agents worldwide.

IARC Scientific Publications are also available from
Oxford University Press, Walton Street, Oxford, UK OX2 6DP
(Fax: +44 1865 267782).